Edited by Sanjay K. Sharma

Green Corrosion Chemistry and Engineering

Related Titles

Krzyzanowski, M., Beynon, J. H., Farrugia, D. C. J.

Oxide Scale Behavior in High Temperature Metal Processing

2010
Hardcover
ISBN: 978-3-527-32518-4

Kreysa, G., Schütze, M. (eds.)

Corrosion Handbook - Corrosive Agents and Their Interaction with Materials
13 Volume Set

2009
Hardcover
ISBN: 978-3-527-31217-7

Heimann, R. B.

Plasma Spray Coating
Principles and Applications

2008
Hardcover
ISBN: 978-3-527-32050-9

Revie, R. W.

Corrosion and Corrosion Control

2010
Hardcover
ISBN: 978-0-471-73279-2

Roberge, P. R., Revie, R. W.

Corrosion Inspection and Monitoring

2007
Hardcover
ISBN: 978-0-471-74248-7

Ghali, E., Sastri, V. S., Elboujdaini, M.

Corrosion Prevention and Protection
Practical Solutions

2007
Hardcover
ISBN: 978-0-470-02402-7

Edited by Sanjay K. Sharma

Green Corrosion Chemistry and Engineering

Opportunities and Challenges

With a Foreword by Nabuk Okon Eddy

WILEY-VCH Verlag GmbH & Co. KGaA

The Editor

Prof. Sanjay K. Sharma
Professor of Chemistry
Department of Chemistry & Environmental Engineering
Jaipur Engineering College & Research Center
JECRC Foundation
Jaipur (Rajasthan)
India

Library of Congress Card No.: applied for

British Library Cataloguing-in-Publication Data
A catalogue record for this book is available from the British Library.

Bibliographic information published by the Deutsche Nationalbibliothek
The Deutsche Nationalbibliothek lists this publication in the Deutsche Nationalbibliografie; detailed bibliographic data are available on the Internet at <http://dnb.d-nb.de>.

Cover Design Grafik-Design Schulz, Fußgönheim
Typesetting Laserwords Private Limited, Chennai, India
Printing and Binding Fabulous Printers Pte Ltd, Singapore

Printed in Singapore
Printed on acid-free paper

Print ISBN: 978-3-527-32930-4
ePDF ISBN: 978-3-527-64180-2
ePub ISBN: 978-3-527-64179-6
Mobi ISBN: 978-3-527-64181-9
oBook ISBN: 978-3-527-64178-9

This book is for

My Parents Dr. M.P. Sharma and Smt. Parmeshwari Devi on their 'Golden Jubilee'.

. . . as they are the "real force"

behind all my success.

Foreword

In spite of the fact that several corrosion inhibitors have been synthesized and utilized for corrosion control, the search for newer inhibitors is not yet a fulfilled mission. The journey started with inorganic compounds and has successfully captured heteroatom(s)-rich organic compounds along its route. So far, the journey has not ended but has captured the extract of living organism into its route. Recently, computer modeling has been a subject matter and has yielded positive and definite results.

One of the major concerns on the industrial utilization of raw materials and other products involves a task that will ensure that the quality of the environment is not negatively altered. We have only one global village and that is the world. Therefore, our action or inaction should not be targeted toward the initiation or extension of adverse environmental impact. Corrosion is an essential process involving the electrochemical conversion of metals into its original form. Corrosion is one of the processes nature has adopted to recycle its content. We cannot stop corrosion but the rate at which metals corrodes can be reduced by using various methods.

I have gone through the contents of this book and I am satisfied that the book has convincingly addressed the major problems associated with corrosion and the various green control methods that can be adopted to reduce its impact. The authors are sound academicians in the field and have translated their basic knowledge of corrosion into a book form.

I hereby recommend the book for use by all science and engineering students of tertiary institutions as well as those who want to gain good insight into the chemistry of corrosion.

Dr. Nabuk Okon Eddy, MRSC
Computational and Corrosion Chemist
Department of Chemistry,
Ahmadu Bello University, Zaria
Kaduna State
Nigeria

Contents

Foreword *VII*
Preface *XIX*
Acknowledgments *XXI*
About The Editor *XXIII*
List of Contributors *XXV*

1 Basics of Corrosion Chemistry *1*
Norio Sato
1.1 Introduction *1*
1.2 Metallic Corrosion *1*
1.2.1 Basic Processes *1*
1.2.2 Potential-pH Diagram *2*
1.2.3 Corrosion Potential *3*
1.2.4 Anodic Metal Dissolution *4*
1.2.5 Cathodic Oxidant Reduction *6*
1.3 Metallic Passivity *7*
1.3.1 Passivity of Metals *7*
1.3.2 Passivation of Metals *9*
1.3.3 Passive Films *11*
1.3.4 Chloride-Breakdown of Passive Films *12*
1.4 Localized Corrosion *13*
1.4.1 Pitting Corrosion *13*
1.4.2 Crevice Corrosion *16*
1.4.3 Potential–Dimension Diagram *18*
1.5 Corrosion Rust *19*
1.5.1 Rust in Corrosion *19*
1.5.2 Ion-Selective Rust *20*
1.5.3 Electron-Selective Rust *22*
1.5.4 Redox Rust *24*
1.6 Atmospheric Corrosion *24*
1.6.1 Atmospheric Corrosion Chemistry *24*
1.6.2 Weathering Steel Corrosion *26*
1.6.3 Anticorrosion Rust *28*

1.7 Concluding Remarks 29
References 29

2 Corrosion and Electrochemistry *33*
Vedula Sankar Sastri
2.1 Introduction *33*
2.2 Thermodynamics and the Stability of Metals *40*
2.3 Free Energy and Electrode Potential *41*
2.4 Electrode Potential Measurements *44*
2.5 Equilibrium Electrode Potentials *45*
2.6 Use of Pourbaix Diagrams *49*
2.7 Dynamic Electrochemical Processes *49*
2.8 Concentration Polarization *61*
References *69*
Further Reading *69*

3 Application of Microelectrochemical Techniques in Corrosion Research *71*
Y. Frank Cheng
3.1 Introduction *71*
3.2 Scanning Vibrating Electrode Technique *72*
3.2.1 The Technique and Principle *72*
3.2.2 Local Dissolution Behavior of the Welding Zone of Pipeline Steel *73*
3.2.3 Effects of Mill Scale and Corrosion Product Deposit on Corrosion of the Steel *79*
3.3 Localized Electrochemical Impedance Spectroscopy *81*
3.3.1 The Technique and Principle *81*
3.3.2 Corrosion of Steel at the Base of the Coating Defect *82*
3.3.3 Microscopic Metallurgical Electrochemistry of Pipeline Steels *86*
3.3.4 Characterization of Local Electrochemical Activity of a Precracked Steel Specimen *88*
3.4 Scanning Kelvin Probe *89*
3.4.1 The Technique and Principle *89*
3.4.2 Monitoring the Coating Disbondment *91*
3.5 Conclusive Remarks *94*
Acknowledgments *95*
References *95*

4 Protective Coatings: an Overview *97*
Anand Sawroop Khanna
4.1 Introduction *97*
4.2 Selection of Paint Coatings *97*
4.3 Classification of Various Coatings *98*

4.4 Chemistry of Resins *99*
4.4.1 Alkyd Resins *99*
4.4.2 Modified Alkyds *101*
4.4.3 Epoxies *101*
4.4.4 Urethanes *105*
4.4.5 Isocyanates *106*
4.4.6 Aliphatic Isocyanates *106*
4.4.7 Polyols *107*
4.4.8 Acrylic Urethanes *107*
4.4.9 Moisture-Cured Polyurethanes *107*
4.4.10 Zinc-Based Coatings *108*
4.5 High-Performance Coatings *109*
4.5.1 The 100% Solventless Epoxies *110*
4.5.2 Concept of Underwater Coatings *111*
4.5.3 Polyvinylidenedifluride Coatings *113*
4.5.4 Polysiloxane Coatings *114*
4.5.5 Fire-Resistant Coatings *119*
4.5.6 Organic–Inorganic Hybrid (OIH) Waterborne Coatings *119*
4.6 Surface Preparation *121*
4.7 Paint Application *121*
4.8 Importance of Supervision, Inspection, and Quality Control during Paint Coatings *122*
4.9 Training and Certification Courses *123*
4.10 Summary *123*
References *123*

5 New Era of Eco-Friendly Corrosion Inhibitors *125*
Niketan Patel and Girish Mehta
5.1 Introduction *125*
5.2 Anodic (Passivating or Film-Forming) Inhibitors *126*
5.2.1 Mechanism *127*
5.2.1.1 Generalized Film Theory *127*
5.2.1.2 Adsorption Theory *127*
5.3 Cathodic (Adsorption-Type) Inhibitors *129*
5.3.1 Mechanism *129*
5.4 Mixed Inhibitors *130*
5.5 Precipitation Inhibitors *131*
5.6 Vapor Phase Inhibitors *131*
5.7 Toxicity of Inhibitors *139*
5.7.1 Irritants *140*
5.7.2 Asphyxiants *141*
5.7.3 Anesthetics and Narcotics *141*
5.7.4 Systemic Poisons *141*
5.7.5 Sensitizers *142*
5.7.6 Carcinogens *142*

5.7.7 Mutagens *142*
5.7.8 Teratogens *142*
References *148*

6 Green Corrosion Inhibitors: Status in Developing Countries *157*
Sanjay K. Sharma and Alka Sharma
6.1 Introduction *157*
6.2 Protection against Corrosion *160*
6.3 Inhibitors *162*
6.3.1 Mechanism of Inhibition *162*
6.3.2 Choice of Inhibitors *163*
6.4 Natural Products as Green Corrosion Inhibitors *166*
6.5 Green Corrosion Inhibition: Research and Progress *169*
6.5.1 The Proposed Mechanism for the Inhibitory Behavior of the Extracts *170*
6.6 Green Corrosion Inhibition in Developing Countries *173*
6.6.1 Usage of Metals and Present Corrosion Management: Practice and Prevention *173*
6.6.2 Summary of Researchers' Work to Develop Green Inhibition Science *174*
Acknowledgments *176*
References *176*

7 Innovative Silanes-Based Pretreatment to Improve the Adhesion of Organic Coatings *181*
Michele Fedel, Flavio Deflorian, and Stefano Rossi
7.1 Introduction *181*
7.2 Hybrid Silane Sol–Gel Coatings *182*
7.2.1 Basic Chemistry of the Silicon Alkoxides and Organofunctional Silicon Alkoxides *183*
7.2.2 Dip Coating *192*
7.2.3 Interaction between Silicon Alkoxides and Metallic Substrates *194*
7.2.4 Interaction between Silicon Alkoxides and an Organic Polymeric Material *199*
7.3 Corrosion Protection by Sol–Gel Coatings *202*
7.3.1 Corrosion Protection Properties of Organofunctional Sol–Gel Coatings *202*
7.3.2 Experimental Methods of Investigation of the Properties of the Silicon Alkoxide Sol–Gel Coatings as Coupling Agents *203*
7.3.3 Practical Examples of Corrosion Protection by Silicon-Based Sol–Gel Coatings *204*
References *207*

8 **Corrosion of Austenitic Stainless Steels and Nickel-Base Alloys in Supercritical Water and Novel Control Methods** *211*
Lizhen Tan, Todd R. Allen, and Ying Yang
8.1 Introduction *211*
8.1.1 Supercritical Water and Its Applications *211*
8.1.2 Austenitic Stainless Steels and Ni-Base Alloys and Their General Corrosion Behavior *214*
8.1.2.1 Austenitic Stainless Steels and Ni-Base Alloys *214*
8.1.2.2 General Corrosion Behavior *215*
8.2 Thermodynamics of Alloy Oxidation *216*
8.3 Corrosion of Austenitic Stainless Steels and Ni-Base Alloys in SCW *220*
8.3.1 Weight Change *221*
8.3.2 Surface Morphology *223*
8.3.3 Oxide Layer Structure *225*
8.4 Novel Corrosion Control Methods *227*
8.4.1 Microstructural Optimization *227*
8.4.2 Grain Size Refinement *231*
8.4.3 Performance Comparison of the Corrosion Control Methods *233*
8.5 Factors Influencing Corrosion *234*
8.5.1 Test Conditions *234*
8.5.2 Effect of Thermodynamics and Kinetics *235*
8.5.3 Effect of Microstructure *236*
8.5.4 Effect of Grain Size *237*
8.6 Summary *239*
References *240*

9 **Metal–Phosphonate Anticorrosion Coatings** *243*
Konstantinos D. Demadis, Maria Papadaki, and Dimitrios Varouchas
9.1 Introduction *243*
9.2 The Scope of Green Chemistry and Corrosion Control *245*
9.3 Metal–Phosphonate Materials: Structural Chemistry *247*
9.3.1 Phosphonobutane-1,2,4-Tricarboxylic Acid (PBTC) *247*
9.3.2 Ethylenediamine-Tetrakis(Methylenephosphonic Acid) (EDTMP) *249*
9.3.3 Dimethylaminomethylene-Bis(Phosphonic Acid) (DMABP) *249*
9.3.4 Magnesium (or Zinc)-(Amino-Tris-(Methylenephosphonate)), Mg(or Zn)-AMP *251*
9.3.5 Calcium-(Amino-Tris-(Methylenephosphonate)), Ca-AMP *252*
9.3.6 Strontium-(Amino-Tris-(Methylenephosphonate)), Sr-AMP *253*
9.3.7 Barium-(Amino-Tris-(Methylenephosphonate)), Ba-AMP *255*
9.3.8 Zinc-Hexamethylene-Diamine-Tetrakis(Methylenephosphonate), Zn-HDTMP *255*
9.3.9 Strontium or Barium-Hexamethylene-Diamine-Tetrakis(Methylenephosphonate), Sr or Ba-HDTMP *257*

9.3.10 Zinc-Tetramethylene-Diamine-Tetrakis(Methylenephosphonate), Zn-TDTMP *259*
9.3.11 Strontium and Calcium-Ethylene-Diamine-Tetrakis(Methylene Phosphonate), Sr-EDTMP and Ca-EDTMP *259*
9.3.12 Barium-Phosphonomethylene-Imino-Diacetate, Ba-PMIDA *260*
9.3.13 Tetrasodium-Hydroxyethyl-Amino-Bis(Methylenephosphonate), Na4-HEABMP *260*
9.3.14 Calcium-Phosphonobutane-1,2,4-Tricarboxylate, Ca-PBTC *263*
9.3.15 Calcium-Hexamethylene-Diamine-Tetrakis(Methylenephosphonate), Ca-HDTMP *264*
9.3.16 Calcium-Hydroxyphosphonoacetate, Ca-HPAA *264*
9.3.17 Strontium and Barium-Hydroxyphosphonoacetate, Sr and Ba-HPAA *267*
9.3.17.1 $Sr[(HPAA)(H_2O)_3]\cdot H_2O$ *267*
9.3.17.2 $M(HPAA)(H_2O)_2$ (M = Sr, Ba) *268*
9.3.18 Calcium-Phosphonomethylene-Imino-Diacetate, Ca-PMIDA *269*
9.4 Metal–Phosphonate Anticorrosion Coatings *269*
9.5 A Look at Corrosion Inhibition by Metal–Phosphonates at the Molecular Level *276*
9.5.1 Anticorrosion Coatings by the Material $\{Zn(AMP)\cdot 3H_2O\}_n$ *276*
9.5.2 Anticorrosion Coatings by the Material $\{Zn(HDTMP)\cdot H_2O\}_n$ *278*
9.5.3 Anticorrosion Coatings by the Material $\{Ca(PBTC)(H_2O)_2\cdot 2H_2O\}_n$ *278*
9.5.4 Anticorrosion Coatings by the Material $Ca_3(HPAA)_2(H_2O)_{14}$ *279*
9.5.5 Anticorrosion Coatings by the Materials $\{M(HPAA)(H_2O)_2\}_n$ (M = Sr, Ba) *281*
9.5.6 Anticorrosion Coatings by the Materials $\{M(PMIDA)\}_n$ (M = Ca, Sr, Ba) *282*
9.5.7 A Comparative Look at the Inhibitory Performance by Metal–Phosphonate Protective Films *285*
9.6 Conclusions/Perspectives *287*
References *287*

10 Metal-Matrix Nanocomposite Coatings Produced by Electrodeposition *297*
Caterina Zanella, Stefano Rossi, and Flavio Deflorian
10.1 Introduction *297*
10.2 Electrodeposition of Composite Coatings–Theoretical Remarks *298*
10.2.1 Suspension of Solid Particles in Electrolytes *298*
10.2.2 Mechanism of Codeposition *299*
10.2.3 Process Parameters Influencing the Incorporation *302*
10.3 Electrodeposition of Composite Coatings *305*
10.4 New Insight in the Electrodeposition of Composite Coatings *313*

10.4.1 Ultrasonic Vibrations *313*
10.4.2 Magnetic Field *314*
References *315*

11 Adsorption Studies, Modeling, and Use of Green Inhibitors in Corrosion Inhibition: an Overview of Recent Research *319*
Sanjay K. Sharma, Ackmez Mudhoo, and Essam Khamis
11.1 Introduction *319*
11.2 Adsorption Mechanisms in Corrosion Inhibition *321*
11.3 Hybrid Coatings *323*
11.4 Modeling Aspects *326*
11.5 Green Inhibitors *328*
11.5.1 Natural Derivatives as Green Inhibitors *328*
11.5.2 Research Orientations *329*
11.5.2.1 Organic-Based Green Inhibitors *329*
11.5.2.2 Amino Acids–Based Green Inhibitors *330*
11.5.2.3 Plant Extracts–Based Green Inhibitors *330*
11.5.2.4 Rare Earth Elements–Based Green Inhibitors *333*
11.6 Conclusions *334*
Acknowledgments *335*
References *335*

12 Indian Initiatives for Corrosion Protection *339*
Anand Sawroop Khanna
12.1 Introduction *339*
12.2 Scenario of the Indian Industry *342*
12.3 Corrosion Protection Scenario in India *343*
12.4 Corrosion Education *345*
12.4.1 Why Corrosion Education Is the Need of the Hour? *345*
12.4.2 Pioneers of Corrosion Education in India *346*
12.4.3 Corrosion Science and Engineering, IIT Bombay *346*
12.4.4 The Central Electrochemical Research Institute (CECRI) *347*
12.4.5 NACE India Section *347*
12.4.6 The Society for Surface Protective Coatings India *347*
12.4.7 The National Corrosion Council of India (NCCI) *348*
12.5 An Overview of Highly Corrosion-Prone Industries in India *348*
12.5.1 Oil and Gas Industry *348*
12.5.2 Process Chemical and Petrochemical Industry *349*
12.5.3 Pulp and Paper Industry *350*
12.5.4 Power Plants *351*
12.6 Conclusions *352*
12.7 Recommendations *352*
12.7.1 For Government or Relevant Ministry Responsible for Controlling Industrial Discipline *352*
12.7.2 For Academic Institutes and Research Organizations *353*

12.7.3 Roles and Responsibilities of the Industry *353*
References *353*
Further Reading *354*

13 Protective Coatings: Novel Nanohybrid Coatings for Corrosion and Fouling Prevention *355*
S. Anandakumar and R. Savitha
13.1 Introduction *355*
13.2 Background *356*
13.3 Fouling *356*
13.4 Marine Fouling *357*
13.4.1 Stages of Marine Fouling *357*
13.4.2 Consequences of Marine Fouling *357*
13.4.3 Methods Used for Fouling Prevention *357*
13.4.3.1 Antifouling Paints *358*
13.5 Corrosion *359*
13.5.1 Consequences of Corrosion *359*
13.5.2 Methods Used for the Prevention of Corrosion *359*
13.5.3 Characteristics of a Good Coating *359*
13.5.4 Evaluation of Corrosion Resistance of Coatings *360*
13.5.5 Electrochemical Impedance Studies (EISs) *360*
13.6 Epoxy Resin Coatings *360*
13.6.1 Advantages of Epoxy Resin *361*
13.6.2 Disadvantages of Epoxy Resin *361*
13.6.3 Justification *362*
13.6.4 Need for Nanotechnology *362*
13.6.5 Polymer Nanomaterials *363*
13.6.6 Different Types of Nanoparticles *363*
13.6.7 Polyhedral Oligomeric Silsesquioxanes (POSSs) *364*
13.6.8 Need for Nanocontainer (Nanozeolite) *364*
13.6.9 Self-Repairing Multifunctional Coatings *366*
13.6.9.1 Fabrication of Nanocontainers *366*
13.6.9.2 Biocide Release Mechanism from Nanocontainer *366*
13.6.9.3 Selection of Natural Products as Biocides *367*
13.7 Scope and Objectives *368*
13.8 Experimental: Synthesis and Structural Characterization of the Nanohybrid Coatings *368*
13.8.1 Materials *368*
13.8.2 Surface Preparation of the Mild Steel Specimens *369*
13.8.3 Synthesis of Phosphorus-Containing Polyurethane Epoxy Resin *369*
13.8.4 Synthesis of Amine-Functionalized POSS (POSS-NH2) *369*
13.8.5 Preparation of Biocides *370*
13.8.6 Loading of Biocides *370*
13.8.7 Preparation of Tris(*p*-Isocyanatophenyl)-Thiophosphate-Modified Epoxy Nanocoatings *371*

13.8.8 Test Methods *371*
13.9 Results and Discussion *373*
13.9.1 Structural Characterization of Tris(*p*-Isocyanatophenyl)-Thiophosphate-Modified Epoxy Resin *373*
13.9.2 Structural Characterization of MCM-41 *375*
13.9.3 Composition of Natural Products *378*
13.9.3.1 M. champaca *378*
13.9.3.2 Neem Oil *379*
13.9.4 Confirmation of Loading of Biocide *379*
13.9.5 Evaluation of Corrosion Resistance by EIS *381*
13.9.6 Colorimetric and Gravimetric Analyses *383*
13.9.6.1 pH Analysis *385*
13.9.6.2 Salt Spray Test Results *386*
13.9.6.3 Seawater Immersion Test *386*
13.9.6.4 Antifouling Studies by Scanning Electron Microscopy (SEM) *389*
13.10 Summary and Conclusion *389*
Acknowledgment *390*
References *391*
Further Reading *392*

Index *393*

Preface

> Green Chemistry represents the pillars that hold up our sustainable future. It is imperative to teach the value of Green Chemistry to tomorrow's Chemists.
>
> *Daryle Busch (ACS President, 1999–2001)*

The mighty words of Daryle Busch are the need of the day and that is why editing this book has been a very special experience for me... because of its theme and essence. Green Chemistry is a 14 years old philosophy given by the brilliant duo Anastas and Warner (1998); which is now the choice of billions of researchers world wide. I am also one of them, who are thrilled by this new concept of thinking and mind-set especially at a time when we are all facing severe environmental disorders and extremely dangerous threats such as air pollution and global warming.

The fast growing industrialization and development activities cause many problems such as water pollution, noise pollution, soil pollution, air pollution, and so on. At the same time, these pollutions cause damage, deformation, destruction, and decay of materials and metals, which is commonly known as *Corrosion*. It is one of the most dangerous industrial problems world wide that must be confronted for safety, environmental, and economic reasons. It also incurs heavy maintenance costs and environmental impacts of billions of dollars.

Green Chemistry provides many environmentally friendly corrosion inhibitors, called "*Green inhibitors.*" Several efforts have been made using corrosion preventive practices. Use of corrosion inhibitors and anti-corrosion coatings are some of them. The theme – Green Corrosion Chemistry and Engineering – involves all such genuine efforts which may reduce the maintenance costs and save the environment.

This book is a sincere effort to address the problem of corrosion and to discuss preventive measures with the help of eco-friendly (green) alternates including protective coatings, use of green inhibitors, application of micro-electrochemical techniques, use of nanocomposites and pre-treatments, and much more.

I hope this book provides an insightful text on the corrosion preventive techniques and processes that are being studied, optimized, and developed to sustain our environment.

I sincerely welcome feedback from my valuable readers and critics.
Happy Reading!

Sanjay K. Sharma
drsanjay1973@gmail.com

Acknowledgments

It is the time to express my gratitude to my friends, supporters, and well wishers to make them know that I am deeply obliged to have them and their valuable co-operation during the journey of the completion of the present book *Green Corrosion Chemistry and Engineering: Opportunities and Challenges.*

First of all, I feel greatly indebted to Prof. Paul Anastas and Prof. John Warner, because they are the key persons who ignited the fire of "Green Chemistry" in my heart. Specially Prof. Warner, who appreciated my work in the field of green corrosion inhibitors during a personal meet at Mumbai.

I also acknowledge Prof. Nabul Eddy for his moral support and best wishes, which I need most in this phase of writing-editing.

All our esteemed contributors to this book deserve special thanks for contributing their work, without which this book could not be possible in this form.

My teachers Dr. R.K. Bansal, Dr. R.V. Singh, Dr. R.K. Bhardwaj, and Dr. Saraswati Mittal, deserve special mention here as they are the *Gurus* behind all my academic achievements, publications etc.

I acknowledge the active interest and useful suggestions of the one and only Ackmez Mudhoo (co-author in many of my works), University of Mauritius, Mauritius. His prompt and precise suggestions are always useful to me. Thanks Ackmez. My friends, Dr. Rashmi Sanghi, Dr. V.K. Garg, Dr. R.V. Singh, Dr. Pranav Saxena, Dr. Alka Sharma, and Aruna were also of moral support in this journey.

I deeply acknowledge my parents Dr. M.P. Sharma and Mrs. Parmeshwari Devi, wife Dr. Pratima Sharma and other family members for their never ending encouragement, moral support, and patience during the course of this book.

I also express my gratitude to Mr. Amit Agarwal and Mr. Arpit Agarwal (Directors, JECRC) for giving me an opportunity to work with them. It is wonderful experience to work under so energetic and young team leaders.

My kids Kunal and Kritika also deserve special attention as their valuable moments were mostly stolen owing to my busy schedules.

I am also thankful to many others whose names I have not been able to mention but whose guidance value has not been less in any way.

Last, but not least I am thankful to all my valuable readers and critics for encouraging me to do more and more work.

Think Green!

Sanjay K. Sharma
drsanjay1973@gmail.com

About The Editor

Prof. (Dr.) Sanjay K. Sharma is a very well-known author and editor of many books, research journals, and hundreds of articles from the past 20 years. His recently published books are "Handbook on Applications of Ultrasound: Sonochemistry and Sustainability," "Green Chemistry for Environmental Sustainability" (both from CRC Taylor & Francis Group, LLC, Florida, Boca Raton, USA), and "Handbook of Applied Biopolymer Technology: Synthesis, Degradation and Applications" (From Royal Society of Chemistry, UK).

He has also been appointed as the Series Editor by Springer's London for their prestigious book Series "Green Chemistry for Sustainability." His work in the field of Green Corrosion Inhibitors is very well recognized and praised by the international research community. Other than this, he is known as a person who is dedicated to educate people about environmental awareness, especially for rain water harvesting.

Presently, he is working as Professor of Chemistry at Jaipur Engineering College & Research Centre, JECRC Foundation, Jaipur (Rajasthan), India where he is teaching Engineering Chemistry and Environmental Engineering Courses to B. Tech. students and pursuing his research interests. Dr. Sharma has delivered many guest lectures on different topics of applied chemistry in various reputed institutions. His students appreciate his teaching skills and hold him in high esteem.

He is a member of the American Chemical Society (USA), International Society for Environmental Information Sciences (ISEIS, Canada), and Green Chemistry Network (Royal Society of Chemists, UK) and is also a life member of various international professional societies, including the International Society of Analytical Scientists, Indian Council of Chemists, International Congress of Chemistry and Environment, and Indian Chemical Society.

Dr. Sharma has 12 books on chemistry from national–international publishers and over 40 research papers of national and international repute to his credit.

Dr. Sharma is also serving as the Editor-in-Chief for four international research journals: the "RASAYAN Journal of Chemistry," "International Journal of Chemical, Environmental and Pharmaceutical Research," "International Journal of Water

Treatment & Green Chemistry,'' and ''Water: Research & Development.'' He is also a reviewer for many other international journals including the prestigious Green Chemistry Letters & Reviews.

List of Contributors

Todd R. Allen
University of Wisconsin-Madison
Department of Engineering
Physics
1500 Engineering Drive
Madison, WI 53706
USA

S. Anand Kumar
Anna University
Department of Chemistry
Sardar Patel Road
Gundy
Chennai 600025
Tamilnadu
India

Y. Frank Cheng
University of Calgary
Department of Mechanical
Engineering
2500 University drive NW
Calgary
Alberta, T2N 1N4
Canada

Flavio Deflorian
University of Trento
Department of Materials
Engineering and Industrial
Technologies
Via Mesiano 77
38123 Trento
Italy

Konstantinos D. Demadis
University of Crete
Department of Chemistry
Crystal Engineering
Growth and Design Laboratory
Voutes Campus
P.O. Box 2208
Heraklion Crete 71003
Greece

Michele Fedel
University of Trento
Department of Materials
Engineering and
Industrial Technologies
Via Mesiano 77
38123 Trento
Italy

Essam Khamis
Alexandria University
Faculty of Science
Mohram Bey
Alexandria
Egypt

Ananad Sawroop Khanna
Indian Institute of Technology
Corrosion Science & Engineering
Adi Shankaracharya Marg, Powai
Bombay 400076
Maharashtra
India

Girish Mehta
S V National Institute of
Technology
Department of Applied Chemistry
Ichchanath
Dumas Road
Surat 395007, Gujarat
India

Ackmez Mudhoo
University of Mauritius
Department of Chemical &
Environmental Engineering
Faculty of Engineering
Réduit
Mauritius

Maria Papadaki
University of Crete
Department of Chemistry
Crystal Engineering
Growth and Design Laboratory
Voutes Campus
P.O. Box 2208
Heraklion Crete 71003
Greece

Niketan Patel
S V National Institute of
Technology
Department of Applied Chemistry
Ichchanath
Dumas Road
Surat 395007, Gujarat
India

Stefano Rossi
University of Trento
Department of Materials
Engineering and
Industrial Technologies
Via Mesiano 77
38123 Trento
Italy

Vedula Sankar Sastri
Sai Ram Consultant
1839 Greenacre Crescent
Ottawa
Ontario, K1J 6S7
Canada

Norio Sato
Hokkaido University
Graduate School of Engineering
Kita-13
Nishi-8
Kita-ku
Sapporo 060-8628
Hokkaido
Japan

R. Savitha
Anna University
Department of Chemistry
Sardar Patel Road
Guindy
Chennai 600025
Tamilnadu
India

Alka Sharma
University of Rajasthan
Department of Chemistry
JLN Marg
Jaipur 302044
Rajasthan
India

Sanjay K. Sharma
Jaipur Engineering College &
Research Center
Department of Chemistry &
Environmental Engineering
JECRC Foundation
Jaipur 302022
India

Lizhen Tan
Oak Ridge National Laboratory
One Bethel Valley Road
Oak Ridge
Tennessee 37831-6151
USA

Dimitrios Varouchas
University of Crete
Department of Chemistry
Crystal Engineering
Growth and Design Laboratory
Voutes Campus
P.O. Box 2208
Heraklion Crete 71003
Greece

Ying Yang
CompuTherm LLC
437 S. Yellowstone
Dr. Suite 217
Madison, WI 53719
USA

Caterina Zanella
University of Trento
Department of Materials
Engineering and
Industrial Technologies
Via Mesiano 77
38123 Trento
Italy

1
Basics of Corrosion Chemistry

Norio Sato

1.1
Introduction

Metallic materials in practical use are normally exposed to corrosion in the atmospheric and aqueous environments. Metallic corrosion is one of the problems we have often encountered in our industrialized society; hence it has been studied comprehensively since the industrial revolution in the late eighteenth century. Modern corrosion science was set off in the early twentieth century with the local cell model proposed by Evans [1] and the corrosion potential model proved by Wagner and Traud [2]. The two models have joined into the modern electrochemical theory of corrosion, which describes metallic corrosion as a coupled electrochemical reaction consisting of anodic metal oxidation and cathodic oxidant reduction. The electrochemical theory is applicable not only to wet corrosion of metals at normal temperature but also to dry oxidation of metals at high temperature [3].

Metallic materials corrode in a variety of gaseous and aqueous environments. In this chapter, we restrict ourselves to the most common corrosion of metals in aqueous solution and in wet air in the atmosphere. In general, metallic corrosion produces in its initial stage soluble metal ions in water, and then, the metal ions develop into solid corrosion precipitates such as metal oxide and hydroxide. We will discuss the whole process of metallic corrosion from the basic electrochemical standpoint.

1.2
Metallic Corrosion

1.2.1
Basic Processes

The basic process of metallic corrosion in aqueous solution consists of the anodic dissolution of metals and the cathodic reduction of oxidants present in the solution:

$$M_M \rightarrow M^{2+}_{aq} + 2e^-_M \quad \text{anodic oxidation} \tag{1.1}$$

Green Corrosion Chemistry and Engineering: Opportunities and Challenges, First Edition.
Edited by Sanjay K. Sharma.

$$2Ox_{aq} + 2e^-_M \rightarrow 2Red(e^-_{redox})_{aq} \quad \text{cathodic oxidation} \tag{1.2}$$

In the formulae, M_M is the metal in the state of metallic bonding, M^{2+}_{aq} is the hydrated metal ion in aqueous solution, e^-_M is the electron in the metal, Ox_{aq} is an oxidant, $Red(e^-_{redox})_{aq}$ is a reductant, and e^-_{redox} is the redox electron in the reductant. The overall corrosion reaction is then written as follows:

$$M_M + 2Ox_{aq} \rightarrow M^{2+}_{aq} + 2Red(e^-_{redox})_{aq} \tag{1.3}$$

These reactions are charge-transfer processes that occur across the interface between the metal and the aqueous solution, hence they are dependent on the *interfacial potential* that essentially corresponds to what is called the *electrode potential* of metals in electrochemistry terms. In physics terms, the electrode potential represents the energy level of electrons, called the *Fermi level*, in an electrode immersed in electrolyte.

For normal metallic corrosion, in practice, the cathodic process is carried out by the reduction of hydrogen ions and/or the reduction of oxygen molecules in aqueous solution. These two cathodic reductions are *electron transfer* processes that occur across the metal–solution interface, whereas anodic metal dissolution is an *ion transfer* process across the interface.

1.2.2
Potential-pH Diagram

Thermodynamics shows that an electrode reaction is reversible at its equilibrium potential, where no net reaction current is observed. We then learn that the anodic reaction of metallic corrosion may occur only in the potential range more positive than its equilibrium potential and that the cathodic reaction of oxidant reduction may occur only in the potential range more negative than its equilibrium potential. Moreover, it is known that metallic corrosion in aqueous solution is dependent not only on the electrode potential but also on the acidity and basicity of the solution, that is, the solution pH.

The thermodynamic prediction of metallic corrosion was thus illustrated by Pourbaix [4] in the form of *potential–pH diagrams*, as shown for iron corrosion in Figure 1.1. The corrosion of metallic iron may occur in the potential–pH region where hydrated ferrous ions Fe^{2+}, ferric ions Fe^{3+}, and hydroxo-ferrous ions $Fe(OH)^-_3$ are stable. No iron corrosion occurs in the region where metallic iron is thermodynamically stable at relatively negative electrode potentials. In the regions where solid iron oxides and hydroxides are stable, no iron corrosion into water is expected to develop and the iron surface is covered with solid oxide films. In the diagram, we also see the equilibrium potentials of the hydrogen and oxygen electrode reactions. Atmospheric oxygen may cause iron corrosion in the potential range more negative than the oxygen equilibrium potential, E_{O_2/H_2O}, while hydrogen ions in aqueous solution may carry iron corrosion in the potential range more negative than the hydrogen equilibrium potential, E_{H^+/H_2O}.

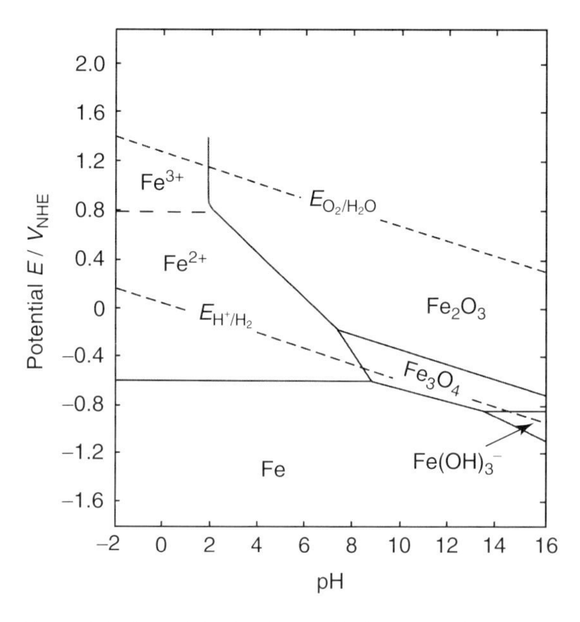

Figure 1.1 Potential–pH diagram for iron corrosion in water at room temperature. E_{O_2/H_2O} is the equilibrium potential for the oxygen electrode reaction, E_{H^+/H_2O} is the equilibrium potential for the hydrogen electrode reaction, and V_{NHE} is volt on the normal hydrogen electrode scale [4].

We note that the potential–pH diagram normally assumes metallic corrosion in pure water containing no foreign solutes. The presence of foreign solutes in aqueous solution may affect the corrosion and anticorrosion regions in the potential–pH diagram. We see in the literature a number of potential–pH diagrams for metallic corrosion in the presence of foreign solutes such as chloride and sulfides [4, 5].

1.2.3 Corrosion Potential

An electrode of metal corroding in aqueous solution has an electrode potential, which is called the *corrosion potential*. As a matter of course, the corrosion potential stands somewhere in the range between the equilibrium potential of the anodic metal dissolution and that of the cathodic oxidant reduction. It comes from the kinetics of metallic corrosion that at the corrosion potential, the anodic oxidation current of the metal dissolution is equal to the cathodic reduction current of the oxidant. The corrosion kinetics is usually described by the electrode potential versus reaction current curves of both the anodic oxidation and the cathodic reduction, as schematically shown in Figure 1.2, which electrochemists call the *polarization curves* of corrosion reactions. We see in Figure 1.2 that the intersecting point of the anodic and cathodic polarization curves represents the state of corrosion, namely, the corrosion potential and the corrosion current. We then realize that the corrosion

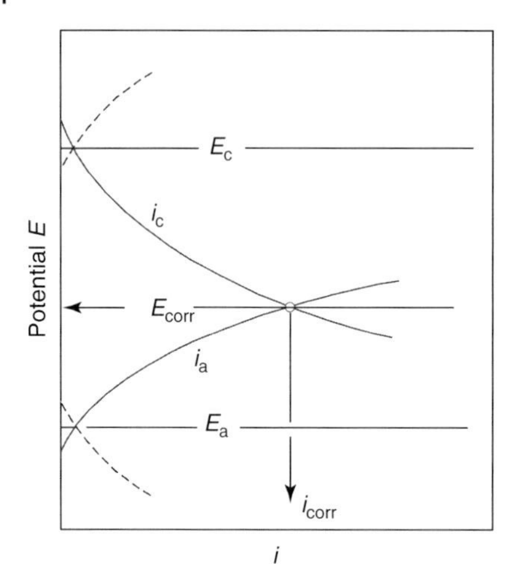

Figure 1.2 Conceptual potential–current curves of anodic and cathodic reactions for metallic corrosion; i_a is the anodic reaction current, i_c is the cathodic reaction current, i_{corr} is the corrosion current, E_a is the equilibrium potential of the anodic reaction, E_c is the equilibrium potential of the cathodic reaction, and E_{corr} is the corrosion potential.

potential is determined by both the anodic and cathodic polarization curves of the corrosion reactions.

The corrosion rate of metals may be controlled by either the anodic or the cathodic reaction. In most cases of metallic corrosion, the cathodic hydrogen ion reduction controls the rate of metallic corrosion in acidic solution, while in neutral solution, the cathodic oxygen reduction preferentially controls the corrosion rate. If the corrosion potential comes out far away from the equilibrium potential of the cathodic reaction, the corrosion rate will be controlled by the cathodic reaction. In practice, we see that metallic corrosion is often controlled by the oxygen diffusion toward the corroding metal surface, in which the corrosion potential is far more negative than the oxygen equilibrium potential.

1.2.4
Anodic Metal Dissolution

Electrochemical kinetics gives the reaction current, i_a, of anodic metal dissolution as an exponential function of the electrode potential, E, of the metal as follows:

$$i_a = K_a \exp\left(\frac{\alpha_a E}{kT}\right) \tag{1.4}$$

In Eq. (1.4), K_a and α_a are parameters. The anodic dissolution current of metallic iron, in fact, increases exponentially with the anodic electrode potential in acid solution as shown in Figure 1.3 [6].

Anodic metal dissolution depends not only on the electrode potential but also on the acidity and foreign solutes present in the aqueous solution. It is a received

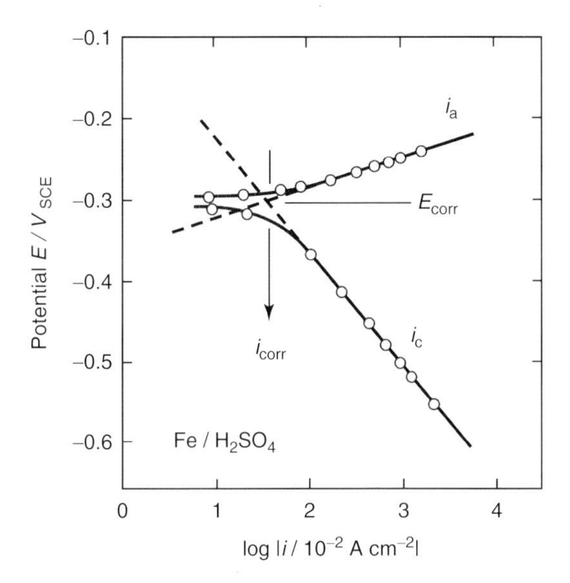

Figure 1.3 Potential–current curves for corroding iron in sulfuric acid solution at pH 1.7 at 25 °C; V_{SCE} is the volt on the saturated calomel electrode scale [6].

understanding that the anodic dissolution current of metallic iron depends on the concentration of hydroxide ions in the solution. Hydrated anions other than hydroxide ions also have some effects on the anodic metal dissolution. Hydrated hydroxide ions are found to accelerate the anodic iron dissolution in weekly acidic solution; whereas they decelerate (hydrated hydrogen ions accelerate) the iron dissolution in strongly acidic solution [7]. Furthermore, in acidic solution, hydrated chloride ions accelerate the anodic iron dissolution in relatively concentrated chloride solution, while they inhibit the iron dissolution in relatively dilute chloride solution [7]. These facts suggest that anions of different sorts compete with one another in participating in the process of anodic metal dissolution probably through their adsorption on the metal surface forming activated intermediates, such as $FeOH^{+}_{ad}$ and $FeCl^{+}_{ad}$, which will determine the metal dissolution rate [7].

It is worth noting that, in general, the surface of metal is *soft acid* in the Lewis acid–base concept [8] and tends to adsorb ions and molecules of *soft base*, forming covalent bonds between the metal surface and the adsorbates. It is also noteworthy that as the anodic metal potential increases in the more positive direction, the Lewis acidity of the metal surface may gradually turn from *soft acid* to *hard acid*. The anions of *soft base* adsorbed in the range of less positive potentials are then replaced in the range of more positive potentials by anions of *hard base* such as hydroxide ions and water molecules. In general, iodide ions I^-, sulfide ions S^-, and thiocyanate ions SCN^- are the *soft bases*, while hydroxide ions OH^-, fluoride ions F^-, chloride ions Cl^-, phosphate ions PO_4^{3-}, sulfate ions SO_4^{2-}, and chromate ions CrO_4^{2-} are the *hard bases*. Bromide ions Br^- and sulfurous ions SO_3^{2-} stand somewhere between the *soft base* and the *hard base*.

1.2.5 Cathodic Oxidant Reduction

The cathodic current, i_c, of oxidant reduction is also an exponential function of the electrode potential, E, of the metal as follows:

$$i_c = K_c \exp\left(\frac{-\alpha_c E}{kT}\right) \tag{1.5}$$

For metallic iron in acid solution, where the hydrogen ion reduction carries the cathodic reaction of corrosion, the cathodic current increases exponentially with increasing cathodic electrode potential in the more negative direction as shown in Figure 1.3.

The cathodic reaction for usual metallic corrosion is carried by the hydrogen reaction and/or the oxygen reaction. For the hydrogen reaction, there are two cathodic processes that produce hydrogen gas at metal electrodes: one is the reduction of hydrogen ions and the other is the reduction of water molecules.

$$2H^+_{aq} + 2e^-_M \rightarrow H_{2,gas} \tag{1.6}$$

$$2H_2O + 2e^-_M \rightarrow H_{2,gas} + 2OH^-_{aq} \tag{1.7}$$

Hydrogen ion reduction preferentially occurs in acidic solution, and its reaction current increases with increasing hydrogen ion concentration. Water molecule reduction occurs in weakly acidic and neutral solutions. The cathodic polarization curves (potential–current curves) of these two reactions are different from each other. The cathodic current of hydrogen ion reduction increases with increasing cathodic electrode potential until it reaches a limiting current of hydrogen ion diffusion toward the metal surface, and beyond a certain cathodic potential, the cathodic current of water molecule reduction emerges and increases with the cathodic electrode potential [9].

For oxygen reduction, similarly, there are two cathodic processes: one involves hydrogen ions and the other requires water molecules.

$$O_{2,gas} + 4H^+_{aq} + 4e^-_M \rightarrow 2H_2O \tag{1.8}$$

$$O_{2,gas} + 2H_2O + 4e^-_M \rightarrow 4OH^-_{aq} \tag{1.9}$$

The oxygen reduction involving hydrogen ions takes place preferentially in acidic solution with the reaction current limited not only by the oxygen diffusion but also by the hydrogen ion diffusion toward the metal surface; whereas the oxygen reduction requiring water molecules develops in neutral and alkaline solutions and its reaction current is limited by the oxygen diffusion [9]. It is most likely in practical environments that the corrosion of metals is controlled by the diffusion of oxygen toward the surface of corroding metals.

1.3 Metallic Passivity

1.3.1 Passivity of Metals

Metallic passivity was discovered in 1790 by Keir [10], who found that metallic iron violently corroding in the active state in concentrated nitric acid solution suddenly turned into the passive state where almost no corrosion was observed. It was not until 1960s that we confirmed the presence of an oxide film several nanometers thick on the surface of passivated metals [11]. Latest overviews on metallic passivity may be referred to in the literature of corrosion science [12].

We illustrate metallic passivity with the potential–current curve of anodic metal dissolution for metallic iron, nickel, and chromium in acid solution as shown in Figure 1.4. Anodic metal passivation occurs at a certain potential, called the *passivation potential*, E_P, beyond which the anodic current of metal dissolution drastically decreases to a negligible level. It is an observed fact that the passivation potential depends on the solution acidity, lineally shifting in the more positive direction with decreasing solution pH. This fact thermodynamically suggests that metallic passivity is caused by the formation of an oxide film on the metal, which is extremely thin and invisible to the naked eye.

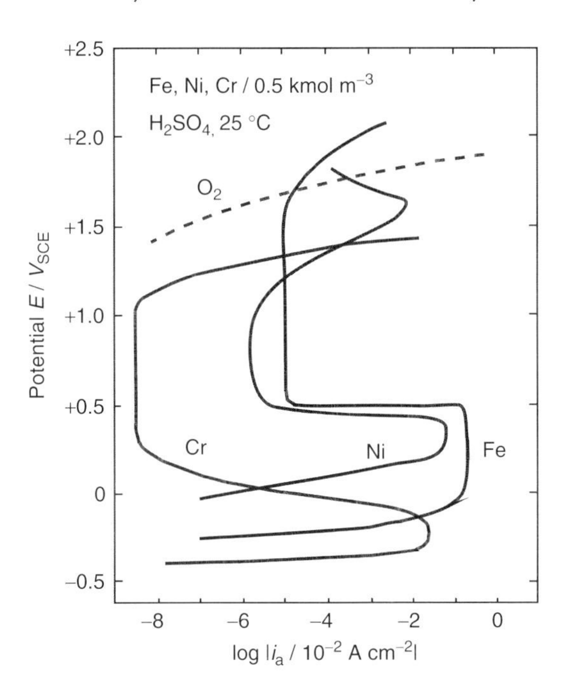

Figure 1.4 Potential–current curves for anodic dissolution of iron Fe, nickel Ni, and chromium Cr in 0.5 mol m^{-3} sulfuric acid solution at 25 °C; O_2 is anodic oxygen evolution current.

In the potential range of passivity, a nanometer-thin oxide film is formed on the metal, which is called the *passive film*. The film grows in thickness with increasing anodic potential at the rate of 1–3 nm V^{-1} equivalent to an electric field of 10^6–10^7 V cm^{-1} across the film [13]. For most of the iron group metals, the passive film is less than several nanometers in thickness in the potential region where it is stable. For some metals such as aluminum and titanium, the passive oxide film can be made thick up to several hundred nanometers by increasing anodic potential; the thick oxide is frequently called the *anodic oxide*.

In the potential range where the passive state is stable, as shown in Figure 1.4, the metal anode normally holds an extremely small, potential-independent metal dissolution current, which is equivalent to the dissolution rate of the passive film itself. We see that the anodic metal dissolution current in the passive state is controlled by the dissolution rate of the passive film. The potential-independent dissolution of passive metals results from the fact that the interfacial potential between the passive oxide film and the solution, which controls the film dissolution rate, remains constant independent of the anodic potential, although depending on the solution pH [12, 14, 15].

For some metals, the passive state turns beyond a certain potential to be what is called the *transpassive* state, where the anodic dissolution current for the most part increases exponentially with the anodic potential as shown in Figure 1.4. The anodic metal dissolution in the transpassive state is thus controlled by the potential-dependent dissolution of the transpassive film. In the transpassive potential range, the interfacial potential between the film and the solution is not constant but depends on the anodic potential of the metal. The passive film is stable as long as the Fermi level (the electrode potential) of the metal anode is within the band gap of electron energy between the conduction and valence bands of the film, a situation which realizes the nonmetallic nature of the interface and hence makes the interfacial potential independent of the metal potential [12, 14, 15]. As the anodic potential increases, the Fermi level finally reaches the valence band edge of the film at the film–solution interface and the quasi-metallization (electronic degeneracy) is realized at the same. The film–solution interfacial potential, hence, turns to be dependent on the anodic metal potential, and as a result, potential-dependent transpassive dissolution occurs beyond the transpassivation potential, E_{TP}.

Let us see the anodic metal dissolution of iron, nickel, and chromium in acid solution as shown in Figure 1.4. While going into solution as hydrated ferrous ions in the active state, metallic iron in the passive state dissolves in the form of hydrated ferric ions, indicating that the passive film is ferric oxide, Fe_2O_3.

$$2Fe + 3H_2O \rightarrow Fe_2O_3 + 6H^+_{aq} + 6e^-_M \quad \text{passive film formation} \tag{1.10}$$

$$Fe_2O_3 + 6H^+_{aq} \rightarrow 2Fe^{3+}_{aq} + 3H_2O \quad \text{passive film dissolution} \tag{1.11}$$

These two reactions proceed simultaneously, maintaining the passive film at a steady thickness, which increases with increasing anodic potential.

Anodic nickel dissolution produces divalent nickel ions both in the active and passive states, implying that the passive film is divalent nickel oxide.

$$\mathrm{Ni} + \mathrm{H_2O} \rightarrow \mathrm{NiO} + 2\mathrm{H^+_{aq}} + \mathrm{e^-_M} \quad \text{passive film formation} \tag{1.12}$$

$$\mathrm{NiO} + 2\mathrm{H^+_{aq}} \rightarrow \mathrm{Ni^{2+}_{aq}} + \mathrm{H_2O} \quad \text{passive film dissolution} \tag{1.13}$$

In the transpassive state, as mentioned earlier, the film dissolution rate is potential dependent in contrast to the passive state in which it is potential independent. Since the dissolution current increases with the anodic potential, the thickness of the transpassive film seems to decrease with the anodic potential in the steady state. Beyond the transpassive potential range of nickel, the transpassive divalent oxide film is assumed to change near the oxygen evolution potential into a trivalent oxide film causing a decrease in the anodic nickel dissolution current [11].

For chromium, the anodic dissolution produces divalent chromium ions in the active state and the passivation occurs forming an extremely thin, trivalent chromium oxide film on the metal surface.

$$2\mathrm{Cr} + 3\mathrm{H_2O} \rightarrow \mathrm{Cr_2O_3} + 6\mathrm{H^+_{aq}} + 6\mathrm{e^-_M} \quad \text{passive film formation} \tag{1.14}$$

$$\mathrm{Cr_2O_3} + 6\mathrm{H^+_{aq}} \rightarrow 2\mathrm{Cr^{3+}_{aq}} + 3\mathrm{H_2O} \quad \text{passive film dissolution} \tag{1.15}$$

$$\mathrm{Cr_2O_3} + 5\mathrm{H_2O} + 6\mathrm{e^-_M} \rightarrow 2\mathrm{CrO_{4,aq}}^{2-} + 10\mathrm{H^+_{aq}} \quad \text{transpassive film dissolution} \tag{1.16}$$

The transpassive dissolution of metallic chromium is the oxidative dissolution of trivalent chromic oxide into soluble hexavalent chromate ions in acidic solution; hence the anodic transpassive dissolution proceeds through the formation of a chromic oxide film on the metal surface.

1.3.2
Passivation of Metals

A metallic electrode may be made passive if its corrosion potential is held in the potential range of passivity. The corrosion potential is determined, as mentioned earlier, by both the anodic metal dissolution current and the cathodic oxidant reduction current. As shown in Figure 1.5, the corrosion potential remains in the active state as long as the cathodic current is less than the maximum current of anodic metal dissolution; whereas it goes to the passive potential range when the cathodic current exceeds the anodic dissolution current. An unstable passive state arises if the cathodic potential–current curve crosses the anodic potential–current curve at two potentials, one in the passive state and the other in the active state. A metallic electrode in the unstable passive state, once its passivity breaks down, never repassivates because the cathodic current of oxidant reduction is insufficient in magnitude for the activated metal to clear its anodic dissolution current peak.

It has been observed that metallic nickel corrodes in acidic solution but passivates in basic solution: the transition from the active corrosion to passivation occurs at pH 6 in sulfate solution [16]. The corrosion of nickel in weakly acidic and neutral solutions is controlled by cathodic oxygen reduction whose current is limited by

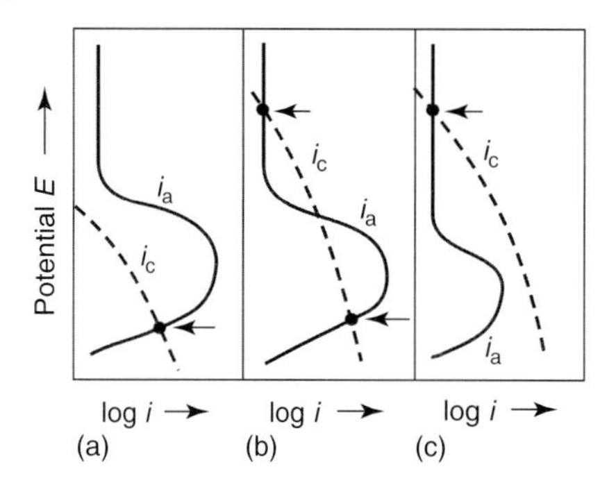

Figure 1.5 Passivation of metals and its stability. (a) Active corrosion, (b) unstable passivity, (c) stable passivity; i_a is the anodic current and i_c is the cathodic current.

the oxygen diffusion toward the metal surface. The nickel dissolution current peak in the active state decreases with increasing solution pH and becomes less than the cathodic oxygen diffusion current beyond pH 6. Consequently, metallic nickel passivates in a solution more basic than pH 6, which is called the *passivation* pH of nickel.

Let us see the oxidizing agent of nitrite ion NO_2^- that provides a cathodic reaction for the corrosion and passivation of metallic iron in weakly acidic and neutral solutions.

$$NO^-_{2,aq} + 5H_2O + 6e^-_M \rightarrow NH_3 + 7OH^-_{aq} \quad \text{cathodic reaction} \tag{1.17}$$

$$2Fe \rightarrow 2Fe^{2+}_{aq} + 4e^-_M$$

$$2Fe + 3H_2O \rightarrow Fe_2O_3 + 6H^+_{aq} + 6e^-_M \quad \text{anodic reaction} \tag{1.18}$$

Metallic iron remains in the active state if the cathodic reaction current is insufficient for the metal to passivate; whereas it turns out to be passive forming a surface film of ferric oxide if the cathodic reaction current surmounts the metal dissolution current peak. Owing to its relatively more-positive redox potential and its greater reaction current, an amount of nitrite salt readily brings metallic iron into the passive state, and thus, it provides an effective passivating reagent for iron and steel.

It is also noted that chromate ions, CrO_4^{2-}, oxidize metallic iron to form a passive film of chromic–ferric mixed oxides on the metal surface.

$$CrO_{4,aq}{}^{2-} + 10H_2O + 6e^-_M \rightarrow Cr_2O_3 + 5H_2O \quad \text{cathodic reaction} \tag{1.19}$$

$$2Fe + 3H_2O \rightarrow Fe_2O_3 + 6H^+_{aq} + 6e^-_M \quad \text{anodic reaction} \tag{1.20}$$

Chromate is one of the strongest oxidants to passivate metallic materials. In the same way, molybdate and tungstate may also make metallic iron passive, although their oxidizing capacity may not always be sufficient.

1.3.3
Passive Films

The passive oxide film on metals is very thin, of the order of several nanometers, and hence, sensitive to the environment in which it is formed. In the formation and growth processes of the film, the oxide ions migrate from the solution across the film to the metal–oxide interface forming an inner oxide layer, while the metal ions migrate from the metal to the oxide–solution interface to react with adsorbed water molecules and solute anions forming an outer oxide layer, which occasionally incorporates anions other than oxide ions into itself. The anion incorporation occurs only when the migrating metal ions react with the adsorbed anions. The ratio of the thickness of the outer anion-incorporating layer to the overall layer is expressed by the transport number, τ_M, of the metal ion migration during the film growth. The transport number was found to be $\tau_M = 0.7–0.8$ for an anodic oxide film 65 nm thick formed on aluminum in phosphate solution [17].

The passive film is mostly amorphous, but as the film grows thicker, it may turn to be crystalline. For the passivity of metallic titanium in sulfuric acid solution, the passive film appears to change from amorphous to crystalline beyond the anodic potential of about 8 V, probably because of the internal stress created in the film [18]. The passive film is either an insulator or a semiconductor. For metallic iron, titanium, tin, niobium, and tungsten, the passive film is an n-type semiconductor with donors in high concentration. Some metals such as metallic nickel, chromium, and copper make the passive film a p-type semiconductor oxide. The passive films on metallic aluminum, tantalum, and hafnium are insulator oxides.

We may classify the passive oxide films into two categories: the network former (glass former) and the the network modifier [19]. The network former, which includes metallic silicon, aluminum, titanium, zirconium, and molybdenum, normally forms a single-layered oxide film. On the other hand, the network modifier, which includes metallic iron, nickel, cobalt, and copper, tends to form a multilayered oxide film, such as a cobalt oxide film, consisting of an inner divalent oxide layer and an outer trivalent oxide layer ($Co/CoO/Co_2O_3$). High-valence metal oxides normally appear to be more corrosion resistive than low-valence metal oxides. The anodic formation of network-forming oxides is most likely carried through the inward oxide ion migration to the metal–oxide interface and hence will probably produce a dehydrated compact film containing no foreign anions other than oxide ions; whereas, network-modifying oxides appear to grow through the outward metal ion migration to the oxide–solution interface forming a more or less defective film occasionally containing foreign anions.

It is worth noting that since the passive film is so extremely thin, electrons easily transfer across the film by the quantum mechanical electron tunneling mechanism, irrespective of whether the passive film is an insulator or a semiconductor. By contrast, however, no ionic tunneling is allowed to occur across the passive film. The passive film thus constitutes a barrier layer to ion transfer but not to electron transfer. Any redox electron transfer reaction is therefore allowed to occur on the passive film-covered metal surface just like on the metal surface without any film.

1.3.4
Chloride-Breakdown of Passive Films

The passive film on metals may break down in the presence of aggressive ions, such as chloride ions, in solution, and the breakdown site may trigger a localized corrosion of the underlying metals. The chloride-breakdown of passive films, in general, occurs beyond a certain potential, which we call the *film breakdown potential*, E_b. Either repassivation or pitting corrosion then follows at the breakdown site, as schematically shown in Figure 1.6. Pitting corrosion is also characterized by a threshold potential, called the *pitting potential*, E_{pit}, above which pitting grows but below which pitting ceases to occur. These two potentials, E_b and E_{pit}, are influenced by the concentrations of chloride ions and hydrogen ions in the solution.

There is a marginal chloride concentration below which no film breakdown occurs. For chloride-breakdown of the passive film on metallic iron, the concentration of chloride ions required for the film breakdown depends on the film thickness, defects in the film, the electric field intensity in the film, and pH of the solution [20]. It is also found that the passive film locally dissolves and thins down before the underlying metal begins pitting at the film breakdown site [21, 22]. It is then likely that the film breakdown results not from a mechanical rupture but from a localized mode of film dissolution because of the adsorption of chloride ions.

It is frequently seen that the passive film preferentially breaks down at the sites of crystal grain boundaries, nonmetallic inclusions, and flaws on the metal surface. For stainless steels, the passivity breakdown and pit initiation most likely occur at the site of nonmetallic inclusions of MnS. It is noted that any localized phenomena are nondeterministic and in general, somehow stochastic. Chloride-breakdown of passive films on stainless steels was found to come out in accordance with a stochastic distribution [23, 24].

The pitting potential, E_{pit}, at which pitting begins to grow, arises at a potential either more positive (more anodic) or less positive (more cathodic) than the film

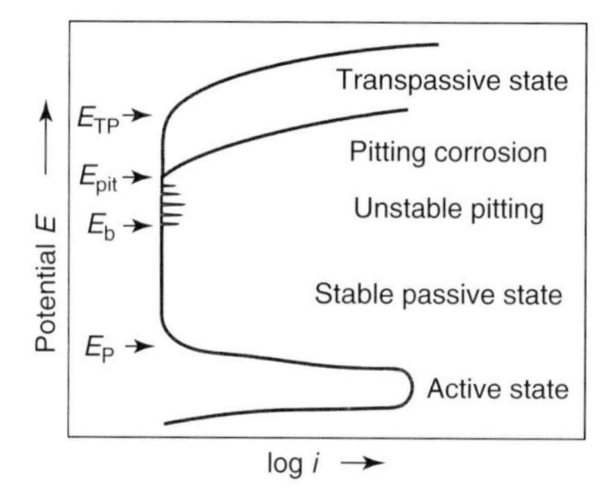

Figure 1.6 Schematic potential–current curves for metallic passivation, passive-film breakdown, pitting dissolution, and transpassivation; E_b is the film breakdown potential, E_{pit} is the pitting potential, E_P is the passivation potential, and E_{TP} is the transpassivation potential.

breakdown potential, E_b. When the film breakdown potential is less-positive than the pitting potential, the breakdown site repassivates, as we observe for some stainless steels in acid solution [25]. On the other hand, pitting corrosion follows film breakdown when the film breakdown potential emerges more positive than the pitting potential, as we observe for metallic iron in acid solution [22].

For the mechanism of chloride-breakdown of passive films, one of the currently prevailing models is the ionic point defect model that assumes injection of metal ion vacancies into the passive film at the adsorption site of chloride ions. The ionic point defects thus injected migrate to and accumulate at the metal–film interface, finally creating a void there to break the film down [26]. Another model is the electronic point defect model that assumes injection of an electronic defect level localized at the film–solution interface. The electronic interfacial state thus injected causes local quasi-metallization at the adsorption site of chloride ions, finally resulting in a local mode of film dissolution [14, 15, 27]. There have also been several chemical models for passivity breakdown, which more or less assume the formation of a soluble chloride complex at the chloride adsorption site.

We note at the end that the passivity breakdown is different from the pitting that follows: the former is a process associated with the passive film itself, whereas the underlying metal is responsible for the latter.

1.4 Localized Corrosion

1.4.1 Pitting Corrosion

Pitting corrosion of metals in general occurs in the potential range more positive than the *pitting potential*, E_{pit} [28]. For usual stainless steel in 1 mol dm^{-3} sodium chloride solution, the pitting potential arises around +0.3 V on the normal hydrogen electrode (NHE) scale [29]. Once metallic pitting sets in, the anodic metal dissolution current increases with the anodic potential as schematically shown in Figure 1.6. The pit initially grows in a semispherical shape with the pit solution that acidifies and concentrates in soluble metal salts. For 304 stainless steels in neutral chloride media at $pH = 6 \sim 8$, the local pH in the pit falls down to $pH = 1 \sim 2$ [29]. The kinetics of pit dissolution appears to be different from that of metal dissolution in the active state. The pitting dissolution current density, i_{pit}, is given by an exponential function of the anodic potential, E.

$$E = a + b \log i_{pit} \tag{1.21}$$

In acid solution, the coefficient b (Tafel constant) is 0.20 V for metallic iron [30] and 0.30 V for stainless steel [29]. These Tafel constants are much greater than those ($0.03 \sim 0.1$ V) normally observed for anodic metal dissolution in the active state. The metal dissolution in the pit proceeds in an acidified, concentrated salt

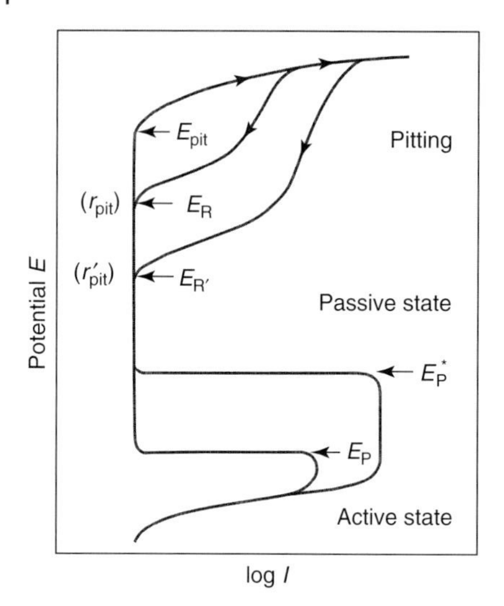

Figure 1.7 Schematic potential–current curves for metallic pitting dissolution, and pit repassivation; r_{pit} is the pit radius ($r'_{pit} > r_{pit}$), E_P^* is the passivation–depassivation potential for the critical pit solution, and E_P is the passivation–depassivation potential for the solution outside the pit.

solution with an electropolishing mode of dissolution, which is different from the usual mode of metal dissolution in the potential range of the active state.

As the electrode potential of pitting metal is made more negative, we come to a certain potential at which the pitting metal dissolution ceases to occur, resulting in repassivation of the pit, a certain potential which we call the *pit-repassivation potential*, E_R. The pit-repassivation potential is found to move more negative with increasing pit size, as schematically shown in Figure 1.7. We may then assume that a pit embryo takes a certain critical size at the pitting potential, E_{pit}, at which the pitting starts to occur. For stainless steel in acidic chloride solution, the smallest pit size for the initiation of pitting was estimated at 0.01–0.02 mm [29].

From the kinetics of pitting dissolution and mass transport in a semispherical pit, it is deduced that the pit-repassivation potential E_R is a logarithmic function of pit radius r_{pit} [29].

$$E_R = a + b \log r_{pit} \tag{1.22}$$

We in fact observe that a logarithmic dependence of the pit-repassivation potential on the pit radius is held for stainless steel in acid solution [31].

It appears that the pitting mode of metal dissolution requires highly concentrated acidic chloride in the pit solution, whose chloride ions must be more concentrated than a certain critical value, $c^*_{Cl^-}$. For usual stainless steel in acid solution, the critical chloride concentration was estimated to be $c^*_{Cl^-} = 1.8\ \text{kmol m}^{-3}$ [29]. The pit solution is also highly acidified to keep running a polishing mode of pitting dissolution and preventing the pit from repassivation at potentials more positive than E_R. The critical chloride concentration, $c^*_{Cl^-}$, is thus accompanied with a critical level of acidity, $c^*_{H^+}$, above which no pit repassivation occurs. The critical chloride–hydrogen ion concentration, $c^*_{H^+}(c^*_{Cl^-})$, will then determine the pit stability. Pitting corrosion continues occurring if $c_{H^+}(c_{Cl^-}) \geq c^*_{H^+}(c^*_{Cl^-})$, but repassivation occurs if $c_{H^+}(c_{Cl^-}) < c^*_{H^+}(c^*_{Cl^-})$.

Metallic passivity in general occurs beyond the passivation potential, E_P, which becomes more positive with increasing solution acidity as mentioned earlier. It may hence be accepted that even for the critically acidified pit solution, the passivation potential may occur, which we call the critical *passivation–depassivation potential*, E^*_P, for the acidified pit solution. The E^*_P obviously stands much more positive than the normal passivation potential, E_P, for the solution outside the pit. It appears, then, that as the repassivation potential, E_R, of the pit goes down more negative than the critical depassivation potential E^*_P, pit repassivation never occurs, and that pitting corrosion in the electropolishing mode turns to be the usual mode of active metal dissolution with the metal potential falling down into the range of the active state.

Figure 1.8 schematically shows that the pit-repassivation potential E_R intersects with the critical depassivation potential E^*_P at a certain pit radius, r^*_{pit}, which is the largest size of the pit that is repassivable. A corroding pit never repassivates if it grows greater than its largest repassivable size r^*_{pit}.

The foregoing discussion indicates that both the pit-repassivation potential E_R and the critical depassivation potential E^*_P play a primary role in the stability of pitting corrosion of metals. For pitting corrosion to occur, a pit embryo first breaks out at a potential in the range of passivity, where the metal surface remains passive except for the pit site. As the pit grows in size, its repassivation

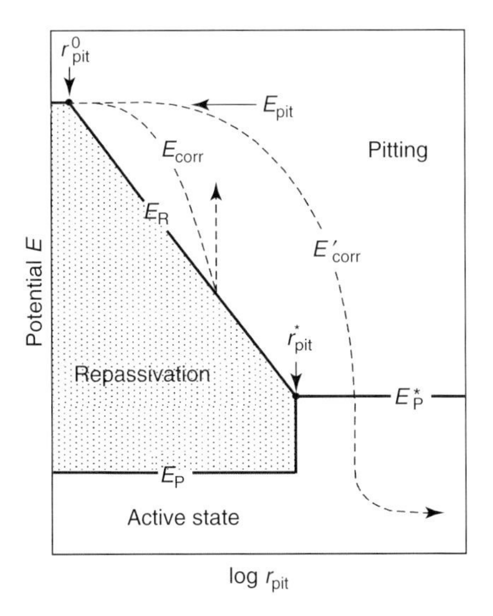

Figure 1.8 Schematic potential–size curves for metallic pitting and pit repassivation: r^*_{pit} is the maximum radius of repassivable pits, r^0_{pit} is the critical size of pit embryos, and E_{corr} is the corrosion potential of pitting metal.

potential E_R goes down more negative, while the critical depassivation potential E_P^* stays constant for the occluded pit solution at the critical chloride–hydrogen ion concentration, $c_{H^+}^*(c_{Cl^-}^*)$. It then appears that a corroding pit repassivates in the potential range between E_R and E_P^* as long as its repassivation potential is more positive than the critical depassivation potential ($E_R > E_P^*$). By contrast, a corroding pit never passivates if its repassivation potential goes down more negative than the critical depassivation potential ($E_R < E_P^*$), and consequently, no pit repassivation is expected to occur for the corrosion pit growing greater than the largest repassivable size, r_{pit}^*. The mode of local corrosion then changes from pitting corrosion to active pit corrosion, namely, from electropolishing to active dissolution.

The corrosion potential, as mentioned earlier, is controlled by the cathodic oxidant reduction. In the case of pitting corrosion, the cathodic reaction mainly occurs on the passive metal surface other than the pitting site. It is also known that the corrosion potential is made more positive and stable with increasing intensity and capacity of the oxidant reduction that supplies the cathodic current for corrosion. With the greater capacity of the cathodic reaction, the corrosion potential remains stable at relatively more positive potentials and the pit may be allowed to grow until its size exceeds the limiting value of repassivable pits, r_{pit}^*, before the potential falls down to its repassivation potential. Hence, no pit repassivation occurs as schematically shown in Figure 1.8. With the smaller capacity of the cathodic reaction, in contrast, the corrosion potential falls more steeply to the pit-repassivation potential before the pit grows over its limiting size of r_{pit}^*, and hence pit repassivation results.

We also note that the pitting potential is made more positive by lowering temperature until it reaches a certain temperature, called the *critical pitting temperature*, below which no pitting occurs [32]. For 316 stainless steels in sea water, the critical pitting temperature was about 30 °C [33]. Criterions of the temperature and potential for pitting corrosion all result from the stability requirement of pitting as described above.

1.4.2 Crevice Corrosion

Crevice corrosion is a type of localized corrosion that arises in structural crevices in the presence of chloride ions in aqueous solution. It normally occurs when the gap of a crevice is thinner than a certain width of the order of micrometers, which was estimated at 30–40 μm for usual stainless steel in acid solution [34]. The anodic metal dissolution in a crevice is usually coupled with the cathodic oxidant reduction that occurs outside the crevice, forming a local corrosion cell that involves the migration of hydrated ions through the crevice. For crevice corrosion to occur, a certain induction period is hence required to create a local corrosion cell between the inside and the outside of the crevice.

In general, crevice corrosion ceases growing in the potential range more negative than a certain critical potential, which we call the *crevice protection potential*, E_{crev}. Figure 1.9 schematically shows that the crevice is protected from corroding as the

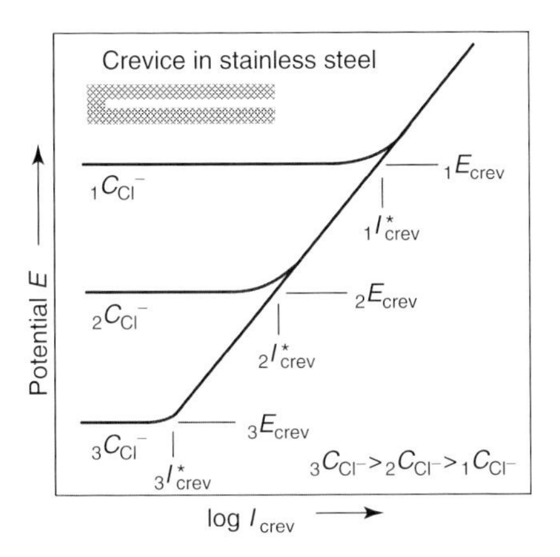

Figure 1.9 Schematic potential–current curves for anodic dissolution of a cylindrical crevice in stainless steel in neutral solutions at three different chloride ion concentrations; I_{crev} is the anodic crevice corrosion current, c_{Cl^-} is the chloride ion concentration outside the crevice, E_{crev} is the crevice protection potential, and I^*_{crev} is the crevice corrosion current at E_{crev} [34].

potential of the corroding crevice is made more negative than the crevice protection potential. The protection potential, E_{crev}, goes more negative with increasing chloride ion concentration, c_{Cl^-} outside the crevice [34].

$$E_{crev} = E^0_{crev} - \alpha \log c_{Cl^-} \tag{1.23}$$

The threshold crevice corrosion current, I^*_{crev}, at the protection potential E_{crev} is also found to be a linear function of the chloride ion concentration [34].

$$I^*_{crev} = I^0_{crev} - \beta c_{Cl^-}$$
$$\beta \propto \frac{1}{h_{crev}} \tag{1.24}$$

The coefficient, β, is found to be inversely proportional to the crevice depth, h_{crev} [35].

It is the acidification of the occluded crevice solution that sets off crevice corrosion. The critical acidity that sets off crevice corrosion is equal to what is called the *passivation–depassivation pH*, above which the crevice passivates spontaneously and below which it remains in the active state. The passivation potential of the crevice metal at the passivation–depassivation pH provides the characteristic protection potential, E^*_{crev}, within the crevice. The actual potential that we measure involves an *IR* drop, ΔE_{IR}, caused by the ionic current through the crevice. From the kinetics of metal dissolution and mass transport, it is revealed that the crevice protection potential, E_{crev}, for a cylindrical crevice is a logarithmic function of the crevice depth, h_{crev} [35, 36].

$$E_{crev} = E^*_{crev} + \Delta E_{IR} = a - b \log h_{crev} \tag{1.25}$$

We see that a linear relationship holds between the crevice protection potential, E_{crev}, and the logarithm of the crevice depth, $\log h_{crev}$, for a corrosion crevice of stainless steel in acid solution [35, 36].

No crevice corrosion occurs below a certain temperature called the *critical crevice corrosion temperature*, a situation which equates with the critical pitting temperature described earlier. The critical crevice temperature for stainless steel in seawater appears to be lower than the critical pitting temperature [37]. The crevice corrosion temperature has actually been used as a measure to evaluate metallic materials for their susceptibility to crevice corrosion.

1.4.3
Potential–Dimension Diagram

From the foregoing discussion, we illustrate the stability of pitting and crevice corrosion in a potential–dimension diagram made up of the corrosion potential and the dimension of pitting and crevice corrosion, as schematically shown in Figure 1.10 [35, 36]. In general, as mentioned earlier, a corrosion pit first breaks out on the passivated surface of metals at the corrosion potential more positive than the pitting potential, E_{pit}, and it then grows in its size with a fall of the corrosion potential in the more negative direction.

If the supply of the cathodic current is insufficient for corrosion, the corrosion potential is rather unstable and readily falls down in the more negative direction with a relatively small rate of pit growth. Since the repassivation potential, E_R, of the small pit stands more positive than the critical depassivation potential E_P^* for the pit solution, the falling corrosion potential reaches E_R before arriving at E_P^*, and consequently, the small pit repassivates into the noncorrosive state. In contrast, if a sufficient supply of the cathodic current is available for pitting corrosion, the pit grows rather fast at relatively positive potentials until its size

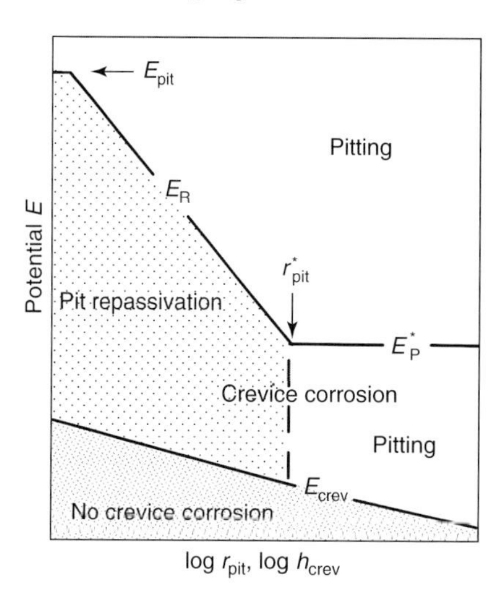

Figure 1.10 Schematic potential–dimension diagram for localized corrosion of stainless steel in aqueous solution [35, 36].

exceeds the critical value, r^*_{pit}, beyond which no pit repassivation is expected to occur. The corrosion potential then gradually falls down to the critical depassivation potential, E^*_P, before reaching the pit-repassivation potential, E_R, and hence the pitting corrosion finally changes to the active mode of pit corrosion. We then see in the potential–dimension diagram that pitting corrosion progresses in the region in which the corrosion potential is more positive than both the pit-repassivation potential E_R and the critical depassivation potential E^*_P, whereas it repassivates back to the passive state in the region in which the corrosion potential stands more negative than the pit-repassivation potential E_R and more positive than the critical depassivation potential E^*_P.

In general, crevice corrosion breaks out in the region where the corrosion potential, E_{corr}, is more positive than the crevice protection potential, E_{crev}, at which the crevice solution is acidified to the passivation–depassivation pH of the crevice metal. We see in Figure 1.10 that no crevice corrosion is expected to break out in the region where the corrosion potential is more negative than the crevice protection potential. A corroding crevice may be made passive by shifting its corrosion potential into the region more negative than the crevice protection potential E_{crev}.

1.5 Corrosion Rust

1.5.1 Rust in Corrosion

Metallic corrosion in general produces metal ions oxidized in the hydrated form and then precipitates a mass of corrosion rust in the solid form. Hydrated metal ions are mostly hard Lewis acids and tend to combine with chemical species of hard Lewis bases such as hydroxide ions, chloride ions, sulfate ions, and phosphate ions to form somewhat covalent or ionic compounds in the soluble or insoluble state. Both soluble and insoluble corrosion products influence the subsequent progress in metallic corrosion, causing either corrosion acceleration or inhibition [38–40]. We examine in this chapter insoluble corrosion products, that is, corrosion rust.

Localized corrosion normally involves the transport of hydrated ions in the corrosion cell: anions migrate from the cathodic reaction site to the anodic reaction site, and cations migrate in the reverse direction. In the presence of a substantial mass of corrosion rust on the metal surface, the ionic migration occurs through the rust precipitate, and the occluded solution directly in contact with the metal under the rust may change its concentration in aggressive ions if the ions migrate selectively across the rust. The ion-selective nature of the rust, therefore, plays an influential role in the corrosion of rust-covered metals.

Corrosion rust in general is more or less ion selective, and its ion selectivity is determined by the fixed ionic charge in the more-or-less porous rust. The rust is anion selective if the fixed charge is positive, while it is cation selective if the fixed charge is negative. Aluminum oxide is anion selective in neutral sodium chloride solution and turns to be cation selective in basic sodium hydroxide solution [41]. Oxides such as hydrous ferric oxide, nickel oxide, and chromic oxide are anion selective in neutral chloride solution [42–44]. The ion selectivity of corrosion rust is normally estimated in terms of the transference number, τ, of migrating ions through the rust. It is found with a hydrous ferric oxide layer that $\tau_{Cl^-} = -0.94$ and $\tau_{Na^+} = 0.06$, indicating that ferric oxide is anion selective in neutral sodium chloride solution [45]. The transference number of water, τ_{H_2O}, also estimates the osmotic flow of water through the rust [46].

Normally, metal hydroxides are anion selective in acidic solution and turns to be cation selective beyond a certain pH, called the *isoelectric point* pH_{iso}; $pH_{iso} = 10.3$ for ferric oxide, and $pH_{iso} = 5.8$ for ferric–ferrous oxide [47]. Because of its effect on the ionic fixed charge, the adsorption of multivalent ions exerts a significant role on the ion selectivity of hydrous metal oxides. We see, for instance, that hydrous ferric oxide that is anion selective in neutral chloride solution turns to be cation selective with the adsorption of multivalent anions such as divalent sulfate ions, divalent molybdate ions, and trivalent phosphate ions [44, 47].

Metallic corrosion, as mentioned earlier, requires not only ionic transfer and migration but also electronic transfer at the surface of metals and rust deposits. It appears that corrosion rust mostly falls in the category of electronic semiconductors, whose electronic energy band structure differs from that of metals. For metals, the Fermi level, ε_F, (the electrode potential, E,) is in the conduction band where electrons are freely mobile, whereas for semiconductors, the Fermi level, ε_F, is located in the forbidden band gap where no freely mobile electrons are allowed to be present. The cathodic reaction involving electron transfer, therefore, proceeds differently on the metal and on the semiconductor. Metallic corrosion receives, as a result, an influence from the electron-selective nature of the rust. It is also apparent that the electrode potential of the rust formed on the metal may be either more positive or more negative than the corrosion potential of the metal, depending on whether the rust is n-type or p-type. Normally, iron oxide and zinc oxide are n-type oxides, while nickel oxide and copper oxide are p-type oxides. The presence of semiconducting rust in contact with metals may have an electronic influence to a certain extent on the metallic corrosion. We will discuss in the following sections how the ionic and the electronic character of corrosion rust affect metallic corrosion.

1.5.2 Ion-Selective Rust

Let us suppose that anodic metal corrosion is in progress under an anion-selective rust layer in aqueous solution containing chloride ions, with the cathodic reaction occurring somewhere away from the anodic corrosion site. The anodic metal dissolution current in the local corrosion cell carries chloride ions from the outside solution through the corrosion rust layer into an occluded solution under the layer, as schematically shown in Figure 1.11. Accumulation of chloride ions in the occluded solution under the rust layer finally reaches a stationary level at which

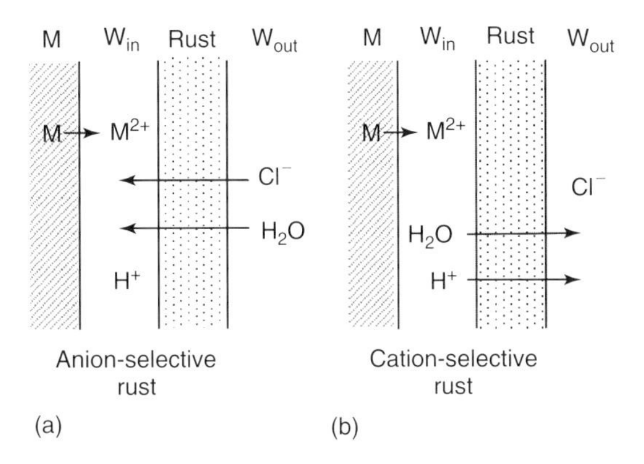

Figure 1.11 Anion-selective and cation-selective rust layers on corroding metals: (a) anion-selective rust makes the occluded water acidify and (b) cation-selective rust makes the occluded water basify; M is the metal, W_{in} is the occluded water under the rust, and W_{out} is the water outside the rust.

the inward migration equals the outward diffusion of the ions. Furthermore, the inward chloride-ionic current carries water molecules by an electro-osmotic flow into the occluded solution. The final steady concentration of chloride ions under the anion-selective rust layer is thus determined by the ratio of the transference number, τ_{Cl^-}, of chloride ions to the transport number, τ_{H_2O}, of water molecules for the layer [48].

As the local chloride ion concentration increases under the anion-selective rust layer, the occluded solution acidifies and eventually accelerates the corrosion of the underlying metal. In order to prevent metallic corrosion due to the enriched chloride ions from occurring under the anion-selective rust, we need to make the rust less anion selective or turn it cation selective by some means such as the adsorption of multivalent anions on the rust, as mentioned earlier.

We next consider a precipitate of cation-selective rust that covers over the surface of corroding metal. The anodic ion transport carries no chloride ions inward through the rust layer, but mainly carries hydrated hydrogen ions (protons) migrating outward from the occluded solution into the outside solution. No accumulation of both chloride and hydrogen ions is thus expected to occur in the occluded solution, where the depression of hydrogen ions causes the basification of the solution and stimulates rust precipitation. Furthermore, the electro-osmotic outward water flow along with the hydrogen ion migration counteracts the inward water diffusion into the occluded solution. Dehydration of the occluded solution may thus occur, resulting in the depletion of water required for anodic metal dissolution, which may suppress the corrosion of the underlying metal.

Metallic corrosion sometimes produces a bipolar rust layer consisting of an anion-selective inner layer and a cation-selective outer layer. The bipolar rust layer may be realized by means of the adsorption of multivalent anions, which turns an outer rust layer cation selective on the inner anion-selective rust layer. The ionic bipolar layer, just like a p–n electronic junction that adjusts electronic transport in semiconductors, rectifies the ionic current across the bipolar junction, suppressing in the anodic direction the ionic migration across the rust layer [44]. Furthermore, the high barrier for ionic transport generates at the bipolar junction a high electric field and makes the bipolar layer dehydrate into a layer of compact corrosion-resistive rust. We thus assume that the bipolar ion-selective rust layer is likely to inhibit metallic corrosion, bringing the metal into the passive state. It is an observed fact that pitting corrosion in stainless steel containing molybdenum is inhibited by the molybdate ion adsorption that makes an outer rust layer cation selective on the original, anion selective rust deposit [49, 50].

It is worth noting at the end that the ion-selective nature of corrosion rust remains valid as long as the concentration of migrating ions is comparable with or less than that of the fixed charge in the rust. If the rust is so coarsely porous that the migrating ion may receive little influence from the fixed charge, no ion-selective migration is expected to occur. In contrast, if the rust is so dense that no hydrated ions may penetrate into it, the normal solid-state diffusion of dehydrated ions will take over only under an extremely intense electric field.

1.5.3 Electron-Selective Rust

We suppose that metallic corrosion produces n-type semiconductor rust of metal oxides or hydroxides on the metal surface. The corrosion potential, E_{corr}, of an isolated metal is usually different from the electrode potential of an isolated semiconducting oxide of the metal. Semiconductor oxide in general sets its electrode potential at around its flat band potential, E_{fb}, where no space charge arises in the semiconductor. For most metals and their oxides in practical use, the flat-band potential of n-type oxides is more negative (cathodic) than the corrosion potential of the metals. The n-type rust oxide in contact with the metal consequently shifts the corrosion potential more negative to equate the Fermi level, ε_F, of the metal with that of the semiconductor oxide, thus giving rise to a space charge layer in the semiconductor oxide, as schematically shown in Figure 1.12. As a result of shifting the corrosion potential in the more negative direction, the n-type semiconductor rust appears to reduce the metallic corrosion, whose rate normally increases with the more positive potential. Furthermore, under the sunlight that injects photo-exited mobile holes (electron vacancies) in the valence band of electron energy levels in the semiconductor, the potential of the space charge layer is reduced in the semiconductor and the corrosion potential consequently shifts further in the more negative direction, resulting in a further decrease in the corrosion rate.

Metallic corrosion in the active state usually increases as the corrosion potential shifts in the positive (anodic) direction, and in the passive state also, pitting corrosion occurs at more positive potentials than the pit-initiation potential, E_{pit}. It is then advantageous for protecting metals from general and localized corrosion that the corrosion potential is made more negative by placing a mass of n-type semiconductor rust on the metal surface. It is also worth noting that for some n-type semiconductor oxides under photo-excitation, anodic water oxidation may

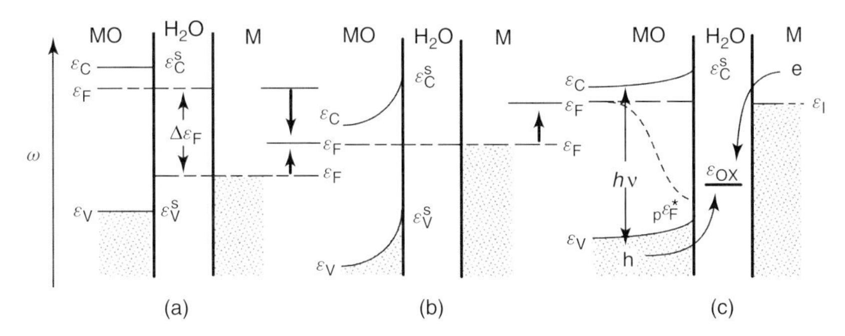

Figure 1.12 Electron energy diagrams for metal and *n*-type oxide in contact with each other in aqueous solution: (a) prior to contact; (b) posterior to contact; and (c) under photo-excitation; M is the metal, MO is the metal oxide, ε_F is the Fermi level, ε_F^* is the quasi-Fermi level for photo-excited holes in oxide, ε_C is the conduction band edge level, ε_V is the valence band edge level, and ε_{OX} is the Fermi level of oxygen electrode reaction.

occur producing gaseous oxygen even at potentials much more negative (cathodic) than the oxygen equilibrium potential. The anodic photo-excited water oxidation on the semiconductor, if coupled with cathodic oxidant reduction on the metal, may take over the anodic metal dissolution, and hence the metal may be protected from corrosion. We see that metallic copper and stainless steel were prevented from corrosion when they were brought into contact with n-type titanium oxide [50, 51]. The effectiveness of corrosion protection due to the n-type semiconductor, by its nature, depends on the ratio of the surface area of the metal to the semiconductor oxide. For stainless steel in contact with an n-type semiconductor of titanium oxide, effective corrosion inhibition was observed at the metal–oxide area ratio less than one tenth [51].

For most metal oxides and hydroxides of p-type semiconductors, in contrast, the flat band potential, E_{fb}, is more positive (anodic) than the corrosion potential, E_{corr}, of the metals. The corroding metal hence shifts its corrosion potential in the more positive (anodic) direction when it is brought in contact with a mass of p-type semiconductor oxide. Photo-excitation, in addition, further shifts the corrosion potential in the more positive direction. Metallic corrosion may then be accelerated if some p-type semiconductor rust is formed on the metal surface. In particular, pitting corrosion that occurs beyond the pit-initiation potential may break out in the presence of p-type semiconductor oxides on the metal. We also note that the photo-excitation of electrons from the valence band into the conduction band of p-type semiconductor oxide makes it thermodynamically possible for the cathodic hydrogen ion reduction to occur on the photo-excited p-type oxide even in the potential range much more positive (anodic) than the equilibrium hydrogen potential. It appears, as a result, that metallic copper, which suffers no corrosion without oxygen gas, may corrode even in the absence of gaseous oxygen, with the cathodic hydrogen ion reduction occurring on a mass of p-type copper oxides in contact with the metal under sunlight or radioactive rays. Hydrogen ion reduction involves radioexcited electrons in the conduction band of the p-type copper oxide.

1.5.4
Redox Rust

Corrosion rust is occasionally subject to redox reactions on metals. We consider iron rust consisting of hydrous ferrous and ferric oxides, which are sensitive to reduction–oxidation reaction on the corroding iron surface.

$$Fe(OH)_3 + H^+_{aq} + e^-_M = Fe(OH)_2 + H_2O \tag{1.26}$$

For the redox reaction to occur, electrons and hydrogen ions will have to migrate through the hydrous oxide toward the reaction partners. In the cathodic reaction, hydrogen ions and electrons both migrate to ferric hydroxide to form ferrous hydroxide releasing water. The anodic reaction, inversely, draws both hydrogen ions and electrons out of ferrous hydroxide, with hydrogen ions migrating outward to the aqueous solution and electrons migrating inward to the metal surface. The reaction rate may thus be controlled by either hydrogen ion migration or electronic conduction in the redox rust.

The electrode potential of the ferric–ferrous redox reaction is normally more positive than the corrosion potential of metallic iron. The redox oxide in contact with the metal hence makes the corrosion potential of the metal more positive, and it also provides the metal corrosion with the cathodic reaction as follows:

$$Fe \rightarrow Fe^{2+}_{aq} + 2e^-_M \quad \text{anodic reaction} \tag{1.27}$$

$$2Fe(OH)_3 + 2H^+_{aq} + 2e^-_M \rightarrow 2Fe(OH)_2 + 2H_2O \quad \text{cathodic reaction} \tag{1.28}$$

The ferrous oxide thus reduced may then be oxidized by atmospheric gaseous oxygen to form ferric oxide on the metal.

$$2Fe(OH)_2 + 2H_2O \rightarrow 2Fe(OH)_3 + 2H^+_{aq} + 2e^-_M \quad \text{anodic reaction} \tag{1.29}$$

$$\frac{1}{2}O_2 + 2\,H^+_{aq} + 2e^-_M \rightarrow H_2O \quad \text{cathodic reaction} \tag{1.30}$$

The rate of the cathodic reduction of hydrous ferric oxide may be greater than the rate of the cathodic reduction of gaseous oxygen that normally occurs on the metal surface, hence the corrosion rate of metallic iron may be greater in the presence of ferric–ferrous redox oxide than in its absence on the metal surface. It is received understanding that the presence of redox rust may accelerate the corrosion of metals.

1.6
Atmospheric Corrosion

1.6.1
Atmospheric Corrosion Chemistry

Metallic iron and steel corrode in the moist atmosphere. The corrosion initially produces hydrated ferrous ions at the anode part and hydrated hydroxide ions at the cathode part where air–oxygen is reduced to hydroxide ions, OH^{-}_{aq}. Hydrated ferrous ions in general arise in a variety of forms, such as Fe^{2+}_{aq}, $Fe(OH)^{+}_{aq}$, $Fe(OH)^{0}_{2,aq}$, and $Fe(OH)^{-}_{3,aq}$, depending on the pH of the solution. As the ferrous ion concentration exceeds its solubility limit, gel-like ferrous hydroxide, $Fe(OH)_{2,gel}$, precipitates out on the metal surface. According to thermodynamic calculations, gel-like ferrous hydroxide in pure water remains in equilibrium at pH 9.31 with its solubility at 1.27×10^{-5} mol dm^{-3} [52].

The hydrated ferrous ions initially formed in corrosion then air-oxidize into the solid precipitate of ferric hydroxide, $Fe(OH)_{3,solid}$.

$$4\left[Fe^{2+}_{aq} + 2OH^{-}_{aq}\right] + O_2 + 2H_2O \rightarrow 4Fe(OH)_{3,solid} \tag{1.31}$$

The term $\left[Fe^{2+}_{aq} + 2OH^{-}_{aq}\right]$ denotes the initial corrosion products at both anode and cathode parts. In the presence of a fair amount of gel-like ferrous hydroxide, $Fe(OH)_{2,gel}$, which maintains the solution at pH 9.31, the corrosion-produced ferrous ions in the hydrated form readily air-oxidize into the solid precipitate of ferric hydroxide, $Fe(OH)_{3,solid}$, and simultaneously, the pH of the solution changes to 7 with the solubility limit of ferric hydroxide at 1.20×10^{-8} mol dm^{-3} [52].

The ferric hydroxide precipitate thus formed gradually dehydrates into a mass of ferric oxyhydroxide.

$$Fe(OH)_{3,solid} \rightarrow FeOOH_{solid} + H_2O \tag{1.32}$$

Solid ferric oxyhydroxide $FeOOH_{solid}$ occurs in a variety of compounds such as α-FeOOH, β-FeOOH, γ-FeOOH, and amorphous FeOOH. A mass of α-FeOOH is the most stable, β-FeOOH is formed in the presence of chloride ions, γ-FeOOH is relatively stable, and an aggregate of amorphous FeOOH consists of extremely fine α-FeOOH particles. The ferric oxyhydroxide thus formed in the early stage of corrosion tends to aggregate into a mass of dense solid precipitate in the pH range of 7–9.31, in which the oxyhydroxide retains almost no surface charge on it and hence readily coagulates into a mass of its precipitate [53]. The solubility limit of α-FeOOH is 1.51×10^{-13} mol dm^{-3} and is much smaller than that of ferric hydroxide $Fe(OH)_{3,solid}$ [52].

Furthermore, the gel-like ferrous hydroxide of $Fe(OH)_{2,gel}$ formed in an early stage of corrosion air-oxidizes and dehydrates into magnetite, $Fe_3O_{4,solid}$, whose solubility limit is 1.10×10^{-13} mol dm^{-3} at pH 9.31 [52].

$$6Fe(OH)_{2,gel} + O_2 \rightarrow 2Fe_3O_{4,solid} + 6\,H_2O \tag{1.33}$$

The magnetite thus formed may be more or less electrochemically reactive, absorbing and desorbing hydrogen ions and electrons as well.

For metallic iron and steel containing alloying transition metal, M, atmospheric corrosion oxidizes the alloying metal into ferrite, MFe_2O_4, which then takes part in the rusting process. The ferrite may come out not only from the alloying elements but also from surface-coating substances on the metal.

$$2\,M(OH)_{2,solid} + 4Fe(OH)_{2,solid} + O_2 \rightarrow 2MFe_2O_{4,solid} + 4\,H_2O \tag{1.34}$$

The solubility of transition metal ferrites, such as $ZnFe_2O_4$, is as small as that of α-FeOOH [52]. We see that, although occupying a minor concentration, transition metal ferrites play a significant role in developing anticorrosion rust in weathering steels as mentioned in the following sections [54].

Atmospheric corrosion normally proceeds in the cycles of wet and dry environments. According to what is called the *Evans model* [55], the anodic metal dissolution in the wet period is coupled with the cathodic reduction of ferric oxyhydroxide into reactive magnetite.

$$\mathrm{Fe} \rightarrow \mathrm{Fe}^{2+}_{\mathrm{aq}} + 2\mathrm{e}^{-}_{\mathrm{M}} \quad \text{anodic reaction} \tag{1.35}$$

$$8\mathrm{FeOOH} + \mathrm{Fe}^{2+}_{\mathrm{aq}} + 2\mathrm{e}^{-}_{\mathrm{M}} \rightarrow 3\mathrm{Fe_3O_4} + 4\mathrm{H_2O} \quad \text{cathodic reduction} \tag{1.36}$$

In the dry stage, the reactive magnetite air-oxidizes into ferric oxyhydroxide.

$$4\mathrm{Fe_3O_4} + 8\mathrm{H_2O} \rightarrow 12\mathrm{FeOOH} + 4\mathrm{H}^{+}_{\mathrm{aq}} + 4\mathrm{e}^{-}_{\mathrm{M}} \quad \text{anodic oxidation} \tag{1.37}$$

$$\mathrm{O_2} + 2\mathrm{H_2O} + 4\mathrm{H}^{+}_{\mathrm{aq}} + 4\mathrm{e}^{-}_{\mathrm{M}} \rightarrow 4\,\mathrm{H_2O} \quad \text{cathodic reaction} \tag{1.38}$$

Combining the wet process with the dry process, we obtain the overall wet–dry corrosion process.

$$4\mathrm{Fe} + 3\mathrm{O_2} + 2\mathrm{H_2O} \rightarrow 4\mathrm{FeOOH} \tag{1.39}$$

It is mainly iron oxyhydroxide rust that occurs in the atmospheric corrosion of metallic iron and steel.

In general, atmospheric corrosion frequently develops the rust into a multilayered structure composed of a variety of iron oxides, hydroxides, and oxyhydroxides mentioned above. It is also observed that as the rusting progresses the corrosive hydroxide rust gradually turns into the more corrosion-resistive rust consisting mainly of α-FeOOH and amorphous FeOOH. The aged rust grows in an imbricate pattern, and the received wisdom is that the smaller imbricate scale provides weathering steels with more corrosion-resistive character.

1.6.2
Weathering Steel Corrosion

Air pollutants such as chloride and sulfate, if present in the atmospheric moisture, penetrate the rust layer and create an anode channel for local corrosion. As the anodic metal dissolution progresses, the chloride and sulfate ions tend to accumulate in the anode channel where the aqueous solution acidifies. The concentrated ferrous chloride ions at 1 mol dm^{-3} solution acidifies the solution at pH 4.75, where no ferrous hydroxide is allowed to precipitate in the solid state because of its solubility greater than 1 mol dm^{-3}. Hydrated ferrous chloride, $\left[\mathrm{Fe}^{2+}_{\mathrm{aq}} + 2\mathrm{Cl}^{-}_{\mathrm{aq}}\right]$, dissolved in local corrosion then gradually air-oxidizes and precipitates into ferric hydroxide rust in the anode channel, simultaneously producing hydrochloric acid $\left[\mathrm{H}^{+}_{\mathrm{aq}} + \mathrm{Cl}^{-}_{\mathrm{aq}}\right]$.

$$4\left[\mathrm{Fe}^{2+}_{\mathrm{aq}} + 2\mathrm{Cl}^{-}_{\mathrm{aq}}\right] + \mathrm{O_2} + 10\mathrm{H_2O} \rightarrow 4\mathrm{Fe(OH)}_{3,\mathrm{solid}} + 8\left[\mathrm{H}^{+}_{\mathrm{aq}} + \mathrm{Cl}^{-}_{\mathrm{aq}}\right] \tag{1.40}$$

The acidified anode channel itself, however, still maintains a significant concentration of hydrated ferrous and ferric ions. These hydrated ions gradually leak out of the anode channel and air-oxidize into corrosion rust outside the anode channel. A model calculation estimates for the total concentration of corroded iron at 1 mol dm^{-3} that air-oxidation finally gives rise to the solution that contains 0.590 mol dm^{-3} of Fe^{+3}_{aq} and 0.096 mol dm^{-3} of $FeOH^{+2}_{aq}$ at the acidic level of pH 1.41 [52]. In the calculation, one-third of the corroded iron precipitates as gel-like ferric hydroxide, $Fe(OH)_{3,solid}$, in the anode channel at pH 1.41. Owing to its substantial surface charge in acidic solution, the gel-like ferric oxide assumes almost no aggregation into solid precipitates but disperses itself in the acidified anode channel, gradually dehydrating into β-FeOOH. The gel-like β-FeOOH then diffuses out forming a flowing mass of nonprotective yellow rust on the metal surface. The remaining two thirds of corroded iron remains in the form of soluble complexes of ferric ions and diffuses out of the anode channel to precipitate into several varieties of fresh corrosion rust in the neutral solution outside the aged rust layer.

In the anode channel containing acidified chloride ions and ferric ions, metallic iron corrodes with the cathodic oxidant reduction of hydrogen ions and ferric ions.

$$Fe + \left[2\,H^{+}_{aq} + 2Cl^{-}_{aq}\right] \rightarrow \left[Fe^{2+}_{aq} + 2Cl^{-}_{aq}\right] + H_2 \tag{1.41}$$

$$Fe + \left[2Fe^{3+}_{aq} + 6Cl^{-}_{aq}\right] \rightarrow \left[3Fe^{2+}_{aq} + 6Cl^{-}_{aq}\right] \tag{1.42}$$

The acidic corrosion reactions continue occurring as long as the anode channel keeps its solution sufficiently acidic.

Weathering steels are normally subject to cyclic wet–dry corrosion in the atmosphere. In the wet period, they corrode locally, producing yellow rust that flows out of the aged rust layers on the steels. In the dry period, on the other hand, the freshly formed rust dehydrates and air-oxidizes into a dense rust layer. The corrosion rust develops, as a result, into a multilayered structure successively consisting of dense and coarse rust layers formed in the dry and wet periods.

Long-term observations of atmospheric corrosion of steels show that a linear bilogarithmic relation holds between the corrosion weight loss or penetration, w_{corr}, and the time, t, of corrosion for 20–30 years [56, 57].

$$\begin{aligned} w_{corr} &= k_1 t^{k_2} \\ \log w_{corr} &= A + B \log t \end{aligned} \tag{1.43}$$

Parameters k_1, k_2, A, and B are dependent on the type of steel materials and on the corrosion environments. No theoretical background has so far been made clear of Eq. (1.36), which, however, can be used to predict atmospheric corrosion for longer periods.

1.6.3
Anticorrosion Rust

It comes out of the foregoing discussion that the key point for weathering steel corrosion is to make the soluble product of ferrous ions air-oxidize as soon as possible into insoluble ferric hydroxide, which eventually turns to be anticorrosion rust on the metal surface. In general, the rate of air-oxidation of hydrated ferrous ions is much greater in neutral solution than in acid solution where no air oxidation of the ions virtually occurs. Once gel-like ferric hydroxide and ferrous hydroxide are formed in water, the solution remains around pH 7–9.31 where the air oxidation rate of ferrous ions is great. In the presence of chloride ions, however, the solution acidifies in the anode channel and hence reduces the rate of the air-oxidation of ferrous ions, eventually counteracting the formation of anticorrosion rust.

It is found that the presence of cupric ions, Cu^{2+}, and phosphate ions, PO_4^{3-}, catalytically accelerates the air-oxidation of ferrous ions even in acid solution [58]. Weathering steels usually contain copper and phosphorus as alloying elements, which dissolve in the form of hydrated ions. The dissolved cupric and phosphate ions increase the rate of the air oxidation of ferrous ions into solid ferric hydroxide in the acidified anode channel, and they thus contribute to the formation of anticorrosion rust.

For some reason not yet made clear, cupric ions Cu^{2+} also influence the coagulation of hydroxide and prevent ferric hydroxide from crystallizing into coarse nonresistive aggregates [59, 60]. The rust then remains amorphous and grows into a dense, void-free barrier layer of anticorrosive rust. In general, transition metal ions such as Ti^{4+}_{aq}, Co^{2+}_{aq}, Cr^{3+}_{aq}, and Ni^{2+}_{aq}, which all come from alloying elements in weathering steels, are also found to catalyze the formation of amorphous iron rust [61, 62]. It is the received understanding that the amorphous rust of ferric oxyhydroxide is much more anticorrosive than the coarse crystalline oxide rust.

The ion-selective character of rust may also affect the corrosion of weathering steels. As mentioned earlier, the rust layer of cation-selective character prevents the occluded solution in the anode channel from being acidified and thus contributes to the formation of corrosion-resistive rust. The rust layer of anion-selective character, on the other hand, tends to make the anode channel acidify, thus retarding the formation of anticorrosion rust. It is, as mentioned earlier, the fixed charge on the inner surface of the rust that generates the ion-selective nature of the rust. The fixed charge depends on various factors such as adsorbed acid–base hydroxyl groups, nonstoichiometric surface composition, and specific adsorption of multivalent ions. In general, the surface charge of iron oxides and hydroxides is positive in the acidic range of water and negative in the basic range of water. The border pH, called the *isoelectric point* pH_{iso}, differs with different sorts of iron rust. It appears that the more acidic is the isoelectric point the more corrosion resistive is the rust. It is noted that the specific adsorption of multivalent anions such as PO_4^{3-} and $CrO^{2-}_{4,aq}$, which come from alloying elements P and Cr in weathering steels, makes the isoelectric point pH, pH_{iso}, of the iron rust more acidic and hence contributes to the formation of anticorrosive rust layers on the steels.

1.7 Concluding Remarks

Corrosion science has made clear that metallic corrosion is in essence an electrochemical process combined with Lewis acid–base reactions at the interface between the metal and the environment, hence the corrosion potential plays a primary role in the process of corrosion. We have, however, a number of crucial issues to be studied to understand further the electrochemical and Lewis acid–base character of metallic corrosion at the molecular level.

In this chapter we have discussed metallic corrosion only under normal conditions. We note that corrosion also occurs under mechanical stresses leading to environmental cracking, stress corrosion cracking, erosion corrosion, and cavitation corrosion. Furthermore, materials other than metals, such as semiconductors, ionic solids, ceramics, and organic polymers, are also more or less all subject to corrosion [63].

References

1. Evans, U.R. (1937) *Metallic Corrosion and Protection*, Edward Arnold, London.
2. Wagner, C. and Traud, W. (1938) Über die Deutung von Korrosionsvorgängen durch Überlagerung von elektrochemischen Teilvorgängen und über die Potentialbildung an Mischelektroden. *Z. Elektrochem.*, **44** (7), 391–402.
3. Wagner, C. (1974) Corrosion in aqueous solution and corrosion in gases at elevated temperature–analogies and disparities. *Werkst. Korros.*, **25** (3), 161–165.
4. Pourbaix, M. (1966) *Atlas of Electrochemical Equilibria in Aqueous Solutions*, Pergamon Press, Oxford.
5. Garreles, R.M. and Christ, C.L. (1965) *Solutions, Minerals and Equilibria*, Harper & Row, London, p. 172.
6. Kaesche, H. (1979) *Die Korrosion der Metalle*, 2 Auflage, Spring-Verlag, New York, p. 122.
7. Sato, N. (1989) Toward a more fundamental understanding of corrosion processes. *Corrosion*, **45**, 354–368.
8. Jensen, W.B. (1980) *The Lewis Acid-Base Concepts*, John Wiley & Sons, Inc., New York, pp. 112–336.
9. Okamoto, G., Shibata, T., and Sato, N. (1969) *Reports of Asahi Glass Foundation for Industrial Technology*, vol. 15, Asahi Glass Foundation for Industrial Technology, Tokyo, pp. 207–230.
10. Keir, J. (1790) *Philos . Trans.*, **80**, 259.
11. Sato, N. and Okamoto, G. (1981) Electrochemical passivation of Metals, in *Comprehensive Treatise of Electrochemistry*, vol. 4 (eds J.O'M.B. Bockris and R.E. White), Plenum Publishing Corporation, New York, London, pp. 193–245.
12. Sato, N. (1990) An Overview on the Passivity of Metals, in *Passivity of Metals Part I*, Corrosion Science, Vol. 31 (supplement) (eds N. Sato and K. Hashimoto), Pergamon Press, pp. 1–19.
13. Sato, N. (1976) Passivity of Iron, Nickel, and Cobalt – General Theory of Passivity, in *Passivity and its Breakdown on Iron and Iron Base Alloys* (eds R. Staehle and H. Okada), NACE, Houston, pp. 1–9.
14. Sato, N. (1982) Anodic breakdown of passive films on metals. *J. Electrochem. Soc.*, **129** (2), 255–260.
15. Sato, N. (2001) The stability and breakdown of passive oxide films on metals. *J. Indian Chem. Soc.*, **78**, 19–26.
16. Okamoto, G. and Sato, N. (1959) Self-passivation of nickel in aerated aqueous solution. *J. Jpn. Inst. Met.*, **23** (12), 725–728.

17. Takahashi, H., Fujimoto, F., Konno, H., and Nagayama, M. (1984) Distribution of aions and protons in oxide films formed anodically on aluminum in a phosphate solution. *J. Electrochem. Soc.*, **131** (8), 1856–1861.
18. Ohtsuka, T., Masuda, M., and Sato, N. (1985) Ellipsometric study of anodic oxide films on titanium in hydrochloric acid, sulfuric acid and phosphate solution. *J. Electrochem. Soc.*, **132** (4), 787–792.
19. Barr, T.L. (1979) An ESCA study of the termination of the passivation of elemental metals. *J. Phys. Chem.*, **82** (16), 1801–1810.
20. Fushimi, K. and Seo, M. (2001) Initiation of a local breakdown of passive film on iron due to chloride ions generated by a liquid-phase ion-gun for local breakdown. *J. Electrochem. Soc.*, **148** (11), B459–B456.
21. Fushimi, K., Azumi, K., and Seo, M. (2000) Use of a liquid-phase ion-gun for local breakdown of the passive film on iron. *J. Electrochem. Soc.*, **147** (2), 552–557.
22. Heusler, K.E. and Fisher, L. (1976) Kinetics of pit initiation at passive iron. *Werkst. Korros.*, **27** (8), 551–556.
23. Shibata, T. (1990) in *Passivity of Metals Part I*, Corrosion Science, Vol. 31 (supplement) (eds N. Sato and K. Hashimoto), pp. 413–423.
24. Sato, N. (1976) Stochastic process of chloride-pit generation in passive stainless steel. *J. Electrochem. Soc.*, **123** (8), 1197–1199.
25. Hisamatsu, Y., Yoshii, T., and Matsumura, Y. (1974) in *Localized Corrosion* (eds R.W. Staehle, B.F. Brown, J. Kruger, and A. Agrawal), NACE, Houston, TX, pp. 420–436.
26. Macdonald, D.D. (1992) The point defect model for the passive state. *J. Electrochem. Soc.*, **139** (12), 3434–3449.
27. Sato, N. (2001) *Proceedings of the 2nd International Conference on Environment Sensitive Cracking and Corrosion Damage, Hiroshima, October 29–November 3*, Japan Society Corrosion Engineering, pp. 49–54.
28. Brennert, C. (1937) Method for testing the resistance of stainless steels to local corrosive attack. *J. Iron Steel Inst.*, **135**, 101–112.
29. Hisamatsu, Y. (1976) in *Passivity and its Breakdown on Iron and Iron Base Alloys* (eds R.W. Staehle and H. Okada), NACE, Houston, TX, pp. 99–105.
30. Engell, H.J. and Storica, N.D. (1959) Untersuchungen über Lochfrass an passiven Elektroden aus unlegiertem Stahl in chloridenhaltiger Schwefelsaure. *Arch. Eisenhuttenwesen*, **30** (4), 239–248.
31. Sato, N. (1982) The stability of pitting dissolution of metals. *J. Electrochem. Soc.*, **129** (2), 260–264.
32. Laycock, H.J. and Newman, R.C. (1998) Temperature dependence of pitting potentals for austenitic stainless steels above their critical pitting temperature. *Corrosion Sci.*, **40** (6), 887–902.
33. Wallen, B. (2001) Critical Pitting Temperature of UNS S31600 in Different Sea Wateers, in *Marine Corrosion of Stainless Steels*, Pubication No. 33 (ed. D. Feron), European Federation of Corrosion, London, pp. 19–25.
34. Tsujikawa, S., Sono, Y., and Hisamatsu, Y. (1987) in *Corrosion Chemistry within Pits* (ed. A. Turnbull), National Physical Laboratory, London, p. 171.
35. Sato, N. (1995) The stability of localized corrosion. *Corrosion Sci.*, **37** (12), 1947–1967.
36. Sato, N. (2001) Potential-Dimension Diagram of Localized Corrosion, in *Marine Corrosion of Stainless Steels*, Publication No. 33 (ed. D. Feron), European Federation of Corrosion, London, pp. 185–201.
37. Steinsmo, U., Rogno, T., and Drugli, J.M. (2001) Recommended Practice for Selection, Quality Control and Use of High-Alloy Stainless Steels in Sea Water Systems, in *Marine Corrosion of Stainless Steels*, Publication No. 33 (ed. D. Feron), European Federation of Corrosion, London, pp. 115–123.
38. Sato, N. (2001) Ion-Selective Lust Layers and Passivation of Metals, in *Passivity of Metals and Semiconductors*, Proceedings 99-42 (eds B. Ives J.L. Luo, and J.R. Rodda), The Electrochemical Society, Inc., Pennington, pp. 281–287.

39. Sato, N. (2002) Surface oxide films affecting metallic corrosion. *Corrosion Sci. Technol.*, **31** (4), 262–274.
40. Sato, N. (1987) Some concepts of corrosion fundamentals. *Corrosion Sci.*, **27** (5), 354–368.
41. Huber, K. (1955) Über anodisch erzeugte Silberbromidschichten. Ein Beitrag zur Frage nach der Struktur von Elektroden II, Art. *Z. Elektrochem.*, **59** (7/8), 693–696.
42. Suzuki, M., Masuko, N., and Hisamatsu, Y. (1971) Study on the rust layer on atmospheric corrosion resistant low alloy steels, part 4 behavior of electrolyte in rust. *Boshoku Gijutsu (Jpn. Corrosion Eng.)*, **20** (7), 319–324.
43. Yomura, Y., Sakashita, M., and Sato, N. (1979) Ion-selectivity in hydrated iron (II), (II-III), and (III) oxide precipitate membranes. *Boshoku Gijutsu (Jpn. Corrosion Eng.)*, **28** (2), 64–71.
44. Sakashita, M. and Sato, N. (1979) Ion-selectivity of precipitate films affecting passivation and corrosion of metals. *Corrosion*, **35** (8), 351–355.
45. Sakashita, M., Yomura, Y., and Sato, N. (1977) Ion-selectivity of hydrous iron (III) oxide precipitate membranes. *Denki Kagaku (Jpn. Electrochem.)*, **45** (3), 165–170.
46. Sakashita, M. and Sato, N. (1975) Water transference through nickel hydroxide precipitate membrane. *J. Electroanal. Chem.*, **62** (1), 127–134.
47. Sakashita, M. and Sato, N. (1977) The effect of molybdate anion on the ion-selectivity of hydrous ferric oxide films in chloride solutions. *Corrosion Sci.*, **17** (6), 473–486.
48. Sakashita, M., Shimakura, S., and Sato, N. (1984) *Proceedings 9th International Congress Metallic Corrosion*, vol. 1, National Research Council of Canada, pp. 126–131.
49. Clayton, C.R. and Lu, Y.C. (1989) A bipolar model of the passivity of stainless steels–III, the mechanism of MoO_4^{2-} formation and incorporation. *Corrosion Sci.*, **29** (7), 881–898.
50. Yuan, J. and Tsujikawa, S. (1995) Characterization of sol-gel-derived TiO_2 coatings and their photoeffectson copper substrates. *J. Electrochem. Soc.*, **142** (10), 3444–3450.
51. Fujisawa, R. and Tsujikawa, S. (1995) *Mater. Sci. Forum*, **183–189**, 1076–1081.
52. Tamura, H. (2008) The role of rusts in corrosion and corrosion protection of iron and steel. *Corrosion Sci.*, **50**, 1872–1882.
53. Cornell, R.M. and Schwertmann, U. (1996) *The Iron Oxides*, Wiley-VCH Verlag GmbH, Weinheim, p. 224.
54. Tamura, H. and Matijevic, E. (1982) Precipitation of cobalt ferrites. *J. Colloid Interface Sci.*, **90** (1), 100–109.
55. Evans, U.R. (1969) Mechanism of rusting. *Corrosion Sci.*, **9** (12), 813–821.
56. Fukushima, T., Sato, N., Hisamatsu, Y., Matsushima, T., and Aoyama, Y. (1982) Atmospheric Corrosion Testing in Japan, in *Atmospheric Corrosion* (ed. W.H. Ailor), A Wiley-Interscience Publication, John Wiley & Sons, Inc., New York, London, pp. 841–872.
57. Pourbaix, M. (1982) in *Atmospheric Corrosion* (ed. W.H. Ailor), A Wiley-Interscience Publication, John Wiley & Sons, Inc., New York, London, pp. 107–121.
58. Tamura, H., Goto, K., and Nagayama, M. (1976) Effect of anions on the oxygenation of ferrous ion in neutral solutions. *J. Inorg. Nucl. Chem.*, **38** (1), 113–117.
59. Furuichi, R., Sato, N., and Okamoto, G. (1969) Reactivity of hydrous ferric oxide containing metallic cations. *Chimia*, **23** (12), 455–465.
60. Suzuki, I., Hisamatsu, Y., and Masuko, N. (1980) Nature of atmospheric rust on iron. *J. Electrochem. Soc.*, **127** (10), 2210–2215.
61. Ishikawa, T., Uemo, T., Yasukawa, A., Kandori, K., Nakayama, T., and Tsubota, T. (2003) Influence of metal ions on the structure of poorly crystallized iron oxide rusts. *Corrosion Sci.*, **45**, 1037–1045.
62. Ishikawa, T., Maeda, A., Kandori, K., and Tahara, A. (2006) Characterization of rust on Fe-Cr, Fe-Ni, and Fe-Cu

bimary alloys by fourier transform infrared and N_2 adsorption. *Corrosion*, **62**, 559–567.

63. Sato, N. (2010) Fundamental Aspects of Corrosion of Metals and Semiconductors, in *Electrocatalysis: Computational, Experimental, and Industrial Aspects*, Surfactant Science Series, Vol. 149, Chapter 22 (ed. C.F. Zinola), CRC Press, Taylor & Francis Book, Inc., pp. 531–588.

2
Corrosion and Electrochemistry

Vedula Sankar Sastri

2.1
Introduction

In very early times, hominids mainly functioned as bioelectrical machines, converting solar energy contained in food into muscle power through electrochemical reactions. In the course of time, scientific knowledge and inventions extended human limitations to such an extent that nuclear and solar energy could be harnessed with increased efficiency. Progress in civilization is marked by an increase in the use of machine power and an attendant decrease in the use of muscle power.

It follows that machines in the service of humanity must function successfully without progressive deterioration in the terrestrial atmosphere. Materials, mainly metals, used in the fabrication of machines must be stable. If the metals are unstable, machines partly made from metals undergo an undesirable obsolescence. Thus it is clear that an industrial civilization depends on the stability of metals and alloys in moist or impure atmosphere.

It is interesting to note that a metal sample remains stable indefinitely when stored in vacuum. Metals become stable when the metal surfaces are isolated from the normal terrestrial environment. In the absence of isolation from the terrestrial environment, metals become unstable in many ways. Metals may suddenly develop cracks and break on strain. Metals might suffer fatigue, that is, loss of strength, when subjected to periodic stress. Metals may undergo embrittlement and oxide scales may form on the metal, which may peel off or dissolve in the medium. In general, all metals, with the exception of noble metals such as gold, platinum, rhodium, and so on, are unstable in terrestrial atmospheres to a varying degree. The most commonly used metals such as iron, copper, aluminum, nickel, and alloys of these metals lose their valuable mechanical properties in unprotected contact with atmospheric air.

The stability of metals is determined by the processes occurring at the interface between the metal and the environment. The internal strength of a metal depends on the nature of events taking place on the surface. If the surface of the metal is stable, the interior of the metal will also tend to be stable. The deleterious

Green Corrosion Chemistry and Engineering: Opportunities and Challenges, First Edition.
Edited by Sanjay K. Sharma.

transformation of the bulk properties of a metal begins at the metal surface. This, then, forms a link between civilization, surfaces, and electrochemical reactions.

Since metals become unstable when they come in contact with moisture, it is likely that the instability of metals arises from the charge-transfer reactions at the interface. The terrestrial environment is essentially moist air containing carbon dioxide. Marine atmosphere consists of moist air containing sodium chloride. Thus, moisture in contact with the terrestrial atmosphere becomes an ionically conducting medium as well as a corrosive medium.

Owing to the occurrence of charge-transfer reactions, metals become unstable when they come into contact with moisture. This is the reason why the corrosive attack on the metal surface is minimal in the absence of moisture. From this, it readily follows that keeping a metal sample in vacuum is the same as keeping the metal free from an electrolyte and hence preventing charge-transfer reactions. Thus, corrosion of metals occurs because of charge-transfer reactions at the electrified interface between the metal and the moist, CO_2- or NaCl-containing air [1].

Experiments involving direct studies of the rate and products of corrosion of a metal as a function of the electrolytic conduction of the moist film and energy-producing electrochemical cells, such as immersed pieces of copper and zinc in an electrolyte solution, as well as a situation in which impure copper containing zinc decayed on exposure to an electrolyte confirm the view that corrosion of metals involves the exposure of the metal to an electrolyte solution.

Metallic corrosion arises from electrodic charge-transfer reactions at the interface between a metal and its electrolytic environment. But in the case of corrosion, the external source driving the charge-transfer reactions or the external load consuming the current produced by the charge-transfer reactions must be identified. Thus the conceptual relationship between electrochemical cells and corroding metals must be understood.

Consider a sample of zinc and a sample of copper immersed in a solution of zinc sulfate and copper sulfate solution, respectively, as shown in Figure 2.1. Because the equilibrium potential of the reduction of zinc cations (Zn^{2+}) to give zinc metal (Zn^0) is negative with respect to the reduction of cupric ion to give metallic copper, the zinc electrode is negatively charged with respect to the copper electrode. The two electrodes may be connected through an external load of resistance, R_e. Electrons begin to flow from zinc to copper by the following reactions:

$$Zn \rightarrow Zn^{2+} + 2e^- \tag{2.1}$$

$$Cu^{2+} + 2e^- \rightarrow Cu^0 \tag{2.2}$$

To maintain the flow of current, the zinc electrode dissolves to give Zn^{2+} ions and cupric ions are reduced to metallic copper, which deposits on the copper electrode. Thus the Zn–Cu electrode couple acts as an energy producer. The potential difference across cells, such as the Cu–Zn cell, has been analyzed, and it has been found to decrease with the cell current (Figure 2.2). The cell current increases with lower values of resistance, R_e. The increase in cell current with decreasing potential occurs as shown in Figure 2.2.

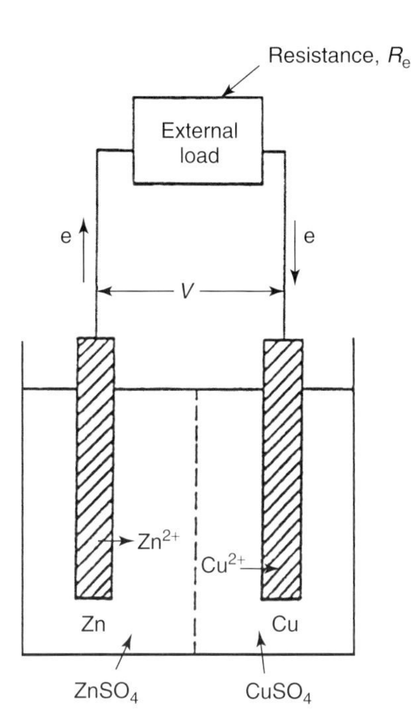

Figure 2.1 Daniel cell.

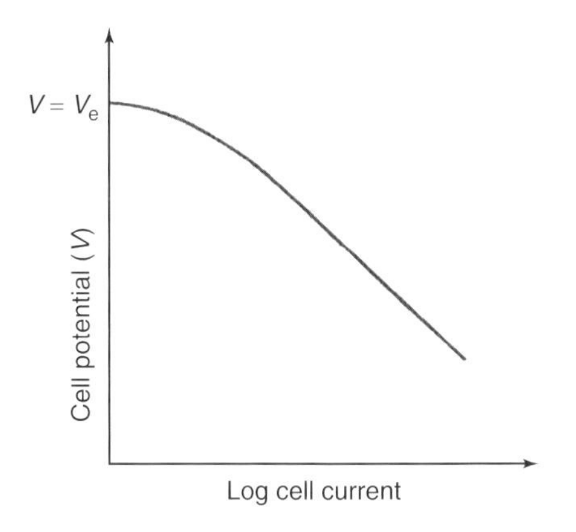

Figure 2.2 Figure showing decreasing potential with increasing current.

When the external resistance is made zero, the copper and zinc electrodes are brought into contact or short circuited. Zinc continues to dissolve at a certain current and copper continues to deposit, but the potential difference across the cell will become zero (Figure 2.3). This phenomenon will occur when a bar of copper welded with zinc is placed in an electrolyte solution containing cupric salt (Figure 2.4). In this situation, zinc continues to dissolve as the copper deposits. Similarly, if iron is welded with another metal and immersed in an electrolyte solution, the dissolution of iron will depend on whether its equilibrium potential is more negative or more positive than that of the other metal.

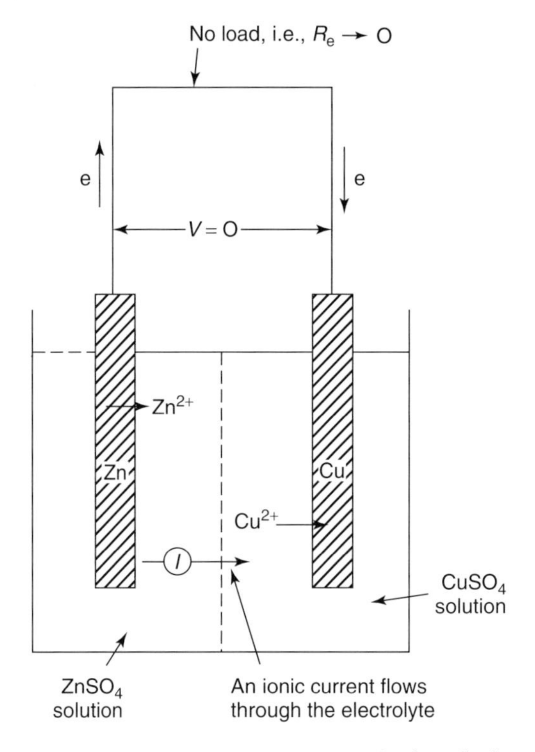

Figure 2.3 Maximum current is given by the cell when the load is removed and the two electrodes are short-circuited.

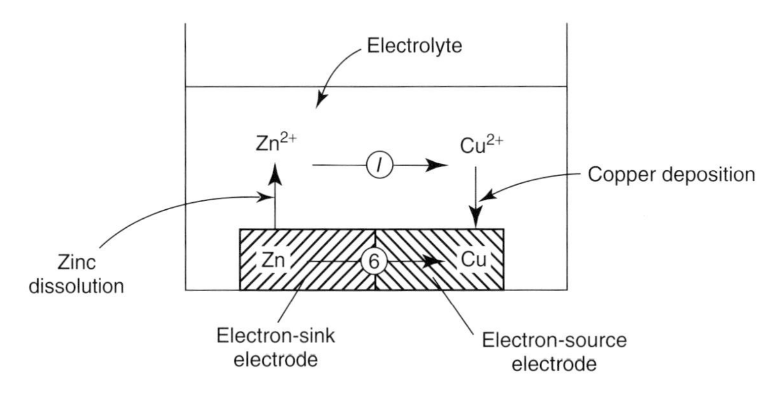

Figure 2.4 Electrodes are short-circuited when two welded bars, one of zinc and one of copper are immersed in an electrolyte of cupric ions.

Let us consider a large number of strips of copper and zinc arranged in alternate layers, that is, a multiband arrangement, and placed in copper sulfate solution. In this situation also, zinc will dissolve along with the deposition of copper.

Consider the case of a bar of zinc containing microscopic inclusions of metallic copper as the impurity. When this bar is placed in a solution containing cupric and

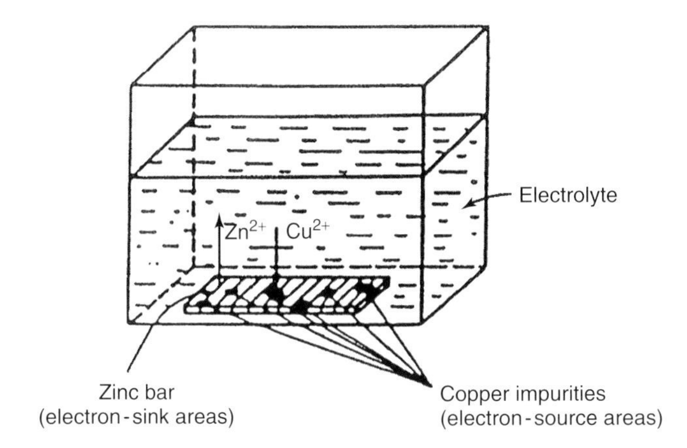

Figure 2.5 Deposition of copper on a zinc bar in cupric ion electrolyte at microscopic inclusions of copper impurities.

zinc ions, dissolution of zinc and deposition of copper on preexisting locations of copper in the metallic bar are observed (Figure 2.5).

In all the situations considered so far, the reduction reaction is one of reduction of cupric ions to metallic copper. Other possible reduction reactions can occur to keep the process going forward, including the reduction of hydrogen or oxygen.

$$2H_3O^+ + 2e^- \rightleftharpoons H_2 + 2H_2O \quad (2.3)$$

$$O_2 + 4H^+ + 4e^- \rightleftharpoons 2H_2O \quad (2.4)$$

The corrosion of zinc in acid solution occurs according to the scheme (Figure 2.6):

$$Zn^0 \rightarrow Zn^{2+} + 2e^- \quad \text{anodic oxidation} \quad (2.5)$$

$$H^+ + e^- \rightarrow H$$

$$H + H \rightarrow H_2 \quad \text{cathodic reaction} \quad (2.6)$$

This model consists of a metal area that can be termed as the *electron-sink area*, where metal dissolution occurs; an ionic conductor to carry the electrons to the

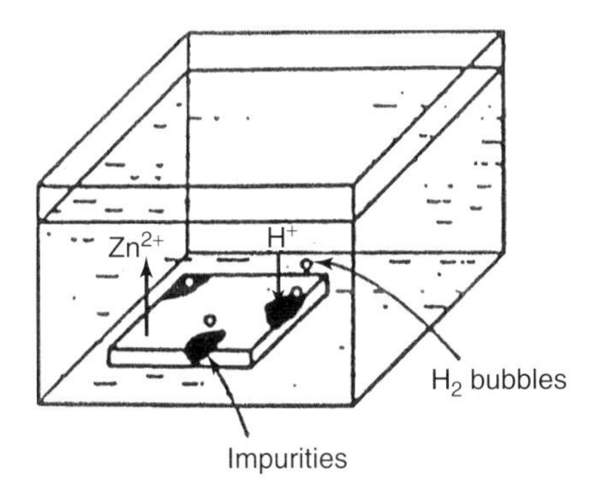

Figure 2.6 Hydrogen evolution reaction that occurs in zinc corrosion in acid solution.

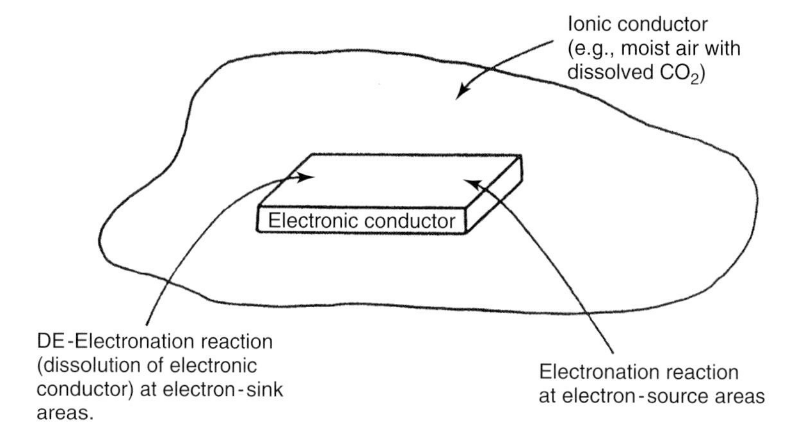

Figure 2.7 Local cell theory of corrosion with separate sites for oxidation and reduction. DE, Dissolution by Electronation.

electron-source area, where the electrons are consumed; and an ionic conductor to keep the flow of ion current as a medium for the electrodic reaction. This model of corrosion is known as the *local cell theory of corrosion* (Figure 2.7).

On the basis of the local cell theory, a highly pure metal devoid of impurity inclusions is expected to be resistant to corrosion. In practice, this is not true since corrosion of highly pure metals has been observed. According to Wagner and Traud, it is not necessary for the presence of impurities that give rise to electron-sink-and electron-source areas on the corroding metal. The necessary and sufficient condition for corrosion is the simultaneous occurrence of metal dissolution reaction and electron consuming reaction at the metal–environment interface. In order for the metal dissolution and consumption of the electrons to be released, the corrosion potential should be more positive than the equilibrium potential of the anodic reaction

$$M^{n+} + ne^- \rightleftharpoons M \tag{2.7}$$

and more negative than the equilibrium potential of the cathodic reaction

$$A + ne^- \rightleftharpoons D \tag{2.8}$$

Thus, when the electron-sink and electron-source are distinct in space and stable in time, the local cell or heterogeneous theory of corrosion is in operation. When the metal dissolution and the cathodic reduction occur randomly over the surface, the Wagner-Traud homogeneous theory of corrosion is in operation.

In practice, heterogeneities of one type or the other are present in the metal as well as in the different phases of the alloy, or in a metal with nonuniform distribution of stress. Thus the local cell or heterogeneous mode of corrosion is more commonly encountered. On the other hand, the homogeneous theory of corrosion involves metals becoming unstable because of different electrodic charge-transfer reactions occurring simultaneously and in opposite directions at the surface.

The corrosion process involves two reactions; namely, the oxidation reaction at the anode

$$M^0 \rightarrow M^{n+} + ne^- \quad \text{anodic reaction} \tag{2.9}$$

and the reduction reaction at the cathode. The three possible cathodic reactions are

$$2H_3O^+ + 2e^- \rightarrow 2H_2O + H_2 \quad \text{(acid solutions)} \tag{2.10}$$

$$O_2 + 4H^+ + 4e^- \rightarrow 2H_2O \quad \text{(acid solutions)} \tag{2.11}$$

$$O_2 + 2H_2O + 4e^- \rightarrow 4OH^- \quad \text{(alkaline solutions)} \tag{2.12}$$

When the corrosive medium contains other species such as a ferric ion and an acid such as nitric acid, the following cathodic reactions may be encountered:

$$Fe^{3+} + e^- \rightarrow Fe^{2+} \tag{2.13}$$

$$3H^+ + NO_3^- + 2e^- \rightarrow HNO_2 + H_2O \tag{2.14}$$

When several cathodic reactions are possible, the cathodic reaction giving rise to the largest corrosion current is considered to be the principal cathodic electronation reaction corresponding to the given potential. This observation is confirmed by the increased corrosion rate of iron in aerated solutions as compared to deoxygenated solutions. The reaction in deaerated acid solutions may be written as

$$2H_3O^+ + 2e^- \rightleftharpoons 2H_2O + H_2 \tag{2.15}$$

and the reaction in oxygenated solutions may be written as

$$O_2 + 4H^+ + 4e^- \rightarrow 2H_2O \tag{2.16}$$

The increase in corrosion rate with an increase in oxygen pressure is clearly illustrated by the data [2] given in Table 2.1.

The type of acid also appears to affect the corrosion rate, as is the case with nitric acid because of the following reaction:

$$3H^+ + NO_3^- + 2e^- \rightarrow HNO_2 + H_2O \tag{2.17}$$

Table 2.1 Effect of oxygen pressure on the corrosion rate of iron in 3.5% NaCl.

Oxygen pressure (atm)	Corrosion rate (mm yr^{-1})
0.2	2.2
1.0	9.3
10.0	86.4
61.0	300

2.2
Thermodynamics and the Stability of Metals

The task of deciding whether a particular metal is suitable as a material of construction in a given environment falls in the realm of thermodynamics and stability of metals. The question or problem on hand is whether the metal or alloy is resistant to corrosion in operating conditions. The real question is whether the oxidation reaction of iron to give ferrous ion (de-electronation reaction) and the electronation reaction (electron accepting reaction), which together constitute the corrosion process, proceed spontaneously or not. This question falls within the realm of equilibrium thermodynamics. In thermodynamics, the free energy change in a reaction is related to the equilibrium potential. The sum of the free energy changes involved in de-electronation and electronation reactions gives the free energy change of the corrosion process,

$$\Delta G = -nFE \tag{2.18}$$

and if the value of the free energy is negative, corrosion of the metal will occur spontaneously.

An alternate approach based on the potential versus pH representation of equilibrium potentials is as follows. Let us suppose the potential of the reaction $M \rightarrow M^{n+} + ne^-$ is independent of pH and can be represented as a straight line parallel to the pH axis (Figure 2.8).

Then we consider the electron acceptor A present in the solution in contact with the metal M and calculate the equilibrium potential for its reactions. Consider the situation in which the proton transfer is involved as well. We may then write

$$xA + mH^+ + ne^- \rightleftharpoons yD + ZH_2O \tag{2.19}$$

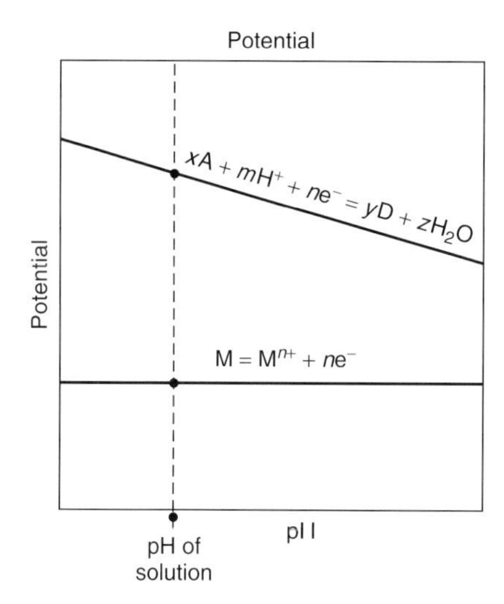

Figure 2.8 Potential–pH diagram indicating the tendency of metal to corrode.

The above reaction involves both electron and proton transfers, and the equilibrium potential will vary with pH and may be represented as a straight line sloping downward in the potential–pH plot.

Now consider drawing a line perpendicular at a chosen value of pH of the solution (Figure 2.9). If this line intersects the line representing

$$M^{n+} + ne^- \rightleftharpoons M \tag{2.20}$$

at a more negative value of potential than

$$xA + mH^+ + ne^- \rightarrow yD + ZH_2O \tag{2.21}$$

line, it can be concluded that the de-electronation reaction will occur readily together with electronation reaction, thus leading to the conclusion that the metal M will corrode readily.

2.3 Free Energy and Electrode Potential

A general reaction for the dissolution of divalent metal in hydrochloric acid may be written as

$$M + 2HCl \rightarrow MCl_2 + H_2 \tag{2.22}$$

The reaction may be written as two half-cell reactions

$$M \rightarrow M^{2+} + 2e^- \tag{2.23}$$

$$2H^+ + 2e^- \rightarrow H_2 \tag{2.24}$$

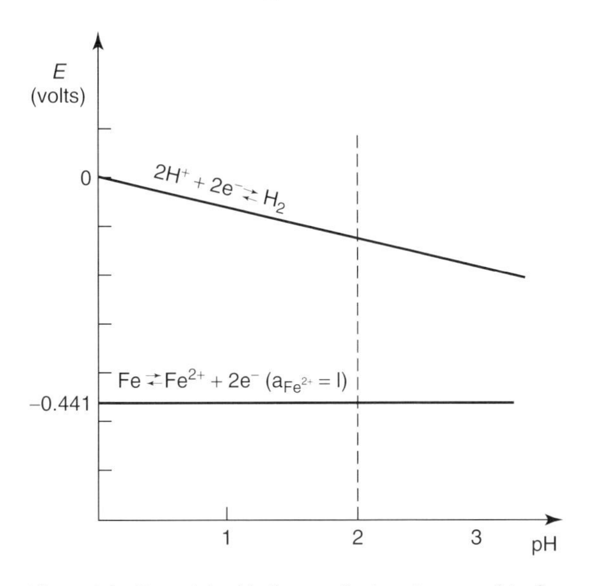

Figure 2.9 Potential–pH diagram for iron immersed in ferrous ion solution of unit activity.

The above reaction is associated with a definite value of the potential, E, which is related to the free energy change

$$\Delta G = -nFE \tag{2.25}$$

where n is the number of electrons involved, $F = 96\,450$ C per equivalent, and E is the potential. Thus, knowing the value of E, the potential, and "n," the number of electrons involved, one can calculate the free energy, ΔG, associated with the reaction. Thus, from the sign of the free energy change involved, it is possible to ascertain whether a postulated corrosion reaction can occur.

The free energy change for a chemical reaction at a temperature, T, is given by

$$\Delta G_T = \Delta G_T^\circ + RT\ln(\pi(a\gamma)\nu_y) \tag{2.26}$$

where ΔG_T° is the standard free energy change for the reaction at the prescribed temperature, π is the multiplication operator, $a\gamma$ is the activity or fugacity, ν_y is the stoichiometric coefficient for y, and R is the gas constant. By convention, the stoichiometric coefficients for reactants are negative, while those for products are positive.

The standard free energy change in a reaction can be calculated using the equation

$$\Delta G^0_{298} = \Sigma\gamma_y M^0_y \tag{2.27}$$

where Σ is the summation operator, M^0_y is the standard chemical potential of the species at 25 °C, and γ_y is the coefficient. Pourbaix [3] has tabulated the standard chemical potentials of most species of interest in corrosion studies.

The free energy change ΔG_T calculated for a particular reaction acts as a definite criterion for the corrosion reaction to occur or not. When the computed free energy gives a negative value, the corrosion reaction will take place, and the corrosion reaction is not likely to occur for positive free energy values.

Let us consider a sample of copper metal immersed in a strong acid solution at 25 °C. Let the activities of H^+ and Cu^{2+} be 10^{-2} and 10^{-4}, respectively. Now, the question on hand is whether corrosion of copper will occur under these conditions. Let the corrosion reaction be

$$Cu + 2H^+ \rightarrow Cu^{2+} + H_2 \tag{2.28}$$

From Pourbaix's *Atlas of Electrochemical Equilibria* [3] and the above reaction we have

Species	M^0_y	γ_y
Cu	0	−1
H^+	0	−2
Cu^{2+}	15.53	+1
H_2	0	+1

Assuming unit fugacity for hydrogen gas and using thermodynamic equations, ΔG can be computed as

$$\Delta G = (-1)(0) + (-2)(0) + 1(15530) + 1(0) + (1.987)(298)\ln\left(\frac{10^{-4}}{10^{-2}}\right)^2 (0.2)$$
$$= +1.553 \times 10^4 \text{ cal mol}^{-1} \text{ of Cu} \quad (2.29)$$

The positive free energy value indicates that the corrosion reaction cannot occur as written.

Let us now consider the effect of oxygen at a fugacity of 0.2 on the corrosion reaction, which may be written as

$$2Cu + 4H^+ + O_2 \rightarrow 2Cu^{2+} + 2H_2O \quad (2.30)$$

From Pourbaix's *Atlas of Electrochemical Equilibria* and inspection of the reaction, we can write

Species	M_y^0 (kcal/mol)	γ_y
Cu	0	−2
H^+	0	−4
O_2	0	−1
Cu^{2+}	15.53	+2
H_2O	−56.59	+2

Assuming unit activity for water and using thermodynamic equations

$$\Delta G = (-2)(0) + -(4)(0) + (-1)(0) + 2(15530) + 2(-56690) + 1.987(298)\ln\frac{(10^{-4})^2}{(10^{-2})^4}(0.2)$$
$$= -8.14 \times 10^4 \text{cal per 2 mol of Cu} \quad (2.31)$$

The ΔG, free energy value, indicates that corrosion of copper can occur according to the postulated reaction.

It is useful to consider a practical situation in which the copper system contains impurities such as 10^{-6} M of ferrous ion and 10^{-7} M of ferric ion. The corrosion reaction in this system will be of the type

$$Cu + 2Fe^{3+} \rightarrow Cu^{2+} + 2Fe^{2+} \quad (2.32)$$

From an inspection of the reaction and reference to Pourbaix's *Atlas of Electrochemical Equilibria* we have

Species	M_y^0 (kcal/mol)	γ_y
Cu	0	−1
Fe^{3+}	−2.53	−2
Cu^{2+}	15.53	+1
Fe^{2+}	−20.30	+2

Using the thermodynamic relationships along with the available data

$$\begin{aligned}\Delta G &= (-1)(0) + (-2)(-2530) + 1(15530) + 2(-20300) \\ &\quad + 1.987(298)\ln 10^{-2} \\ &= -2.27 \times 10^7 \text{cal mol}^{-1} \text{ of Cu}\end{aligned} \tag{2.33}$$

The negative value for free energy indicates the occurrence of corrosion of copper in the presence of ferric or ferrous ion impurities.

2.4 Electrode Potential Measurements

Electrode reactions may be written in the following general form:

$$\text{Reduced form} = \text{oxidized form} + ne^{-1} \tag{2.34}$$

where "n" is the number of electrons involved in the reaction. Electrochemical reactions generally occur at the metal–solution interface. An electric field exists across the metal–solution interface because of the nature of the reaction and the disposition of ionic and dipolar species on the solution side of the interface. This existing electric field cannot be measured directly, but a relative measurement is possible, which yields useful information. The test electrode (TE) is connected to a reference electrode (RE) through a voltmeter (E) with high impedance ($>10^9\ \Omega$).

The RE should be stable with a known electrode potential. By convention, the electrode potential of the system

$$H_2 \rightarrow 2H^+ + 2e^- \tag{2.35}$$

is assigned a value of 0.000 V at unit fugacity of hydrogen gas and hydrogen activity. This represents the standard hydrogen electrode (SHE). In practice, other REs as listed in Table 2.2 are used.

For example, consider the potential of an iron electrode −0.528 V measured against a saturated calomel electrode. This value can be converted to the standard hydrogen scale by using the calomel electrode value of 0.241 V versus the SHE:

$$E = -0.528 + 0.241 = 0.187 \text{ V versus SHE} \tag{2.36}$$

Table 2.2 Reference electrodes.

System	Electrolyte	Electrode potential (V versus SHE)
Calomel		
$2Hg + 2Cl^- = Hg_2Cl_2 + 2e^-$	Saturated KCl	0.241
calomel		
$Cu = Cu^{2+} + 2e^-$	Saturated $CuSO_4$	0.298
$Ag + Cl^- \rightarrow AgCl + e^-$	0.001 M KCl	0.400
	0.01 M KCl	0.343
	0.1 M KCl	0.288
	1.0 M KCl	0.234

2.5 Equilibrium Electrode Potentials

The condition for equilibrium in an electrochemical reaction is

$$\text{Reduced species} = \text{oxidized species} + ne^- \tag{2.37}$$

When the measured potential is negative with respect to the equilibrium value, the reaction favors the reduced form.

Let us consider water, which is an electrochemically active species. The electrochemical reactions involving water are

$$2H_2O = 4H^+ + O_2 + 4e^{-1} \tag{2.38}$$

$$H_2 + 2OH^- = 2H_2O + 2e^{-1} \tag{2.39}$$

For solutions in which the activity of water and the fugacities of oxygen and hydrogen are unity, the equilibrium electrode potentials for the above reactions at 25 °C are given by

$$E_{0,H_2O/O_2} = +1.23\text{–}0.059\ \text{pH} \tag{2.40}$$

$$E_{0,H_2/H_2O} = -0.059\ \text{pH} \tag{2.41}$$

The electrode potentials given are with respect to the SHE.

Pourbaix has shown that plotting electrode potentials of electrochemical reactions against pH of the solution is useful in identifying regions of stability of various chemical species in solution. These plots are known as *Pourbaix diagrams*. These diagrams are extremely useful in corrosion science. Let us consider the Pourbaix diagram for the system H_2O-H_2-O_2-H^+-OH^- (Figure 2.10). The lines (1) and (2) represent plots of $E_{0,H_2O/O_2}$ and $E_{0,H_2/H_2O}$ versus pH, respectively. Inspection of line (1) and the reaction $2H_2O = 4H^+ + O_2 + 4e^-$ shows that water is a stable entity at potentials below this line and, conversely, unstable at potentials above this line. Similar analysis of line (2) and the reaction $H_2 + 2OH^- = 2H_2O + 2e^-$ shows that water is a stable entity at electrode potentials above line (2) and unstable below

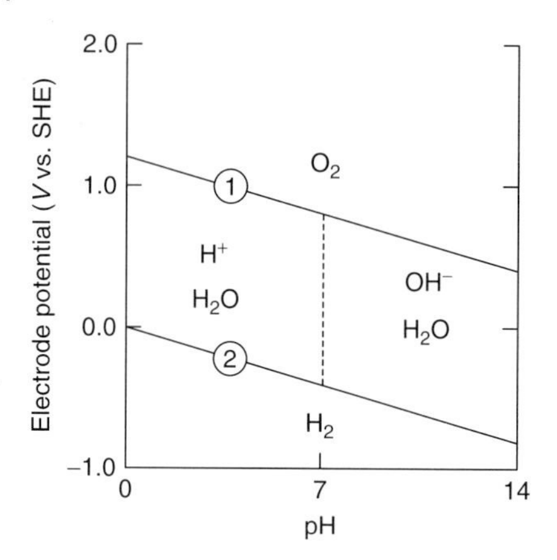

Figure 2.10 Pourbaix diagram for the system H_2O-H_2-O_2-H^+-OH^-. Fugacity of hydrogen and oxygen are unity; activity of water is unity.

this line. The Pourbaix diagram for this system is similar to a phase diagram. This may be considered as a form of electrochemical phase diagram. The region where a species is stable is identified with the chemical symbol of the species.

The Pourbaix diagram is extremely useful in determining stable chemical species for metals in contact with aqueous solutions. When the Pourbaix diagram for water is superimposed on that of the metal of interest, two types of behavior are observed. This behavioral classification of metals is based on the relationship between the metal–metal ion equilibrium reaction and the region of stability of water. In one class of metals, the metal–metal ion equilibrium potential falls within the region of stability of water. In these cases, it is possible to measure the equilibrium electrode potential of metal–metal ion reaction in aqueous solution and devise a method by which the kinetic properties of the reaction may be obtained with minimal kinetic complexity. In the second class of metals, the metal–metal ion equilibrium electrode potential falls below the region of stability of water. These metals form mixed potential systems with solvent water. Therefore, the equilibrium electrode potential of the metal–metal ion reaction cannot be measured in aqueous solution, and the kinetics of the complete reaction cannot be determined with ease. Copper is an example of the first type, and iron is an example of the second class of metals. The Pourbaix diagram for copper and some of its ionic species and compounds in aqueous solution at 25 °C is shown in Figure 2.11.

The equilibrium electrode potential for the copper–cupric ion reaction is located within the region of stability of water represented by dashed lines. Thus, the measurement of the equilibrium electrode potential is possible, and the kinetics of the copper–cupric ion system can be studied without interference from the reactions involving solvent decomposition. All reactions involved in this system are electrochemical in nature with the exception of the reaction

$$Cu^{2+} + H_2O \equiv CuO + 2H^+ \tag{2.42}$$

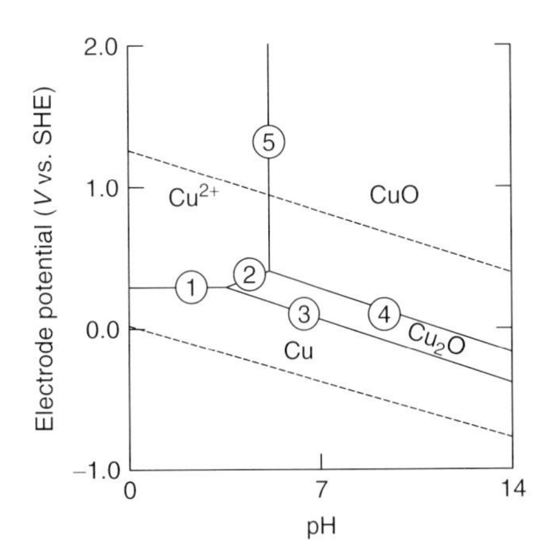

Figure 2.11 Activity of cupric ion is 0.01; dashed lines denote the stability range of water; temperature 25 °C; reactions considered are (1) copper giving cupric ion and two electrons; (2) cuprous oxide reacting with hydrogen to give cupric ion and two electrons; (3) cuprous oxide reacting with water to give cupric oxide, hydrogen ion, and two electrons; (4) copper metal reacting with water to give cuprous oxide, hydrogen ion, and electrons; and (5) cupric ion reacting with water to give cupric oxide and hydrogen ion.

and can be studied by the standard electrochemical techniques. If the measured electrode potential and pH are known, this figure may be used to determine the stable form of copper and its compounds under those conditions.

The reactions considered in the Pourbaix diagram are (i) copper giving cupric ion and two electrons; (ii) cuprous oxide reacting with hydrogen ion to give cupric ion, water, and two electrons; (iii) cuprous oxide reacting with water to give cupric oxide, hydrogen ion, and two electrons; (iv) copper metal reacting with water to give cuprous oxide, hydrogen ion, and electrons; and (v) cupric ion reacting with water to give cupric oxide and hydrogen ion.

The Pourbaix diagram for iron and some of its compounds in an aqueous system at 25 °C is given in Figure 2.12. The equilibrium potential of the reaction

$$Fe^0 = Fe^{2+} + 2e^- \tag{2.43}$$

falls outside the stability region of water, represented by dashed lines. Hence measurement of the equilibrium electrode potential is complicated by the solvent undergoing a reduction reaction, while the iron is undergoing electrochemical oxidation. This is the basis of the mixed potential model of corrosion.

In the potential–pH region where iron metal is the stable species, corrosion cannot occur since the reactions are not thermodynamically favorable. Pourbaix calls these regions "immune" to corrosion. However, in the broader sense of corrosion, iron may corrode in this region, such as fracture of iron alloys in the presence of hydrogen. Hence it is not sufficient for a metal to be "immune" for it to be free of corrosion.

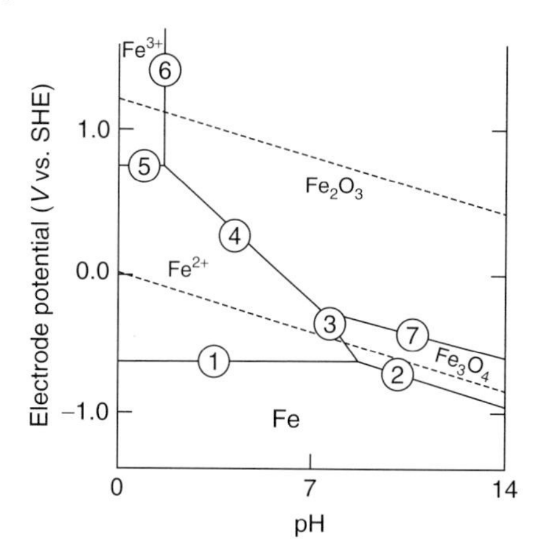

Figure 2.12 Activities of ferrous and ferric ions are 10^{-6}, temp. 25 °C. Reactions considered are (1) iron giving ferrous ion and two electrons; (2) iron reacting with water to give hematite, hydrogen ion, and electrons; (3) ferrous ion reacting with water to give hematite, hydrogen ion, and electrons; (4) ferrous ion reacting with water to give ferric oxide, hydrogen ion, and electrons; (5) ferrous ion giving ferric ion and an electron; (6) ferric ion reacting with water to give ferric oxide and hydrogen ion; and (7) hematite reacting with water to give ferric oxide hydrogen ion and electrons.

In the Pourbaix diagram for iron, the region of stability of iron oxide shows that the film of iron oxide on the metal surface forms a barrier between the metal and the environment. This condition is known as "*passivity*" and is characterized by the measured electrode potentials in the regions where the passive oxide film is stable. Figure 2.13 is a simplified Pourbaix diagram for iron. The figure clearly identifies the regions of immunity, corrosion, and passivity.

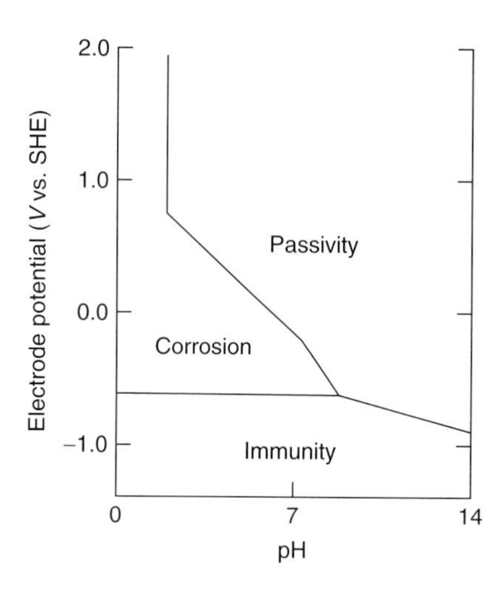

Figure 2.13 Pourbaix diagram for iron in terms of corrosion, passivity, and immunity.

2.6 Use of Pourbaix Diagrams

Let us consider copper metal at +0.150 V versus SCE in an aqueous solution of pH 2.5 and cupric ion activity of 0.01 at 25 °C.

In order to use the Pourbaix diagram, the potential is converted into SHE scale by

$$E = +0.150 + 0.241 = 0.391 \text{ V versus SHE} \tag{2.44}$$

Reference to the Pourbaix diagram for water and copper shows the stable species to be Cu^{2+}, H^+, and H_2O. It appears that corrosion is possible under these conditions. It is important to note here that corrosion of copper has also been observed in deaerated solutions with low values for corrosion rates [4].

Let us now consider the case of iron metal at a potential of −0.750 V versus SCE in a solution of pH 5.0 and a ferrous ion activity of 10^{-6} at 25 °C. The electrode potential with respect to SHE is −0.509 V, and reference to the Pourbaix diagrams for water and iron shows that the stable species are ferrous ion and H_2 and that corrosion is possible under these conditions. Thus, it is clear that Pourbaix diagrams reveal information on the stability of the chemical species and whether corrosion is likely to occur under a given set of conditions.

2.7 Dynamic Electrochemical Processes

The potential series and Pourbaix diagrams involving equilibrium conditions discussed thus far led to determine the feasibility of the corrosion process based on thermodynamics. These concepts do not give any information on the rates of the corrosion processes. In order to obtain any information on the corrosion rates, it may be necessary to understand the intimate dynamical processes occurring at the metal exposed to an electrolyte solution.

Consider the case of an electrode immersed in an electrolyte solution. A potential difference will arise at the interface between the electrode and the surrounding electrolyte solution. The potential difference that is observed is due to charge separation. If the electrons leave the electrode and combine with cations in solution, the electrode will acquire a positive charge, the solution will lose electroneutrality, and anions tend to move closer to the positive electrode. This scenario is depicted in Figure 2.14. Thus we have a pair of positive and negative sheets, which is otherwise known as the *electrical double layer*, named after Helmholtz, who recognized and postulated the presence of the double layer.

The process of contact adsorption of an ion consists of desolvation of the anion and removal of the water molecule from the electrode surface followed by the adsorption of the anion. The steps of anion adsorption in terms of energy are

$$E\ (\text{electrode} - \text{anion}) > E\ (\text{water on electrode}) + E\ (\text{desolvation of anion}) \tag{2.45}$$

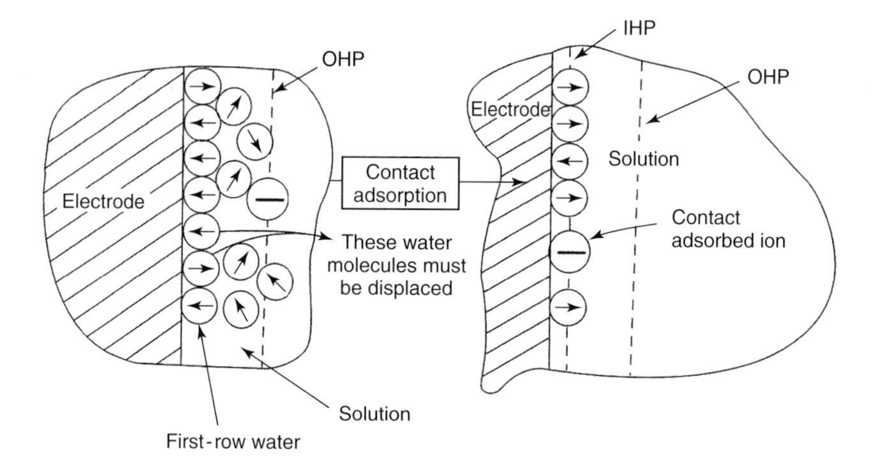

Figure 2.14 Process of contact adsorption in which a negative ion can lose its hydration shell and displace first-row water. The locus of the centers of the ions defines the inner Helmholtz plane. OHP, outer Helmholtz plane; IHP, inner Helmholtz plane.

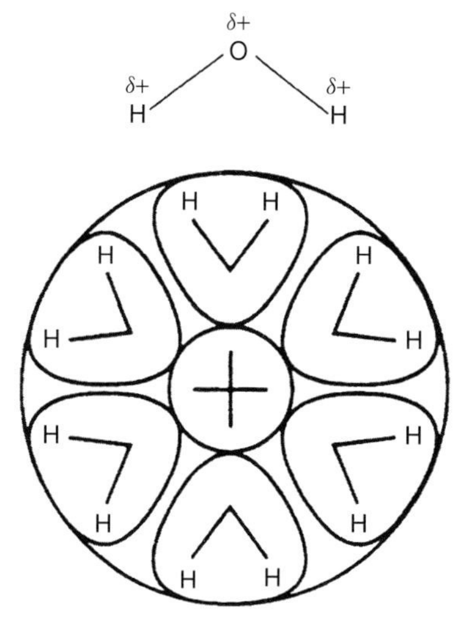

Figure 2.15 Schematic of a positive ion surrounded by a sheet of apex-inward water dipoles.

The disposition of water molecules attached to positive ions and negative ions with metal–oxygen and anion–hydrogen interactions are shown in Figures 2.15 and 2.16, respectively.

The first-row water molecules near the electrode surface play a useful role in the sense that they allow either an anion or a solvated cation to come close to the electrode surface. The locus of the center of the first-row molecules on the electrode surface defines the inner Helmholtz plane, and the distance between the electrode and the inner Helmholtz plane is ~3 Å. Similarly, the locus of the center of the

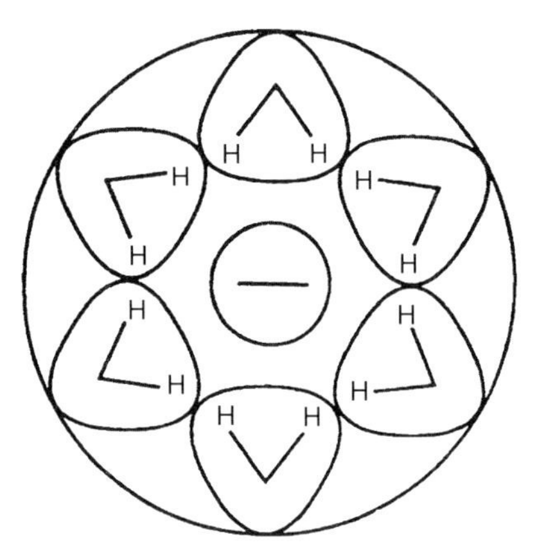

Figure 2.16 Schematic of a negative ion surrounded by a sheet of apex-outward water dipoles.

water molecules adjacent to the first-row water molecules is the outer Helmholtz plane, which is about 5–6 Å from the electrode surface.

The solvation of cations and anions by water molecules is possible due to the dipolar nature of water in the sense that the oxygen and hydrogen in water have partial negative and positive charges, respectively. It is useful to note that the bonding between a cation and a water dipole is stronger than that between an anion and a water dipole.

It is useful to visualize the situation when the electrode is negatively charged as shown in Figure 2.17. The solvated cations are positioned at the outer Helmholtz plane. The interaction forces involved in cation–water interaction are stronger than the negatively charged electrode–water interactions. In terms of mechanism,

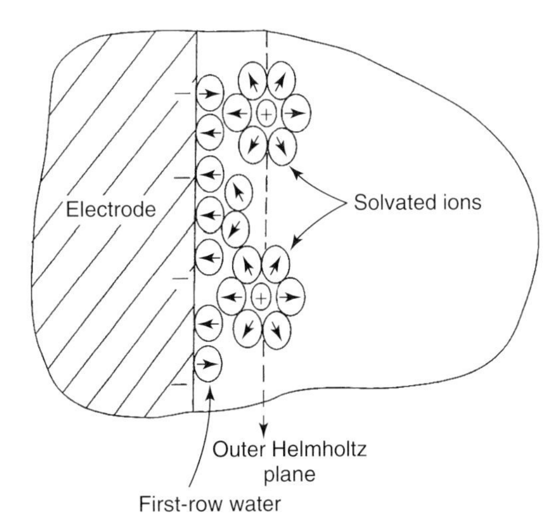

Figure 2.17 A layer of hydrated positive ions, whose hydration sheath cannot be stripped, on the first layer of water molecules. The locus of the centers of the ions define the outer Helmholtz plane.

the water molecules attached to the cation do not readily exchange with the water molecules adsorbed on the electrode surface. The solvated cation is located at the outer Helmholtz plane.

The Helmholtz double-layer model assumes a fixed layer of charges on the electrode and outer Helmholtz plane. This model has been modified by Guoy–Chapman analysis, which assumes that the ions of charge opposite to the charge on the electrode are distributed in a diffuse manner as depicted in Figure 2.18. The variation of potential with distance is exponential according to the Guoy–Chapman model, while it is linear according to the Helmholtz model. Another modification of the double-layer theory by Stern consists of a synthesis of the Helmholtz and Guoy–Chapman models in that both a fixed layer of charges and a region of scattered ions are considered. Stern's model of charge distribution and the variation of potential with distance are shown in Figure 2.19. Considering the layer of charges as a parallel plate capacitor, the relationship between total differential capacity and the capacitance due to Helmholtz (C_H) and Guoy–Chapman (C_G) can be obtained. The variation of potential with distance shows both linear and nonlinear portions, confirming the conclusion that the model embraces a synthesis of Helmholtz and Guoy–Chapman models.

When a test charge at a distance from the electrode approaches the electrode surface after passing through the solution, the potential will vary as shown in Figure 2.20. This trend in the variation of potential is the predicted trend. The

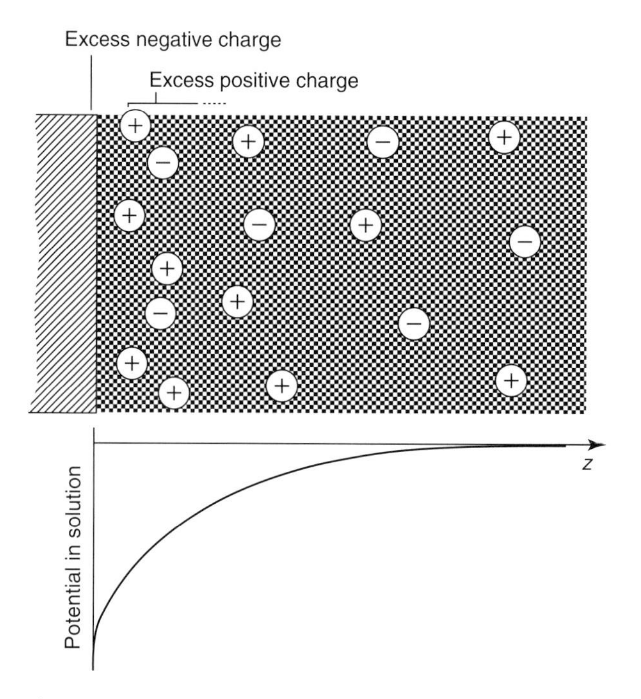

Figure 2.18 Guoy–Chapman model of double layer.

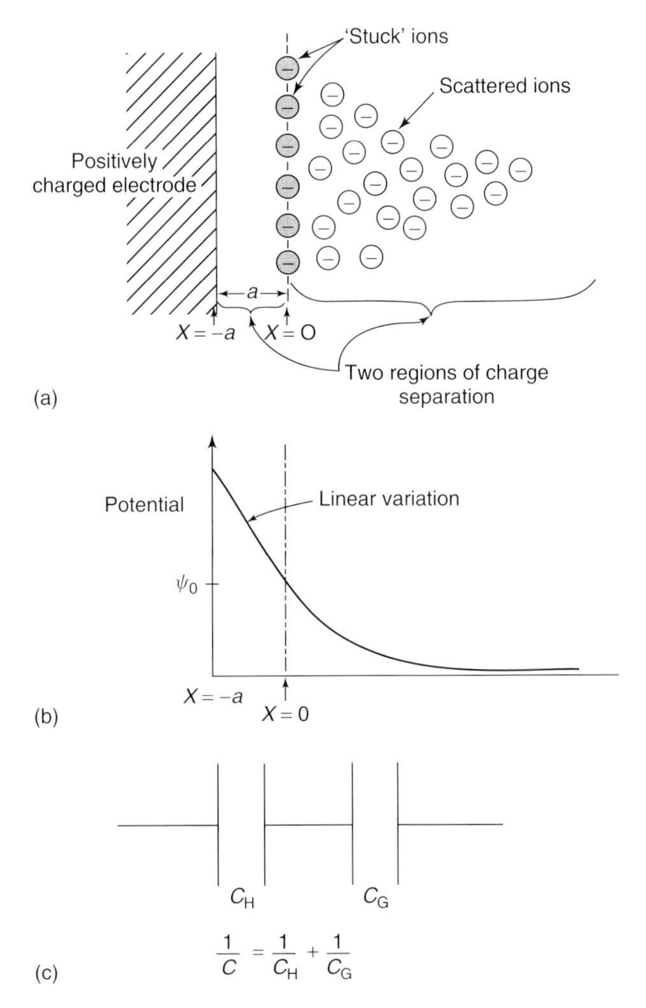

Figure 2.19 Stern model. (a) A layer of ions stuck to the electrode and the remainder in scattered manner; (b) the potential variation according to the model; and (c) the corresponding total differential capacity, C is given by the Helmholtz and Guoy capacities in series.

potential difference between the metal and the solution is given by the relation

$$\Delta \Phi = \Phi_{\mathrm{M}} - \Phi_{\mathrm{S}} \tag{2.46}$$

where Φ_{M} and Φ_{S} are the potentials of metal and the solution, respectively. The potential difference, $\Delta \Phi_{(\mathrm{M,S})}$ is the measured parameter when an electrode is immersed in an electrolyte solution.

In corrosion, the dynamic electrochemical processes play an important role; hence the consequences of perturbation of a system at equilibrium are described. The Daniel cell composed of copper metal immersed in copper sulfate solution

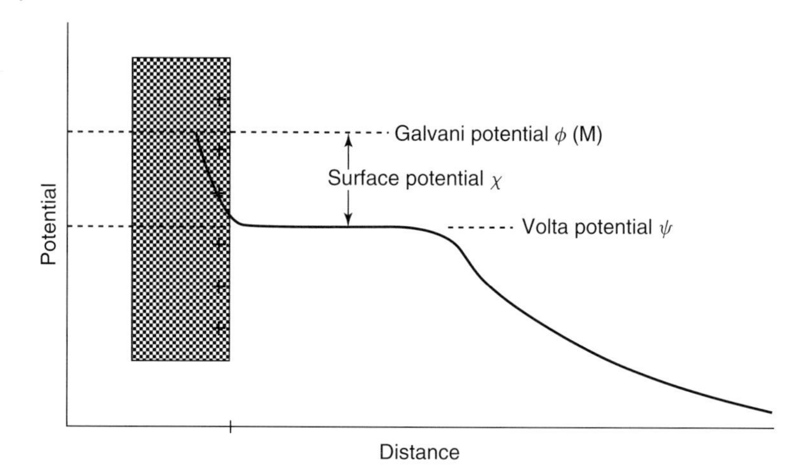

Figure 2.20 Variation of potential with distance from the electrode.

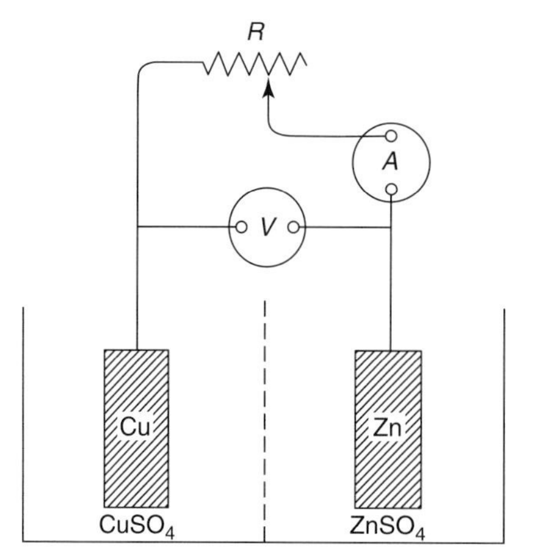

Figure 2.21 Polarized copper–zinc cell.

and zinc metal immersed in zinc sulfate solution, as shown in Figure 2.21, may now be considered.

When there is no current flow, the cell gives an electromotive force of 1.1 V. When a small current flows through the resistance, R, the potential decreases below 1.1 V. When current flows continuously, the potential difference between the two electrodes approaches a value close to zero, and this behavior is termed as the *polarization of the electrodes*. The effect of net current flow on the voltage of the Daniel cell can be represented by plotting the individual potentials of copper and zinc electrodes against the current as shown in Figure 2.22. This plot is known as the *polarization diagram*. When there is no net current flow, we have Φ_{Cu} and Φ_{Zn}, the potential values of copper and zinc, respectively, also known as *open-circuit potentials*. The zinc electrode polarizes along a-b-c and the copper electrode along d-e-f. At a current value of I, the polarization of zinc is the potential Φ at b minus

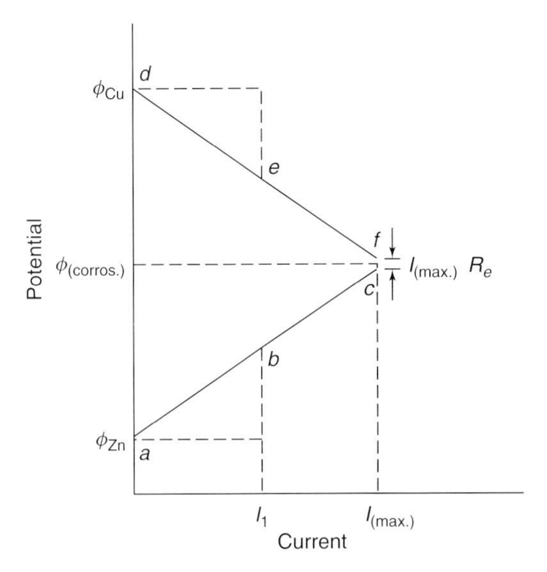

Figure 2.22 Polarization diagram for copper–zinc cell.

Φ_{Zn} or the open-circuit potential Φ_{Zn}^{-}. Similarly, in the case of copper, we have Φ at e minus Φ at d or Φ_{Cu}. The potential difference of the polarized electrodes, that is, zinc at b and copper at e is equal to I_1 $(R_{\mathrm{e}} + R_{\mathrm{m}})$ where R_{e} and R_{m} are electrolytic and metal resistances in series. The current is maximum at point c, and the potential difference is minimum and equal to $I_{\mathrm{max}} R_{\mathrm{e}}$.

The potential corresponding to I_{max} is known as the *corrosion potential*, E_{corr}. Using the equivalent weight of zinc (65.32/2), Faraday constant of 96 500 C equivalent^{-1}, and the value of I_{max}, the amount of zinc corroding in unit time may be calculated. It is useful to note than an equivalent amount of copper will be deposited during the reaction. The lines (a-b-c) and (d-e-f) are known as *cathodic and anodic branches of the polarization diagram.*

The rate of corrosion of zinc in the case of a polarized Daniel cell can be obtained using Faraday's Law

$$Q = ZFM \tag{2.47}$$

where Q is the charge due to ionization of M moles, Z is the valence, and F is the Faraday constant. Differentiating the equation representing Q, we have

$$\frac{\mathrm{d}Q}{\mathrm{d}t} = ZF\frac{\mathrm{d}M}{\mathrm{d}t} \tag{2.48}$$

where $\mathrm{d}M/\mathrm{d}t$ is the flux of the substance, I, the current flowing through the unit area of the cross-section, and i the current density, we can write

$$i = ZFJ \tag{2.49}$$

The flux of the material represents the corrosion rate, and the corrosion rate equals the current density.

Polarization represents a form of perturbation that results in disturbing the equilibrium, producing a dynamic situation. There are three types of polarization, namely, (i) concentration polarization, (ii) activation polarization, and (iii) IR drop.

Consider a situation in which metallic copper is immersed in water giving rise to cupric ions. The energy profile for this system is as shown in Figure 2.23. The environment should be capable of providing sufficient energy for metallic copper to corrode and give rise to cupric ions by overcoming the energy barrier, $\Delta G^{\ddagger}$.

$$\mathrm{Cu} \rightarrow \mathrm{Cu}^{2+} + 2\mathrm{e}^{-} \tag{2.50}$$

With the progress of corrosion, the cupric ion concentration will increase. With time, the tendency of copper to corrode will result in an increase in cupric ion concentration. After some time, the tendency for copper to corrode decreases as the current increases from zero and the value of free energy, ΔG, decreases along with the potential in conformity with Faraday's Law. In due course, the thermodynamic energies of the metal atoms and the metal ions approach each other. After a lapse of time, an equilibrium state is reached when the rate of formation of cupric ions will equal the rate of reduction of cupric ions to form metallic copper. Thus, the equilibrium shown in Figure 2.24 with energy of activation $\Delta G^{\ddagger}$ may be written as,

$$\mathrm{Cu}^{0} \underset{i_c}{\overset{i_a}{\rightleftarrows}} \mathrm{Cu}^{2+} \tag{2.51}$$

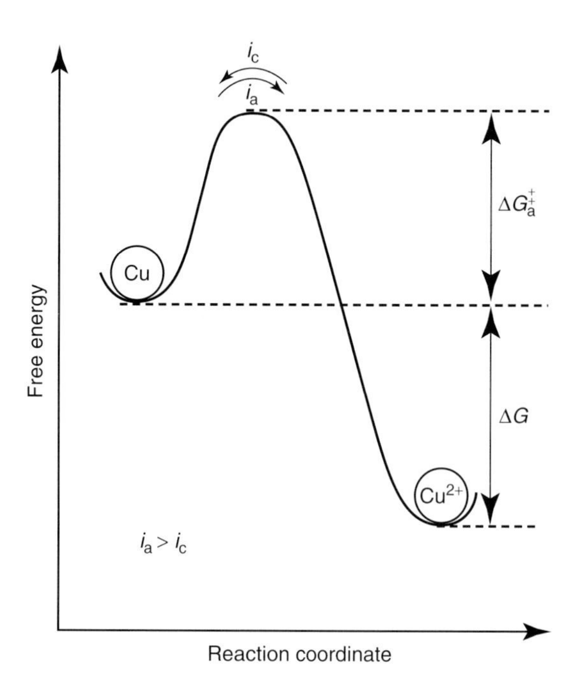

Figure 2.23 Energy profile for oxidation of copper.

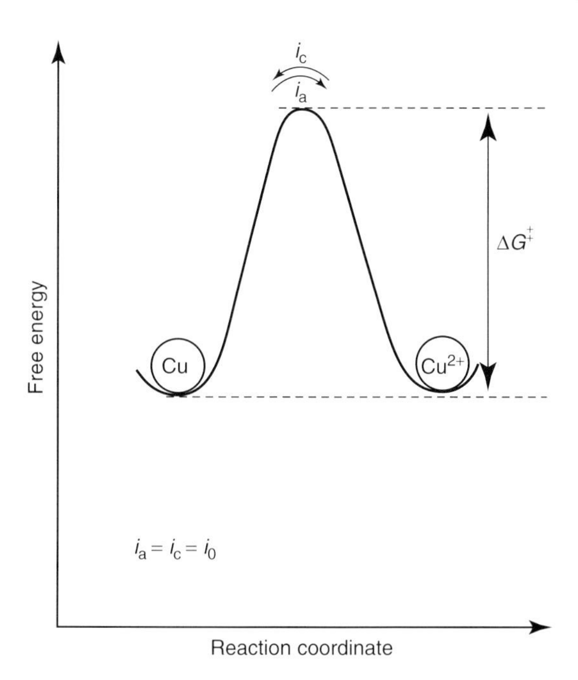

Figure 2.24 Energy profile for copper in equilibrium with a solution of divalent ions; $i_a = i_c = i_o$.

and, in general terms as

$$M \underset{i_c}{\overset{i_a}{\rightleftarrows}} M^{2+} \tag{2.52}$$

At equilibrium, $i_a = i_c$, the measured current density $i_{meas} = i_a - i_c$, and no net current flows, where a is the anode and c is the cathode. Actually, there is current flow, but it is equal and opposite, which cannot be measured. This current is known as *exchange current* I_o and i_o when divided by area.

The current density is a measure of corrosion rate. Corrosion rates are given in mpy (mils per year), ipy (inches per year), ipm (inches per month), and mdd (weight loss in milligrams per square decimeter per day).

Polarization consists of perturbation of a system at equilibrium in terms of potential. The polarization of a system at equilibrium is also known as *overpotential* or *overvoltage* and is denoted by η.

Let us consider a general system

$$M \underset{i_c}{\overset{i_a}{\rightleftarrows}} M^{2+} + 2e^- \tag{2.53}$$

for which the corrosion rate can be expressed as

$$r = k_{corr}\,[\text{reactants}] \tag{2.54}$$

where

$$k_{corr} = Ae^{-\Delta G/RT} \tag{2.55}$$

and A is a constant. We may then write for the corrosion rate, $r = Ae^{-\Delta G/RT}$ [reactants].

Since the system is at equilibrium, $k_f = k_r$ or $i_a = i_c = i_o$.

Since solid metal is corroding, it may be treated as constant

$$i_a = i_o = A_o e^{-\Delta G/RT} \tag{2.56}$$

The energy profile of a system at equilibrium and the anodic polarization of the equilibrium may be visualized as in Figure 2.25.

Reckoning the anodic polarization as $\alpha\eta$ and cathodic polarization as $(1 - \alpha\eta)$, we may write [4]

$$\begin{aligned} i_a &= A_0 e^{-\Delta G^{\ddagger} + \alpha\eta F/RT} \\ &= A_0 e^{-\Delta G^{\ddagger}/RT} Z e^{\alpha\eta ZF/RT} \end{aligned} \tag{2.57}$$

$$i_a = i_0 e^{\alpha\eta ZF/RT} \tag{2.58}$$

$$i_c = i_0 e^{(1-\alpha)\eta ZF/RT} \tag{2.59}$$

Since bulk current flow $i_{measured} = i_a - i_c$

$$I_{meas} = i_0 [e^{\alpha\eta ZF/RT} - e^{(1-\alpha)\eta ZF/RT}] \tag{2.60}$$

This relationship is known as the *Butler–Volmer equation*. Denoting

$$\alpha ZF/RT = A' \tag{2.61}$$

we can write

$$i_a = i_0 e^{(A'\eta)} \tag{2.62}$$

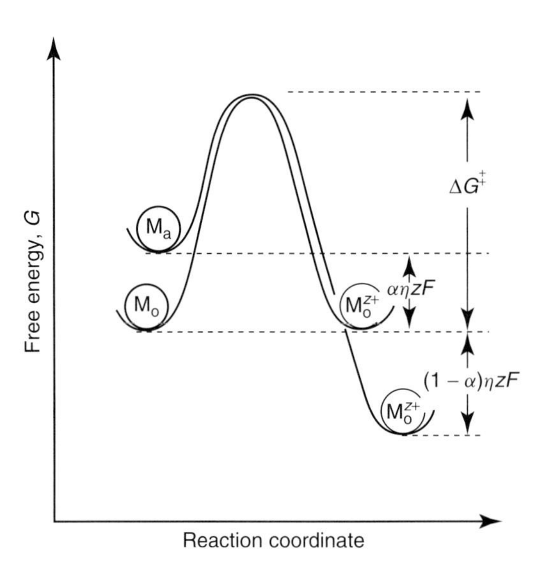

Figure 2.25 Energy profile for anode at equilibrium and for anodic activation polarization.

In logarithmic form, the equation takes the form

$$\ln i_a = \ln i_0 + A'\eta \tag{2.63}$$

$$\ln\left(\frac{i_a}{i_o}\right) = A'\eta \tag{2.64}$$

$$\eta = \frac{2.303}{A'}\log\left(\frac{i_a}{i_o}\right) \tag{2.65}$$

$$\text{Let } \beta = \frac{2.303}{A'} = \frac{2.303RT}{\alpha ZF} \tag{2.66}$$

We then have

$$\eta a = \beta a \log\left(\frac{i_a}{i_0}\right) \tag{2.67}$$

In general terms, we can write $\eta = c\log i + D$, where D is the diffusion coefficient, c is the constant, which is known as the *Tafel equation*. For the anodic and cathodic processes we have

$$\eta_a = \beta_a \log i_a - \beta_a \log i_o \tag{2.68}$$

$$\eta_c = \beta_c \log i_c - \beta_c \log i_o \tag{2.69}$$

where β_a and β_c represent $\frac{2.303RT}{\alpha ZF}$ and $\frac{2.303RT}{(1-\alpha ZF)}$, respectively.

In the Tafel equations, β_a and β_c are known as the *anodic and cathodic Tafel constants*.

Let us consider a sample of iron or steel polarized 300 mV anodically and 300 mV cathodically from the corrosion potential, E_{corr}, at a potential scan rate of 0.1–1.0 mVs^{-1}.

The resulting current is plotted on a logarithmic scale. The resulting hypothetical plot is shown in Figure 2.26. The corrosion current, i_{corr}, is obtained from the plot by extrapolation of the linear portions of the anodic and cathodic branches of the polarization curves to the corrosion potential, E_{corr}. The corrosion current value may then be used to calculate the corrosion rate using the equation

$$\text{Corrosion rate (mpy)} = \frac{0.13 i_{corr}\,(Eq.\ wt.)}{d} \tag{2.70}$$

where $i_{corr.}$ is the corrosion current density (Acm^{-2}), "d" the density of the corroding metal (gcm^{-3}), and *Eq. wt.* is the equivalent weight of the corroding metal in grams.

The linear polarization technique is rapid and gives corrosion rate data, which gives good agreement with the weight-loss method. Linear polarization technique involves scanning through 25 mV above and below the corrosion potential, and the resulting current is plotted against the potential as shown in Figure 2.27. The corrosion current i_{corr} is related to the slope of the line.

$$\frac{\Delta E}{\Delta i} = \frac{\beta_a \beta_c}{2.303(i_{corr})\beta_a + \beta_c} \tag{2.71}$$

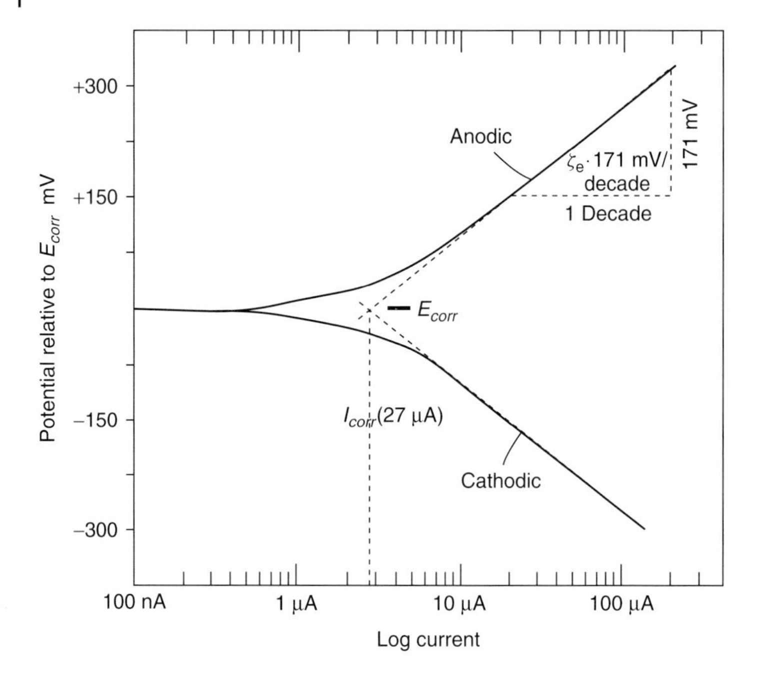

Figure 2.26 Experimental Tafel plot.

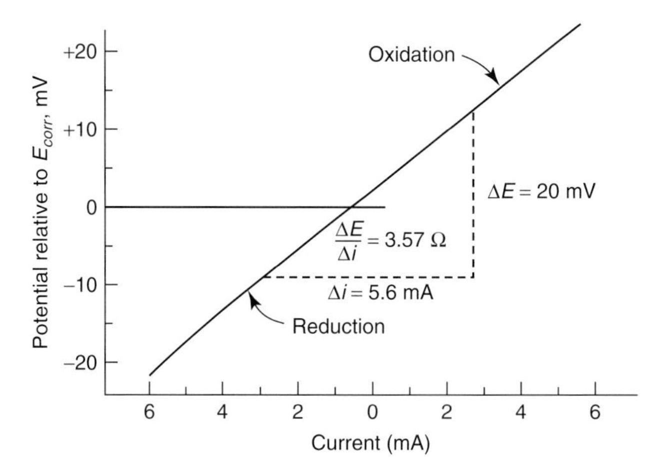

Figure 2.27 Experimental polarization resistance plot.

$$i_{\mathrm{corr}} = \frac{\beta_a \beta_c}{2.303(\beta_a + \beta_c)} \frac{\Delta i}{\Delta E} \tag{2.72}$$

$$\text{Corrosion rate (mpy)} = \frac{0.13 i_{\mathrm{corr}}\ (Eq.\ wt.)}{d} \tag{2.73}$$

A potentiostat is used to carry out electrochemical polarization measurements. The experimental arrangement consists of a cell containing a working electrode, an RE, and a counter electrode. The counter electrode is used to apply a potential on the working electrode both in the anodic and cathodic directions, and the resulting currents are measured. The electrochemical arrangement used in these measurements is shown in Figure 2.28.

2.8 Concentration Polarization

Concentration polarization may be illustrated by a copper cathode immersed in copper sulfate solution. The oxidation potential is given by the Nernst equation

$$E_1 = -0.337 - \frac{0.059}{2} \log \left[Cu^{2+} \right] \tag{2.74}$$

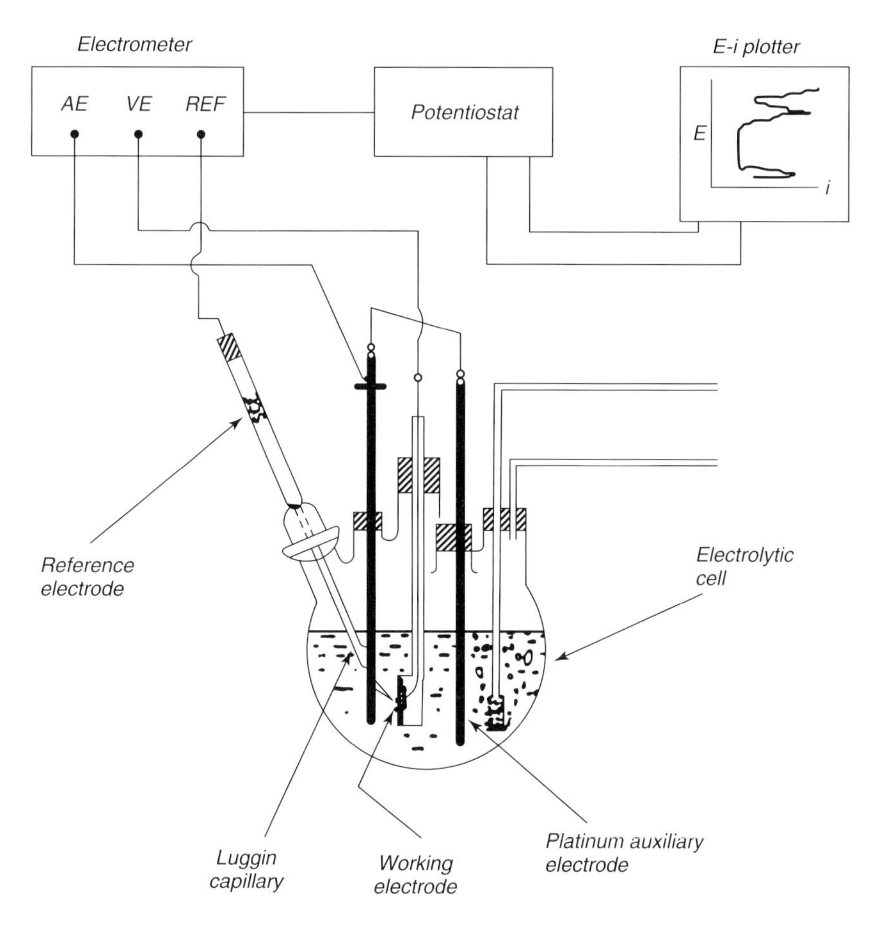

Figure 2.28 Experimental arrangement for potentiodynamic polarization studies.

On passage of current, the amount of cupric ions is reduced to $[Cu^{2+}]_s$

$$E_2 = -0.337 - \frac{0.059}{2} \log \left[Cu^{2+}\right]_s \tag{2.75}$$

and the difference in potential

$$E_2 - E_1 = \frac{0.059}{2} \log \frac{\left[Cu^{2+}\right]}{\left[Cu^{2+}\right]_s} \tag{2.76}$$

More the current smaller is the $[Cu]_s$ value and larger is the polarization, which is known as *concentration polarization*. As the value of $[Cu]_s$ approaches zero, the observed current density is known as *limiting current density*, i_L. When the limiting current density, i_L, is reached, we have

$$E_2 - E_1 = \frac{-RT}{nF} \ln \frac{i_L}{i_L - i} \tag{2.77}$$

and the limiting current density is obtained from

$$i_L = \frac{DnF}{\delta t} c10^{-3} \tag{2.78}$$

where D is diffusion coefficient, F is Faraday, δt is thickness of the electrode layer, and "n" is the number of electrons. The Helmholtz double layer plays a significant part in concentration polarization since the concentration of the ions on the electrode surface and the diffusion of ions from the bulk of the solution into the Helmholtz plane are contributing factors to the limiting current density. This scenario is illustrated in Figure 2.29.

Activation polarization involves a slow step in the electrode reaction. The reduction of hydrogen at the cathode consists of

$$H^+ + e^- \rightarrow H \tag{2.79}$$

$$H + H \rightarrow H_2 \tag{2.80}$$

which is also known as *hydrogen overvoltage*.

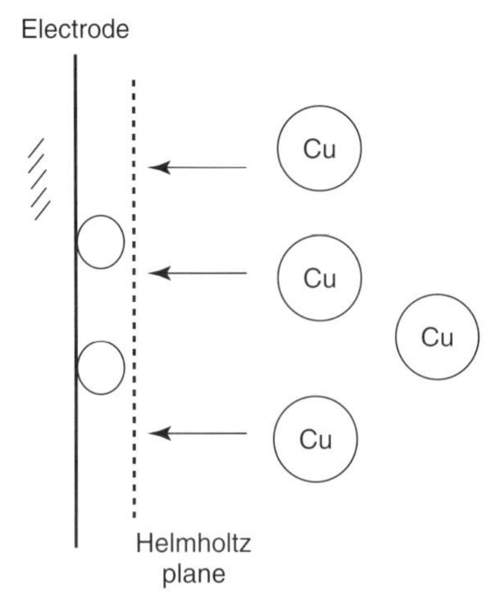

Figure 2.29 Diffusion of ions from bulk solution into Helmholtz plane.

The reaction of hydroxyl giving rise to oxygen may be termed as *oxygen overvoltage*.

$$2OH^- \rightarrow \frac{1}{2}O_2 + H_2O + 2e^- \quad (2.81)$$

Activation polarization η in general terms is given by:

$$\eta = \beta \log \frac{i}{i_0} \quad (2.82)$$

where β and i_0 are constants for a particular medium and a metal.

Since acids attack most metals, it is useful to know the exchange current densities and the hydrogen overpotentials. They are given in Tables 2.3 and 2.4, respectively.

The smaller the exchange current density the greater is the polarizability. In the first instance, the equilibrium oxidation potential determines whether corrosion is likely to occur or not. Then the hydrogen overpotential will determine the corrosion rate.

Let us consider a system consisting of zinc and iron in an acid solution. The two factors to be considered are the anodic polarization of the metal and the exchange current density for hydrogen evolution on the metal. The position of zinc in the galvanic series leads to the conclusion that zinc should corrode and not iron. In

Table 2.3 Hydrogen reaction.

Metal	Exchange current density (Acm^{-2})
Pb, Hg	10^{-13}
Zn	10^{-11}
Sn, Al, Be	10^{-10}
Fe	10^{-6}
Ni, Ag, Cu, and Cd	10^{-7}
Pd, Rh	10^{-4}
Pt	10^{-2}

Table 2.4 Overvoltage data [5].

Metal	i_o (Acm^{-2})	η, V (1 $mAcm^{-2}$)
Pt	10^{-3}	0.0
Pd	2×10^{-4}	0.02
Ni	8×10^{-7}	0.31
Fe	10^{-7}	0.40
Cu	2×10^{-7}	0.44
Al	10^{-10}	0.70
Sn	10^{-8}	0.75
Zn	1.6×10^{-11}	0.94
Pb	2×10^{-13}	1.16

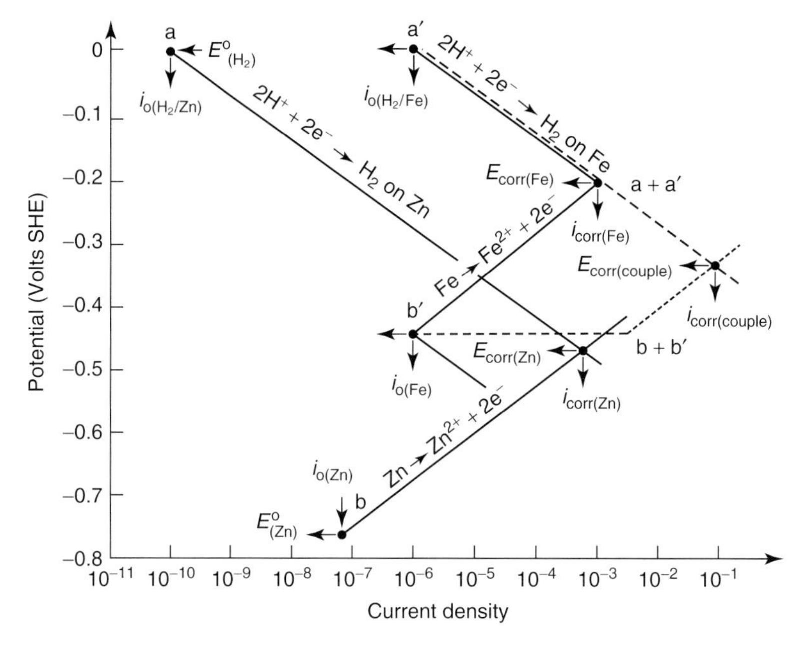

Figure 2.30 Mixed potential plot for iron and zinc.

this system, corrosion of iron occurs and not zinc because the exchange current density for hydrogen evolution is higher on iron than on zinc. A mixed potential diagram for iron and zinc is depicted in Figure 2.30. The lines a and b refer to zinc alone and a″ and b″ are those of iron alone corroding in an isolated condition. The lines a + a″ and b + b″ refer to a mixed electrode system of iron and zinc. The mixed electrode system of iron and zinc considered refers to an acid solution. The same metallic couple studied in neutral or alkaline solution involves oxygen reduction as the cathodic reaction

$$2H_2O + O_2 + 4e^- \rightarrow 4OH^- \tag{2.83}$$

and the rate of diffusion of oxygen to the surface of the metal plays an important role.

The potentiodynamic polarization technique used in corrosion studies consists of using a sample specimen as the working electrode, calomel electrode as the RE, and platinum as the counter or auxiliary electrode. A Luggin capillary is placed close to the working electrode to avoid or minimize the IR drop. The entire electrode assembly is immersed in the corrosive medium, and the corrosion potential and current are recorded. The potential is applied in the positive direction at a suitable scan rate (0.1 mVs^{-1}) and the resulting current recorded. Then the potential is applied in the negative direction to the corrosion potential, and the resulting current is measured. The resulting potentiodynamic polarization curves are as depicted in Figure 2.31. From the potentiodynamic curves, the corrosion potential; corrosion current; and regions corresponding to activation polarization,

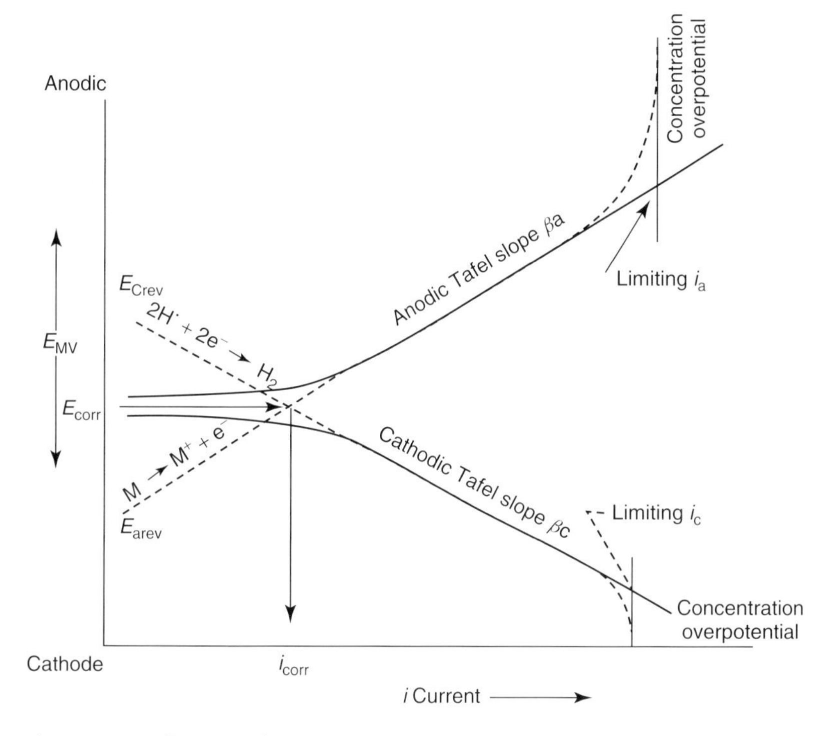

Figure 2.31 Polarization diagram.

concentration polarization, and resistance polarization are obtained. The software allows the recording of both anodic and cathodic polarization branches of the curve at a suitable scan rate. The software performs the calculations and gives the data for the corrosion potential and corrosion current density for the system under investigation.

Rapid developments in electrochemical techniques together with the associated instrumentation electrochemical impedance, electrochemical potential noise, and current noise are being used in a common manner in corrosion studies.

The kinetics and mechanism of corroding systems can be conveniently studied by electrochemical impedance spectroscopy. The advantages of AC techniques over DC techniques may be enumerated as follows:

1) AC techniques use very small excitation amplitudes in the range of 5–10 mV peak-to-peak voltage.
2) Data on electrode capacitance and charge-transfer kinetics may be used to obtain some information on the mechanism.
3) AC techniques can be applied to low-conductivity solutions, while DC techniques are prone to serious errors in these media.

Suppose a small sinusoidal potential ($\Delta E \sin \omega t$) is applied on a corroding sample. This results in a signal along with current flow of harmonics 2ω, 3ω, and so on. Then the impedance $\Delta I \sin(\omega t + \Phi)$ is the relation between $\Delta E/\Delta I$ and phase Φ.

In the case of corrosion studies, the sample is made part of a system known as *equivalent circuit*, which consists of solution resistance, R_S, charge-transfer resistance, R_{CT}, and the capacitance of the double layer C_{dl}. The measured impedance plot appears as a semicircle, also known as a *Nyquist plot*. The equivalent circuit and the Nyquist plot appear as shown in Figure 2.32. The experimental arrangement consisting of an AC impedance analyzer, the electrochemical cell, and the computer to acquire the experimental data over a period of time is shown in Figure 2.33. The

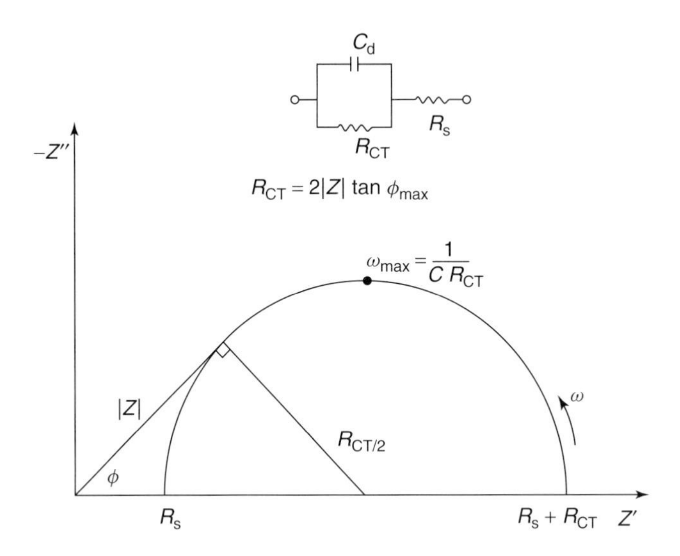

Figure 2.32 Representation of Nyquist plot.

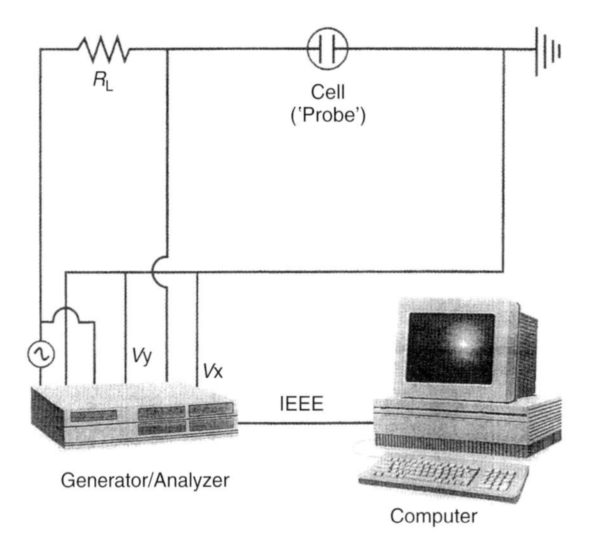

Figure 2.33 Experimental arrangement for AC impedance studies.

polarization resistance data obtained may be used to obtain the corrosion rate of the sample.

Polarization resistance of the corroding sample may be monitored over an extended duration. Thus, AC impedance may be used for on-line monitoring of the performance of corrosion inhibitors as shown in Figure 2.34. The polarization

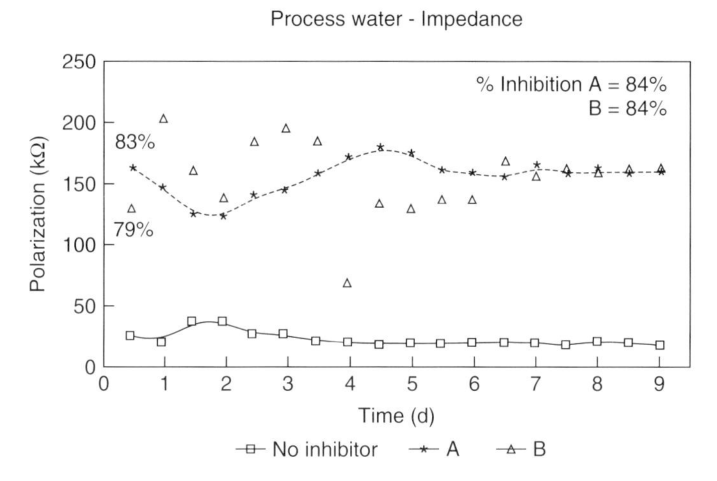

Figure 2.34 Polarization resistance as a function of time.

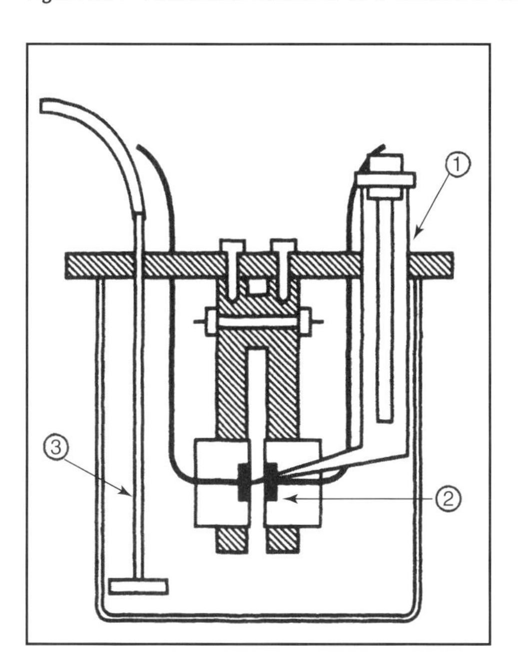

Figure 2.35 Electrochemical cell. (1) Calomel reference electrode; (2) steel specimen; (3) gas purge.

resistance of two inhibitors is plotted as a function of time. The data showed both the inhibitors to be equally effective.

Electrochemical potential noise and current noise are becoming popular for monitoring corrosion especially when localized corrosion, such as pitting corrosion, is encountered. The electrochemical cell and the experimental arrangement for potential noise measurement are depicted in Figures 2.35 and 2.36, respectively. The technique is quite simple as seen in the figures. It is especially suited for on-line monitoring of corrosion processes for long periods, and the typical data obtained in the evaluation of the performance of corrosion inhibitors in a field study is shown in Figure 2.37. The AC impedance data given in Figure 2.33 and the potential noise data given in Figure 2.37 refer to the same system and give an inhibition efficiency of 79–84%.

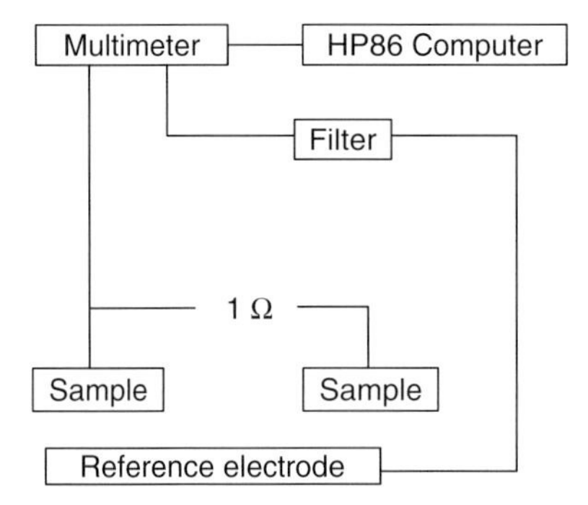

Figure 2.36 Experimental arrangement for potential noise measurements.

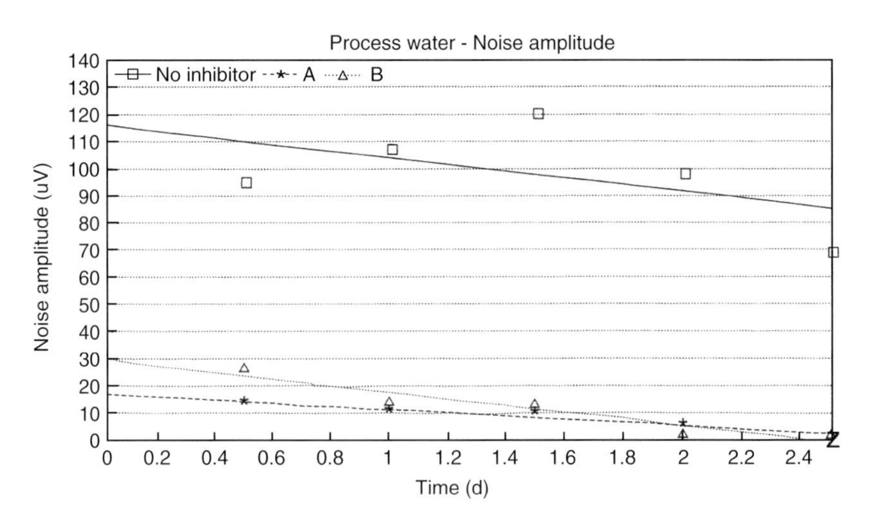

Figure 2.37 Potential noise amplitude as a function of time.

References

1. De La Rive, A. (1830) *Ann. Chem. Phys*, **43**, 423.
2. LaQue, F.L. and Copson, H.R. (eds.) (1963) *Corrosion Resistance of Metals and Alloys*, Reinhold, New York.
3. Pourbaix, M. (1966) *Atlas of Electrochemical Equilibria*, Pergamon, Oxford.
4. Trethewey, K.R. and Chamberlain, J. (1995) *Corrosion for Science and Engineering*, Longman.
5. Conway, B.E. (1952) *Electrochemical Data*, Elsevier, New York.

Further Reading

Bockris, J.O.M. and Conway, B.E. (1954) *Modern Aspects of Electrochemistry*, Butterworths, London.

3
Application of Microelectrochemical Techniques in Corrosion Research

Y. Frank Cheng

3.1
Introduction

Corrosion of metals is electrochemical in nature [1]. Corrosion scientists and engineers have used various electrochemical measurement techniques to investigate corrosion phenomena, obtaining mechanistic information about the corrosion reactions. These techniques, such as potentiodynamic polarization curve, linear polarization resistance, electrochemical impedance spectroscopy (EIS), electrochemical noise, and so on, are quite convenient to use [2]. Compared to the classic weight-loss method to obtain corrosion rate, the electrochemical techniques are quite time saving.

In the past decades, corrosion researchers have attempted to overcome the "spatial resolution" limitation of conventional electrochemical techniques, one of the most essential problems disabling the understanding of corrosion phenomenon at a microscopic, and thus a more mechanistic level. For example, the EIS technique has been used extensively to study the coating performance and corrosion of steel under the coating [3]. However, the measured impedance result is attributed to the electrochemical response of the whole electrode, reflecting an "averaged" behavior of the macroscopic electrode. In case a coating contains microdefects, such as pinholes, the electrochemical process occurring locally at these microdefects is "averaged" out [4].

Development of various microelectrochemical techniques has enabled the conventional electrochemical measurements to study corrosion phenomena with a high spatial resolution. These techniques, including scanning electrochemical microscopy (SECM), scanning reference electrode technique (SRET), scanning vibrating electrode technique (SVET), localized EIS (LEIS), scanning Kelvin probe (SKP), electrochemical scanning tunnel microscopy (ECSTM), SKP force microscopy (SKPFM), and so on, have been used extensively in corrosion research in the last two decades. This chapter summarizes the author's recent research progress in the application of thc SVET, LEIS, and SKP in pipeline corrosion and stress corrosion cracking (SCC) areas.

Green Corrosion Chemistry and Engineering: Opportunities and Challenges, First Edition.
Edited by Sanjay K. Sharma.

3.2
Scanning Vibrating Electrode Technique

3.2.1
The Technique and Principle

SVET measures the changes in microelectrochemical activity in proximity of the specimen surface, allowing the measurement of localized variations of potential and current flow over the surface of an electrochemically active specimen. The measurement relies on potential variations produced in an electrolyte by ionic currents associated with corrosion occurring locally.

Figures 3.1 and 3.2 are schematic diagrams of the measuring principle and the experimental setup of an SVET system, respectively. During the SVET measurement, a fine-tipped scanning probe, which is usually made of Pt–Ir alloy and in microns scale, is used to measure changes in microgalvanic activity close to the specimen surface. The microprobe vibration is controlled by a piezoceramic displacement device. Vibration amplitudes range from 1 to 30 μm in a direction perpendicular to the electrode surface. The potential of the microprobe is proportional to its position in the vibrating plane. The difference of potentials when the microprobe is located at the vibrating peak and valley, ΔE, is measured by an electrometer. The solution resistance between the vibrating peak and valley, R, is determined by $R = d/k$, where d is the vibrating amplitude of the microprobe and k is the solution conductivity. The SVET current, I, is then obtained by $I = \Delta E/R$. A flat SVET current density diagram means that there is uniform electrochemical activity of the test electrode, while a fluctuating SVET diagram is associated with an electrode with nonuniform electrochemical activity.

The operational parameters that affect the SVET signal output include

- the output time constant of the lock-in amplifier;

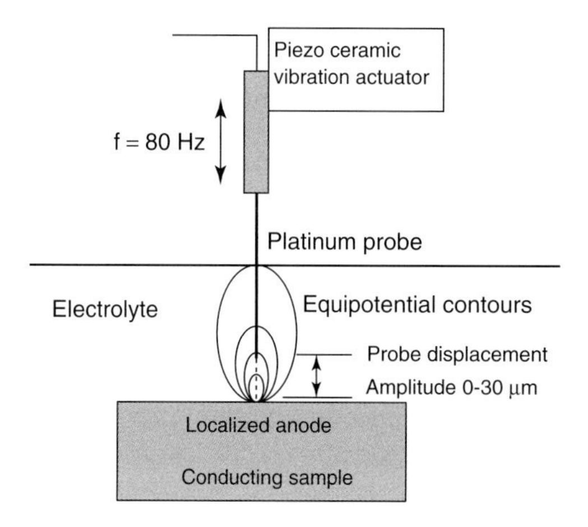

Figure 3.1 Schematic diagram of the measuring principle of SVET.

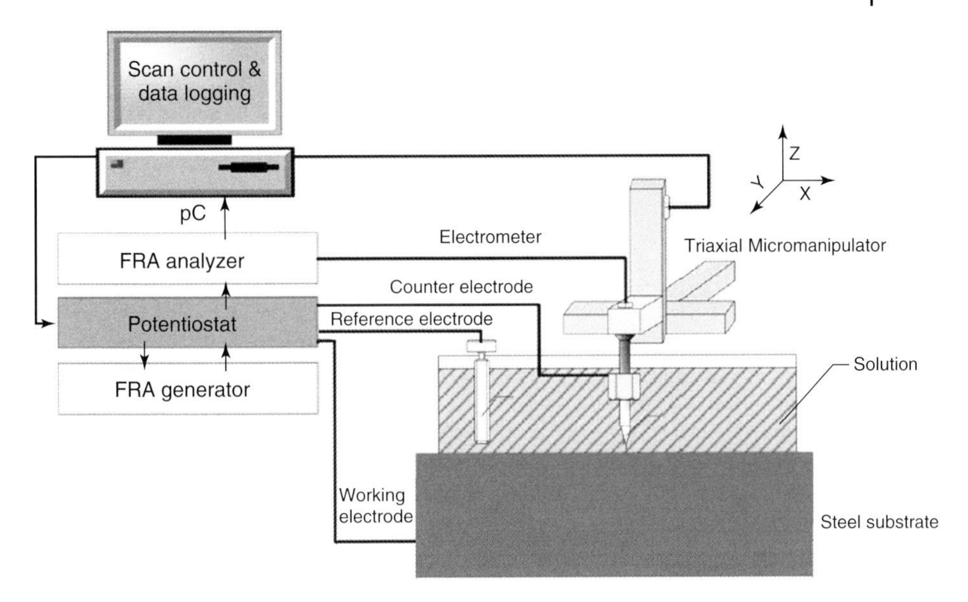

Figure 3.2 The experimental setup of an SVET system, where FRA is frequency response analyzer.

- the microprobe scanning speed in the *x*-direction;
- vibration amplitude of the microprobe;
- microprobe–specimen distance; and
- the microprobe scanning mode: continuous sweep or step.

A detailed description of the working principle of SVET can be found in Rossi *et al.*'s paper [5].

3.2.2
Local Dissolution Behavior of the Welding Zone of Pipeline Steel

It has been evidenced [6] that pipeline SCC occurred frequently at welding and in the adjacent area. Samples were cut from a spirally welded X70 pipeline steel and were machined into flat tensile specimens with the weld metal located in the center, as shown in Figure 3.3 [7]. The specimen was sealed with an epoxy resin except the working faces, that is, gauge section, and was immersed in an NS4 solution, which has been widely used to simulate the dilute electrolyte trapped between the coating and the pipeline, with a chemical composition (gl^{-1}): 0.483 $NaHCO_3$, 0.122 KCl, 0.181 $CaCl_2{\cdot}2H_2O$, and 0.131 $MgSO_4{\cdot}7H_2O$. Before each test, the solution was purged with 5% CO_2 balanced with N_2 gas for 2 hours to achieve an anaerobic and near-neutral pH condition (pH $\sim$ 6.8).

The SVET current mapping on the specimen containing the weld metal, heat-affected zone (HAZ), and the X70 base steel is shown in Figure 3.4. It is

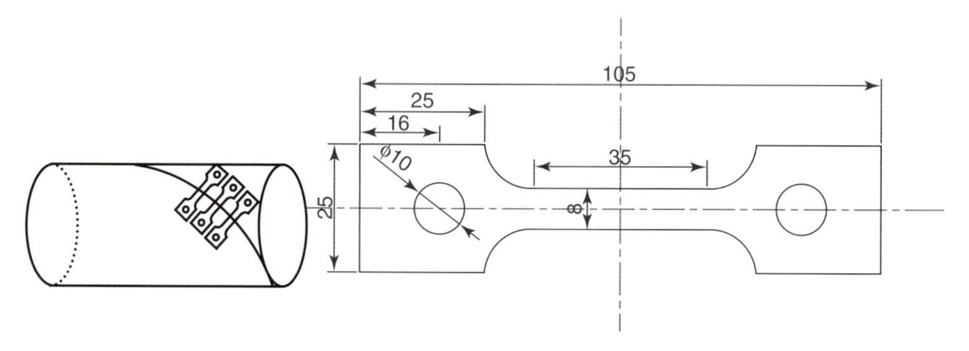

Figure 3.3 Schematic diagram of the tensile specimen taken from a welded X70 steel pipe for SVET measurement (unit: mm).

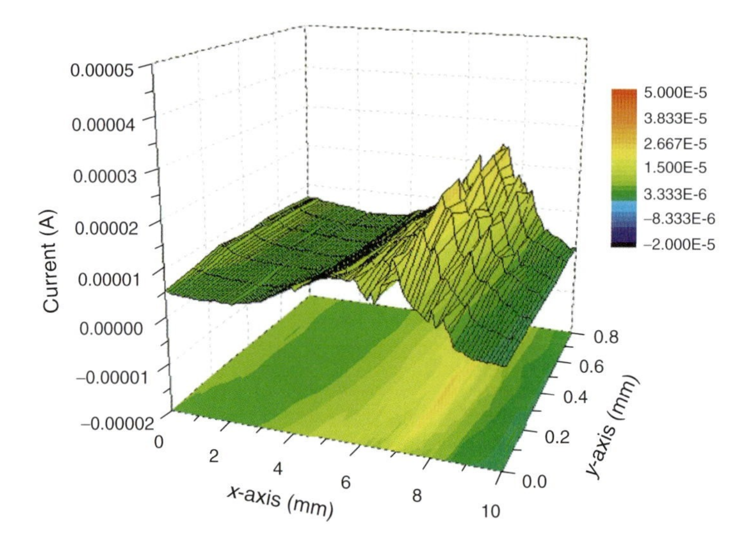

Figure 3.4 SVET measurement on the welded steel specimen containing the weld metal, HAZ, and base steel zones in NS4 solution (no hydrogen charging and no stress).

seen that the anodic current of the weld metal is about 7.4 μA. The current increased gradually and reached a maximum of approximately 25 μA at the HAZ. There was the smallest current in the base steel.

When the specimen is hydrogen charged, the dissolution current of the specimen increases. Figure 3.5 shows the SVET current map measured on various zones in the specimen in NS4 solution at various hydrogen-charging current densities for 2 hours. It is seen that the anodic current of all zones increases with the increase in the charging current density. The maximum current is at the HAZ. Similarly, when the specimen is under tensile stress, the anodic activity also increases, as seen in Figure 3.6, in which the SVET current map measured on the various zones in the tensile specimen at various stressing levels in NS4 solution is shown. It is

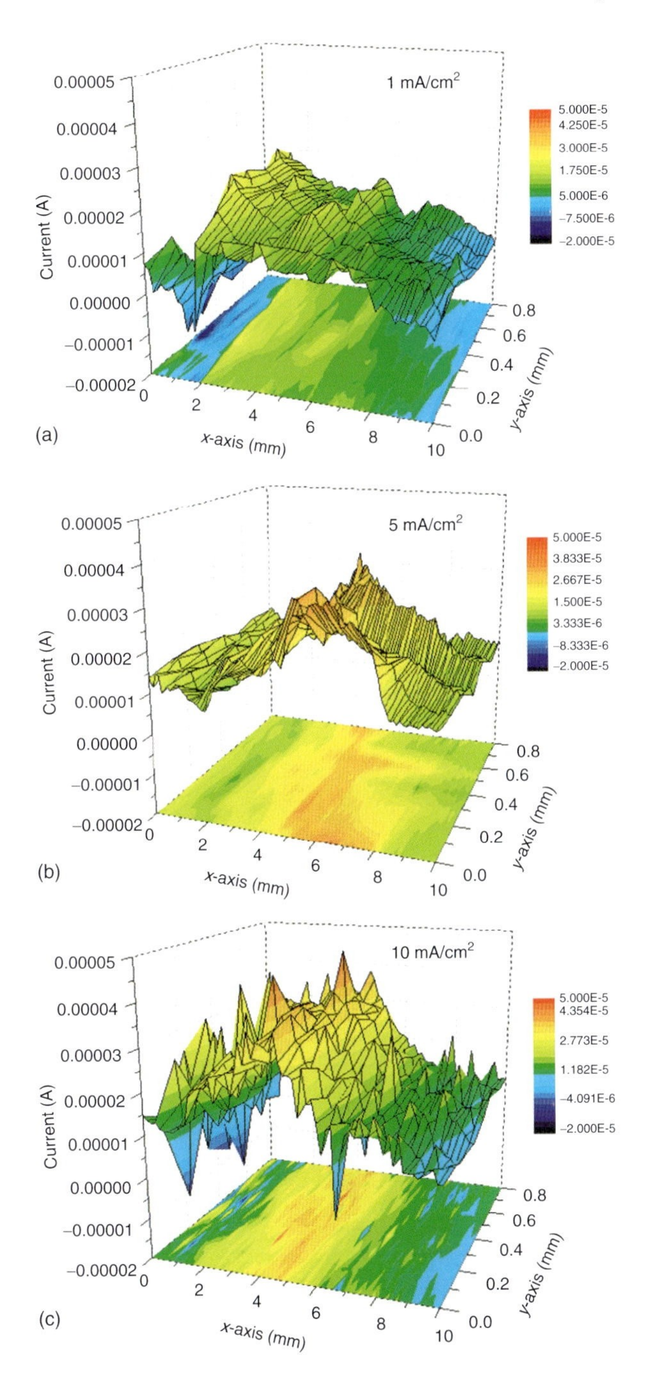

Figure 3.5 (a)–(c) SVET measurements on the specimen containing weld metal, HAZ, and base steel in NS4 solution with various hydrogen-charging current densities for 2 hours (no stress).

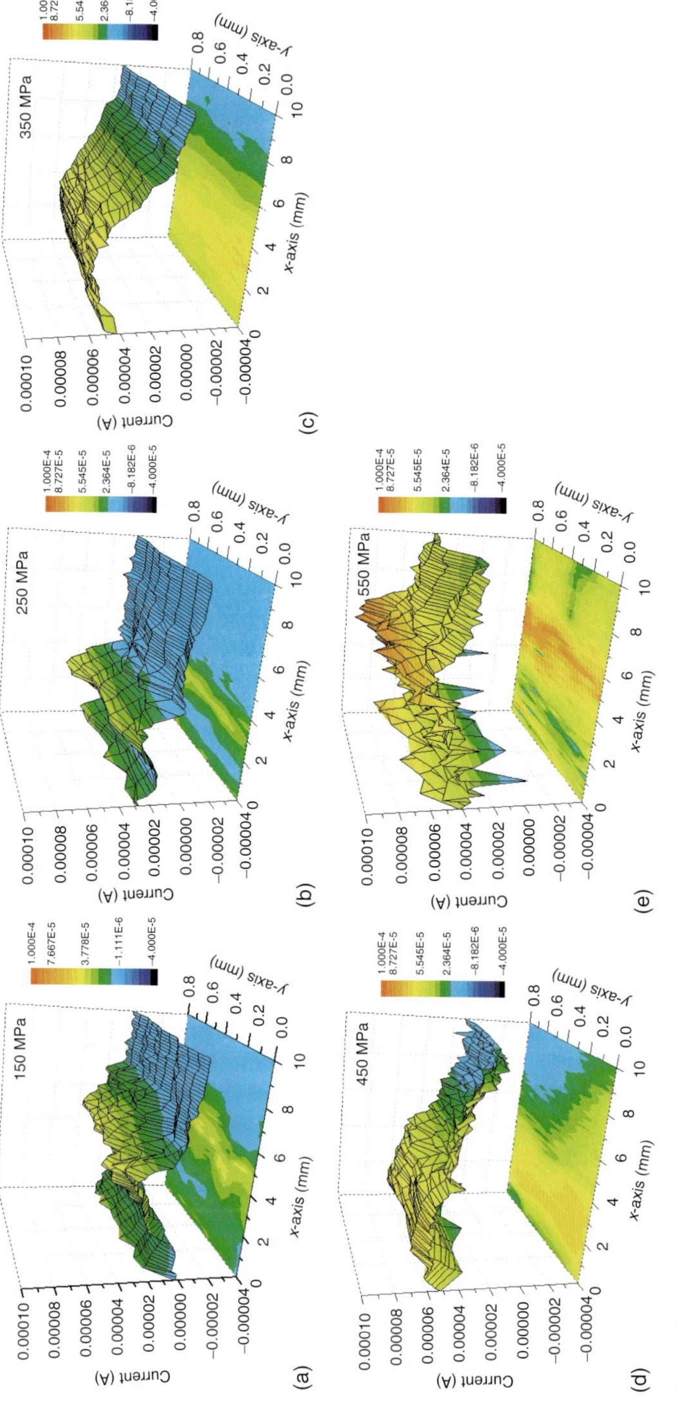

Figure 3.6 (a)–(e) SVET measurements on the specimen containing weld metal, HAZ, and base steel in NS4 solution under various stresses (no hydrogen charging).

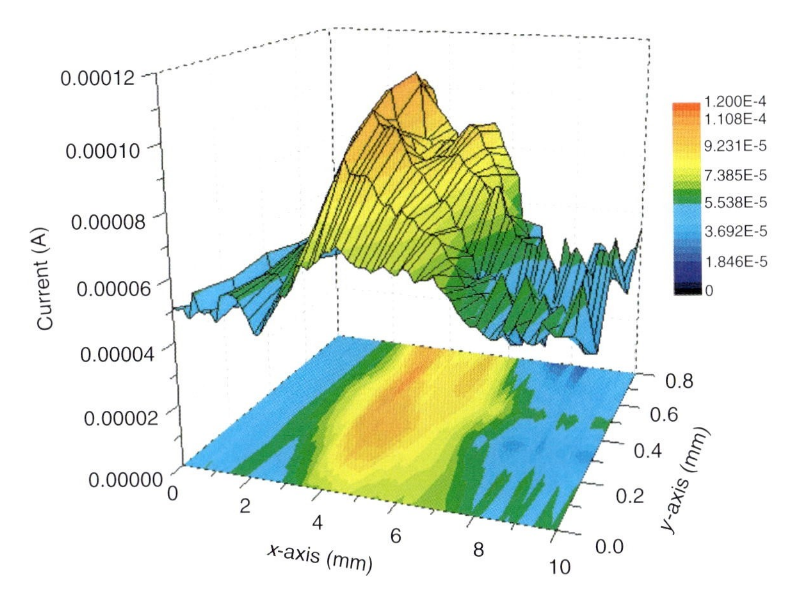

Figure 3.7 SVET measurement on the specimen with application of a 10 mAcm^{-2} hydrogen charging for 2 hours and 550 MPa.

obvious that the currents in all zones increase with the increase of applied stress. The maximum current is observed at the HAZ.

Figure 3.7 shows the SVET current map measured on the specimen under the synergism of hydrogen charging and stressing. It is seen that the current in all zones is higher than that with hydrogen charging (Figure 3.5) or under applied stress (Figure 3.6) individually. The maximum current is also observed at the HAZ.

The SVET measurements demonstrate that hydrogen charging and applied stress as well as their synergism contribute to an enhanced anodic dissolution current of the steel. Moreover, the highest dissolution activity is at the HAZ, compared to the weld metal and base steel. On hydrogen charging, the chemical potential and exchange current density of the steel is altered, resulting in an increasing activity. As the hydrogen concentration in steel is proportional to the hydrogen-charging current density [8], the anodic dissolution current increases with the latter. The hydrogen charge would diffuse toward and accumulate at the HAZ because of the presence of hardening phases and the high density of lattice defects.

Furthermore, the applied stress introduces strain energy to steel, resulting in activation of the steel. With the increase of the stress level, the local deformation strengthens, causing more dislocations to be activated. Consequently, more permanent slip bands form on the steel, enhancing the activity of the steel. Thus, the anodic activity of the steel increases with the applied stress.

For stressed steel, hydrogen charging would increase the energy resulting from the interaction between the hydrogen-induced lattice strain and the externally

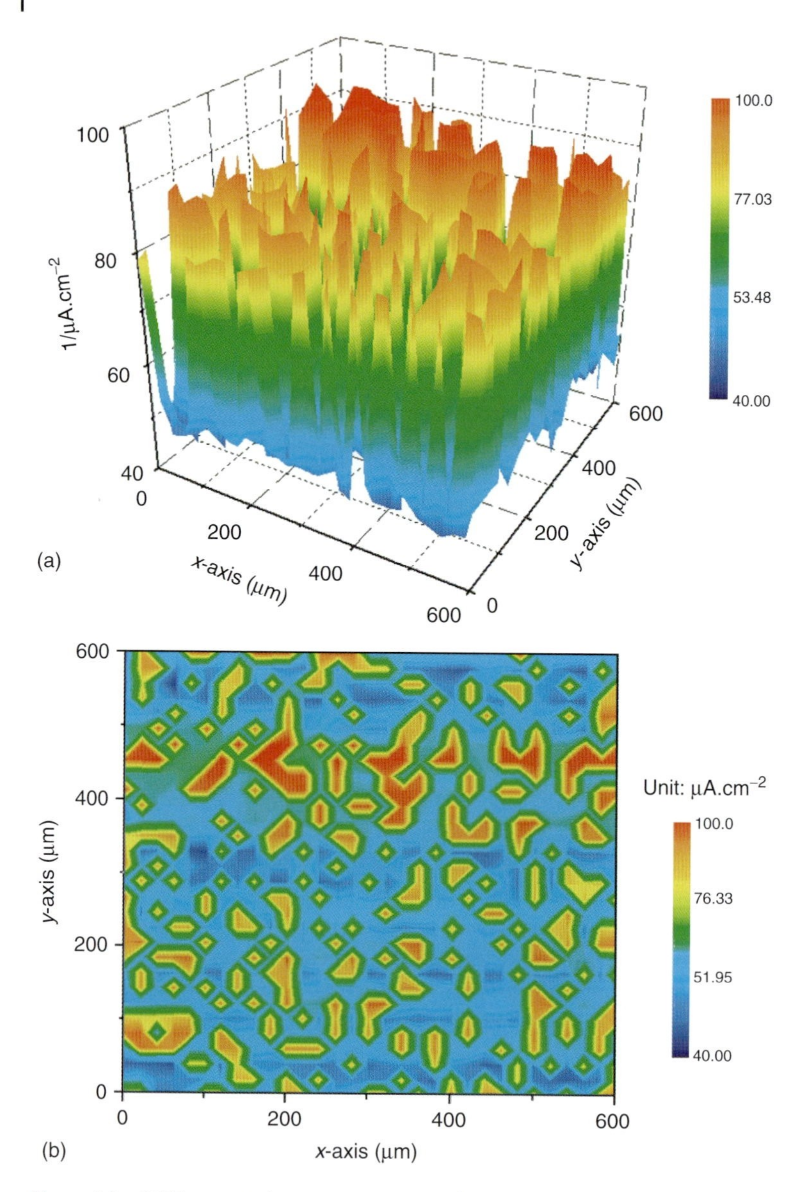

Figure 3.8 SVET current density map measured on an X70 steel specimen covered with a corrosion deposit layer in NS4 solution: (a) 3D map and (b) 2D contour map.

applied stress field, resulting in the enhancement of anodic dissolution of the steel. Similarly, for charged steel, the applied stress would increase the hydrogen content in the steel because of the enhanced stress-assisted hydrogen diffusion.

This work demonstrates that the HAZ is the most sensitive to the enhancement of dissolution reaction upon hydrogen charging, applied stress, and their synergistic effect. It provides an essential insight into the preferential site for initiation and propagation of stress corrosion cracks in the steel. The field reports show that most of the cracks are concentrated around the weld, especially the HAZ.

3.2.3 Effects of Mill Scale and Corrosion Product Deposit on Corrosion of the Steel

It has been demonstrated previously [9] that the near-neutral pH environment is capable of generating catalytic surface effect on hydrogen evolution, which is proposed to be associated with the presence of a layer of porous corrosion product that could separate cathodic and anodic reactions. Qin *et al.* also speculated [10] that the deposit layer on the steel surface is always characterized by a porous structure that could increase the probability of the local separation of anodic and cathodic sites and lead to the initiation of localized dissolution at the base of the pores. However, to date, there has been limited experimental evidence to provide direct confirmation of this speculation.

Figure 3.8 shows the SVET current density map measured on an X70 steel electrode covered with a corrosion deposit layer in NS4 solution, where Figure 3.8a is the 3D current density map and Figure 3.8b is the projected 2D contour map [11]. It is shown that the current density distribution was not uniform, with the maximum current density of about 90 $\mu A\, cm^{-2}$ and the minimum value of about 45 $\mu A\, cm^{-2}$, indicating the porous and noncompact structure of the corrosion product deposit formed on the electrode surface.

Figure 3.9 shows the SVET current density map measured on the steel electrode with one half of the area covered with a corrosion product layer and the other half of bare steel. The boundary line between the two areas is at $x = 550\ \mu m$ (parallel to the y-axis) and is marked as line l in Figure 3.9b. The area located left of l is the surface covered by a corrosion deposit layer, and the other area is the bare steel. It is apparent that the SVET current density map contains three different zones (A, B, and C in Figure 3.9b). The current density measured in zone A, where the electrode surface is covered with the corrosion deposit layer, is much higher (about 70–80 $\mu A\, cm^{-2}$) than in the other two zones. Zone B, which is adjacent to both zones A and C, is characterized by an intermediate current density of about 55 $\mu A\, cm^{-2}$. The smallest current density of about 35–40 $\mu A/cm^{-2}$ is in zone C.

The current density mapping over the deposit layer in this work (Figure 3.8) shows that the dissolutive activity of the deposit layer is not uniform, indicating a porous, noncompact structure of the deposit. It is reasonable to assume that anodic reaction of the steel is concentrated mainly at the pores that are associated with a high dissolution current density and that the cathodic hydrogen evolution occurs primarily on the deposit surface. A small anode/big cathode geometry is generated to accelerate the localized dissolution of the steel. The generated electrons are consumed by the cathodic reaction. Consequently, the cathodic reaction is enhanced in order to maintain the charge neutrality. Therefore, the porous deposit layer enhances the hydrogen generation.

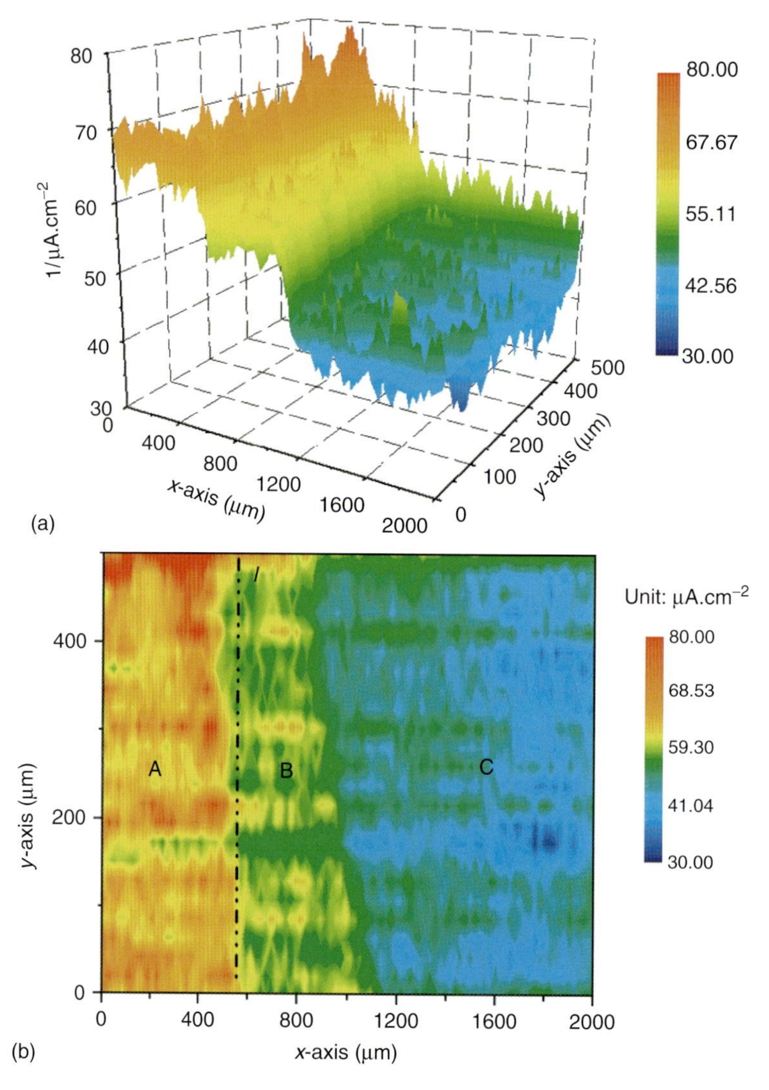

Figure 3.9 SVET current density map measured on the electrode surface with one half covered with a corrosion product layer and the other half of bare steel. (a) 3D map, and (b) 2D contour map, where A, B and C refer to the corrosion product-covered zone, transition zone and bare steel zone, respectively.

The SVET measurements also show that the presence of a layer of corrosion product deposit on the electrode surface greatly enhances the anodic dissolution rate of the steel. A transition zone is observed in the SVET measurement. The anodic dissolution current density of the transition zone (zone B in Figure 3.9) is less than that of the deposit-covered zone A and higher than that of the freshly exposed zone C. It is believed that the steel adjacent to the deposit layer is somewhat oxidized to generate a deposit scale, where the adsorption of intermediate species is enhanced, but not as strong as the deposit zone. As a consequence, the "self-catalytic" dissolution effect is less than zone A but much higher than zone C.

Occurrence of pipeline corrosion and SCC under a disbonded coating is usually associated with the presence of corrosion product deposit and scale on the pipe steel surface. This work shows that the deposit would accelerate corrosion of the steel. In particular, the small anode/big cathode geometry resulting from the porous structure of deposit could result in pitting corrosion, as indicated by the quite-high local dissolution current density at individual points over electrode surface (Figure 3.8). Corrosion pits might be the most common sites for crack initiation. Therefore, in the presence of deposit layer on the pipe steel surface, pipeline corrosion, especially pitting corrosion, is expected to be enhanced. Stress corrosion cracks could initiate from the corrosion pits that are developed under deposit.

3.3 Localized Electrochemical Impedance Spectroscopy

3.3.1 The Technique and Principle

LEIS is an evolution of the conventional EIS, with a measuring principle similar to that employed in EIS. It enables the measurement of the electrochemical impedance specific to a local, microscopic site. During LEIS measurement, an electrochemical microprobe electrode, usually made of platinum, is scanned over the surface of a target specimen, measuring the local current density in the electrolyte by the following principle. The potential difference between the platinum microprobe and another platinum ring electrode, ΔV_{local}, is measured via an electrometer, and the local AC current density, i_{local}, is calculated for a known solution conductivity, κ, by

$$i_{\text{local}} = \frac{\Delta V_{\text{local}}}{d} \times \kappa \tag{3.1}$$

where d is the distance between the two platinum electrodes.

The ratio of the AC voltage perturbation applied on the specimen, V_{appl}, to i_{local} then gives the value of local impedance, Z_{local}, by

$$Z_{\text{local}} = \frac{\Delta V_{\text{applied}}}{i_{\text{local}}} \tag{3.2}$$

The LEIS scanning microprobe is operated in two modes. The first mode is used for point-to-point LEIS measurements. The microprobe is set directly above the specimen to measure the typical impedance responses at individual points over a certain frequency range. The second mode is for LEIS mapping. The microprobe is stepped over a designated area of the specimen. The scanning takes the form of a raster in the x–y plane. The impedance distribution at a fixed frequency within the scanning area is then measured.

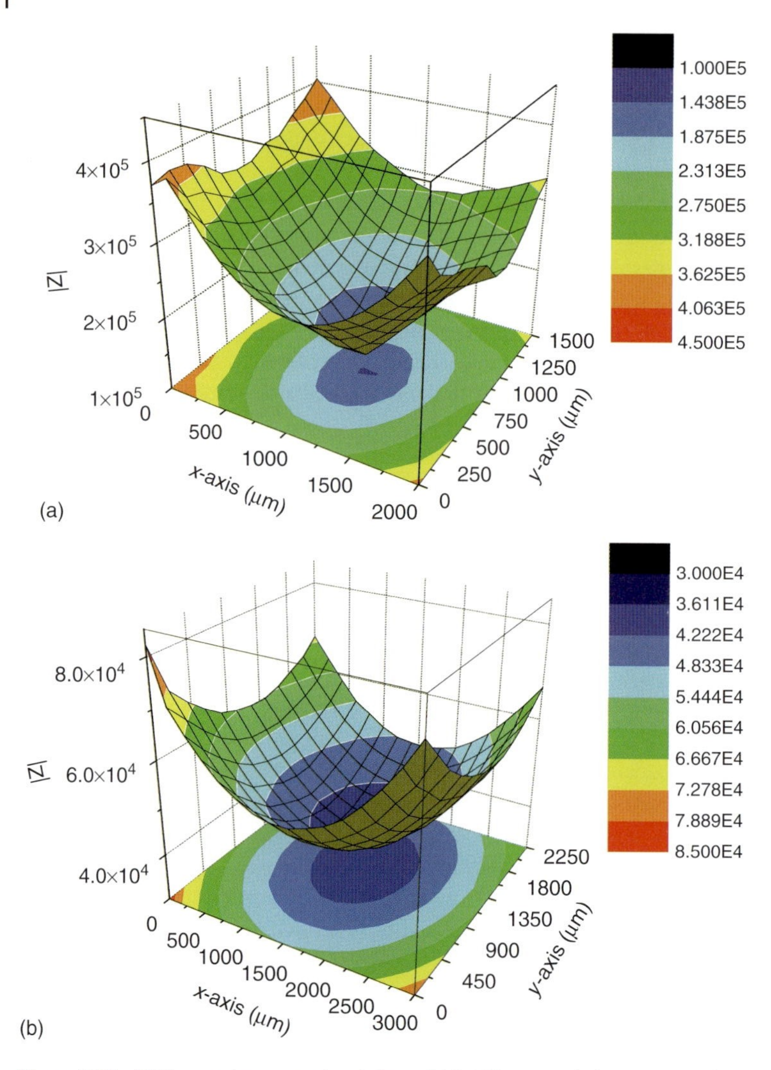

Figure 3.10 LEIS mapping around a defect of (a) 200 μm and (b) 1000 μm diameter in a coated steel electrode in NS4 solution.

3.3.2
Corrosion of Steel at the Base of the Coating Defect

Figure 3.10 shows the LEIS maps measured at 50 Hz on the high-performance composite coating (HPCC) X65 steel electrode with a central 200 and 1000 μm defect in NS4 solution [4]. It is clear that the defect area in the electrode is associated with the lowest impedance value in the map. Furthermore, Figure 3.10b shows a larger area with a low impedance value than Figure 3.10a, indicating that the former electrode contains a bigger defect.

Figure 3.11 shows the LEIS Nyquist diagrams measured at a defect of 200 μm at various immersion times in NS4 solution. It is seen that after 3 hours of immersion, the LEIS plot exhibits a capacitive loop over the whole frequency range (Figure 3.11a). With 24 hours of immersion, the LEIS plot (Figure 3.11b) is characterized by two depressed semicircles. Upon 48 hours of immersion, the impedance plot (Figure 3.11c) contains two semicircles followed by a straight line with an approximately 45° slope in the low-frequency range. When the immersion time increases to 168 hours, a linear impedance dominates the low-frequency range and a semicircle exists in the high-frequency range, as seen in Figure 3.11d.

As a comparison, Figure 3.12 shows the Nyquist diagrams measured by a conventional EIS on a coated steel electrode containing a central 200 μm defect at different immersion times (3, 24, 48, and 168 h). It is apparent that the impedance plot measured at individual time is quite different from the LEIS plot. A single capacitive loop is observed at 3 hours of immersion. On further immersion, the impedance plots are characterized by two semicircles in the high- and low-frequency ranges. With the increase of the immersion time, the sizes of both semicircles reduce.

When a coating contains small pinholes, such as the defect of 200 μm in the diameter in this work, corrosion of steel at the defect experiences different mechanistic changes with the immersion time, as identified by the LEIS plots. At the early stage of immersion (3 h), the LEIS plot shows a capacitive behavior with a high low-frequency impedance. At this stage, owing to the small size of the defect, there is not sufficient time for the steel to be fully exposed to the electrolyte. Therefore, the impedance response is dominated by the dielectric property of the coating. After 24 hours of immersion, two time constants are observed in the LEIS plot (Figure 3.11b). Moreover, the low-frequency impedance decreases dramatically compared to that measured at 3 hours of immersion. It is the typical impedance behavior measured on steel under a degraded coating [3]. The presence of two time constants indicates that there are two interfacial reaction processes occurring over the measuring frequency range. The high-frequency semicircle is associated with the pore impedance at the defect (about $1.5 \times 10^5\ \Omega$), while the low-frequency semicircle is the impedance response from the corrosion reaction occurring at the defect base, with a low-frequency impedance value of about $10^6\ \Omega$.

After 48 hours of immersion, three frequency-dependent time constants are observed in the Nyquist diagram, as seen in Figure 3.11c. Similarly, the high-frequency semicircle is from the pore impedance at the defect, with an impedance of approximately $1.5 \times 10^5\ \Omega$. The two time constants in the low-frequency range indicate that corrosion of steel at the defect is experiencing an activation–diffusion mixed-control process. The straight line with a slope close to 45° in the low-frequency range is a Warburg diffusive impedance. The overlapped Warburg impedance with a semicircle in the low-frequency range shows that the diffusion process contributes to the corrosion reaction at the base of the defect. Owing to the small size of the defect and a relatively long pathway of about 1.1 mm for corrosive species to diffuse from bulk solution to the defect base and, furthermore, owing to the deposit of corrosion product blocking the diffusive path, the diffusion step becomes dominant

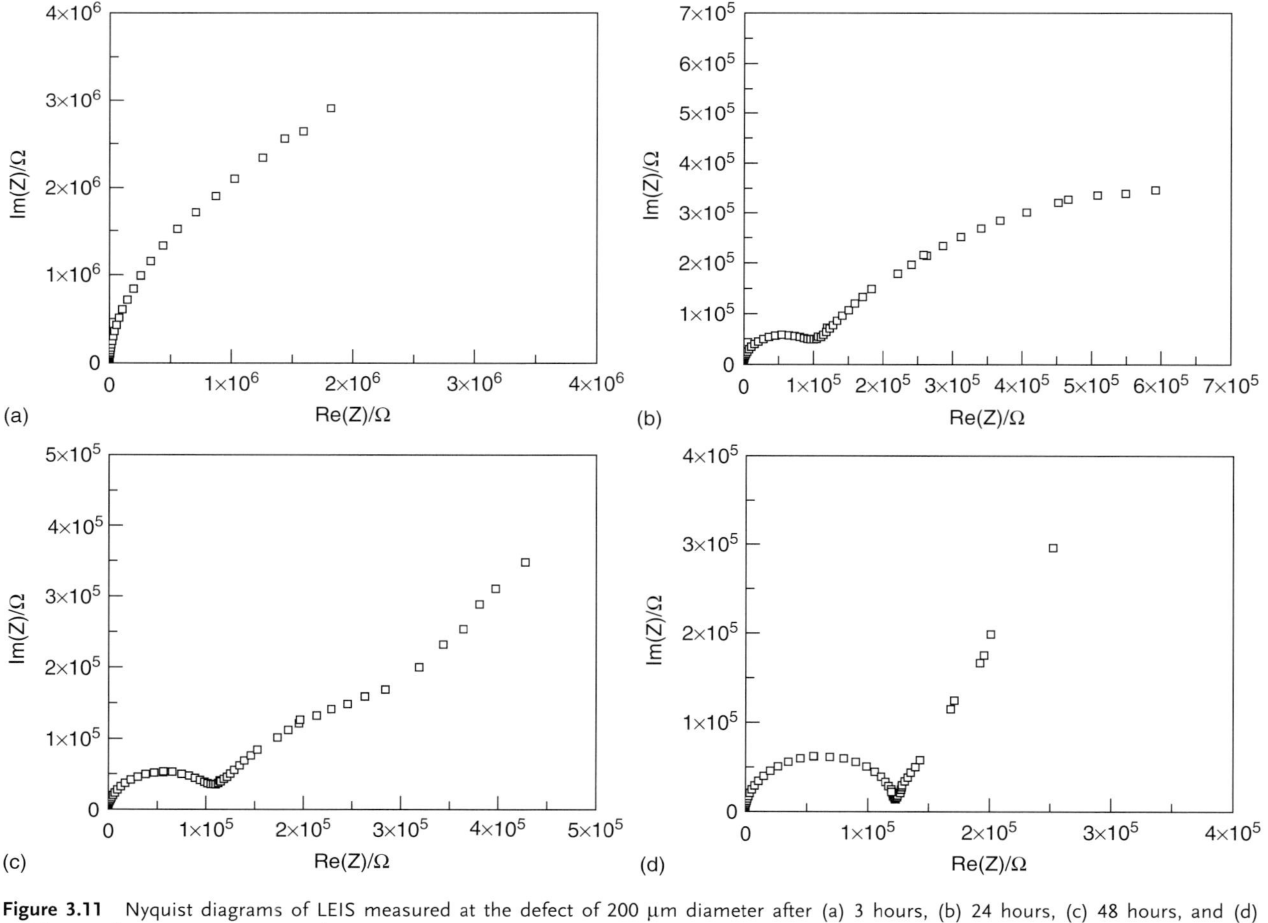

Figure 3.11 Nyquist diagrams of LEIS measured at the defect of 200 μm diameter after (a) 3 hours, (b) 24 hours, (c) 48 hours, and (d) 168 hours of immersion in NS4 solution.

over the corrosion process. With 168 hours of immersion, the measured LEIS plot contains two time constants again, with one semicircle at high frequency and a diffusive Warburg straight line at low frequency, as shown in Figure 3.11d. The high-frequency semicircle is from the pore impedance, and the low-frequency Warburg impedance behavior shows that diffusion dominates the corrosion process. The diffusion-controlled effect is primarily due to the block effect of the corrosion product deposited in the small defect.

The conventional EIS measurement on a macroscopic, coated steel electrode containing a same-sized defect provides results distinctly different from those provided by LEIS measurement. In the early stage of immersion, the EIS plot measured after 3 hours of immersion (Figure 3.12) is similar to the LEIS plot measured at the defect with the same immersion period (Figure 3.11a), that is, a capacitive behavior with a one time constant over the measuring frequency range. The electrolyte has not yet reached the steel surface to result in corrosion of the steel. Therefore, the measured impedance is dominated by the coating property at this stage. After 24, 48, and 168 hours of immersion, the impedance diagrams are featured with two semicircles in the high- and low-frequency ranges, representing the typically reported EIS feature measured on a defect-containing coating system [12]. The high-frequency semicircle is attributed to the impedance response from coating, while the low-frequency semicircle is associated with the corrosion reaction occurring on the steel/coating interface. Compared with the LEIS plots measured at the coating defect, it is apparent that EIS measurements

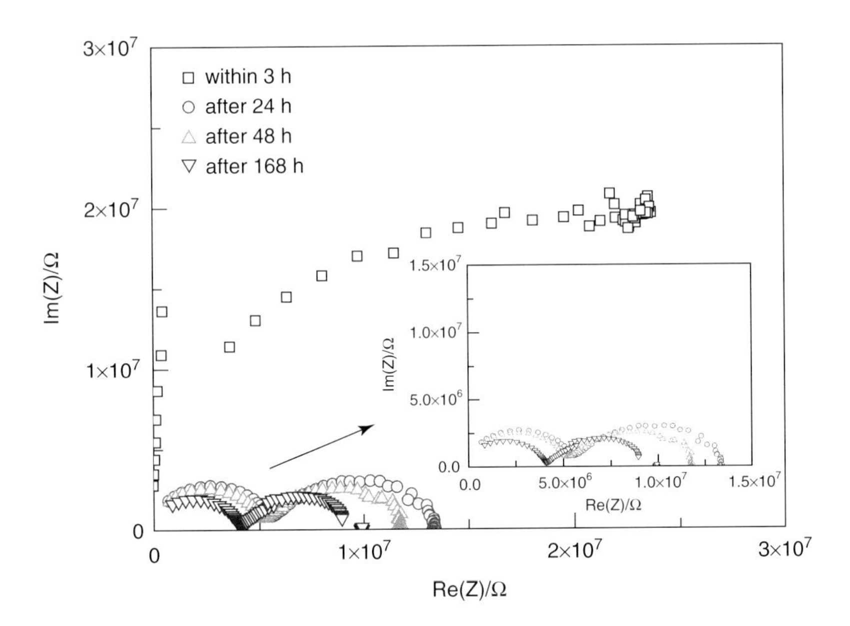

Figure 3.12 Nyquist diagrams measured by a conventional EIS on a macroscopic specimen (HPCC-coated X65 steel electrode with a central 200 μm defect), where Im(Z) and Re(Z) refer to imaginary and real impedance, respectively.

on the macroscopic electrode miss some important information that is directly associated with the electrochemical corrosion processes occurring at the defect. For example, the conventional EIS plots do not reveal the mass transport process and the block effect of corrosion product as well as the resultant change of corrosion mechanism. An "averaged" effect of EIS measured on both coating and the defect completely eliminates the localized corrosion processes occurring at the defect of the coating.

3.3.3
Microscopic Metallurgical Electrochemistry of Pipeline Steels

An accurate characterization of the electrochemical activity of various metallurgical defects, such as inclusions, is essential to understand mechanistically the initiation and propagation of localized corrosion and stress corrosion cracks. The characterization of microscopic electrochemistry of the inclusions in steel is complex, and the complexity arises from the low spatial resolution of the conventional electrochemical measurement techniques. However, LEIS enables the detection of local dissolution current or local impedance at a microscopic site and thus characterizes the microscopic metallurgical electrochemistry of the steel [13].

Figure 3.13 shows three metallurgical microdefects, which are Si-enriched inclusion, microvoid, and carbide inclusion, existing in X100 pipe steel and the LEIS measuring result on the defects. It is seen that there are lower local impedances around the Si-enriched inclusion (point A) and microvoid (point B)

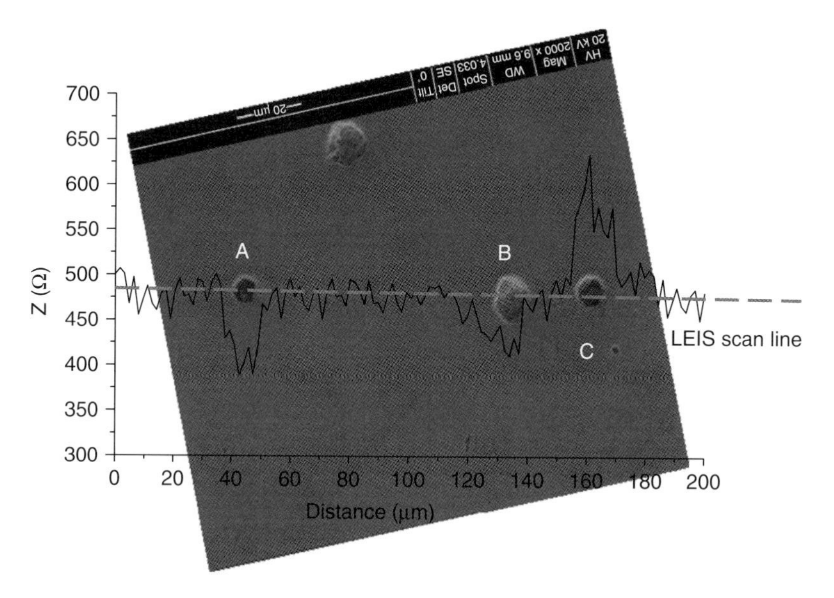

Figure 3.13 Localized electrochemical impedance measured on the X100 steel electrode across three micrometallurgical defects A, B, and C.

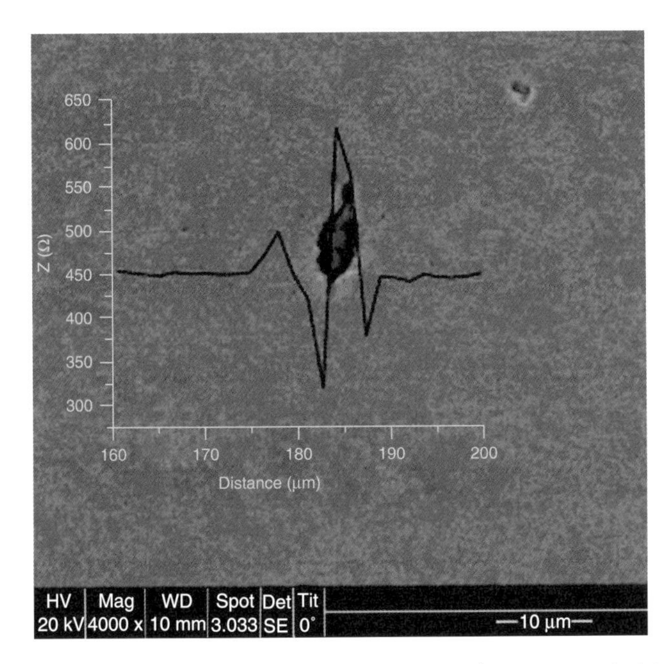

Figure 3.14 The LEIS map measured across an aluminium-enriched inclusion.

than that measured on the steel matrix, whereas the impedance measured at the carbide inclusion (point C) is higher than that of the steel matrix.

Figure 3.14 shows another LEIS scanning on the electrode surface, where an inclusion is identified as an aluminum-enriched one. It is seen that the impedance fluctuations are associated with the inclusion. Moreover, the impedance is approximately constant on the steel matrix. It drops at the boundary between the steel matrix and inclusion, and then increases at the inclusion itself. The impedance drops again at the interface between inclusion and the steel and is then back to the constant value.

In a previous work [14], it was found that the presence of inclusions in the steel serves as hydrogen traps, initiating hydrogen-induced cracks. This work demonstrates that there exists an electrochemical heterogeneity between inclusion and the adjacent steel matrix. It is thus expected that these inclusions would contribute to the locally preferential dissolution (or corrosion) of the steel.

The local electrochemical activity of the inclusion depends on its composition. The low local impedance associated with the Si-enriched inclusion (Figure 3.13) indicates that there is a high electrochemical activity at the inclusion in the test solution. Therefore, a galvanic couple forms between the inclusion and the adjacent steel matrix, where the former serves as an anode and the latter as a cathode. The preferential dissolution of the Si-enriched inclusion generates a local microvoid, which will continue to dissolve because of its high electrochemical activity, as demonstrated by the low local impedance. A pit is consequentially formed and could be the site to initiate stress corrosion crack in the presence of applied stress.

However, an aluminum-oxide-enriched inclusion presents a higher local impedance than that obtained on the steel substrate, as seen in Figure 3.14. Thus, compared to the steel, the inclusion is more stable. In the galvanic couple formed between the aluminum-oxide-enriched inclusion and the adjacent steel matrix, the former serves as a cathode and the latter as an anode. Preferential dissolution would occur on the steel matrix, resulting in the drop of local impedance at the interfaces between inclusion and the steel. The local dissolution of the steel causes the "drop-off" of the inclusion, and consequently, generates a microvoid. Further dissolution of the microvoid results in the generation of a corrosion pit.

3.3.4
Characterization of Local Electrochemical Activity of a Precracked Steel Specimen

In a high-pH concentrated carbonate/bicarbonate solution, the properties of passive film formed on the steel surface are crucial to the stress corrosion crack initiation and propagation. Furthermore, the solution electrochemistry and stress state at the crack tip are quite different from the region ahead of the crack, and the passive film formed at crack tip would be different from that formed on the surrounding region. The LEIS technique is used to characterize the electrochemical activity of an X70 steel at the crack tip and the region away from the crack (point A).

Figure 3.15 shows the LEIS plots measured at the crack tip and point A of an X70 steel specimen under a 1500 N tensile load and polarized at $-0.3\ V_{SCE}$ (saturated calomel electrode, SCE) in a concentrated carbonate/bicarbonate solution [15]. It is seen that the size of the semicircle measured at the crack tip is much smaller than that at point A.

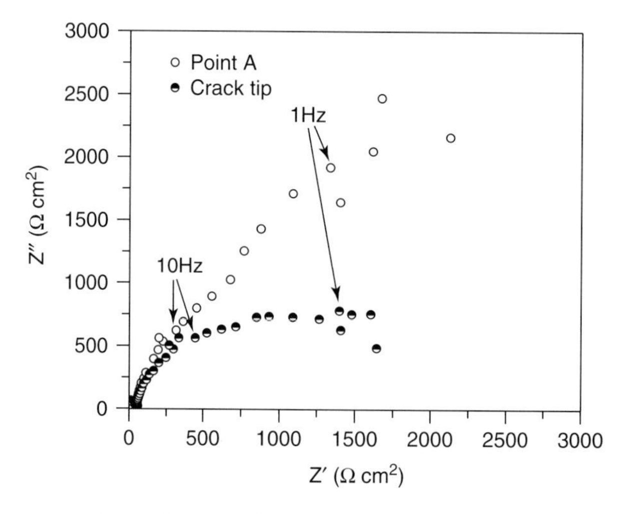

Figure 3.15 LEIS Nyquist diagrams measured on the pre crack X70 steel specimen at the crack tip and point A under a 1500 ΩN force and the polarization potential at $-0.3\ V_{SCE}$, where Z' and Z'' refer to real and imaginary impedance, respectively.

Figure 3.16 shows the LEIS maps measured on the precracked specimen under various loads. It is seen that there is an obvious impedance valley at the crack tip under individual load. With the increase of the load, the average value of the impedance decreases, especially at the crack tip.

It is demonstrated in this work that with the increase in applied load (stress), the electrochemical impedance at both the crack tip and the region away from the crack increases, as seen in Figure 3.16. Moreover, the local impedance measurements (Figure 3.15) show that there is a lower impedance at the crack tip than in the surrounding region. Therefore, applied stress is capable of enhancing the electrochemical activity of the steel, which is further increased at the crack tip because of the local stress concentration. It was proposed [16] that the stress concentration at the crack tip introduces additional strain energy to the steel, increasing the internal energy and lattice strain and resulting in an increasing anodic dissolution rate locally. The remarkable difference of the electrochemical impedance measured at the crack tip and the region ahead of the crack is due to the presence of the additional local strain at the crack tip. Moreover, the stress concentration would considerably increase the local stress intensity factor [17].

As the charge-transfer resistance is inversely proportional to the dissolution rate of the steel, the stress effect factor on anodic dissolution, k_σ, could be determined quantitatively by calculating the ratio of the charge-transfer resistance of an unstressed steel to that of a stressed steel by

$$k_\sigma = \frac{R_{ct}^0}{R_{ct}^\sigma} \tag{3.3}$$

where R_{ct}^0 and R_{ct}^σ are the charge-transfer resistances of the unstressed and stressed steel specimens, respectively. The charge-transfer resistance is derived from the impedance data fitting with an electrochemical equivalent circuit $R_s(QR_{ct})$, where R_s is the solution resistance, Q is the constant-phase element of the double-charge layer, and R_{ct} is the charge-transfer resistance. Generally, k_σ increases with the applied force and also the local stress intensity factor. Furthermore, the increase of k_σ at the crack tip is higher than that away from the crack. When the applied force reaches 1500 N, the stress effect factor at the crack tip is up to 6.9.

3.4 Scanning Kelvin Probe

3.4.1 The Technique and Principle

SKP is a noncontact and nondestructive technique that enables the measurement and mapping of the work function difference between a sample and a reference probe. The major advantage of the Kelvin probe in comparison to conventional electrochemical devices is that the Kelvin probe measures electrode potential without touching the surface under investigation across a dielectric medium of high resistance.

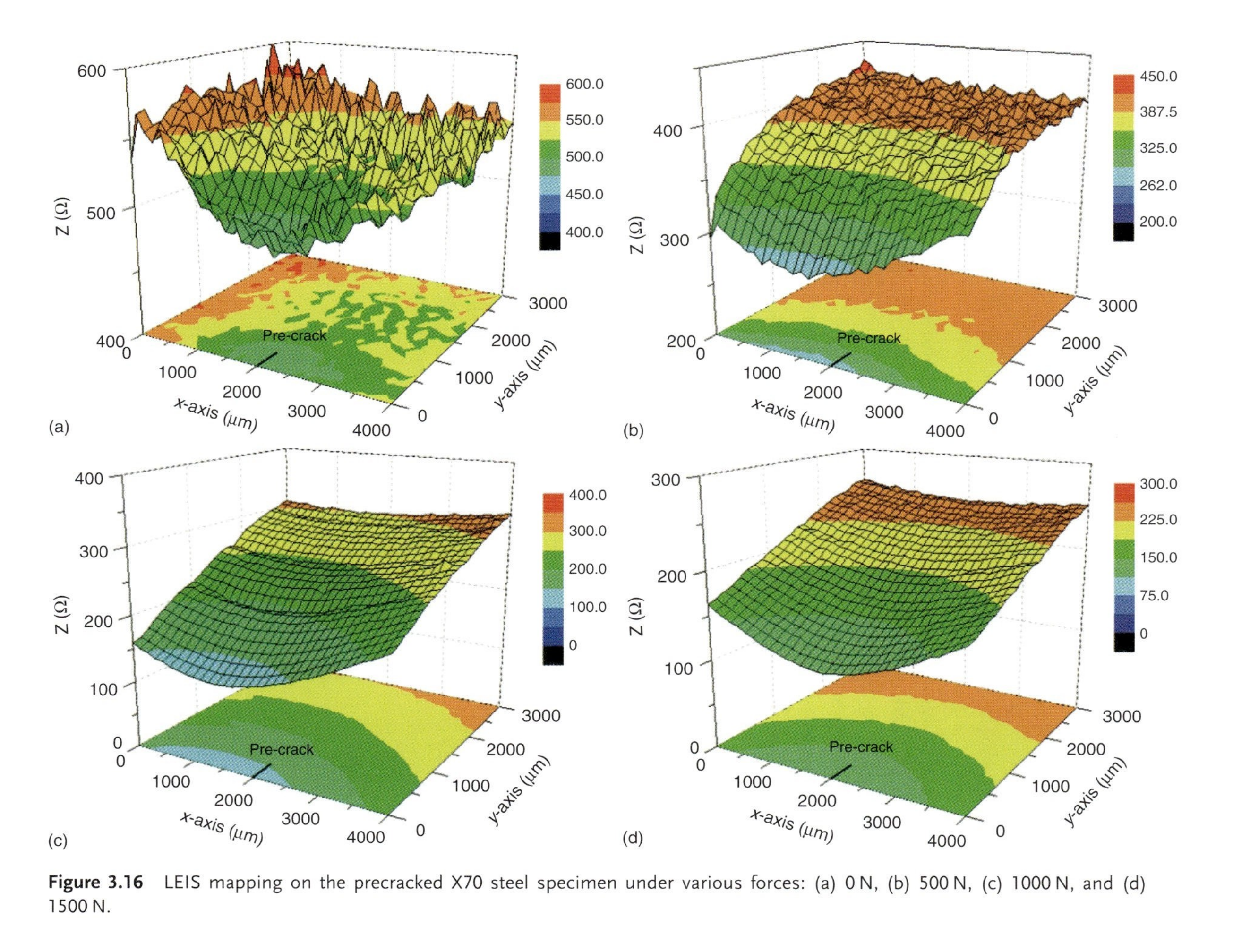

Figure 3.16 LEIS mapping on the precracked X70 steel specimen under various forces: (a) 0 N, (b) 500 N, (c) 1000 N, and (d) 1500 N.

When the target electrode is in a certain environment, there is a work function difference between it and the reference electrode (Kelvin probe). During SKP measurement, the two electrodes are electrically connected by an external electric circuit, and the electrons will distribute themselves, establishing an equilibrium of charge. The work function, Φ, of a solid material at a solid–liquid interface can be divided into two components [18]

$$\Phi = \Psi + \chi \tag{3.4}$$

where Ψ is the contact potential and χ is surface potential. As electrons have to move from the working electrode through several interfaces to the reference electrode, the work function is extremely sensitive to surface conditioning and is affected by surface phenomena such as adsorption or absorption, surface charging, oxidization, surface/bulk contamination, and corrosion.

Under certain circumstances, the work function is determined by the electrode potential, and therefore, the Kelvin probe is able to measure local corrosion potential, E_{corr}, by [19]

$$E_{corr} = \text{Constant} + V_b \tag{3.5}$$

where V_b is the measured work function difference between the Kelvin probe and the sample.

3.4.2 Monitoring the Coating Disbondment

Figure 3.17 shows the Kelvin potential profile measured on an fusion bonded epoxy (FBE) coated X70 steel specimen in a concentrated carbonate/bicarbonate solution as a function of time, where the blue and green potential regions refer to the disbonded and intact areas, respectively [20]. It is seen that there is a similar feature for all potential maps, that is, a distinct potential difference exists between the "intact" and disbonded areas, and the potential of the "intact" area is less negative than that of the disbonded area. Furthermore, potentials measured on both sides are shifted negatively with time. For example, the Kelvin potential of the "intact" area near the disbonding boundary line is about −750 mV (tungsten, W) at day 1, which is shifted to a more negative value of −1150 mV (W) at day 9. The Kelvin potential of the disbonded area is decreased from −1000 mV (W) at day 1 to −1400 mV (W) at day 9.

Figure 3.18 shows the SKP measurements on the coated steel specimen (open to air) experiencing wet–dry cycles, where the low-potential region in the central part of the scanning area is the disbonded area. With evaporation of the trapped solution, the solution layer changes from "thick" to relatively "thin" in thickness. Correspondingly, the Kelvin potential is shifted negatively.

When the SKP measurement is conducted on a disbonded coating under which corrosion of steel occurs in the trapped electrolyte, the recorded Kelvin potential

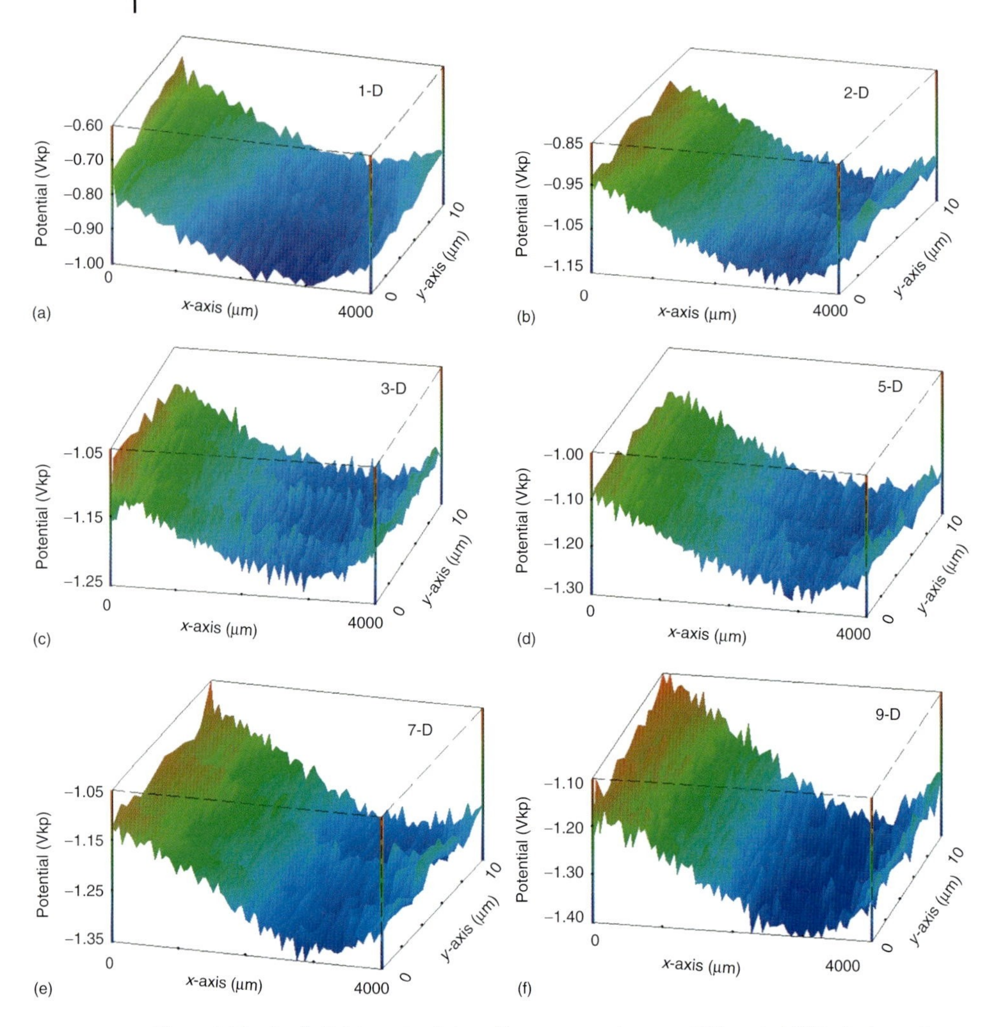

Figure 3.17 (a–f) Kelvin potential profiles measured on an FBE-coated X70 steel specimen in a concentrated carbonate/bicarbonate solution as a function of time.

actually contains several components, which can be defined as:

Intact coating:

$$\Delta\psi_{1\text{coating}}^{\text{probe}} = \Delta\phi_{\text{coating}}^{\text{steel}} + \chi_{\text{coating}} - \frac{1}{F}(W^{\text{probe}} + \mu_{\text{e}}^{\text{steel}}) \quad (3.6)$$

Disbonded coating without corrosion product:

$$\Delta\psi_{2\text{coating}}^{\text{probe}} = \Delta\Phi_{\text{electrolyte}}^{\text{steel}} + \Delta\Phi_{\text{coating}}^{\text{electrolyte}} + \chi_{\text{coating}} + \Delta\Phi_{\text{Donnan}} - \frac{1}{F}(W^{\text{probe}} + \mu_{\text{e}}^{\text{steel}}) \quad (3.7)$$

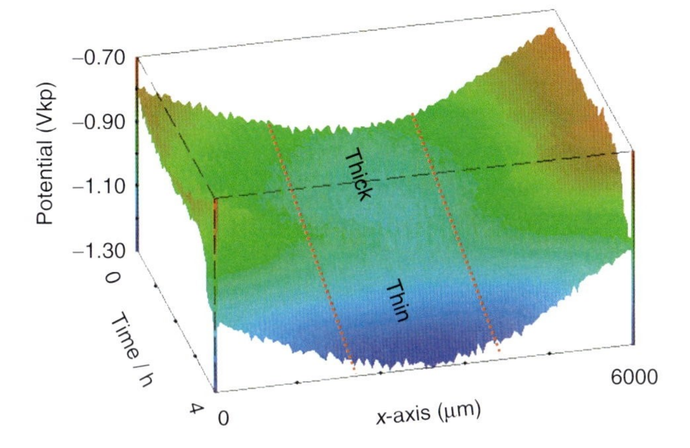

Figure 3.18 The Kelvin potential measured on the coated steel specimen during wet–dry cycles.

Disbonded coating with corrosion product (oxide):

$$\Delta\psi^{\text{probe}}_{3\text{coating}} = \Delta\Phi^{\text{steel}}_{\text{oxide}} + \Delta\Phi^{\text{oxide}}_{\text{electrode}} + \Delta\Phi_{\text{oxide}} + \Delta\Phi^{\text{electrolyte}}_{\text{coating}} + \chi_{\text{coating}} + \Delta\Phi_{\text{Donnan}} - \frac{1}{F}(W^{\text{probe}} + \mu^{\text{steel}}_{\text{e}}) \quad (3.8)$$

where $\Delta\psi^{\text{probe}}_{\text{coating}}$ is the Kelvin potential measured by SKP, $\Delta\Phi^{i}_{j}$ is the contact potential at the individual interface, χ_{coating} is the surface potential of the coating, $\Delta\Phi_{\text{Donnan}}$ is the Donnan potential that is generated as a result of the presence of two solution phases separated by the disbonded coating membrane, F is the Faraday constant, W^{probe} is the work function of the reference probe, and $\mu^{\text{steel}}_{\text{e}}$ is the chemical potential of the steel electrode.

This work shows a gradual increase of Kelvin potential from the disbonded area to the "intact" area, as seen in Figure 3.17, and there is no distinct boundary line separating the two areas, as recorded by Leng *et al.* [18]. Therefore, there is no true "intact" area for the coating used in this work as electrolyte keeps penetrating through the coating gradually. Furthermore, the Kelvin potential measured on the "intact" area is shifted negatively with time, which is probably due to the alteration of the electrolyte concentration and/or electrochemical reaction rate under coating. The Kelvin potential of the "intact" area decreases remarkably in the first three days, and then maintains a relatively steady value in the following days, indicating that the continuous solution intake occurs mainly within three days. After that, the trapped solution reaches a saturation status. Upon solution intake, the initial steel–coating interface is replaced with the steel–electrolyte and electrolyte–coating interfaces. With the increase in the amount of solution, it is expected that the electrolyte concentration and electrochemical reaction rate change, resulting in a significant decrease of the interfacial potential. Moreover, the coating surface potential and Donnan potential also decrease remarkably with the increase in the water amount. With continued immersion, the double-charge layer achieves a relatively steady state, so does the steel–electrolyte interfacial potential. Thus, there is a slight change in the Kelvin potential in the following days.

This work also shows that the Kelvin potential measured on the disbonded area is shifted negatively with time. Moreover, the Kelvin potential is more negative on the disbonded area than that on the "intact" area, which is attributed to corrosion of steel occurring under the disbonded coating.

During the wet–dry cycle, the thickness of the solution layer trapped under a disbonded coating decreases because of evaporation of water. The reduction of the solution thickness facilitates the oxygen diffusion, and simultaneously, the concentration of the solution increases, resulting in a reduction of the oxygen solubility. While the former could result in the steel oxidation and thus an increase of electrode potential, the latter is always related to a negative shift of potential because of less oxygen involved in the corrosion of steel. There exists a competition between the two effects. This work shows that the Kelvin potential decreases during wet–dry cycles, indicting that the effect associated with the reduction of the oxygen solubility is favored over that related to the oxygen diffusion and reduction. Furthermore, the SKP measurement results show that the apparent potential drop occurs only after a certain time period of evaporation. For example, the less negative potential of −1.1 V (W) is observed over a certain area of disbondment, that is, Kelvin potential is at a relatively steady value in this area, although the solution layer gradually thins because of the water evaporation. It is thus reasonable to assume that there exists a critical thickness of solution layer below which the oxygen solubility is sufficiently low to support the electrochemical corrosion reaction of the steel.

3.5 Conclusive Remarks

While conventional electrochemical techniques have been used extensively in corrosion research, the microelectrochemical measurement techniques provide promising alternatives for characterizing *in situ* the corrosion phenomena occurring at a microscopic scale, such as pitting corrosion and SCC, and those with a high geometrical limitation, such as corrosion of steel in a thin layer of electrolyte under coating. The microelectrochemical measurements would enable the understanding of the corrosion process at a microscopic scale, and thus at a more mechanistic level.

The future trend in the advanced electrochemical measurement techniques will be the development of micro- or nano-sized multifunctional probe, allowing simultaneous measurements of electrochemical, chemical, and topographical changes of the target material. Moreover, a more stable, higher signal-to-noise ratio and a lower detection limit will be desired.

Acknowledgments

The author appreciates the supports from Canada Research Chairs Program, Natural Science and Engineering Research Council of Canada (NSERC), Canadian Foundation of Innovation (CFI), and the University of Calgary. A number of postdoctoral fellows and graduate students had participated in this research and contributed to the chapter. They are Drs. Guoan Zhang, Xiao Tang, Anqing Fu, Cheng Zhong, and Guozhe Meng as well as Mr. Taiyan Jin.

References

1. Stansbury, E.E. and Buchanan, R.A. (2000) *Fundamentals of Electrochemical Corrosion*, ASM International, Materials Park.
2. Frankel, G.S. (2008) Electrochemical techniques in corrosion: status, limitations, and needs. *J. ASTM Int.*, **5** (2), 1–27.
3. Mansfeld, F. (1995) Use of electrochemical impedance spectroscopy for the study of corrosion protection by polymer coatings. *J. Appl. Electrochem.*, **39**, 187–202.
4. Zhong, C., Tang, X., and Cheng, Y.F. (2008) Corrosion of steel under the defected coating studied by localized electrochemical impedance spectroscopy. *Electrochim. Acta*, **53**, 4740–4747.
5. Rossi, S., Fedel, M., Deflorian, F., and Del Carmen Vadillo, M. (2008) Localized electrochemical techniques: theory and practical examples in corrosion studies. *C.R. Chim.*, **11**, 984–994.
6. Baker, M. (2004) Stress Corrosion Cracking Studies. Integrity Management Program, Department of Transportation, Office and Pipeline Safety DTRS56-02-D-70036.
7. Zhang, G.A. and Cheng, Y.F. (2009) Micro-electrochemical characterization of corrosion of welded X70 pipeline steel in near-neutral pH solution. *Corros. Sci.*, **51**, 1714–1724.
8. Oriani, R.A., Hirth, J.P., and Smialowski, M. (1985) *Hydrogen Degradation of Ferrous Alloys*, Noyes Publications, Park Ridge.
9. Cheng, Y.F. and Niu, L. (2007) Mechanism for hydrogen evolution reaction on pipeline steel in near-neutral pH solution. *Electrochem. Commun.*, **9**, 558–562.
10. Qin, Z., Demko, B., Noel, J., Shoesmith, D., King, F., Worthingham, R., and Keith, K. (2004) Localized dissolution of mill scale-covered pipeline steel surfaces. *Corrosion*, **60**, 906–914.
11. Meng, G.Z., Zhang, C., and Cheng, Y.F. (2008) Effects of corrosion product deposit on the subsequent cathodic and anodic reactions of X-70 steel in near-neutral pH solution. *Corros. Sci.*, **50**, 3116–6122.
12. Mansfeld, F. (1993) Models for the impedance behaviour of protective coatings and cases of localized corrosion. *Electrochim. Acta*, **38**, 1891–1897.
13. Jin, T.Y. and Cheng, Y.F. (2011) In-situ characterization by localized electrochemical impedance spectroscopy of the electrochemical activity of microscopic inclusions in an X100 steel. *Corros. Sci.*, **53**, 850–853.
14. Jin, T.Y., Liu, Z.Y., and Cheng, Y.F. (2010) Effects of non-metallic inclusions on hydrogen-induced cracking of API5L X100 steel. *Int. J. Hydrogen Energy*, **35**, 8014–8021.
15. Zhang, G.A. and Cheng, Y.F. (2010) Micro-electrochemical characterization of corrosion of pre-cracked X70 pipeline steel in a concentrated carbonate/bicarbonate solution. *Corros. Sci.*, **52**, 960–968.
16. Hirth, J.P. (1980) Effects of hydrogen on the properties of iron and steel. *Metall. Trans. A*, **11**, 861–890.
17. Rolfe, S.T. and Barsom, J.M. (1977) *Fracture and Fatigue Control in*

Structures-Application of Fracture Mechanics, Prentice-Hall, Inc., Englewood Cliffs.

18. Leng, A., Streckel, H., and Stratmann, M. (1999) The delamination of polymeric coatings from steel. Part 1. Calibration of the Kelvin probe and basic delamination mechanism. *Corros. Sci.*, **41**, 547–578.
19. Furbeth, W. and Stratmann, M. (2001) The delamination of polymeric coatings from electrogalvanised steel – a mechanistic approach. Part 1. Delamination from a defect with intact zinc layer. *Corros. Sci.*, **43**, 207–227.
20. Fu, A.Q. and Cheng, Y.F. (2009) Characterization of corrosion of X65 pipeline steel under disbonded coating by scanning Kelvin probe. *Corros. Sci.*, **51**, 914–920.

4
Protective Coatings: an Overview

Anand Sawroop Khanna

4.1
Introduction

Corrosion results in material degradation, leading to loss of load-bearing capability of a material and many forms of surface deformations, such as color fading, loss of gloss, pitting, blisters, and so on. Many a times, it requires material replacement, discarding the whole component or machinery, production losses, accidents, and perhaps plant shutdown. However, corrosion is a phenomenon that cannot be prevented but can be avoided or minimized using one of the corrosion protection technologies, such as better use of materials, use of inhibitor chemicals, use of cathodic and anodic protection, and use of protective coatings. It is the last one that has a share of about 75% of all the protective methods used.

It is a myth that protective coating is the simplest of all the corrosion control methods. For many, it is considered just a simple application of paint using a brush on the surface to be protected. Although it appears simple, it is actually not so in practice. It requires the use of scientific principles for selection of paint coatings from the thousands of available formulations. It is also an art as the skill of the applicator is the most important in prolonging the life of a paint coating. The issues in paint coatings are proper selection of binder chemistry; suitable formulation to achieve chemical, mechanical, and weathering properties; skilled application strategy, which involves proper surface preparation selection of primer, intermediate, and/or topcoat application; and maintaining proper coating intervals, drying requirements, and environmental conditions. The last part is the necessity of supervision, which can assure integrity of coating.

4.2
Selection of Paint Coatings

Paint coatings are known from their generic names, based on the binder chemistry, which in turn are usually classified on the basis of the severity of the environment. These are more generally described as mild, moderate, and aggressive, based

Green Corrosion Chemistry and Engineering: Opportunities and Challenges, First Edition.
Edited by Sanjay K. Sharma.

Table 4.1 Classification of various environments as per International Organization for Standardization.

C1	Mild environment with low humidity
C2	Low RH and condensation
C3	Urban industrial with SO_2
C4	Industrial and Coastal–chemical plants
C5-1	Industrial zone with high RH
C5-M	Marine offshore with high salinity

on the *relative humidity* (RH) of the environment as well as on the severity of industrial pollutants such as sulfur dioxide, carbon dioxide, or marine pollutants such as chloride. Thus, for a mild environment with low humidity and almost no pollutants, simple oil-based resins (alkyds) are sufficient. But for more aggressive environments, better resin chemistry such as epoxies, polyurethanes, polyesters, and so on, are required. Table 4.1 gives the classification of various environments as per International Organization for Standardization (ISO). Further, using organic paint coatings, it is possible to create only a barrier coating, which is able to protect the substrate from environment, but in case of any coating damage, the substrate directly comes in contact with the environment and results in corrosion. Thus, the use of a primer coat, such as organic zinc-rich primer or inorganic zinc silicate primer, helps in providing additional corrosion protection because of cathodic protection of zinc in the primer coat. So for the selection of coatings, it is first necessary to understand the chemistry of various binders.

A paint system consists of four main components:

- resin or binder;
- solvent;
- pigments;
- additives.

Paint is nothing but a blend of these items in the correct proportion to achieve the following properties: adhesion, impermeablity, toughness or flexiblibility, chemical resistance, and good mechanical and ultraviolet (UV) properties. In order to achieve this, there are large numbers of combinations of various resins with different concentrations of pigments and additives. Pigments provide corrosion resistance, color and opacity, and mechanical strength or barrier properties to the coatings.

4.3 Classification of Various Coatings

1) Conventional coating systems, usually those with a combination of the above four components:
   a. oil-based coatings (alkyd resin);

   b. epoxies;
   c. polyurethanes;
   d. polyesters;
   e. vinyl;
   f. polyurea;
   g. rubber-based (chlorinated) coatings;
   h. zinc-based coatings.
2) High-performance coatings:
   a. usually solventless, thus give higher coating thickness per coat;
   b. can be applied in highly aggressive environment, even with 90% humidity;
   c. long durability – usually achieved by the addition of special pigments;
   d. high chemical resistance.
3) Waterborne coatings:
   a. use water as solvent;
   b. can be formed either by water soluble resins or by sol–gel route.

4.4 Chemistry of Resins

The best way to describe and compare the behavior of various resins is to look into their chemical origin and their curing mechanism. The simplest of all are the alkyd resins, which are derived from various oils. They are good only in mild environment and deteriorate fast in aggressive environments such as coastal environment. They cure by reacting with oxygen and therefore take a longer time to dry.

4.4.1 Alkyd Resins

Alkyd resins are the reaction products of polyhydric alcohols and polybasic acids (Figure 4.1). The properties of alkyd coatings are derived from the properties of the drying oil used in the manufacture of the alkyd resin. The drying time, hardness, color, and moisture sensitivity all depend on the characteristics of the drying oil, its type, and the degree of unsaturation. Soybean oil has been shown to give good drying rates and has good color retention. However, linseed oil generally dries faster but darkens on exposure to sunlight. Castor and coconut oils have good color-retention properties and are used as plasticizing resins because of their nonoxidizing characteristics. The amount of oil combined with the resin influences the protective capability of the applied alkyd coating. High-solubility oil resins ($2\ L\,kg^{-1}$) have poor moisture and chemical resistance and have longer drying times. Low-oil alkyds ($0.80\ L\,kg^{-1}$) are fast-drying coatings and have good moisture and chemical resistance, but these are relatively hard and brittle [1].

All alkyd coating systems initially dry by solvent evaporation mechanism and cure by auto-oxidative cross-linking of the oil constituent. Because of the presence of the

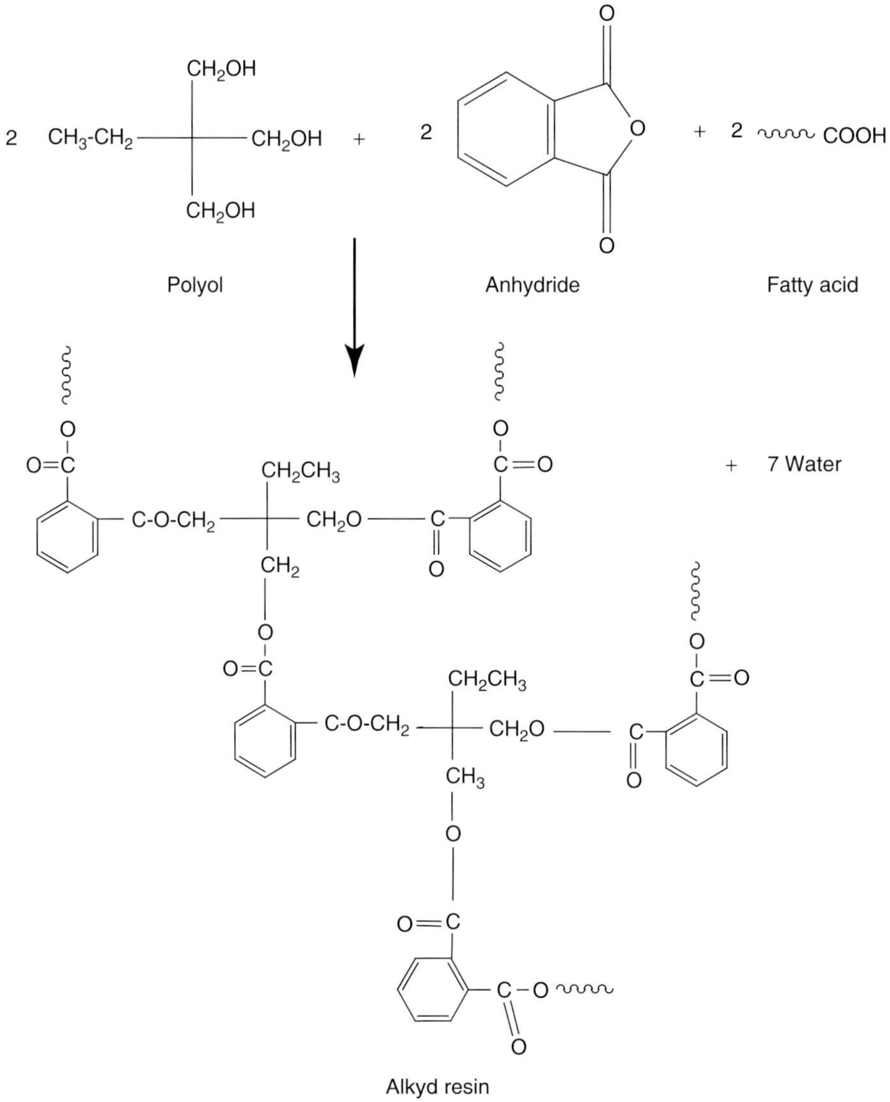

Figure 4.1 Schematic for the synthesis of an alkyd resin.

drying oil alkyd coating systems have limited chemical and moisture resistance, cannot be used in highly chemical environments (acid or alkali), and are not resistant in immersion or near-immersion condensing conditions. However, their relatively low cost, ease of mixing and application, and excellent ability to penetrate and adhere to relatively poorly prepared, rough, dirty, or chalked surfaces make them the coating system of choice on steel exposed to nonchemical atmospheric services.

4.4.2
Modified Alkyds

Alkyds have relatively poor properties, such as low chemical resistance, poor mechanical properties, and low UV resistance. However, it is possible to enhance the properties of alkyds by grafting them with some other resins, which although increases the cost, improves its properties, thus compensating the increased cost. For example, modification with a phenolic resin improves gloss retention and water and alkali resistance. Alkyd resins with vinyl modification are commonly formulated as universal primers. These primers can generally be topcoated with the most generic types of intermediates and topcoats. The alkyd constituent improves the adhesion, film buildup, and thermal resistance; the vinyl modification enhances recoatability and chemical and moisture resistance. These coatings are frequently used as shop primers or as tie coats between different generic coatings (e.g., over inorganic, zinc-rich primers or between alkyd primers and epoxy topcoats). Another important modification is the addition of silicone (up to 30%), which improves corrosion protection with greatly improved durability, gloss retention, and heat resistance. Further, improvement in moisture resistance helps its use as marine and maintenance paints [1].

4.4.3
Epoxies

Alkyd resins can be combined with epoxy resins to produce coatings with improved chemical and moisture resistance. Epoxy ester coatings are similar to alkyds, and they are used when improved performance is required. Epoxy resin-based coatings have better adhesion, moisture, and chemical resistance than alkyds, although they are slightly expensive. These coatings are used as baking and air-dry paints on appliances, machinery, pumps, valves, and so on. Uralkyd or urethane oil coating is formed by the isocyanate reaction, which decreases the drying time of the coating and provides enhanced resistance to chemicals, moisture, weathering, and abrasion. Uralkyd coatings are used as marine coatings of wooden boat hulls, machinery enamels, and an upgrade to an alkyd coating.

The combination of bisphenol-A (BPA) and epichlorhydrin (ECH) results in the formation of an epoxide group, generally called an *epoxy resin*. Although, such an epoxy resin when dissolved in a solvent can give corrosion protection, such protection is not long lasting. The protective properties of epoxies result when such an epoxy resin is combined with a hardener, which reacts with these epoxide groups. Cross-linking takes place preferentially through the terminal epoxy groups and then through the midchain hydroxyl groups. Figure 4.2 shows the formation of a basic epoxy resin monomer, diglycidyl ether of BPA, from the starting raw materials ECH and BPA, while Figure 4.3 shows the formation of diglycidyl ether of BPF from the starting raw materials ECH and bisphenol F (BPF). These two epoxy monomers are the base for a large number of epoxy-based coatings, which are being used for corrosion protection. Figure 4.4 shows the role of various components

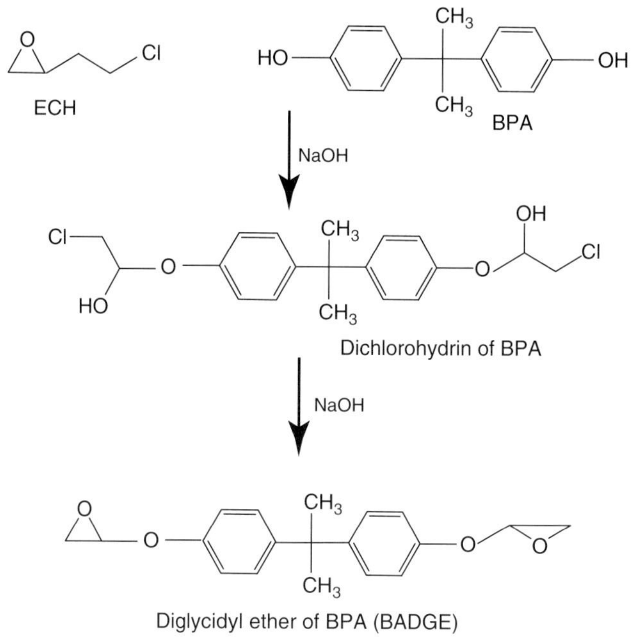

Figure 4.2 Schematic for the synthesis of an epoxy resin bis-phenol A (BPA).

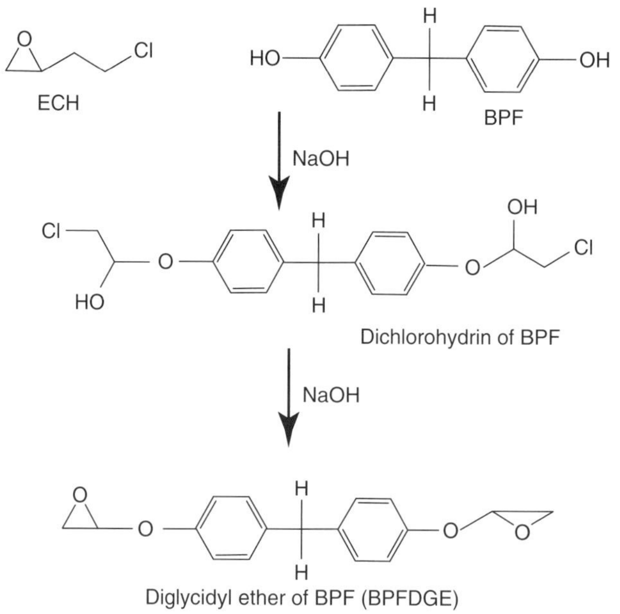

Figure 4.3 Schematic for the synthesis of an epoxy resin bis-phenol F (BPF).

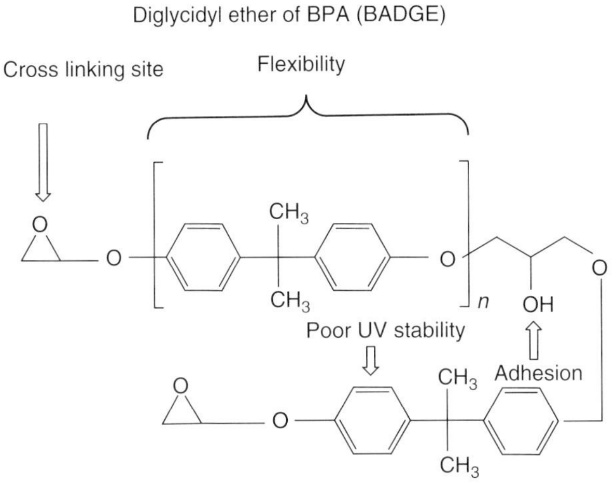

Figure 4.4 Showing the main properties of epoxies, based upon the structure of epoxy, such as relationship of flexibility with the chain length, poor UV stability due to phenolic group and good adhesion due to polar –OH group.

of epoxies in providing functional improvement in the properties. For example, adherence of epoxy coating is provided by polar groups such as –OH, stretchability, and flexibility of coating comes from the chain length of the polymer, cross-linking takes place at epoxide groups (Figure 4.5), and UV degradation of epoxy coatings is due to the phenolic group.

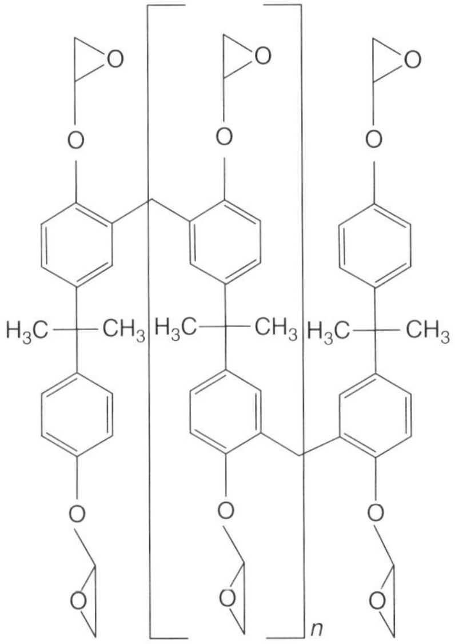

Figure 4.5 Schematic of epoxy with multi epo-oxide groups, which help in stronger cross-linking giving better rigidity and lower permeability.

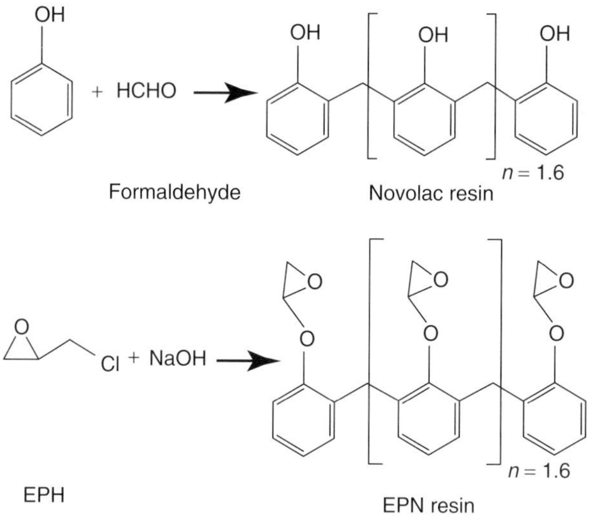

Figure 4.6 Schematic of multi epo-oxide epoxy called Novolac epoxy used for internal linings in pipelines.

Other developed epoxy resins, including cycloaliphatic epoxies, offer improvements in light stability and UV light degradation, but these epoxy resins do not exhibit the same extent of adhesion, chemical resistance, and flexibility.

Further improvements in the properties of epoxies can be achieved by still strong cross-linking. For this, more than two epoxide groups are created. This can be achieved by a combination of formaldehyde with phenol (Figure 4.6), which gives cresol (novolac epoxy). These coatings provide high temperature resistance and great chemical resistance but at the expense of brittleness and a lack of toughness and flexibility [1].

Polyamines (i.e., diethylenetriamine, hydroxyethyldiethylenetriamine, bishydroxydiethylenetriamine) are relatively small molecules with a low molecular weight compared to the epoxy resin, lead to tight cross-linking, and give high chemical and moisture resistance. Generally, polyamine cross-linked-epoxy coatings have excellent alkali resistance and the greatest chemical resistance of the epoxies. They also have good moisture and water resistance. These epoxies are the most brittle and the least flexible, and they have a strong tendency to degrade on UV light exposure, resulting in chalking. Because of their high cross-link density (achieved as a result of the small molecular size of most of the amines used as coreactants), amine-cured epoxies are the epoxies of choice in atmospheric or immersion environments of high- and low- (pH 3–12) hydrolyzing chemicals.

Polyamide curing agents improve flexibility, gloss, and flow and have excellent water resistance and good chemical resistance. However, polyamide-cured coatings have less solvent and alkali resistance than amine-cured epoxies. Because fatty acids have a water repellent tendency, polyamide-cured epoxies are said to be tough, flexible, water repellent coatings. They are perhaps the most widely used systems and have wide application in the protection of steel and concrete in freshwater and saltwater immersions. Polyamide-cured epoxies have the best exterior weathering resistance and the best ability of the epoxies to recoat after extended periods. Polyamide epoxies are used to protect substrates exposed to condensation and high humidity. Specially formulated polyamide-cured epoxies have the ability to displace water from a surface. These coating materials can be applied and cured under water to form corrosion-resistant coatings.

Coal tar epoxy is a combination of coal tar and an epoxy resin. The epoxy resin is usually packaged separately from the curing agent, which is frequently combined with the coal tar resin. The curing agent may be an amine or a polyamide. The cross-linking reaction is the same as that previously described, with active hydrogens from the amine nitrogen providing a cross-linking site to the epoxide groups and, in some situations, to the hydroxyl groups of the epoxy resin. The coal tar acts as a filler within the cross-linked epoxy matrix, and the resulting film has the toughness, adhesion, UV resistance, and thermal stability of the epoxies combined with the extremely high moisture resistance afforded by the coal tar. The amine-cured coal tar epoxies generally have great chemical and moisture resistance but are more brittle and harder to apply than the amine adduct and polyamide-cured coal tar epoxies. The polyamide-cured coal tar epoxies are more water resistant, flexible, easier to topcoat, and more tolerant of application variables than the other epoxies.

4.4.4 Urethanes

Another class of widely used resin, is the urethane resin, which is formed by the reaction of an isocyanate (–N=C=O) with a polyol (–OH) (Figure 4.7). Cross-linking occurs because of the high reactivity and affinity of the isocyanate group for the active hydrogen of the polyolhydroxyl or any other active hydrogen atom attached to a nitrogen or oxygen atom.

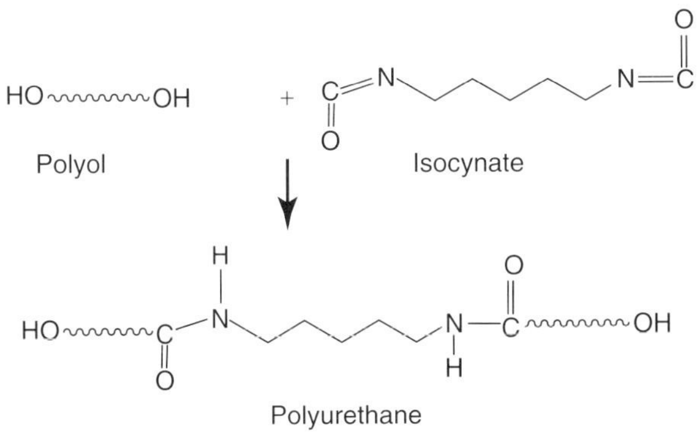

HO~~~~OH + O=C=N~~~~N=C=O

Polyol Isocynate

Polyurethane

Figure 4.7 Schematic for the synthesis of polyurethane resin.

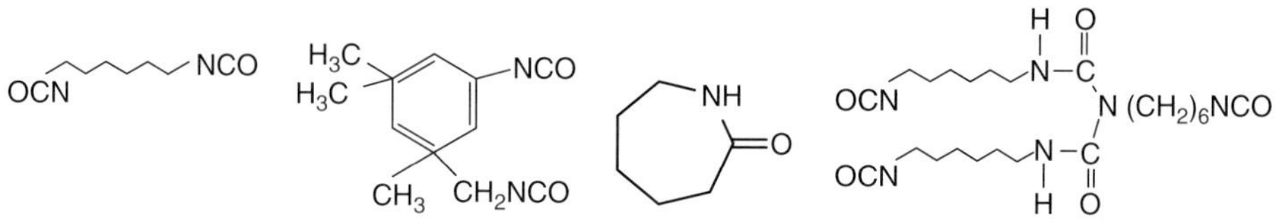

Figure 4.8 List of various isocynates like, diisocyanate (TDI), 4,4-diphenylmethane diisocyanate (MDI), and 1,6-hexamethylene diisocyanate (HDI) used for synthesis of various polyurethane resins with different properties.

The rate of this cross-linking reaction depends on a number of factors, such as the type and configuration of both the isocyanate and polyol materials, temperature, and so on. However, the reaction is such that with most formulations, the cross-linking can occur at temperatures as low as 18 °C or less [1].

4.4.5 Isocyanates

The isocyanate reactant can be either aromatic (containing the benzene ring) or aliphatic (straight chain or cyclical) hydrocarbons (Figure 4.8). Aromatic polyurethanes are prone to darkening and yellowing on exposure to sunlight because of the chromophoric nature of the benzene ring. Because aliphatic polyurethanes, by definition, do not contain the benzene ring, they do not yellow or darken and are preferred for exterior use.

The most important monomeric diisocyanates used for coatings are toluene diisocyanate (TDI), 4,4-diphenylmethane diisocyanate (MDI), and 1,6-hexamethylene diisocyanate (HDI). They are normally converted into isocyanate-terminated polymers or adducts of polyols such as hydroxyl-terminated polyesters and polyethers. The molecular weight of these isocyanates can be increased by self-reaction in the presence of catalysts to form dimers and/or trimers.

4.4.6 Aliphatic Isocyanates

They react more slowly and are considerably more expensive than the aromatic isocyanates, but they allow the formulation of nonyellowing, light-stable, high gloss finish coats. The appearances of polyurethane coatings, formulated with aliphatic isocyanates are well superior to any of the epoxies, acrylics, or other coating materials. One of the most important aliphatic isocyanates is HDI. In its monomeric form, HDI is an irritant, as is true with TDI and MDI. However, HDI can be reacted (commonly with water) to obtain a higher molecular weight modification that is less volatile and safer. When HDI or its higher molecular weight modifications are reacted with a suitable polyol in the presence of certain metal catalysts (tin, bismuth, zinc, iron, cobalt), a urethane coating with excellent resistance to discoloration, hydrolysis, and heat degradation is produced.

4.4.7 Polyols

Polyols coreact with isocyanates to form a polyurethane film. A polyol consists of large molecules (commonly acrylics, polyesters, polyethers, epoxies, vinyls, and alkyds) that have been reacted with di- or polyfunctional alcohols such as propylene glycol, trimethylolpropane, pentaerythritol, and others. The hydroxyl-terminated polyol materials are packaged separately from the isocyanate, and the packaging usually includes appropriate solvents and pigments. On application, the isocyanate and polyol constituents are mixed, and cross-linking proceeds via the isocyanate–hydroxyl functions and liberates carbon dioxide gas. To prevent bubbles and voids in the coating cross-section as a result of the carbon dioxide gas inclusion, all polyurethane coatings must be applied relatively thin (0.038–0.05 mm per coat). This allows the gas to pass easily from the coating before the coating cures and hardens.

4.4.8 Acrylic Urethanes

Acrylic urethanes are perhaps the most widely used urethanes for corrosion protection and atmospheric service. When properly formulated, these materials have excellent weatherability, gloss, and color retention and good chemical and moisture resistance. They can be tinted easily and pigmented to provide a variety of deep and pastel colors at a lower cost than the next most popular class, the polyester urethanes. Acrylic urethanes are not used for water immersion service and, for the most part, they do not have the chemical resistance of the polyester urethanes. However, they have excellent weathering and color retention properties, when an aliphatic isocyanate coreactant is used. These are the most popular aliphatic polyurethanes; they are widely used as topcoats over epoxy primers and intermediate coats in most nonchemical atmospheric environments. Many water tanks, bridges, railroad cars, aircraft, and other highly visible structures are coated with these light-fast, glossy, aesthetically appealing coatings.

4.4.9 Moisture-Cured Polyurethanes

Isocyanates can react with the hydroxyl group in water (H–OH) to form a unique class of coatings known as *moisture-cured urethanes*. Single-package moisture-cured urethanes use an isocyanate prepolymer that, when applied, reacts with the humidity in the air to form a tough, hard resinous film. Because of their rapid rate of reaction, aromatic isocyanates are used almost exclusively in moisture-cured urethanes. The pigments added must be essentially nonreactive with the isocyanate. Although it is possible to use a number of pigments, aluminum leaf is commonly used. When properly formulated and applied, urethanes have excellent adhesion to blast-cleaned structural steel surfaces.

When spray applied, urethanes form a tough, glossy, highly protective chemical and solvent-resistant film. Because of their high cross-link density, the recoating interval is less than 24 hours. However, within the range of 24 hours to a month or more, subsequently applied polyurethane topcoats may exhibit disbonding or poor adhesion.

Moisture-cured urethanes are used as primers under some epoxies, as full system coating on steel and nonferrous metals, and as primers on marginally cleaned steel. Frequently, the moisture-cured polyurethane primer and/or intermediate coats are topcoated with nonmoisture-cured aliphatic polyurethane to minimize yellowing and darkening.

4.4.10
Zinc-Based Coatings

Zinc-based coatings are perhaps one of the most important coatings required for many industrial applications. As the name suggests, zinc-based coatings provide corrosion protection to the otherwise pure organic-based coatings, which are termed *barrier coatings*. Presence of zinc-based coating as a primer along with epoxy as the main coating on top enhances the life of coating several times. Zinc-based coatings are mainly classified into the following two categories:

1) organic zinc-rich coatings;
2) inorganic zinc silicate coating.

They are called *organic* and *inorganic* mainly because by definition, organic zinc-rich coating is a dispersion of pure zinc powder in an organic resin. Thus, the organic part, the resin, forms a bond with the metal substrate when such a coating is applied, like many other organic coatings. On the other hand, an inorganic zinc-rich coating has a chemical bond with the substrate. This happens because in this coating, pure zinc powder is added to an active organic resin such as ethyl silicate, which forms a bond with zinc as well as with the steel substrate by −Si−O−Fe− bond (Figure 4.9). Two main differences between the organic and organic zinc rich-coatings are

- An organic zinc-rich primer is simply a dispersion of pure zinc in an organic resin. Purity of zinc and its concentration (minimum 75% by weight) is very important. Such a coating forms a mechanical bond with a fully cleaned (Sa2 $^1/_2$ surface) or partially cleaned surface (st2/st3 surface), and in case of any damage of the topcoat, zinc provides cathodic protection to the steel substrate. Corrosion resistance (cathodic protection) is decided by the concentration of zinc in dry film. The life of such a coating with an intermediate coat of epoxy (100 μm) and a topcoat of polyurethane (50 μm) is a minimum of 4–6 years, if it is applied as per the specified application procedure.
- An inorganic zinc silicate coating, on the other hand, is a mixture of 82% or more of pure zinc dust in an active resin such as ethyl silicate. Such a resin provides a chemical bond with the clean steel substrate (surface finish Sa2 $^1/_2$ or better), and with a topcoat of 100 μm epoxy intermediate and top polyurethane

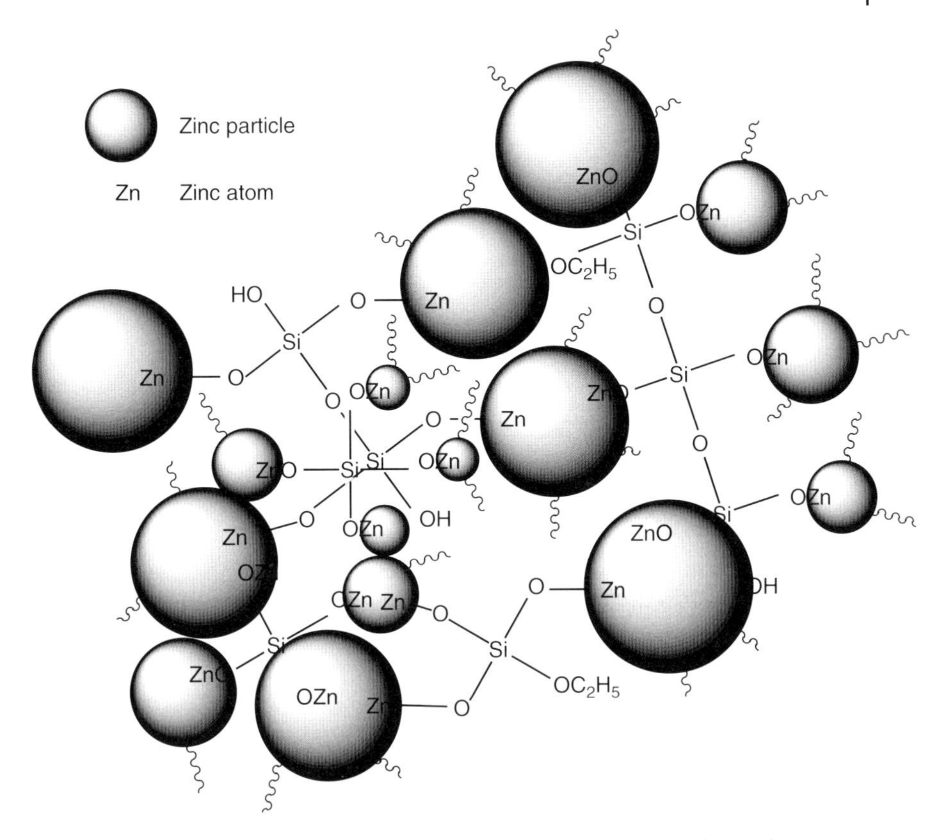

Figure 4.9 Structure of a inorganic zinc rich coating, showing the bond of Zn with Si through –O–.

coat, it can give a life of 6–8 years in a moderate environment. In case of highly aggressive environment, the intermediate coating thickness can be doubled. The only precaution to be taken during inorganic zinc silicate (IOZ) application is that the curing of IOZ requires high moisture. In case the RH is less than 45%, a requirement of organic-rich coating, it must be enhanced to more than 75% [1].

4.5 High-Performance Coatings

A high-performance coating can have one or more of the following characteristics:

- High corrosion resistance in highly aggressive environment.
- It can be applied at high RH (more than 90%).
- High thickness dry film thickness (DFT) can be achieved per coat.
- Provides resistance to weathering, especially by UV light.
- Good mechanical properties such as high impact, abrasion resistance, and hardness.

Some high-performance coatings can be can applied on splash zone of the offshore structures, under water in sea and rivers, damp surfaces, highly acidic and alkaline environments such as pickling units, battery chamber of a ship or submarine, sewage disposal unit, seawater transport through jetties, and so on.

Requirements of such paint coatings are

- high chemical resistance chemistry such as epoxies;
- solventless systems so that large thickness can be obtained per coat;
- long life and structural stability that can be obtained by addition of pigments such as glassflakes and fibers.

As an example, we focus on the properties of a solventless epoxy system in detail.

4.5.1 The 100% Solventless Epoxies

The 100% solid epoxies can be formulated from the low-molecular-weight polyfunctional liquid epoxy resins. The viscosity of these resins can be lowered even further by the use of compatible reactive diluents with an epoxy functionality of 1; they do not contribute to cross-linking but are chemically bound into the final cross-linked film. The liquid epoxy resin system is cross-linked by a liquid polyamine or polyamide without the addition of any solvent. A tertiary amino phenolic catalyst, such as tri (dimethylaminomethyl) phenol may be added to produce polymerization of the epoxy resin with itself. This and other phenols act as accelerators in the curing reaction. Silicone resins may be added as flow agents, and dibutyl phthalate can be added as a plasticizer. Thus, the entire liquid paint can be converted to a cross-linked coating that becomes a 100% solids epoxy. These materials have little, if any, volatile organic content (VOC), so they are VOC compliant.

Coatings formulated in this manner show typical epoxy finish properties; but they are less flexible than other epoxies because the films are thicker and the close spacing of the reacting groups leads to a high cross-link density. The film is tough and relatively nonbrittle. Solventless epoxy systems have low internal stress and are less brittle because there is negligible volume contraction on curing to a solid.

The 100% solids epoxies generally have short pot lives because the coreactants are not diluted. Some formulations require the use of special twin-feed airless spray equipment for external mixing. When aromatic polyamines are used for curing, hardening times of 4–12 hours can be attained even at temperatures as low as freezing. Although these coating systems are relatively expensive, they are used primarily as corrosion-resistant linings for storage tanks (e.g., oil tankers) both on land and in marine vessels. Because of their low-molecular-weight liquid formulation, these materials also can be used as self-leveling epoxy flooring systems. Generally, these coating systems are clear or high gloss and, when applied, have a water-like consistency that hardens to a smooth, glossy flooring.

Further improvement in their life can be achieved by adding glassflakes, which not only improve their strength but also help in increasing the tortuous path, as shown in Figure 4.10, and thus enhances the life of the coating. Such glassflake epoxies are used as internal linings for the sea water pipelines from jetties to power plants for condenser cooling. As we know, sea water is very corrosive and also quite abrasive because of its high speed of travels. Such glassflake epoxies provide more than 10 years of life to these pipes if the coating is applied properly up to a thickness of 1–2 mm. Figure 4.11 shows how glassflakes in epoxy matrix increase the tortuous path of moisture or other pollutants and thus enhance coating durability.

Such a glassflake epoxy coating is, however, not very suitable when applied on splash zone on an offshore structure because of two reasons: (i) relatively longer drying times and (ii) their degradation in UV light. Thus in order to incorporate these two properties in coating, a suitable polyester coating can be a better choice. Isophthalic-based polyester system or polyester modified with vinyl ester is the best example for such application. Isophthalic-modified polyesters, modified with glassflakes are being used for the splash zone for several offshore platforms in the Arabian sea. About 72 platforms of Oil and Natural Gas Corporation (ONGC) are coated with this system.

For underwater applications and for applications on damp surfaces, a hydrophobic system of epoxy is usually recommended. The principle of such a coating is that when it is applied under water, it replaces water very quickly and sticks on the steel surface. However, the full curing takes place within 24 h in water. Such coatings have been found to have a long life under water [2]. A typical underwater coating being applied in a trough underwater is shown in Figure 4.12.

4.5.2 Concept of Underwater Coatings

The adhesion of paint is due to the strong polar bonds formed by the resin with the substrate (usually mild steel). On dry surfaces, the bond between the surface and the epoxy (resin, commonly used in paints) displaces the air. The same is also true for underwater coating as the polar bond attraction is strong enough to displace the water (air in dry surface), and produces a strong bond with the substrate under water. Thus, coating underwater, in theory, is no different than that in the water. Figure 4.13 shows how the extra lone pair of electrons on amine hardener, cross-links with the resin to build a network that helps in repelling water from the surface of the substrate. For this, amine or modified amine hardeners or curing agents, with lots of additives, fillers, and accelerators, are required to make underwater paints.

The cross-linking reaction of epoxies should be independent of the surrounding environment (Figure 4.14). For the displacement of water, the curing agent should be such that it has high hydrophobicity to the water molecules on the surface of substrate.

(a)

(b)

Figure 4.10 Picture showing (a) offshore platform requiring high performance coatings such as solvent less coatings for splash zone (b).

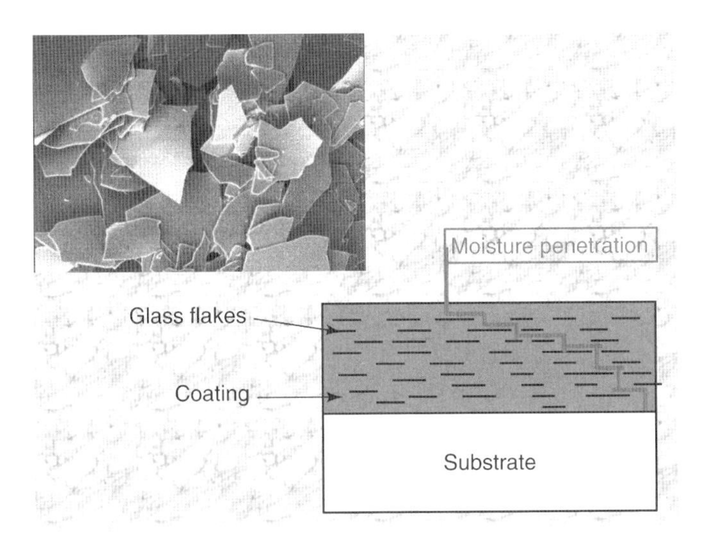

Figure 4.11 Mechanism of glassflakes in epoxy matrix, showing the enhanced torturous path, leading to high durability.

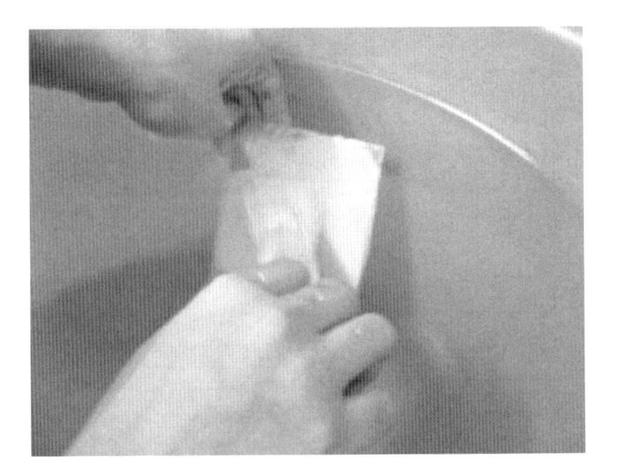

Figure 4.12 Showing the application of special coating to underwater objects.

The properties of a typical formulation made in our laboratory are given in Table 4.2.

4.5.3
Polyvinylidenedifluride Coatings

For highly acidic and alkaline environments, there cannot be any better coating than polyvinylidenedifluoride (PVDF) coatings, which resist strong acids and alkalis and

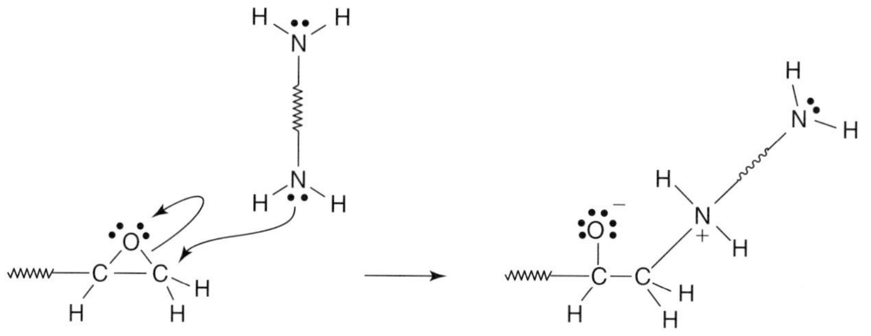

The diamine's electrons attack the carbon atom next to the epoxide oxygen, giving us a negative charge on the oxygen, and a positive charge on the nitrogen.

Figure 4.13 Mechanism of underwater coating, showing the extra lone pair of electrons on amine which repels water.

also have strong UV resistance. PVDF coatings are now being used in external wall claddings on aluminum as aluminum composite panels, and a life of 25–30 years is assured with no maintenance. These high-performance properties are due to the strong –C–F– bond which is not broken by any strong alkali, strong acid, or even sunlight (Figure 4.15).

Basically, it is a composite coating with 70% PVDF and 30% acrylic coating. Both the ratio of the two, as well as the concentration of fluorine in the PVDF decide the properties of the coating. Resistance to acids is affected if any of these requirements are not met in the composite coating. Figure 4.16 shows the best elongation at 70 : 30 ratio [3, 4].

PVDF coatings also have long gloss and color retention properties. This again is due to the fact that a strong –C–F– bond has a much larger bond strength that cannot be easily broken by UV light during exposure to sunlight. Figure 4.17 compares the gloss retention of PVDF with several other coatings in a Florida exposure test for more than 10 years.

4.5.4 Polysiloxane Coatings

Silicon, which in many ways is analogous to carbon, forms interesting polymers with properties more superior to carbon-based polymers. Si-based polymers form coatings that have stronger resistance to oxidation and low surface energy, which results in coatings that are not degraded in sunlight (UV light), and can be used for much higher temperatures. Low surface energy helps in coatings with low surface tension, which are also water repellent and can be used as hydrophobic coatings [5].

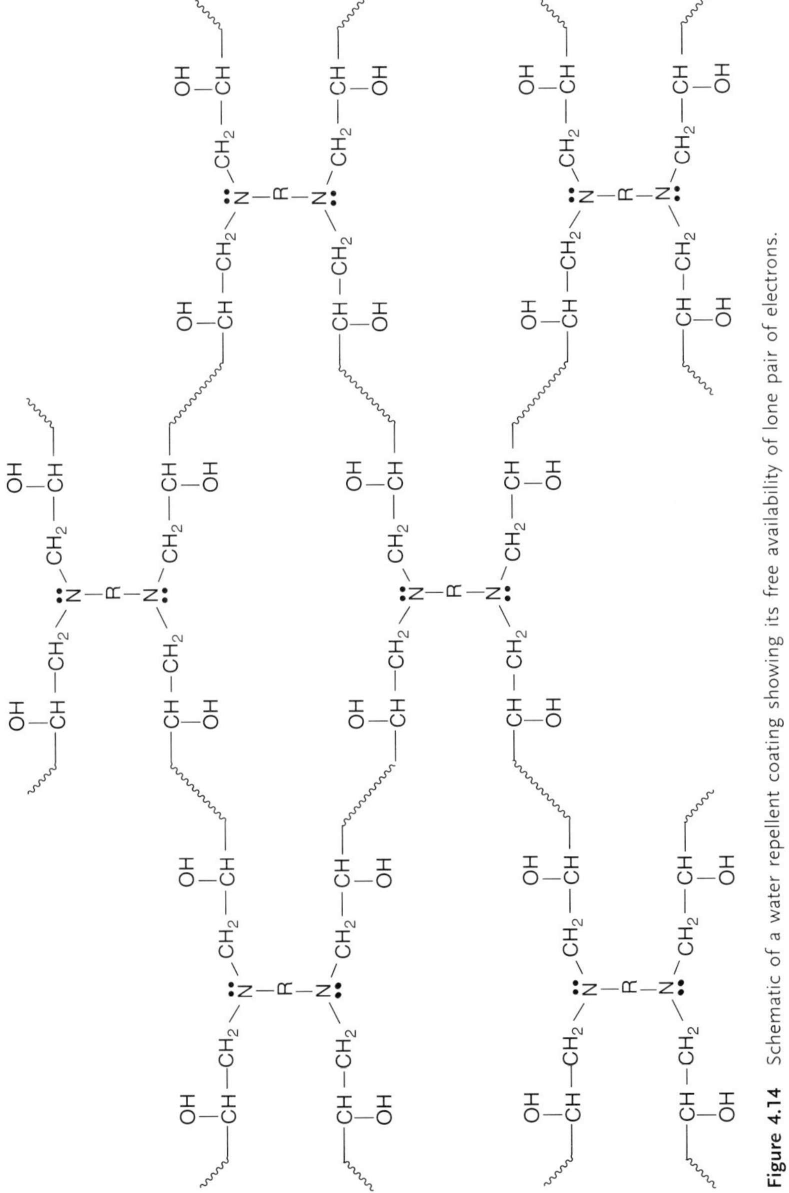

Figure 4.14 Schematic of a water repellent coating showing its free availability of lone pair of electrons.

Table 4.2 Properties of the underwater coating prepared in our laboratory.

Properties	Result
Pot life	2 h
Hard dry	7–8 h
Ease of application	Good
Impact resistance	27 J
Cross hatch adhesion	5 A
Pull-off adhesion (ASTM D4541)	2.2 MPa
Cylindrical mandrel flexibility (ASTM D522)	190°
Abrasion resistance (ASTM D4060)	5 mg
Tensile strength (ASTM D882)	18.3 MPa

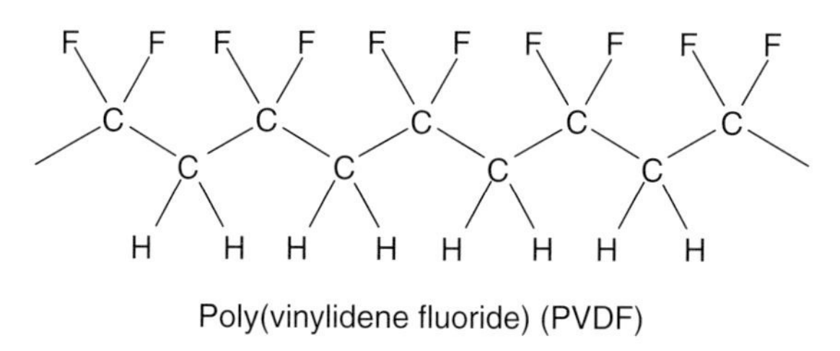

Figure 4.15 Schematic of a polyvinylidene (PVDF) coating.

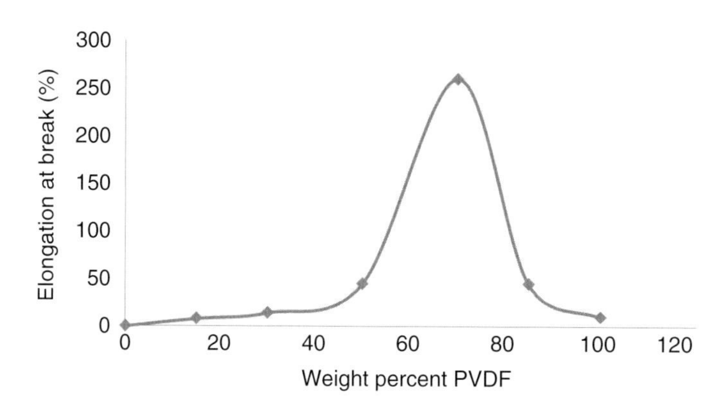

Figure 4.16 Showing the best corrosion and UV properties when fluorine based resin is 70% with 30% acrylic.

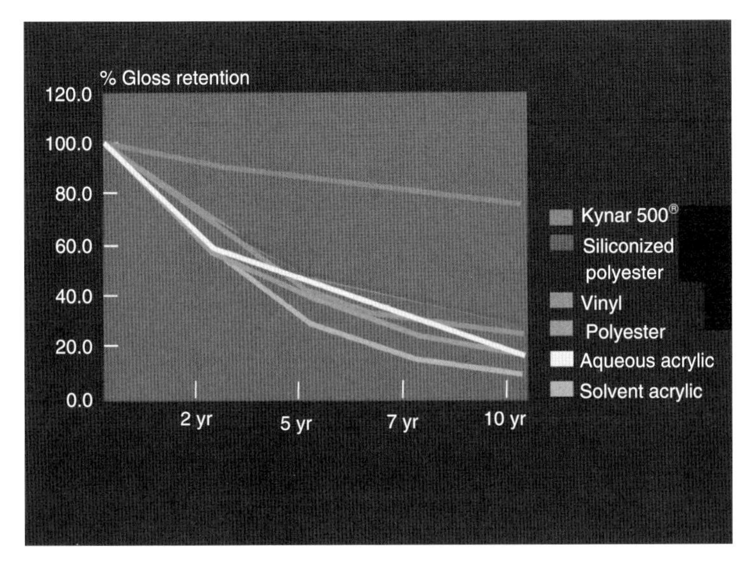

Figure 4.17 Comparison of gloss retention of PVDF Coating (Kynar 500) with other coatings, exposed in Florida open environment for 10 years.

The starting material for Si-based coatings can be chlorosilanes, which can be converted to the respective silanes by hydrolysis:

$$Cl-\underset{R}{\overset{R}{\underset{|}{\overset{|}{Si}}}}-Cl \xrightarrow[-2HCl]{+H_2O} OH-\underset{R}{\overset{R}{\underset{|}{\overset{|}{Si}}}}-OH \tag{4.1}$$

The silanes formed from the hydrolysis reaction are then condensed together to form siloxane-linked polymers by releasing water:

$$--O-\underset{R}{\overset{R}{\underset{|}{\overset{|}{Si}}}}-OH + OH-\underset{R}{\overset{R}{\underset{|}{\overset{|}{Si}}}}---O \rightarrow --O-\underset{R}{\overset{R}{\underset{|}{\overset{|}{Si}}}}-O-\underset{R}{\overset{R}{\underset{|}{\overset{|}{Si}}}}-O-- + H_2O \tag{4.2}$$

Reaction with silanes leads to linear or complex structure siloxane:

$$4RSi(OH)_3 \longrightarrow HO-\underset{R}{\overset{R}{Si}}-O-\underset{R}{\overset{R}{Si}}-O-\underset{R}{\overset{R}{Si}}-O-\underset{R}{\overset{R}{Si}}-OH$$

$$\downarrow$$

$$\begin{array}{ccccccccc} & R & & R & & R & & R & \\ & | & & | & & | & & | & \\ - & Si & -O- & Si & -O- & Si & -O- & Si & -O- \\ & | & & | & & | & & | & \\ & O & & O & & O & & O & \\ & | & & | & & | & & | & \\ - & Si & -O- & Si & -O- & Si & -O- & Si & -O \\ & | & & | & & | & & | & \end{array} \quad + \quad 6H_2O \tag{4.3}$$

Disilanes give linear polymers having controlled degree of polymerization and are known as *silicon oils.* They have low surface tension and are widely used as lubricants. Silicon resins for coatings are produced from a mixture of di- and trisilanols. Their synthesis is long and tedious, and the yield is not very high; hence these coatings are costlier.

R in Eq. (4.1) can be methyl, ethyl, epoxy, polyester, or acrylic. The performance of a silicon resin deteriorates with an increase in the size of "R." While methyl silicon films resist cracking up to 200 °C for more than 10 years, amyl Si films will crack within 24 h at this temperature. Methyl silicons are more reactive and hence cure faster than phenyl resins. Methyl resins have better water resistance, chemical and solvent resistance, and good physical properties such as hot hardness. Phenyl resins, on the other hand, have better resistance to oxidation and are more stable at high temperatures. It is possible to enhance the high-temperature properties of methyl silicons by pigmenting them with aluminum flakes, which makes them capable to hold up to 650 °C, while addition of ceramic frits enhance temperature resistance to as high as 800 °C.

Polysiloxane coatings can have excellent gloss retention for up to 15 years and can have a little effect on the color fastness in addition to their excellent heat resistance. This is due to a strong –Si–O– bond, which is not affected by sunlight; therefore, such coatings retain gloss for a very long time. The bond strength of –Si–O– bond is 193 kcal $mole^{-1}$, compared to 145 kcal $mole^{-1}$ for –C–C– bond.

Polysiloxane coatings are thus a unique class of coatings that offer outstanding properties, such as low VOC (can be as low as 100 $g\,L^{-1}$), outstanding color and gloss retention, extremely good abrasion resistance, anti-dirt pickup, and graffiti resistance properties. When used in combination with inorganic zinc silicate or organic epoxy zinc or other surface tolerant coatings, it gives better corrosion resistance at lower dry film thickness.

4.5.5
Fire-Resistant Coatings

Another area of great concern is the fire safety of the structures, especially, the public utilities such as shopping malls, multiplexes, multistoried buildings, and so on. The main concern in case of fire is how fast can the evacuation process be carried out to save maximum people and important documents and accessories. The concept of providing intumescent coatings of good rating is the need of the hour.

What is an intumescent coating? An intumescent coating is that which when applied on a surface, on burning, can give a stable char of thickness several times the thickness of the applied coating, which can insulate the steel structure from crumbling. The principle of intumescent coating is given in Figure 4.18.

Intumescent coatings range from simple cementitious coatings to organic and inorganic coatings. The main principle of these coatings is to generate an acidic medium, a material which burns into a char, and a medium which accelerates the thickness of the char, that is, blower. Several phosphate-based inorganic coatings are available and are in commercial use; however, they have a fragile char. Silicate-based coatings are cited in the literature but are difficult to convert into good coating. Recently, a modified silica composition with expandable graphite as the source of stable char was developed in our laboratory. It has a thickness rating of 60 minutes for 2 mm coating thickness and 180 minutes for 6 mm thickness.

4.5.6
Organic–Inorganic Hybrid (OIH) Waterborne Coatings

Organic–inorganic hybrids (OIHs) are a new class of materials having important technological potential in the coating industry. OIHs are molecules containing

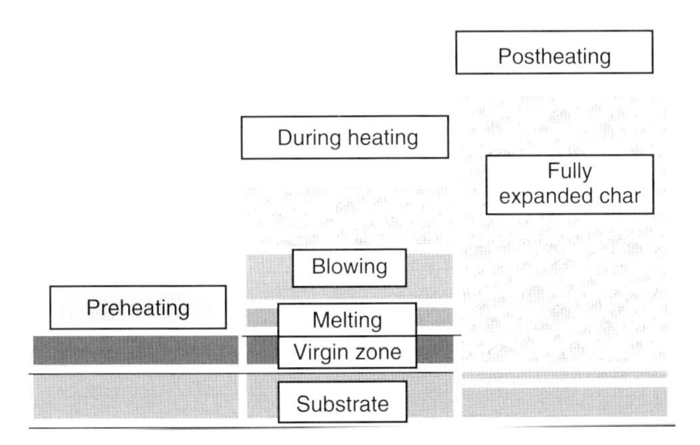

Figure 4.18 Schematic of a intumescent coating.

a metal core bonded to reactive alkoxy groups and/or organic groups. From the coating prospective, the precursors for OIH coatings can be divided into two major categories: (i) metal alkoxide, M(OR)n, where hydrolyzable alkoxy groups are only bonded to the core metal and (ii) functional metal alkoxide R′nM(OR)$x - n$, where a hydrolyzable alkoxy group (OR) and an organofunctional group (R′) are bonded to the core metal. Figures 4.19 and 4.20 show some metal alkoxides and functional silanes, respectively. Metal alkoxides based on silicon, aluminum, zirconium, titanium, and cerium have been commonly used in coating formulation. Extensive coatings market as well as research and development in the twenty-first century is directed toward the development of silicon-based OIH coatings. The silicon–oxygen backbones of functional alkoxysilanes, provide the hybrid with

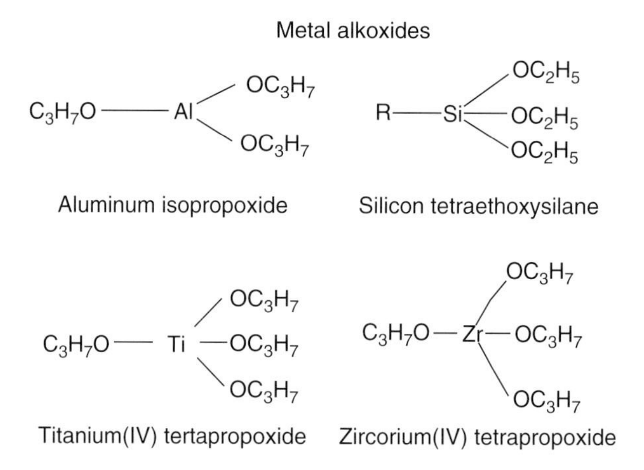

Figure 4.19 Structure of four common metal-alkoxides.

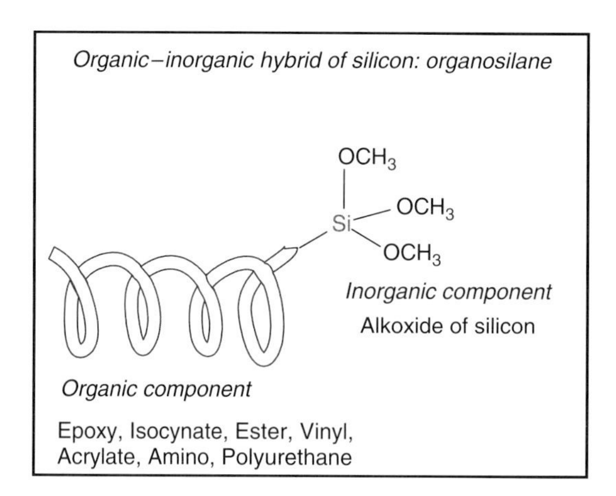

Figure 4.20 Schematic of an inorganic-organic hybrid coating.

a variety of attractive properties. Owing to the hybrid character, they possess high optical transparency, superior weathering and corrosion resistance, excellent abrasion and impact resistance, low surface energy, and better adhesive properties than traditional organic and inorganic coatings [6].

4.6 Surface Preparation

It is a fact that more than 75% of prematured paint failures are due to poor surface preparation. What is surface preparation and what is its significance? Why is it important to adhere to a specified surface preparation as suggested in the paint specifications? These are pertinent questions which are often overlooked. Surface preparation basically means creating a clean surface, free from any contamination, corrosion products, salt, or dust along with a suitable surface roughness, which anchors the paint coating. It is thus mandatory that the substrate be properly cleaned as per the required standards, such as NACE, The Society for Protective Coatings SSPC, or ISO. Confirmation by available techniques of both the required surface cleanliness and anchor profile is a must before proceeding for coating application. Some of the surface preparation standards are summarized in Table 4.3.

4.7 Paint Application

Paint application is no longer a brush or roller application. When application is carried out on large structures, paint and rollers are not efficient methods. Coatings by spray techniques is the most appropriate. The various coating spray techniques and their advantages are given in Table 4.4. The principle of spray painting is to convert a liquid paint into small droplets that are then forced on to a substrate by a stream of compressed air. In air spray, which is more suitable for decorative applications, solvent-borne coatings are usually used. Machines are usually cheap, and the pressure required is about 60 psi; however, the efficiency of the process is

Table 4.3 Various standards describing the surface preparation before paint coating.

SSPC	Swedish SIS055900	British Std BS4232	NACE Std
White metal SP-5	Sa3	First quality	NACE-1
Near white SP-10	Sa2 $^1/_2$	Second quality	NACE-2
Commercial SP-6	Sa2	Third quality	NACE-3
Sweep blasting	Sa1	–	NACE-4
Solvent cleaning SP1	–	–	–
Hand tool cleaning SP2	–	–	–
Power tool cleaning SP11	–	–	–

Table 4.4 Transfer efficiency of various paint spray processes.

Application method	Transfer efficiency (%)
Air spray	30
Airless spray	45–50
Air-assisted spray	60
Electrostatic spray	75

just 30%, but the quality of paint finish is excellent. Therefore, all good-appearing coatings such as auto coatings are carried out using air spray. However, when the requirement is high protective properties and high bond strength but low surface finish, airless spray technique is used. In airless spray system, a very high pressure of 2000 psi is created using a hydraulic system, and the efficiency of the process is as high as 45–50%. Efficiency can be further improved by 10% by using a hybrid system of compressed air and hydraulic pressure.

The best efficiency is obtained when we use electrostatic guns, in which the paint is charged, using high voltage, before it gets out of the gun nozzle and hits on a surface, which is oppositely charged, thus increasing the efficiency to as high as 75%. Many powder coatings are carried out using electrostatic guns.

There are several new developments in paint applications, especially the application of very fast curing paints such as polyurea, which has a curing time of 3–12 seconds. Very sophisticated machines are available in which two components are added from two different streams, allowed to meet just before the gun nozzle, and immediately sprayed on the substrate. Polyurea coatings are well known today as "*spray and walk*" coatings, which have all the properties of a superior epoxy and polyurethane coating with fast curing, not affected by temperature.

4.8 Importance of Supervision, Inspection, and Quality Control during Paint Coatings

In India, the main cause of premature paint failure is the lack of supervision, inspection, and quality control. In the section above, it was told that 75% paint failures are due to poor surface preparation or overlooking the requirement of surface cleanliness and its anchor profile. The remaining 25% of failures may be due to nonadherence, improper drying times, wrong coating intervals, and poor paint application practices. Even a high-quality paint will perform miserably when applied on a poorly prepared surface and using a poor application technique. It is high time that users insist on employing coating supervisors and inspectors and on following quality assurance procedures, such as intermediate inspection, maintaining daily log books, inspection plan, calibration of equipments, and so on. This is important not only in maintaining good coating quality but also in case of any premature failure, enough evidence for arbitration. Contractors must employ trained supervisors, inspectors, and skilled persons with a basic knowledge of the paint to be used, surface preparation standards, application techniques.

4.9 Training and Certification Courses

In order to achieve excellence in the four parameters mentioned above, it is very important to undergo training and certification programs, which not only impart quick, relevant information that one requires to enhance the coating durability but also trains for several practical techniques used during paint coatings. Several such kinds of courses are now being organized in India by NACE International, British Gas (BGAS), SSPC India, and Indian Institute of Technology (IIT) Bombay. Courses organized by NACE International and BWG are costly (~Rs 75 000–Rs 90 000 per person). The information provided is very mundane, and owing to the high cost, these courses are usually taken by managers and senior people and not by the persons who really work at site. On the other hand, courses conducted by SSPC India and IIT Bombay are within the reach (Rs 15 000 per person) and are very useful, with excellent theoretical and practical training.

4.10 Summary

The critical issues in the use of paint coating in many industrial applications are basic knowledge of resin chemistry for selection of paint coatings for various environments, knowledge of surface preparation and its standards, modern application techniques, and thus paint savings. This must be coupled with thorough paint application supervision, inspection, and quality control.

References

1. Khanna, A.S. (2008) Key issues in applying organic paint coatings, in *High Performance Coatings* (ed. A.S. Khanna), Woodhead Publisher, pp. 1–26.
2. Swati, G. (2010) MTech Thesis, Swati Gaur, IIT Bombay.
3. Wood, K., Tanaka, A., Zheng, M., and Garcia, D. (1996) 70% PVDF Coatings for Highly Weatherable Architectural Coatings, *www.arkema-inc.com/pdf/techpoly/kurt%20 paper.pdf.*
4. Wood, K.A., Cypcar, C., Hedhli, L., and Hare, C.H. (2000) Predicting the exterior durability of new fluoropolymer coatings. *J. Fluor. Chem.*, **104**, 63–71.
5. Hare, C.H. (1994) Protective Coatings, SSPC 94-17, chapter 17.
6. Pathak, S.S. and Khanna, A.S. (2010) Organic Inorganic hybrids as functional coatings, in *Advances in Solid Hybrid Materials and Membranes* (eds T. Xu and C. Wu), Transworld Research Network, pp. 1–26.

5
New Era of Eco-Friendly Corrosion Inhibitors

Niketan Patel and Girish Mehta

5.1
Introduction

Corrosion is damage caused to a metal by chemical, electrochemical, or even biological reactions between the metal and the surrounding medium. Most human activities are affected by this phenomenon, considering the use of metals and their alloys in all domains. The knowledge of the causes of corrosion is essential to develop a control technology and to improve the means of protection.

Corrosion is recognized as one of the most serious problems in the modern society. The resulting losses are in hundreds of billions of dollars per year, and with further development, the losses too are increasing exponentially. The annual corrosion costs range from approximately 1 to 5% of the gross national product (GNP) of each nation.

Among the practical methods available for control of corrosion, the most significant ones are (i) selection of materials, (ii) application of coatings, (iii) cathodic protection, (iv) anodic protection, and (v) addition of corrosion inhibitors.

A corrosion inhibitor is a chemical additive, which when added in small quantity to a corrosive environment reduces the rate of corrosion remarkably. Inhibitors function by adsorption or absorption of ions or molecules onto the metal surface.

Hundreds of chemicals – inorganic and organic – have been studied and recommended as inhibitors of corrosion for a range of metals in a variety of environments – aqueous, nonaqueous, molten salts, and dry atmospheres. Inhibitor properties are reported at various temperatures – ranging from very low to high. Most inhibitors are developed by trial and experiment. A lot of these are proprietary in nature; their compositions are not known.

The corrosion of iron and mild steel (MS) is a fundamental academic and industrial concern that has received considerable amount of attention [1]. A study of the mechanism of action of corrosion inhibitors has relevance both from the point of view of the search for new inhibitors and of their effective use [2].

A very large number of inhibitors are reported in the literature for different corroding systems. A useful summary in this regard is provided in the checklist of inhibitors. One striking feature of these inhibitors is their selectivity. An inhibitor

Green Corrosion Chemistry and Engineering: Opportunities and Challenges, First Edition.
Edited by Sanjay K. Sharma.

that is extremely efficient in one system may perform very badly in another system. This selectivity is attributed to the preferential adsorption of the inhibitor on the metal surface.

The parameters to be considered for the inhibitors are mainly

1) solubility;
2) compatibility with the corroding system;
3) stability over long time under stationary and flowing conditions at different temperatures and pH of the medium;
4) undesired effects of the effluents containing the inhibitor – pollution;
5) cost.

Corrosion inhibitors can be classified as follows:

1) anodic (passivating or film forming) inhibitors;
2) cathodic (adsorption type) inhibitors;
3) mixed inhibitors;
4) precipitation inhibitors;
5) vapor phase inhibitors (VPIs).

5.2 Anodic (Passivating or Film-Forming) Inhibitors

An anodic inhibitor interferes with the anodic process of corrosion. These substances reduce the anode area by acting on the anodic sites, usually by the formation of protective oxide layer, and polarize the anodic reaction, as shown for steel. They are mostly inorganic compounds added in neutral or alkaline media.

$$Fe \rightarrow Fe^{2+} + 2e^- \tag{5.1}$$

Generally, anodic inhibitors are salts that contain anions that form sparingly soluble compounds with ions of the metal under consideration. These are ions of transition elements with a high content of oxygen and are capable of limiting the corrosion reactions taking place at the anode. They form a passive film of corrosion product at the surface that is corroded, and this film of corrosion product limits further corrosion. Oxidizing anions as such have the ability to passivate metal in the absence of oxygen. Typical oxidizing anions are chromate, nitrite, and nitrate. Nonoxidizing ones such as phosphate, tungstate, and molybdate require oxygen to perform passivation. Compounds of orthophosphate, nitrite, tungstate, venadate, chlorate, permanganate, ferricyanide, and silicates inhibit corrosion by producing passive material on the metal surface, particularly on Fe, stainless steel (SS), and other alloys that show active–passive transition. These are known as *passivators* or *anodic inhibitors* and are also often called *dangerous inhibitors*. If all anodic sites are not protected because of insufficient concentration of such inhibitor, leading to an incomplete surface adsorption or partial film formation, or if the protective films are nonadherent at a few points, corrosion sets in at the exposed sites. This corrosion, if supported by an adequate cathodic process, can become high since the area of corrosion is small. This intensive corrosion can become deep rooted and lead to catastrophic failures.

5.2.1 Mechanism

The mechanism of action of the anodic inhibitors (passivators) has been the subject of considerable study and long controversy. Many theories of passivity have been proposed. The earliest seems to be Faraday's [3], who considered that the passivity of iron in concentrated HNO_3 is due to the formation of protective oxide film. The oxide film theory has been generalized and is the one most widely held at present.

5.2.1.1 Generalized Film Theory

According to this theory, the passive film is always a diffusion barrier layer of reaction products (e.g., metal oxide or other compounds) that separates a metal from its surroundings and slows down the corrosion rate. This reasoning is the basis of what is generally referred to as the *oxide film theory*. Evans [4] stated that passivity is due to the formation of a protective film on the surface. This film may not always be an oxide film. Hedges [5] and Mears [6] supported this view. Glasstone [7] concluded that iron in concentrated HNO_3 produces a thin adherent film of ferric oxide or a basic salt, which protects the metal. Evans [8] has stated that anions of sodium or potassium, which form a sparingly soluble salt with the metal ion, act as inhibitors, while Hoar and Evans [9] have shown that chromate ions react with ferrous ions and precipitate an adherent protective film of hydrated ferric and chromic oxides on the anodic areas of the metal surface. Moreover, the amount of chromate ions reduced on iron surface has been measured [10–14] by residual radioactivity using ^{51}Cr isotope in environment, and it was found that Cr^{3+} or hydrated Cr_2O_3 remained adsorbed on the metal surface along with iron oxide. The presence of γ-Fe_2O_3 has been disclosed using electron diffraction technique by Mayne and Pryor [15]. Graham and Cohen [16] used Mossbauer spectroscopy for study of passive film. The properties of the film formed on the iron surface in the presence of chromate, phosphate, molybdate, tungstate, arsenate, and nitric acid have been studied [17].

5.2.1.2 Adsorption Theory

This theory is also known as *electron configuration theory*. Langmuir [18] pointed out that passive film on tungsten was because of adsorption of oxygen. Tammann [19] also suggested that passivities of Fe, Ni, Co, and Cr were due to adsorbed oxygen rather than oxide. Several other theories have been proposed, most of them in opposition to the oxide film theory.

Russell [20], Swinne [21], and Sborgi [22] correlated the electron configuration of metal atoms and passivity. Uhlig and Wulff [23] and Uhlig [24] proposed and extended a theory for critical passivity concentrations in alloys.

According to the electron configuration theory, transition metals having electron vacancies or unpaired electrons in the "d" shells of the atom are responsible for strong covalent bond formation with components of the corrosive environments, especially oxygen (paramagnetic because of unpaired electron), in addition to ionic bonding. These metals have higher heats of formation so that metal atoms tend to remain in their lattice, whereas oxide formation requires metal atoms to leave their lattice. The peculiar higher activation energies for adsorption of oxygen on such metals reveals the chemical bond formation, hence such films are called *chemisorbed films* in contrast to lower-energy films, which are called *physically absorbed films*. On the transition metals, the life of chemisorbed film is much longer. Such chemisorbed layer displaces the normally adsorbed H_2O molecules and slows down the anodic dissolution rate. Alternatively, it can be said that adsorbed oxygen decreases the exchange current density (increases anodic over potential). Even less than a monolayer of the film on the surface is observed to have a passivating effect [25, 26]; hence it is suggested that the film cannot primarily act as diffusion barrier layer. Multilayer chemisorbed passive films react in time with the metal to form compounds such as oxides. These oxides are less important in accounting for passivity than the chemisorbed films that form initially and continue to form on metal exposed at pores in the oxide. It is reported that a film of γ-Fe_2O_3 is in equilibrium with an underlying Fe_3O_4 layer [27]. Uhlig [28] correlated the observed noble value of Flade potential (E_F) for iron with the potential calculated for chemisorbed film of oxygen on the surface of iron and suggested that on metal surface, an initial layer of atomic oxygen is formed, over which molecular oxygen is chemisorbed to give a thick film and the adsorbed passive film can be represented by O_2, O_{ads} on Fe. Oxygen in the passive film is present in a higher energy state [29].

Passivities of Cr and SS can occur by direct chemisorption of oxygen from the air or from aqueous solutions because of their higher affinity for oxygen. Equivalents of oxygen adsorbed are found [30] to be of the same order of magnitude as the equivalents of passive films formed on the iron passivated by anodic polarization, concentrated HNO_3, or chromates.

Flade [31] and Frankenthal [26] observed that the film is stabilized by continued exposure of the metal or alloy to the passivating environment. It is probable that the observed stabilizing effect is the result of positively charged metal ions entering the adsorbed layers of negatively charged oxygen ions and molecules, the coexisting opposite charges tending to stabilize the adsorbed film. This statement is supported by MacRae [32]. The presence of H_2O in the passive film on 18-8 SS [33] and passive iron [34] has been reported. The mode of initiation and lateral growth of chemisorbed atomic and molecular oxygen film on the metal surface is explained by Uhlig [35].

Frankenthal [26] in the electrochemical study of passivity of Fe–Cr alloys in H_2SO_4 as well as by microscopic observations has shown that at least two distinct potential-dependent films are formed. The primary film responsible for the initial passivation is stable only within a few millivolts of the primary activation potential. A secondary film forms at more positive potentials, grows to a thickness greater than 10 Å, and, with increasing potential and time, becomes very stable and resistant to reduction. The primary passivation process is reversible. The thickness of the primary film at the primary activation potential corresponds to less than the equivalent of one oxygen atom per surface metal atom. From the thickness and pH dependence of primary activation potential, he suggested that the primary passivation process forms a film containing Cr and oxygen ions and possibly some constituent of the electrolyte solution also [36].

5.3 Cathodic (Adsorption-Type) Inhibitors

A cathodic inhibitor interferes with the cathodic process of corrosion. Substances that reduce the cathodic area by acting on the cathodic sites, usually by formation of barrier film, and polarize the cathodic reaction, given below for steel in acid, are called *cathodic inhibitors*. They are mostly organic compounds containing hetero atoms such as N, O, S, P, and so on, and are added in acidic media. These hetero atoms in the compounds having higher electron density act as reaction centers and help the compound to get absorbed or adsorbed on the metal surface either in ionic or in molecular form.

For example, the major cathodic reaction in cooling systems is the reduction of oxygen.

$$2e^- + H_2O + \frac{1}{2}O_2 \rightarrow 2OH \qquad (5.2)$$

There are other cathodic reactions and additives that suppress these reactions called *cathodic inhibitors*. They function by reducing the accessible area for the cathodic reaction. This is often achieved by precipitating an insoluble species onto the cathodic sites.

Many organic inhibitors work by an adsorption mechanism. The resultant film of chemisorbed inhibitor is then responsible for protection either by physically blocking the surface from the corrosion environment or by retarding the electrochemical processes. The main functional groups capable of forming chemisorbed bonds with metal surfaces are amino ($-NH_2$), carboxyl ($-COOH$), and phosphonate ($-PO_3H_2$) groups, although other functional groups or atoms also can form coordinate bonds with metal surfaces. These are long-chain organics and polymers that form films, isolating the corroding surface from the corroding solutions.

5.3.1 Mechanism

Interface inhibition presumes a physically powerful interaction between the corroding substrate and the inhibitor. The two-dimensional adsorbed layer can effect the fundamental corrosion reactions in different ways, which may be discussed in terms of the inhibition efficiency (IE), ε, defined by

$$\varepsilon = \frac{i - i_{INH}}{i} \qquad (5.3)$$

where i and i_{INH} represent the current densities of the electrode reaction at constant potential in the absence and presence of inhibitor, respectively. If Eq. (5.3) is to be converted into percentage ε, the right-hand-side term is to be multiplied by 100.

1) Geometrical blocking effect of the electrode surface by an indifferent adsorbate at a relatively high degree of coverage, θ_{INH}.
   In this case, $(1-\theta)\, i = i_{INH}$, therefore

$$\varepsilon = \theta \tag{5.4}$$

   where $\theta = \frac{\tau}{\tau_S}$; τ denotes the potential-dependent surface concentration of inhibitor and τ_s represents its saturation value.
2) Blocking of active surface sites by an indifferent adsorbate at a relatively low degree of coverage.
   In this case, Eq. (5.4) holds again, if τ_s is replaced by a surface concentration $\tau_{a,s}$ necessary for a complete blocking of all active surface sites.
3) The adsorbate is not indifferent but reactive.
   Two different cases may be distinguished. Firstly, the inhibitor acts as a positive or negative electrocatalyst on the corrosion reaction and secondly, the adsorbate itself undergoes an electrochemical redox process. In the latter case, primary and/or secondary inhibition can occur depending on the retardation effects caused by the original adsorbate and its reaction product, respectively.
   In the case of a reactive coverage, ε will be a more complex function of θ than Eq. (5.4) and can also be negative:

$$\text{Inhibition} \quad 0 \le \varepsilon(\theta) \le 1 \tag{5.5}$$

$$\text{Stimulation} \quad \varepsilon(\theta) \le 0 \tag{5.6}$$

These three types of interface inhibition are mostly observed in corrosion systems exhibiting a bare metal surface in contact with the corroding medium. This condition is often realized for active metal dissolution in acid solutions [37].

5.4 Mixed Inhibitors

Substances that effect both the cathodic and anodic reactions of corrosion are called *mixed inhibitors*. In general, these are organic compounds that adsorb on the metal surface and suppress metal dissolution and reduction reaction. In the majority of cases, it appears that mixed inhibitors influence both the anodic and cathodic processes, although in numerous cases, the effect is not the same. If the influence of the cathodic reaction is superior to that of the anodic reaction, the inhibitor can be classified as harmless; if, however, the influence of anodic reaction is greater, it is classified as a hazardous one. Because of the risk of pitting when using anodic inhibitors alone, it becomes common practice to add-in a cathodic inhibitor into the formulation, the performance of which is greater by a combination of the inhibitors than from the sum of the individual performances. This observation is generally referred to as "*synergism*" and demonstrates the synergistic action that exists between Zn^{2+} and $CrO_4{}^{2-}$ ions. When two or more inhibiting substances are added to a corrosive system, the inhibiting effect is sometimes greater than that which would be achieved by either of the two (or more) substances alone. This is called a *synergistic effect*.

5.5 Precipitation Inhibitors

These inhibitors are often film forming in nature, for instance, silicates and phosphates. They are effective at blocking anodic and cathodic sites. They precipitate on the metal surface, forming a protective barrier. Hard water is rich in magnesium and calcium. When these salts precipitate on the metal surface, for example, at the cathode where the pH is higher, they establish a protection layer on the metal. Film-forming types of inhibitors are often distinguished into two classes. The first one works by slowing down the corrosion without stopping it completely. The second ceases the attack completely.

However, the efficiency of this inhibitor depends on the pH value and saturation index. The saturation index is then in turn determined by the water composition and temperature.

5.6 Vapor Phase Inhibitors

They are organic inhibitors that readily vaporize and form a protective layer of the inhibitors on the metal surface. Volatile corrosion inhibitors (VCIs), also called VPIs, are compounds transported in a closed environment to the site of corrosion by volatilization from a source. In boilers, volatile basic compounds, such as morpholine or hydrazine, are transported with steam to prevent corrosion in condenser tubes by neutralizing acidic CO_2 or by shifting surface pH toward less acidic and corrosive values. In closed vapor spaces, such as shipping containers, volatile solids such as salts of dicyclohexylamine, cyclohexylamine, and hexamethylene-amine are used. On contact with the metal surface, the vapor of these salts condenses and is hydrolyzed by any moisture to liberate protective ions. It is desirable, for an efficient VCI, to provide inhibition speedily while lasting for extended periods. Both qualities depend on the volatility of these compounds, fast action demands high volatility while enduring protection requires low volatility [38].

Generally, heterogeneous organic compounds containing hetero atoms such as N, S, and P having higher electron density form the reaction center and are responsible for adsorption or absorption on the metal surface. Some of them also increase the hydrogen over voltage, for example, oxides of arsenic and stibium (antimony) are reduced to the corresponding metals and these metals are deposited on the surface of the substrate metal. Such inhibitors are usually added in acidic corrosives during surface preparation by acid cleaning and acid pickling. Anodic inhibitors are usually inorganic compounds having oxidizing nature, such as permanganates, chlorates, vanadates, and tungstates, which convert metal ions formed because of corrosion on the metal surface into protective oxide films and retard the rate of anodic half-cell reaction of corrosion. Recently, some organic compounds with oxidizing functional groups have also been reported. Although chromates and dichromates have been used as anodic inhibitors for a long time in industrial applications and in paint coatings, because of the carcinogenic effect of Cr-VI species, they are now replaced by the other compounds. Mixed type of inhibitors contain at least two functional groups by which the rates of both the half-cell reactions are retarded, for example, *p*-amino benzoic acid in which the $-NH_2$ group retards the rate of cathodic half-cell reaction and the –COOH group retards that of anodic half-cell reaction. For temporary protection in sealed packing of metal articles, VPIs are included inside the packing, which act when they go into vapor phase from either solid or liquid phase.

Large amounts of acid solutions are used in the chemical industry for removal of the undesired scales and rust. The addition of corrosion inhibitors effectively secures the metal against an acid attack. Inhibitors are generally used in these processes to control metal dissolution [39]. A detailed survey of the literature on acid corrosion indicates there are numerous descriptions regarding application of corrosion inhibitors for a wide range of metals, of which only a few are now applied in practice. This may be because of the fact that when the inhibitors are actually employed, a lot of other desirable factors such as cost, toxicity, availability, and so on, come into picture that extends beyond metal protection. Some examples, from a few published review papers, include a wide range of chemicals that have inhibitive properties. Harsch *et al.* [40] have examined over 70 compounds, and Trabanelli and Carassiti [41] have published and discussed some 150 compounds as corrosion inhibitors. Indian workers have also given the list of organic inhibitors for aluminum and its alloys with 225 references [42] and also for corrosion inhibition of copper with 93 references [43]. Corrosion inhibitors for industrial use were reviewed by Rama Char with 134 references [44]. Walker [45] has given 92 references in discussing the use of benzotriazoles as the inhibitors for copper corrosion. The corrosion of copper and its inhibition by benzotriazole and its derivatives has been widely studied and is being studied and reported by many workers. Huynh *et al.* [46] have studied the influence of coating of alkyl esters of carboxybenzotriazole as inhibitors for copper corrosion. The synergistic effect of a mixture of sodium dodecylsulfate–benzotriazole as an inhibitor for processes on copper–chloridric interfaces has been studied by Villamil *et al.* [47].

Recent developments in the mechanisms of corrosion inhibition have been discussed in reviews regarding acid solutions [48] and neutral solutions [49], and for this study, novel and improvised techniques such as surface-enhanced Raman spectroscopy (SERS) [50], infra red (IR) [51], Auger and X-ray electron spectroscopy [52], and electrochemical impedance spectroscopy (EIS) have been used to study the adsorption, interaction, and reaction of inhibitors at metal surface. In addition to the reviews on inhibition, a number of books are also available such as *Metallic corrosion inhibitors* by Putilova *et al.* [53], *Advances in Corrosion Science and Technology-1* by Trabanelli and Carassiti [54], and *Corrosion inhibitors* by Rozenfeld [55], and so on, which provide very useful information on inhibitors and mechanisms of corrosion inhibition.

Most of the corrosion inhibitors are specific in their action toward particular metal and the environment. The mechanism of corrosion of metal in neutral solutions differs from that in acid solutions in two important aspects: first, in neutral air-saturated solutions, the cathodic half-cell reaction of corrosion is the reduction of dissolved oxygen and H_2 evolution by reduction of H^+ ions in the acidic solutions. Second, in neutral solutions, the metal surface is covered with stable films of oxide/hydroxide/salt, because of the decreased solubility of the species, but in the acid solutions, the metal surfaces become free from the oxides. Because of this difference of the condition, the chemical compounds that may be very effective in acid solutions may not be much effective in neutral ones, with some exceptions; for example, organic compounds of high molecular weight such as gelatin, agar, and dextrin [53] are effective in partly protecting the metal surface in neutral as well as in acidic solutions. The mechanism of corrosion inhibition in neutral solutions includes the formation of protective salts/hydroxides/oxides films by action of the inhibitors on metals. Inhibition can also occur by incorporation of anion in the oxide film as reported in the case of iron and steel. This was further supported by radiotracers and Auger electron spectroscopic studies. Anodic passivation involves the iron in lower valence state along with inhibiting anion [49] before forming the ferric oxide film. The inhibitive anions may be of two types: type I includes carboxylates such as phthalate and acetate, which have little or no inhibitive effect in retarding the active dissolution or facilitating passivation in deaerated solutions. Type II involves more effective inhibitors such as nitrites, molybdate, and substituted benzoates and salicylates that possess inhibitive properties in deaerated solutions. The inhibition of copper corrosion by low-molecular-weight organic compounds such as benzotriazole [56] and 2-mercaptobenzotriazole [57] in neutral solutions has been reported. Benzotriazole is effective in reducing the tarnishing of copper in chloride solutions, and according to the report of Mansfeld *et al.* [58], the inhibitors further retard the anodic dissolution, dissolved oxygen reduction, and oxide film growth, indicating the strong adsorption on the cuprous oxide surface. Similarly, extensive study of other metals such as zinc and aluminum in neutral solutions has been carried out. In general, the bonding of adsorption inhibitors on the metal surface is described in terms of hard–soft acid and bases by Aramaki [59]. In many cases, either in acid or neutral solutions, the substituents (electrophilic/nucleophilic) increase the electron density of the functional group and thus increase the IE by the stronger adsorption forces.

Most of the inhibitors studied in acid solutions are organic compounds containing mainly N, S, and O atoms as discussed earlier. In general, the polar function of these heteroatoms is regarded as the reaction center for the establishment of the process of chemisorption. In such cases, the strength of the adsorption bond is determined by the electron density and polarizability of the atom (N, S, or O) acting as the reaction center.

Corrosion inhibition by organic compounds takes effect generally by the mechanism of adsorption of molecules and ions at metal surface. Adsorption can be of the following types:

- Physisorption due to electrostatic attractive force between the inhibiting ions or dipoles and the electrically charged metal surface.
- Chemisorption caused by the interaction between unshared electron pairs or electrons and metals to form a co-ordinate type of bond. This type of absorption takes place when there are heteroatoms such as P, Se, S, N, and O present with lone pair electrons and/or aromatic rings in the adsorbed molecules [60–64].
- A combination of both.

The level of protection of organic corrosion inhibitors depends on the adsorption, which in turn is determined by many factors such as

- nature and surface charge of the metal;
- mode of adsorption of the inhibitor;
- chemical structure of the inhibitor;
- type of solution the metal is exposed to [65].

The adsorption process can be improved by the presence of triple bonds, aromatic rings, and heteroatoms present in the chemical structure of the corrosion inhibitor. The efficiency of heterocyclic organic compounds of corrosion inhibitor increases in the order of oxygen < nitrogen < sulfur < phosphorus [65–68].

Generally, the acid corrosion inhibitors function by interfering with the reactions that occur on anodic/cathodic sites of the metal surface. They are able to form chemical linkages with the metal by covering the active sites of metallic surface. Normally, inorganic inhibitors such as tungstates, benzoates, nitrites, and so on function by passivating the metals. The organic inhibitors form a barrier at the metal surface either with physical linkages through adsorption or by forming an insoluble compound on the surface of the metal. Kaesche and Hackermann [69] reported that in the acid solutions, organic inhibitors are in equilibrium with cationic species and interfere with the cathodic reactions. Although classical adsorption theory considers large size as the prime factor for effective corrosion inhibition, it is now well accepted that the chemical bonding also has major influence. As discussed in Section 5.1, the structure of the organic compound provides a guideline, but its value as a corrosion inhibitor is proved only if its performance is tested. The field of corrosion inhibition is rapidly growing, and several organic compounds are being tested under various operating conditions by many workers in industries as well as in laboratories of good reputation.

Proton acceptors, generally, are considered as the absorbers on the cathodic site, and they accept H^+ ion and migrate to the cathode, for example, anilines, quinolines, thiourea (TU), and the aliphatic amines. In the same way, electron acceptors are, generally, effective on the anodic sites because of their ability to accept electrons and are most effective for the reactions under anodic control. Some of them even act by passivation; for example, organic peroxides, organic thiols, selenols, inorganic tungstates, and nitrites act as aforesaid. The third category is "ambiodic"/mixed inhibitors that can affect both anodic as well as cathodic half-cell reactions equally; for example, amine benzenethiol, pyridinium benzyl bromide, and so on, act as aforesaid.

The inhibiting action of organic compounds in acidic solutions depends upon the factors such as nature of electrode/electrolyte, environment, pH, and so on. Antrapov [70] studied the behavior of pyridine derivatives as inhibitors for the acid corrosion of iron. After an examination of variations in the coverage as a function of the molecular structure, it was stated that the compounds examined were adsorbed on the metal surface in the form of cations and not as neutral molecules. Thus, the N atom, which might have been considered as being protonated in acidic solution, was directed at the metallic surface in the cationic form. However, other authors such as Trabanelli *et al.* [71], Felloni and Cozzi [72], and Hackermann [73] believed the existence of cations in bulk of the acid but subsequent discharge with the formation of molecules at the interface. These compounds were chemisorbed through an electron pair of the heteroatom, and the mechanism was proposed as follows:

$$R\text{-}NH_3^+ + e^- \rightarrow R\text{-}NH_2 + 1/2\ H_2 \quad \text{(for aliphatic compounds)} \tag{5.7}$$

$$NH^+ + e^- \rightarrow N + 1/2\ H_2 \tag{5.8}$$

These additives can act as transporters of protons and catalyze the electrode reaction by providing lower activation energy for the discharge of a proton, as a function of the structure of the additive. A different opinion was expressed by authors like Ertelk and Horner [74] in the case of quaternary (-onium) organic compounds for which a secondary inhibition action analogous to dibenzyl sulfoxide (DBSO) was considered. Lorenz and Fischer [75] considered the inhibition as "primary" in the case of triphenyl phosphonium/arsonium, which was in contrast to "secondary" of DBSO. Jofa [76] described the inhibition in acidic solutions by many of the organic substances as cationic type and proposed that the IE depends on the nature and charge on the metal surface. According to him, the adsorption of cations from solutions of H_2SO_4 on to the Fe surface was weak. He concluded that nonionic additives, adsorbed on Fe surface, have almost low percentage of IE in the acid media. On the other hand, anionic substances were assumed to be absorbed immediately from the H_2SO_4 solutions, but sometimes they may stimulate the corrosion by formation of $[FeOH]^-$. Thus, in halogen-containing solutions, halogens were chemisorbed onto the metal surface creating dipoles oriented to the surface and may attract cationic inhibitors. This has explained the increase in IE of quaternary compounds in acid solutions in the presence of KI. The bond between chemisorbed halogen and nonionized inhibitor may be covalent and may not be electrostatic. In contrast to the above statements, Cavallaro *et al.* [77], while studying the effect of TU in the presence of KI in the acid solutions, reported that the anion had no influence on the adsorption of (uncharged) neutral inhibitor molecules. Here, the absence of synergistic effect of I^- was attributed to the adsorption of inhibitors in molecular form and not as cations. Such contradictory review was well reflected in the study of Trabanelli *et al.* and Aramaki *et al.* [79] DBSO was studied for the inhibition in acidic media by Trabanelli *et al.* [78], and its secondary inhibiting action by the protonation and reduction was shown as

$$[C_6H_5CH_2]_2SO + H^+ \rightarrow [C_6H_5CH_2]_2\,SOH^+ \qquad (5.9)$$

The above cation was electrochemically reduced and dibenzyl sulfide (DBS) was formed.

$$[C_6H_5CH_2]_2\,SOH^+ + H^+ + 2e^- \rightarrow [C_6H_5CH_2]_2\,S + H_2O \qquad (5.10)$$

A multimolecular layer formation has taken place by accumulation of DBS, and its low solubility has contributed to the high IE. Aramaki *et al.* [79] extended this study by EIS and SERS techniques and suggested that the sulfoxide reduction was possible; but DBS, formed by reduction, was absorbed on the Fe surface in the form of neutral molecule and only the first layer of these absorbed DBS molecules contributed to the inhibition. Thus, IE of DBS in 0.5 M H_2SO_4 was larger by one order compared to that of DBSO at the same concentration. The denotation of primary and secondary inhibitions is discussed later in the following paragraphs.

Riggs and Every [80], using some simple structures of aniline and benzenethiol, have developed mechanics for designing organic molecules, which prevented corrosion of carbon steel in HCl. Numerous investigators have studied the relation between the organic structure of inhibitors and their effectiveness in the acid corrosion inhibition, and these are summarized by Eldridge and Warner in Uhlig's corrosion handbook [81]. This idea was further developed by Douty [82] who classified inhibitor compounds according to elemental composition and functional groups. For doubtful treating situations, a checklist [83] of inhibitors having detailed structures can serve as a guideline and was given for a wide range of metals at various conditions.

Some of the earlier studies in corrosion inhibition are given below. Hackermann along with Hurd, Anand, and Aramaki [84–86] have studied and published a series of papers on polymethylimines as inhibitors of steel corrosion in HCl. The same studies were extended to include polymeric amines by concluding that the polymeric compounds were more efficient corrosion inhibitors than their monomers. The inhibition of corrosion of iron in HCl by polymethylimines was believed to be related to the C–N–C bond angle or strain in the molecule and more effectively because of donation of the unshared π-electron pair of its "N" atom to the metal but not because of molecular size. On the other hand, it was found that increase in the number of repeating methylene unit groups in the molecule increased the inhibition supporting the effect of molecular size concept also. However, the electron density concept for inhibitor to act can account for both the adsorption and/or complexion properties of imines. Moreover, Okamoto *et al.* [87] studied the corrosion of MS in 10% H_2SO_4 and the effect of dibutylthioether by galvanostatic polarization method, and the results were consistent and have mainly shown anodic inhibition.

Chin and Nobe [88] have studied the electrochemical characteristic of iron in H_2SO_4 in the presence of aminoazophenylene (AAP), compared the corrosion rates with adsorption measurement, and found that AAP was strongly absorbed on the iron surface and that this inhibition was enhanced by quaternizing with dimethylbenzyl bromide. Various other compounds such as *N*-alkyl quaternary compounds were studied by Meakins [89] for increase in inhibition with increase in alkyl chain. Moreover, Hoar and Holiday [90] reported quinolines and the same quaternized with dodecylbenzyl chloride to inhibit corrosion of carbon steel in H_2SO_4 at 80 °C and found that the latter was far superior to quinolines alone. It was found by Riggs and Hurd [91] that the major pickling inhibitors of aromatic quaternary amine type such as dodecyl benzyl were able to remove mill scale in the HCl in lesser time and at lower temperatures than normally operational conditions and were efficient in preventing the corrosion of bare steel surface. Inhibitors such as 2, 6-dimethyl quinoline, β-naphthoquinoline, and aliphatic sulfides were studied for their effect on corrosion of iron in H_2SO_4 solutions. These absorbed inhibitors blocked the active sites of metal so that the dissolution did not occur as there might not be any change in mechanism of the reaction, because there was no change in the displacement of Tafel slopes.

The studies of anodic dissolution in the presence of some inhibitors such as aniline and its derivatives, the benzoate ion, and also halide ions indicated the participation of species from the solution either from the inhibitor or from the electrolyte. The dissolution proceeded via the alternative path through the formation of complex of the type $[Fe.\ I]_{ads}$ or $[Fe.\ OH.I]_{ads}$ with less rapidity than that via the $[Fe.\ OH]_{ads}$. This path could inhibit the anodic dissolution efficiently and could increase the Tafel slope. The adsorbed species may also accelerate the rate of anodic dissolution, as in the case of H_2S in the acid solutions that stimulate corrosion of iron through the formation of $[Fe.\ HS^-]_{ads}$ complex. This was reported in the case of some inhibitive S^- compounds such as TU [90], at lower concentrations since H_2S has been identified as the reduction product. In some cases, the absorbed corrosion inhibitors may react, usually by electrochemical reduction, to form a product, which may also be inhibitive. Corrosion inhibition due to added substance is termed *primary inhibition* and that due to the reaction product is termed *secondary inhibition*. In such cases, the efficiency changes with time, and it also depends on whether the secondary or primary inhibitor is more effective or not. Such examples for the secondary inhibition were reported earlier in 1969 by Trabanelli *et al.* [78] in the case of DBSO and DBS on corrosion of iron in 1 N HCl as discussed earlier. The results obtained by these workers [74, 75] from the experimental measurement have confirmed the hypothesis of secondary inhibiting action of DBSO. The sulfoxides were protonated first in an acid solution as shown in Eq. (5.9) and the cation thus formed was electrochemically reduced as shown in Eq. (5.10) at the electrode where DBS was formed. The low solubility of DBS caused its accumulation at the electrode in the form of multimolecular layers resulting in the high IE%. However, by using ^{35}S-labeled DBSO in the radiotracer technique Trabanelli has proved that only the first layer (DBS) contributed to the corrosion inhibition. Hence the sulfides, formed by secondary process, were more efficient inhibitors. Further study of DBSO on bare and iron-coated Ag in 0.5 M HCl was carried out extensively by Aramaki *et al.* in 1993 [73] and is attributed to strong chemisorption of DBS on the iron surface as neutral molecule but not as cation. Likewise, quaternary phosphonium and arsonium compounds [74, 75] could be reduced to phosphine or arsine compounds, respectively. Both the primary and secondary compounds have shown almost similar IE. On the contrary, TU was reduced first to HS^- ion [89] but later acted as stimulator of corrosion. TU has strong inhibition effect of nearly 98% at a lower concentration of about 10^{-3} M. As the concentration of TU or acid was increased, the IE was decreased, and even in extreme cases, the stimulation of corrosion or reversal of inhibition was observed due to the reduction of TU as stated earlier. It was interesting to note that IE was increased with increasing substitution of H atom of TU: larger the number of H atoms substituted greater was the bulkiness of the substituents and inhibition. The stimulation is associated with the formation of protonated SH^+ cation, which when adsorbed can serve as a bridge for a faster reduction of protons. Other authors considered the possibility of decomposition of TU into S^{-2} and HS^- products that stimulate the corrosion [92].

Elkadi *et al.* [93] investigated the inhibition effect of 3,6-bis[2-methoxyphenyl]-1, 2-dihydro-1,2,4,5-tetrazine (2-MDHT) on the corrosion of MS in acidic media by weight loss (WL) and various electrochemical techniques. Results obtained have revealed that this organic compound is a very good inhibitor. 2-MDHT was able to reduce the corrosion of steel more effectively in 1 M HCl than in 0.5 M H_2SO_4. Potentiodynamic polarization studies showed that 2-MDHT was a mixed-type inhibitor in 1 M HCl and cathodic-type inhibitor in 0.5 M H_2SO_4. Surface analyses were also carried out to establish the mechanism of corrosion inhibition of MS in acidic media. The adsorption of this inhibitor on the MS surface in both the acids obeyed the Langmuir absorption isotherm.

Avci [94] investigated corrosion inhibition of indole-3-acetic acid on MS in 0.5 M HCl containing the desired amount of inhibitor at different temperatures by using potentiodynamic polarization, EIS, and linear polarization resistance (LPR) measurements. The experimental results showed that E_{corr} shifted toward a more negative potential region in the presence of indole-3-acetic acid than that of blank solution. According to the results obtained from all the measurements, IE was about 77% with 1.7×10^{-3} M inhibitor concentration, increasing to about 93% at the 1×10^{-2} M inhibitor concentration. Potentiodynamic polarization measurements showed that the current at anodic and cathodic regions gave a smaller value in the presence of inhibitor at almost all potentials than that of the blank solution.

Tebbji *et al.* [95] have studied the inhibition effect of a new bipyrazole derivative namely *N*-benzyl-*N*,*N*-bis[(3,5-dimethyl-1*H*-pyrazol-1-yl)methyl]amine (BBPA) on the corrosion of steel in 1 M HCl which was studied at 308 K by WL measurements, Potentiodynamic polarization, LPR, and EIS methods. The results have shown that BBPA was a good inhibitor and that IE reached 87% at 5×10^{-4} M. The values of the IE calculated from these techniques were reasonably in good agreement. Polarization curves revealed that this organic compound acted as a mixed-type inhibitor.

Achary *et al.* [96] investigated the inhibition effects of 8-hydroxy quinoline (HQ) and 3-formyl-8-hydroxy quinoline (FQ) on corrosion of MS in HCl. It was studied through WL, LPR, and EIS techniques. The results indicated that the IE and extent of surface coverage were increased with increase in inhibitor concentration and decreased with increase in temperature and HCl concentration. The thermodynamic parameters were evaluated for corrosion inhibition process. The inhibitors follow Langmuir adsorption isotherm. The compound FQ showed more IE than that of HQ.

Bentiss *et al.* [97] synthesized and studied a new pyridazine derivative, namely 1,4-bis[2-pyridyl]-5H-pyridazino[4,5-b] indole (PPI) to act against MS corrosion in 1 M HCl solutions by WL and electrochemical techniques such as potentiodynamic polarization curves and EIS. The experimental results suggested that PPI was a good corrosion inhibitor and that the IE increased with the increase of PPI concentration, while the adsorption followed the Langmuir isotherm. The corrosion inhibition was because of the formation of a chemisorbed film on the steel surface.

Although many synthetic compounds have shown good anticorrosive activity, most of them are highly toxic to both human beings and environment. The safety and environmental issues of corrosion inhibitors that arise in industries have always been a global concern. Such inhibitors may cause reversible (temporary) or irreversible (permanent) damage to organ systems such as kidneys or liver, disturb a biochemical process, or disturb an enzyme system at some site in the body. The toxicity may manifest either during the synthesis of the compound or during its applications. Less-toxic corrosion inhibitors can be designed, if one has a reliable method of estimation of toxicity of these compounds before these are actually synthesized. Although the most effective and efficient organic inhibitors are compounds that have π bonds, the biological toxicity of these products, especially organic phosphates, is specifically documented with regard to their harmful environmental characteristics. From the standpoint of safety, at present, the development of nontoxic and effective inhibitors is considered more important and desirable, which are also called *eco-friendly* or *green corrosion* inhibitors.

5.7 Toxicity of Inhibitors

Toxicity is defined as the ability of a chemical to damage an organ system, such as the liver or kidneys, to disrupt a biochemical process, such as blood-forming mechanism, or to disturb an enzyme system at some site in the body [98, 99]. In simple words, toxicity is a property of a chemical that causes damage to the body of a living organism.

Most toxic effects are reversible and do not cause permanent damage, but complete recovery may require a long time. However, some poisons can cause irreversible (permanent) damage. Poisons can have serious effect on just one particular organ system or they may produce generalized toxicity by affecting a number of systems. The respiratory system includes body parts such as the nose, trachea, and lungs. When toxins are present in this system, the organism is likely to experience irritation, coughing, chest pain, and choking. In the case of gastrointestinal organs such as the stomach and intestines, the common syndromes are nausea, vomiting, and diarrhea. Renal organs are kidneys, and the victim will suffer from back pain, difficulty to urinate, and abnormality when urinating. The amount of urine can be dramatically less or more than usual. When the brain and spinal cord are affected by toxins, the victim feels dizzy and is likely to suffer from headache. One will have depression and confusion and has a possibility to progress to coma.

The blood-forming and related system is known as the *hematological system*, and when affected by poison, the production of blood is greatly reduced and altered. Anemia is one of the most visible syndromes. Serious effect may lead to leukemia. The dermatological system covers the skin and eyes, which are prone to irritation, redness, swelling, and itching. Very toxic chemicals such as chromate induce critical damage to the reproductive system. Women often experience miscarriage and abnormal fetus development.

Most of the inorganic/organic inhibitors used are toxic substances. They can be classified as

- irritants;
- asphyxiants;
- anesthetics and narcotics;
- systemic poisons;
- sensitizers;
- carcinogens;
- mutagens;
- teratogens.

5.7.1 **Irritants**

Irritants are substances which have the capacity to cause reddening, inflammation, or chemical burns to the eyes, skin, nose, throat, lungs, and other tissues of the body. Some chemicals such as concentrated acids are irritant and corrosive. They lead to second- and third-degree chemical burns when they come in contact with eyes or skin. Besides acids, some vapors or fumes when inhaled, will lead to severe lung injury. If these chemicals are ingested, the person can have severe damage to the mouth, throat, stomach, and intestinal tract. If inhaled as a gas, vapor, fume, mist, or dust, they may cause severe lung injury, and if ingested, they can seriously damage the mouth, throat, stomach, and/or intestinal tract.

Irritants when exposed to skin have the ability to dissolve natural oils in the skin, causing dermatitis and turning it dry. The dry skin will then crack at some point and will be inflamed, and possible infection may result. The same chemical that is causing irritation to the skin may as well cause irritation of the eyes. If it comes into contact with the membrane that protects the surface of the cornea, which is recognized as epithelium, there is a possibility that one may suffer loss of vision.

Irritation can also affect the lungs and cause pneumonitis and pulmonary edema. Severe consequences may lead to bleeding (hemorrhage) and the death of living tissue (tissue necrosis).

5.7.2 Asphyxiants

An asphyxiant is a gas that can be damaging when it is inhaled in large concentrations. A not very harmful type of asphyxiant is nitrogen in the air. Air consists of only 78% of nitrogen, and it is harmless at this composition. But if there is more than this percentage of nitrogen or any other gas in the air, it will suppress the oxygen content in the air, which is necessary to uphold consciousness and life. A person may suffer from difficulty in breathing because there is insufficient oxygen. Some heavier gases that decompose from chemical reactions sink in the air. At low-lying areas, people will experience difficulty to breathe as the concentration of oxygen decreases.

Chemical asphyxiants are substances that, when once inhaled, prevent the body from consuming the oxygen that has been taken in. A common type of chemical asphyxiant is carbon monoxide emitted from the exhaust of automobiles. Carbon monoxide combines with hemoglobin in the blood and prevents the transport of oxygen from the lungs to other organs of the body. When the lack of oxygen becomes too high, a person may become unconscious and die.

There are four stages of asphyxiation. In the first stage, oxygen is reduced to between 19.5 and 20.9% by volume, with that the pulse and respiration rate increase. This will disrupt the coordination of the muscles. Oxygen is further suppressed to 12–19.5% in the second stage. Rapid fatigue occurs, and a person becomes insensitive to pain with faulty judgment. At the third stage, there is only 6–10% oxygen available; the signs of illness include nausea and vomiting. This is followed by collapse, and the patient will suffer permanent brain damage. At the final stage, the volume of oxygen falls below 6%. At this moment, the patient goes into convulsion and stops breathing, which results in death.

5.7.3 Anesthetics and Narcotics

Some toxic substances have the ability to depress the central nervous system, preventing one from feeling any pain. The first signs on exposure are dizziness, drowsiness, feeling weak, and thus fatigue and gradual loss of coordination. Serious situation will cause one to become unconscious, the respiratory system to be paralyzed, and death.

5.7.4 Systemic Poisons

- Hepatotoxic agents: they are substances that damage liver.
- Nephrotoxic agents: they are substances that damage kidney.
- Neurotoxic agents: they are substances that affect the nervous system and possibly cause neurological damage.

- Hematopoietic agents: they are substances that affect the blood and the blood-forming tissues.
- Agents that damage the lungs or respiratory system.

5.7.5 Sensitizers

There are also some chemicals that will not affect the victims when they are first exposed to them. But when they are exposed again, the victims will experience significant and harmful effects even if the exposed concentration is very low. The victims then become very allergic to the contaminants and those of similar characteristics.

5.7.6 Carcinogens

They are substances that induce cancer within some part of the body. All cancers are characterized by an abnormal and unregulated growth of cells. This growth has a negative effect of destroying the surrounding body tissues and may spread to other parts of the body in a process known as *metastasis*. There are different kinds of cancer, namely skin cancer, lung cancer, brain cancer, breast cancer, colon cancer, leukemia, prostate cancer, ovarian cancer, lymphoma, and others. Apparently, there is no limit to where the cancer may develop or what age is more susceptible to cancer.

Cancer is not contagious, and it is often caused by the damaged genes in a single cell. Cells that are affected by cancer are known as *malignant cells*. These cells have a division rate that is significantly higher than that of other normal cells in the body. Carcinogenic activity is evidenced by the increase of malignant neoplasm, the increase of a combination of malignant and benign neoplasm, and the increase of benign neoplasm if these tumors have the ability to proceed to malignancy.

5.7.7 Mutagens

A mutagen is a substance or agent that causes an increase in the rate of change in genes (subsections of the DNA of the body's cells). These mutations can be passed along as the cell reproduces, sometimes leading to defective cells or cancer.

5.7.8 Teratogens

A substance may induce the risk of nonhereditable birth defects in offspring, if it comes into contact with skin for prolonged and repeated periods. This tends to result in abnormal developments of sperm, eggs, and/or fetal tissue.

These toxic effects have led to the use of natural products such as anticorrosion agents, which are ecofriendly and harmless. In recent days, many alternative eco-friendly corrosion inhibitors have been studied and developed; they range from rare earth elements to organic compounds.

On looking at the nature, one can find very rich resources for substances with wide varieties of chemical structures. So, why rush to synthesize injurious chemicals while the nature around us is full of the safest ones? Plants are great chemical factories that can supply us with the chemicals required to inhibit the corrosion process. Most of the naturally occurring substances are safe and can be extracted by simple and cheap procedures. Recent literature has many researches that have tested different extracts for corrosion-inhibition applications. The examples are numerous, such as henna, olive, shirsh zallouh, vanillin, natural honey, khella, onion, ficus, opuntia, many oils extracted from different parts of different plants, and many others. Many of these naturally occurring substances have proved their ability to act as corrosion inhibitors for the corrosion of different metals and alloys in different aggressive media.

Herbal extracts are an economical, environment-friendly, and highly effective alternative to many of the current inhibitors, which are toxic, polluting, and relatively very expensive. Herbal extracts contain large number of compounds, and it is significant that out of this huge number of compounds available, there are relatively high concentrations of alkaloids, flavonoids, and so on. Even the presence of tannins, cellulose, and polycyclic compounds normally enhances the film formation over the metal surface, thus preventing the metal from corrosion. In other words, these compounds contain active groups that should render them electrochemically active, that is, they may react with the metal surface and inhibit metallic corrosion.

Recent trends of using different herbal extracts as corrosion inhibitors is attributed to their biodegradability, less/nontoxic nature, high solubility, and stability in the acidic solutions. Natural products such as lignin and tannin, cinchona alkaloids, pomegranate alkaloids, eucalyptus leaves, Mahasudarshan churna, *Swertia angustifolia, Emblica officinalis, Terminalia belerica, Terminalia chebula, Sapindus trifolianus, Acacia conciana, Calotropis giganta* latex, *Acacia arabica, Eugenia jambolans,* and *Lawsonia inermis* have been evaluated as very effective acid corrosion inhibitors in standard as well as stringent conditions. The study of these green inhibitors has been even extended in recent times to the cooling water systems as reported by Farooqi and Quraishi [100]. The aqueous extracts of *Azadirachta indica* (neem), *Eucalyptus,* and *Cordia latifolia* were reported as inhibitors for corrosion and scale formation in the cooling waters. The performance of these inhibitors was not affected by the presence of hardness-causing Ca^{2+} or biocidal Cl^- ions, and it was reported that these plant extracts were safe for aquatic life. The environmental parameters such as chemical oxygen demand (COD), biological oxygen demand (BOD), and so on, were reported to be within permissible limits. Earlier, studies on corrosion inhibitor was also proposed by Agarwaala [101] in the case of macromolecules such as porphyrins and their compounds (in which redox groups have been introduced), which could form stable octahedral complexes with Fe[III]. It was proposed that the donor "N" atoms in these compounds can make bond with Fe^{2+} ions.

In 1930, plant extracts (dried stems, leaves, and seeds) of celandine (*Chelidonium majus*) and other plants were used in H_2SO_4 pickling baths. An additive, ZH-1 consisting of finely divided oil cake, a by-product formed in the phytin manufacture was developed for the control of corrosion. Animal proteins (by products of meat and milk industries) were also used for retarding acid corrosion. The additives used in acid included flour, bran, yeast, a mixture of molasses and vegetable oil, starch, and hydrocarbons (tars and oils). "Antra" made by sulfonating anthracene or anthracene oil and "TM" consisting of heavy oils obtained in the fractionation of coal tar were used in Russia [102]. The first patented corrosion inhibitors used were either natural products such as flour, yeast, and so on [103] or by-products of food industries for restraining iron corrosion in acid media [104].

Saleh *et al.* [105] reported that Opuntia extract, *Aloe vera* leaves, and orange and mango peels give adequate protection to steel in 5 and 10% HCl at 25 and 40 °C. Srivatsava and Sanyal studied the performance of caffeine [106] and nicotine [107] in the inhibition of steel corrosion in neutral media. Khamis and Al-Andis [108] have proved the use of herbs (such as coriander, hibiscus, anis, black cumin, and garden cress) as new types of green inhibitors for acidic corrosion of steel.

The extracts of chamomile, halfabar, black cumin, and kidney bean were analyzed for their inhibitive action of corrosion of steel in acid media by Abdel-Gaber *et al.* [109]. El-Hosary *et al.* [110] studied the corrosion inhibition of aluminum and zinc in HCl using *Hibiscus sabdariffa* extract. Srivatsava and Srivatsava [111] found that tobacco, black pepper, castor oil seeds, acacia gum, and lignin can behave as good inhibitors for steel in acid medium.

Muller [112] investigated the effect of saccharides (reducing sugars – fructose and mannose) on the corrosion of aluminum and zinc in alkaline media. Hammouti *et al.* studied the extracts of ginger [113], jojoba oil [114], eugenol, acetyl eugenol [115], artemisia oil [116, 117], and *Mentha pulegium* [118] for corrosion inhibition of steel in acid media. Ethanolic extract of *Ricinus communis* leaves was studied for the corrosion inhibition of MS in acid media by Ananda *et al.* [119]. Aqueous extracts of hibiscus flower and Agaricus have been studied as corrosion inhibitors for industrial cooling system by Minhaj *et al.* [120].

Zucchi and Omar [121] have found that *Papaia, Poinciana pulcherrima, Cassia occidentalis,* and *Datura stramonmium* seeds, *Calotropis procera, Azadirachta indica,* and *Auforpio turkiale* sap are useful as acid corrosion inhibitors.

Cabrera *et al.* found that molasses treated in alkali solution inhibit the corrosion of steel in HCl used in acid cleaning [122]. El-Etre [123] has studied the application of natural honey as corrosion inhibitor for copper in aqueous solution. Similar study has also been conducted on carbon steel [124]. Parikh and Joshi [125] studied the anticorrosion activity of onion, garlic, and bitter gourd for MS in HCl media.

The application of extracts of henna, thyme, bgugaine, and inriine was investigated for their anticorrosion activity [126–129]. The effect of addition of bgugaine on steel corrosion in HCl is patented [130]. Saleh *et al.* studied the peel of pomegranate

[131] and beetroot [132] as corrosion inhibitor for MS in acid media. The anticorrosion effects of *Andrographis paniculata* [133] and tea wastes [134] have also been reported. Kliskic *et al.* analyzed aqueous extract of *Rosmarinus officinalis* [135] as corrosion inhibitor for aluminum alloy corrosion in chloride solution.

Sanghvi *et al.* investigated the anticorrosion activity of *Emblica officinalis, T. chebula, Terminalia belivia* [136], *S. trifolianus,* and *Accacia conicianna* [137]. Corrosion inhibition has also been studied for the extracts of *Malachra capitata* [138], *L. inermis* [139], *E. jambolans* [140], *Pongamia glabra, Annona squamosa* [141], *Accacia arabica* [142], *Carica papaya* [143], *A. indica* [144], and *Vernonia amygdalina* [145] for steel in acid media.

Avwiri *et al.* studied the inhibitive action of *V. amygdalina* on the corrosion of aluminum alloys in HCl and HNO_3 [146]. The inhibition effect of *Zenthoxylum alatum* extract on the corrosion of MS in dilute HCl was investigated by Chauhan and Gunasekaran [147]. *Nypa fruticans* Wurmb [148] leaves were studied for the corrosion inhibition of MS in HCl media. Guar gum was analyzed for its anticorrosion activity by Abdallah [149]. Martinez and Stern have studied the inhibitory mechanism of low-carbon steel corrosion of Mimosa tannin in H_2SO_4 media [150]. Oguzie investigated the efficiency of *Telforia occidentalis* extract as corrosion inhibitor in both HCl and H_2SO_4 media [151].

Patel *et al.* also evaluated the extract of *T. chebula* fruits and found out the extract as an efficient corrosion inhibitor for MS in H_2SO_4 medium [152]. Oguzie studied the inhibitive effect of *Occium viridis* extract [153] on the acid corrosion of MS and *Sansevieria trifasciata* extract [154] on the acid and alkaline corrosion of aluminum alloy. El-Etre *et al.* investigated Khillah extract [155] for the corrosion inhibition of SX 316 steel in acid media, Lawsonia extract [156] was studied for its effect against acid-induced corrosion of metals, Opuntia extract [157] was investigated for the corrosion of aluminum in acid medium and vanillin [158] for the corrosion of MS in acid media. Berberine, an alkaloid isolated from Captis, was studied for its anticorrosion effect for MS corrosion in H_2SO_4 medium [159] by Li *et al.*

The corrosion inhibition activity in many of these plant extracts could be due to the presence of heterocyclic constituents such as alkaloids, flavonoids, and so on. Even the presence of tannins, cellulose, and polycyclic compounds normally enhances the film formation over the metal surface, thus aiding corrosion protection [160].

Verma and Mehta [161] studied the inhibiting influence of acid extract of *C. giganta* latex on the corrosion of MS in 0.1 N H_2SO_4 solutions by DC polarization techniques using potentiogalvanoscan. From the changes in the value of LPR and corrosion current density in the presence of the additives, it was concluded that the extract of *C. giganta* latex inhibited acid corrosion but not much effectively. The inhibition was lowered with gradual decrease in concentration of the acid extract of the latex, and the additive was found to be a mixed-type inhibitor.

Chauhan and Gunasekaran [162] studied the inhibition effect of *Z. alatum* plant extract on the corrosion of MS in 20, 50, and 88% H_3PO_4 by WL and EIS methods. The extract was able to reduce the corrosion of steel more effectively in 88% H_3PO_4 than in 20% H_3PO_4. The effect of temperature on the corrosion behavior of MS in 20, 50, and 88% H_3PO_4 with addition of the plant extract was studied in the temperature range 50–80 °C. Results on corrosion rate and IE have indicated that this extract was effective up to 70 °C in 88% H_3PO_4 medium.

Raja and Sethuraman [163] investigated the corrosion inhibitive effect of the extract of black pepper on MS in 1 M H_2SO_4 media by conventional WL studies (at 303–323 K), electrochemical TI and EIS, and by scanning electron microscopic studies. The results of WL study revealed that black pepper extract acts as a good inhibitor even at high temperatures. The inhibition was through adsorption, which followed Temkin adsorption isotherm. The data of TI method revealed the mixed mode of inhibition of black pepper extract.

De Souza and Spinelli [164] studied the inhibitor effect of the naturally occurring biological molecule caffeic acid on the corrosion of MS in 0.1 M H_2SO_4 by WL, potentiodynamic polarization, EIS, and Raman spectroscopy. The results of these different techniques confirmed the adsorption of caffeic acid onto the MS surface and, consequently, the inhibition of the corrosion process. Caffeic acid acted by decreasing the available cathodic reaction area and by modifying the activation energy of the anodic reaction.

Okafor *et al.* [165] evaluated the inhibitive action of leaves, seeds, and a combination of leaves and seeds extracts of *Phyllanthus amarus* on MS corrosion in HCl and H_2SO_4 solutions using WL and gasometric techniques. The results indicate that the extracts functioned as a good inhibitor in both environments and that IE increased with extracts' concentration. Temperature studies revealed an increase in IE with rise in temperature, and activation energies decreased in the presence of the extract. A mechanism of chemical adsorption of the plants components on the surface of the metal is proposed for the inhibition behavior. The adsorption characteristics of the inhibitor were approximated by Temkin isotherm.

Eddy [166] studied the inhibitive and adsorption properties of ethanol extract of *Colocasia esculenta* for the corrosion of MS in H_2SO_4 using WL, hydrogen evolution, and infrared methods of monitoring corrosion. The results obtained indicates that ethanol extract of *C. esculenta* is a good inhibitor for the corrosion of MS in H_2SO_4 and that its inhibitive action is attributed to its phytochemical constituents that aided its adsorption on the surface of MS. Calculated values of activation energy and IE at 303 and 333 K revealed that the mechanism of adsorption of ethanol extract of *C. esculenta* on MS surface is physical adsorption. Also, the adsorption of the inhibitor on MS surface was found to be spontaneous, endothermic, and consistent with the assumptions of Langmuir adsorption isotherms.

Cheng *et al.* [167] studied the inhibiting influence of carboxymethyl-chitosan (CM-chitosan) on the corrosion of MS in 1 M HCl solution by WL, potentiodynamic polarization, and EIS methods. Polarization measurements show that the CM-chitosan acts essentially as a mixed-type inhibitor. The protection efficiency of this inhibitor increases with the inhibitor concentration to reach 93% at 200 mg L^{-1} but decreases slightly with the rise of temperature. The adsorption of used compound on the steel surface obeys modified Langmuir's isotherm. The efficient inhibition is also characterized by a series of greater activation energies of corrosion reaction in the presence of CM-chitosan at various concentrations.

The drawback of the extensive applications of such green inhibitors may be attributed to the less-enough availability of supporting literature for the inhibition mechanisms involved as they have been recently studied. The details of preparation of the herbal acid extracts, the nature of inhibitor species that are actually effective in decreasing the corrosion, and also, the surface analysis of films/complexes that are formed on the metal surface have not been exclusively reported. Most of such green inhibitors were studied in the laboratory conditions. Such systems can be applied to the practical purpose with ample modifications according to the environment and the need. The near future may witness these inhibitors as better alternates to toxic chemical inhibitors, not only because of the ever-increasing research work going on in this field from various researchers but also because of their ecofriendliness and better stability [92].

It is certain that natural compounds will emerge out as effective inhibitors of corrosion in the coming years because of their biodegradability, easy availability, nontoxic nature, and cheaper cost. Thorough investigation of the literature clearly reveals that the era of green inhibitors has already begun.

Corrosion inhibitors are chemical compounds added to the corrosive medium to reduce the rate of its attack on the metal or alloy. The chemicals that can act as corrosion inhibitors may be inorganic or organic. The inorganic compounds such as chromates inhibit the corrosion process via formation of passive oxide film on the metal surface and thus prevent the corrosive medium from attacking the bar metal. On the other hand, the organic compounds adsorb on to the metal surface forming a barrier between the metal and the corrosive environment. Some structural features of the organic compounds help them to do so. These include the presence of oxygen, nitrogen, or sulfur atoms as well as the presence of double bonds. The lone pair electrons of the mentioned atoms facilitate the adsorption process. Some criteria should be considered when making a choice of chemical compounds for inhibition of corrosion. Inhibition of metallic corrosion is mainly an economical process. Therefore, the first criterion that must be fulfilled by the inhibitors used is their price. The other very important criterion that should be considered when dealing with a corrosion inhibitor is its effect on the humans and environment. Unfortunately, most of the effective corrosion inhibitors are synthetic chemicals with high cost. At the same time, the use of such synthetic compounds can cause harm to humans and environment. On looking around, we'll find very rich resources for substances with wide varieties of chemical structures. So, why do we rush to synthesize harmful chemical while the nature around us is full of the safest ones? Plants are the great chemical factories that can supply us with the chemicals required to inhibit the corrosion process. Most of the naturally occurring substances are safe and can be extracted by simple and cheap procedures. The recent literature is full of researches that test different extracts for corrosion inhibition applications. The examples are numerous, such as henna, olive, shirsh zallouh, vanillin, natural honey, khella, onion, ficus, opuntia, many oils extracted from different parts of different plants, and many others. Many of these naturally occurring substances have proved their ability to act as corrosion inhibitors for different metals and alloys in different aggressive media.

References

1. Uhlig, H.H. and Revie, R.W. (1985) *Corrosion and Corrosion Control*, John Wiley & Sons, Inc., New York.
2. Trabanelli, G. (1991) Inhibitors. An old remedy for a new challenge. *Corrosion*, **47** (6), 410–419.
3. Faraday, M. (1844) *Experimental Researches in Electricity*, vol. II, University of London.
4. Evans, U.R. (1937) *Metallic Corrosion, Passivity and Protection*, Edward Arnold and Company, London.
5. Hedges, E.S. (1937) *Protective Films on Metals*, D Van Nostrand Co., New York.
6. Mears, R.B. (1940) A unified mechanism of passivity and inhibition. *Trans. Electrochem. Soc.*, **77**, 288.
7. Glasstone, S. (1940) *Textbook of Physical Chemistry*, D Van Nostrand Company, New York.
8. Evans, U.R. (1948) *Metallic Corrosion, Passivity and Protection*, Edward Arnold & Company, London.
9. Hoar, T.P. and Evans, U.R. (1932) The passivity of metals. Part VII. The specific function of chromates. *J. Chem. Soc.*, 2476–2481.
10. Uhlig, H. and King, P. (1959) The flade potential of iron passivated by various inorganic corrosion inhibitors. *J. Electrochem. Soc.*, **106** (1), 1–7.
11. Powers, R. and Hackerman, N. (1953) Surface reactions of steel in dilute $Cr^{51}O_4{}^{2-}$ solutions: applications to passivity. *J. Electrochem. Soc.*, **100** (7), 314–319.
12. Cohen, M. and Beck, A. (1958) The passivity of iron in chromate solution I. Structure and composition of the film. *Z. Electrochem.*, **62**, 696–702.
13. Simnad, M. (1953) "Radioisotopes in the study of metal. Surface reactions in solutions," in properties of metal surfaces. *J. Inst. Metals Monogr.*, **13** 23.
14. Brasher, D. and Stove, E. (1952) Nature of the passive film on iron in concentrate nitric acid. *Chem. Ind.*, **8**, 171.
15. Mayne, J.E.O. and Pryor, M.J. (1949) The mechanism of inhibition of corrosion of iron by chromic acid and potassium chromate. *J. Chem. Soc.*, 1831.
16. Graham, M.J. and Cohen, M. (1976) Analysis of iron corrosion products using moessbauer spectroscopy. *Corrosion*, **32** (11), 432–438.
17. Lumsden, J.B. and Szklarska-Smialowska, Z. (1978) Properties of films formed on iron exposed to inhibitive solutions. *Corrosion*, **34** (5), 169–176.
18. Langmuir, I. (1916) Adsorption of Gases by Solids. *J. Am. Chem. Soc.*, **38** (8) 2267–2273; Langmuir, I. (1916) Relation between contact potentials and electrochemical action. *Trans. Electrochem. Soc.*, **29**, 125–129.
19. Tammann, G. (1919) The chemical and galvanic properties of alloy states and their atomic configurations. *Z. Anorg. Chem.*, **107**, 155–156.
20. Russell, A. S. (1925) Passivity of iron and other metals. *Nature*, **115**, 455–456; Russell, A. S.. (1926), Passivity, catalytic action, and other phenomena. *Nature*, **117**, 47–48.
21. Swinne, R. (1925) Das periodischen system der chemischen elementen in laufe des atombaus. *Z. Electrochem.*, **31**, 422.
22. Sborgi, U. (1925) An electronic interpretation is suggested for various anode phenomena. *Atti Accad. Lincei*, **6**, 388; Sborgi, U. (1926) Umberto sul comportaniento anodico del niobio. *Gazz.*, **56**, 532.
23. Uhlig, H.H. and Wulff, J. (1939) The nature of passivity in stainless steels and other alloys. *Trans. Am. Inst. Mining Met. Engs.*, **135**, 494–534.
24. Uhlig, H.H. (1944) Passivity in Copper-Nickel and Molybdenum-Nickel-Iron alloys. *Trans. Electrochem. Soc.*, **85**, 307.
25. Kabanov, B., Burstein, R., and Frumkin, A. (1947) Kinetics of electrode processes on the iron electrode. *Discuss. Faraday Soc.*, **1**, 259–269.
26. Frankenthal, R. (1967) On the passivity of iron-chromium alloys. *J. Electrochem. Soc.*, **114** (6), 542–547.

27. Gohr, H. and Langi, E. (1957) Interpretation of passivity, and especially of the flade standard potential of iron. *Z. Electrochem.*, **61**, 1291.
28. Uhlig, H.H. (1958) The solubilities of gases and surface tension. *Z. Electrochem.*, **62**, 626.
29. Uhlig, H. and Connor, T.O. (1955) Nature of the passive film on iron in concentrated nitric acid. *J. Electrochem. Soc.*, **102** (10), 562–572.
30. Uhlig, H. and Lord, S. (1953) Amount of oxygen on the surface of passive stainless steel. *J. Electrochem. Soc.*, **100** (5), 216–221.
31. Flade, F. (1911) Contributions to the knowledge of passivity. *Z. Phys. Chem.*, **76**, 513–546.
32. MacRae, A. (1964) Adsorption of oxygen on the {111}, {100} and {110} surfaces of clean nickel. *Surface Sci.*, **1** (4), 319–348.
33. Okamato, G. and Shibata, T. (1965) Desorption of tritiated bound-water from the passive film formed on stainless steels. *Nature*, **206**, 1350–1357.
34. Kudo, K., Shibata, T., Okamoto, G., and Sato, N. (1968) Ellipsometric and radiotracer measurements of the passive oxide film on Fe in neutral solution *Corros. Sci.*, **8** (11), 809–814.
35. Uhlig, H.H. (1967) Structure and growth of thin films on metals exposed to oxygen. *Corros. Sci.*, **7** (6), 325–339.
36. Mehta, G.N. (1979) Surat Electrochemical techniques for study of passivity and evaluation of inhibitor. PhD Thesis. South Gujarat University.
37. Lorenz, W.J. and Mansfeld, F. (1983) Interface and interphase inhibition. International Conference on Corrosion Inhibition, NACE, Texas.
38. Fontana, M.G. (1988) *Corrosion Engineering*, 3rd edn, TATA McGraw-Hill Publication, New Delhi.
39. Schmitt, G. (1984) Application of inhibitors for acid media. *Br. Corros. J.*, **19**, 165–176.
40. Harsch, P., Hare, J., Robertson, J.B., and Sutherland, S.M. (1961) An experimental survey of rust preventives in water II- The screening of organic inhibitors. *J. Appl. Chem.*, **11** (7), 251–265.
41. Trabanelli, G. and Carassiti, V. (1970) Mechanism and phenomenology of organic inhibitors in *Advances in Corrosion Science and Technology*, 1st edn, (eds M.G. Fontana and R.W. Staehle), Plenum Press, New York, London, pp. 149–190.
42. Desai, M.N., Desai, S.M., Gandhi, M.H., and Shah, C.B. (1971) Corrosion inhibitors for aluminium and aluminium-based alloys – Part I. *Anti-Corros. Methods Mater.*, **18** (4), 8–13.
43. Desai, M.N., Rana, S.S., and Gandhi, M.H. (1971) Corrosion inhibitors for copper. *Anti-Corros. Methods Mater.*, **18** (2), 19–23.
44. Rama Char, T.L. and Padma, D.K. (1969) Corrosion inhibitors in industry review of iterature of inhibitors minimizing corrosion in different environments from 1965–1968. *Trans Inst. Chem. Eng.*, **47**, 177–182.
45. Walker, R. (1970) The use of benzotriazole as a corrosion inhibitor for copper. *Anti-Corros. Methods Mater.*, **17** (9), 9–15.
46. Huynh, N., Bottle, S.E., Notoya, T., and Schweinsberg, D.P. (2002) Inhibition of copper corrosion by coatings of alkyl esters of carboxybenzotriazole. *Corros. Sci.*, **44** (11), 2583–2596.
47. Villamil, R.F.V., Corio, P., Rubim, J.C., and Agostinho, S.M.L. (2002) Sodium dodecylsulfate-benzotriazole synergistic effect as an inhibitor of processes on copper | chloridric acid interfaces. *J. Electroanal. Chem.*, **535** (1–2), 75–83.
48. Clubley, B.G. (ed.) (1990) *Chemical Inhibitors for Corrosion Control*, Royal Society of Chemistry, Cambridge.
49. Szklarska-Smialowska, Z. (1978) *Passivity of Metals*, Princeton, NJ.
50. Thierry, D. and Leygraf, C. (1985) Simultaneous Raman spectroscopy and electrochemical studies of corrosion inhibiting molecules on copper. *J. Electrochem. Soc.*, **132** (5), 1009–1014.
51. Bockris, J.O.M., Habib, M.A., and Carajal, J.L. (1984) Adsorption of thiourea on passivated iron. *J. Electrochem. Soc.*, **131**, 3032.
52. Augustynki, J., Balsanc, L., and Hinden, J. (1978) X-ray photoelectron

spectroscopic studies of RuO_2-based film electrodes. *J. Electrochem. Soc.*, **125** (7), 1093–1097.
53. Putilova, I.N., Balezin, S.A., and Barannik, V.P. (1960) *Metallic Corrosion Inhibitors*, Pergamon Press, London.
54. Trabanelli, G. and Carassiti, V. (1970) *Advances in Corrosion Science and Technology-1*, Plenum Press, New York, London.
55. Rozenfeld, I.L. (1981) *Corrosion Inhibitors*, McGraw-Hill, New York.
56. Eldakar, N. and Nobe, K. (1981) Effect of tolyltriazole on iron corrosion and the hydrogen evolution reaction in H//2SO//4. *Corrosion*, **32** (6), 238–242.
57. Bonora, P.L., Bolognesi, G.P., Borea, P.A., Zucchini, G.L., and Brunoro, F. (1971) *III European Symposium on Corrosion Inhibitors*, University of Ferrara, Ferrara, p. 685.
58. Mansfeld, F., Smith, T., and Parry, E.P. (1971) Benzotriazole as corrosion inhibitor for copper. *Corrosion*, **28** (7), 289–294.
59. Aramaki, K. (1987) Adsorption and corrosion inhibition effect of anions plus an organic cation on iron in 1*M* $HClO_4$ and the HSAB principle. *J. Electrochem. Soc.*, **134** (8), 1896–1901.
60. Ohsawa, M. and Sue Taka, W. (1979) Spectro-electrochemical studies of the corrosion inhibition of copper by mercaptobenzothiazole. *Corros. Sci.*, **19** (10), 709–722.
61. Penninger, J., Wippermann, K., and Schultze, J.W. (1987) Molecular structure and efficiency of triazole derivatives and other heterocyclics as corrosion inhibitors for copper. *Werkst. Korros.*, **38** (11), 649–659.
62. Schultze, J.W. and Wippermann, K. (1987) Inhibition of electrode processes on copper by AHT in acid solutions. *Electrochim. Acta*, **32** (5), 823–831.
63. El-Rahman, H.A. (1991) Evaluation of AHT as corrosion inhibitor for α-brass in acid chloride solutions. *Corrosion*, **47** (6), 424–428.
64. Singh, M.M., Rastogi, R.B., and Upadhyay, B.N. (1994) Inhibition of copper corrosion in aqueous sodium chloride solution by various forms of the piperidine moiety. *Corrosion*, **50** (8), 620–625.
65. Saleh, R.M. and Shams El Din, A.M. (1981) Efficiency of organic acids and their anions in retarding the dissolution of aluminium. *Corros. Sci.*, **12** (9), 689–697.
66. Tadros, A.B. and Abd-el-Nabey, B.A. (1988) Inhibition of the acid corrosion of steel by 4-amino-3-hydrazino-5-thio-1,2,4-triazoles. *J. Electroanal. Chem.*, **246** (2), 433–439.
67. Thomas, J.G.N. (1980) Some new fundamentals aspects in corrosion inhibition. Proceedings, 5th European Symposium on Corrosion Inhibitors, Ferrara, Italy, p. 453.
68. Donnelly, B., Downic, T.C., Grzeskowiak, R., Hambourg, H.R., and Shori, D. (1977) The effect of electronic delocalization in organic groups *R* in substituted thiocarbamoyl R–CS–NH_2 and related compounds on inhibition efficiency. *Corros. Sci.*, **18** (2), 109–116.
69. Kaesche, H. and Hackermann, N. (1958) Corrosion inhibition by organic amines. *J. Electrochem. Soc.*, **105** (4), 191–198.
70. Antrapov, L.I. (1967) A correlation between kinetics of corrosion and the mechanism of inhibition by organic compounds. *Corros. Sci.*, **7** (9), 607–620.
71. Trabanelli, G., Zucchi, F., and Carassiti V. (1966) Annali dell Universita di Ferrara (N.S.), *2nd European Symposium on Corrosion Inhibitors, Sezione V*, 4, p. 147.
72. Felloni, L. and Cozzi, A. (1966) Annali dell Universita di Ferrara (N.S.), *2nd European Symposium on Corrosion Inhibitors, Sezione V*, Suppl. 4, p. 253.
73. Hackermann, N. (1962) Zero point of charge of solid metals in contact with electrolyte solution. *Science*, **138**, 988–992.
74. Ertelk, H. and Horner, L. (1966) European Symposium on Corrosion Inhibitors, Annali dell Universita di Ferrara (N.S.) Sezione V, Suppl. 4, p. 71.
75. Lorenz, J.W. and Fischer, H. (1966) European Symposium on Corrosion

Inhibitors, Annali dell Universita di Ferrara (N.S.) Sezione V, Suppl. 4, p. 81.

76. Jofa, Z.A. (1966) European Symposium on Corrosion Inhibitors, Annali dell Universita di Ferrara, Suppl. 4, p. 93.

77. Cavallaro, L., Felloni, L., Trabanelli, G., and Pulidori, F. (1964) The anodic dissolution of iron and the behaviour of some corrosion inhibitors investigated by the potentiodynamic method. *Electrochim. Acta*, **9** (5), 485–494.

78. Trabanelli, G., Zucchini, G.L., Zucchi, F., and Carassiti, V. (1969) Inhibition of copper corrosion in chloride. *Br. Corros. J.*, **4**, 267–270.

79. Aramaki, K., Ohno, N., and Uehara, J. (1993) A SERS study on adsorption of dibenzyl disulfide, sulfide, and sulfoxide on an iron-deposited silver electrode in a hydrochloric acid solution. *J. Electrochem. Soc.*, **140** (9), 2512–2519.

80. Riggs, O.L. and Every, R.L. (1962) Study of organic inhibitors hydrochloric acid attack on iron. *Corrosion*, **18**, 262.

81. Eldredge, G.G. and Warner, J.C. (1948) *Inhibitors and Passivators-The Corrosion Handbook*, John Wiley & Sons, Inc., New York.

82. Douty, A. (1953) *Metal Industry*, Pergamon Press, London.

83. Brooke, M. (1962) *Chemical Engineering*, Pergamon Press, London.

84. Hackerman, N., Hurd, R.M., and Armand, R.R. (1962) Some structural effects of organic N-containing compounds on corrosion inhibition. *Corrosion – NACE*, **18**, 37t.

85. Anand, R.R., Hurd, R.M., and Hackerman, N. (1965) Adsorption of monomeric and polymeric amino corrosion inhibitors on steel. *J. Electrochem. Soc.*, **112** (2), 138–144.

86. Aramaki, K. and Hackerman, N. (1969) Inhibition mechanism of medium-sized polymethyleneimine. *J. Electrochem. Soc.*, **116**, 568.

87. Okamoto, G., Nagayama, M., Kato, J., and Baba, T. (1961) Effect of organic inhibitors on the polarization characteristics of mild steel in acid solution. *Corros. Sci.*, **2** (1), 21–27.

88. Chin, R.J. and Nobe, K. (1971) Electrochemical characteristics of iron in H_2SO_4 containing benzotriazole. *J. Electrochem. Soc.*, **118** (4), 545–548.

89. Meakins, R.J. (1963) Alkyl quaternary ammonium compounds as inhibitors of the acid corrosion of steel. *J. Appl. Chem.*, **13**, 339–345.

90. Hoar, T.P. and Holiday, R.D. (1953) The inhibition by quinolines and thioureas of the acid dissolution of mild steel. *J. Appl. Chem.*, **3**, 502–513.

91. Riggs, O.L. and Hurd, R.M. (1968) Effect of inhibitors on scale removal in HCl pickling solutions. *Corrosion*, **24** (2), 45–49.

92. Kalpana, M. (2003) Study of effect of herbal extract on acid corrosion of mild steel by DC polarisation methods. PhD Thesis of South Gujarat University, India.

93. Elkadi, L., Mernari, B., Traisnel, M., Bentiss, F., and Lagrenée, M. (2000) The inhibition action of 3,6-bis(2-methoxyphenyl)-1,2-dihydro-1,2,4,5-tetrazine on the corrosion of mild steel in acidic media. *Corros. Sci.*, **42** (4), 703–719.

94. Avci, G. (2008) Corrosion inhibition of indole-3-acetic acid on mild steel in 0.5 M HCl. *Colloids Surf. A: Physicochem. Eng. Aspects*, **317** (1–3), 730–736.

95. Tebbji, K., Bouabdellah, I., Aouniti, A., Hammouti, B., Oudda, H., Benkaddour, M., and Ramdani, A. (2007) *N*-benzyl-*N,N*-bis[(3,5-dimethyl-1*H*-pyrazol-1-yl)methyl]amine as corrosion inhibitor of steel in 1 M HCl. *Mater. Lett.*, **61** (3), 799–804.

96. Achary, G., Sachin, H.P., Arthoba Naik, Y., and Venkatesha, T.V. (2008) The corrosion inhibition of mild steel by 3-formyl-8-hydroxy quinoline in hydrochloric acid medium. *Mater. Chem. Phys.*, **107** (1), 44–50.

97. Bentiss, F., Gassama, F., Barbry, D., Gengembre, L., Vezin, H., Lagrenee, M., and Traisnel, M. (2006) Enhanced corrosion resistance of mild steel in molar hydrochloric acid solution by

1,4-bis(2-pyridyl)-5H-pyridazino[4, 5-b]indole: Electrochemical, theoretical and XPS studies. *App. Surf. Sci.*, **252** (8), 2684–2691.
98. Ottoboni, A.M. (1997) *The Dose Makes The Poison*, Van Nostrand Reinhold, p. 5.
99. Martin, R.L., Alink, B.A., Braga, T.G., McMahon, A.J., and Weare, R. (1995) Environmentally acceptable water soluble corrosion inhibitors, *Corrosion* **95**, Paper No.36, NACE. 121–125.
100. Farooqi, I.H. and Quraishi, M.A. (2002) Proceedings of the GLOCORR'2002, International Congress on Emerging Corrosion Control Strategies for the new Millennium, NCCI & CECRI, New Delhi, February 2002.
101. Agarwaala, V.S. (1984) A new approach in corrosion inhibition. Proceedings of International Congress on Metallic Corrosion, Toronto, 1, p. 380.
102. Sanyal, B. (1981) Organic compounds as corrosion inhibitors in different environments – A review. *Prog. Org. Coat.*, **9** (2), 165–236.
103. Baldwin, J. (1895) British Patent 2,327.
104. Putilova, N., Balezin, S.A., and Barannik, V.P. (1960) *Metallic Corrosion Inhibitors*, Pergamon Press, Oxford, London.
105. Saleh, R.M., Ismail, A.A., and El Hosary, A.H. (1982) Corrosion inhibition by naturally occurring substances. VII. The effect of aqueous extracts of some leaves and fruit peels on the corrosion of steel, al, zn and cu in acids. *Br. Corros. J.*, **17** (3), 131–135.
106. Srivatsava, B.C. and Sanyal, B. (1973) Proceedings of Symposium of Cathodic Protection, Defence Research Laboratory, Kanpur, India, Paper 1.2.
107. Srivatsava, K. and Sanyal, B. (1973) Proceedings of Symposium of Cathodic Protection, Defence Research Laboratory, Kanpur, India, Paper 1.4.
108. Khamis, E. and Al-Andis, N. (2002) Herbs as new type of green inhibitors for acidic corrosion of steel. *Materialwiss. Werkstofftech.*, **33** (9), 550–554.
109. Abdel-Gaber, M., Abd-El-Nabey, B.A., Sidahmed, I.M., El-Zayaday, A.M., and Saa dawy, M. (2006) Inhibitive action of some plant extracts on the corrosion of steel in acidic media. *Corros. Sci.*, **48** (9), 2765–2779.
110. El-Hosary, A., Saleh, R.M., and Sharns El Din, A.M. (1972) Corrosion inhibition by naturally occurring substances – I. The effect of *Hibiscus subdariffa* (karkade) extract on the dissolution of Al and Zn. *Corros. Sci.*, **12** (12), 897–904.
111. Srivastava, K. and Srivastava, P. (1981) Studies on plant materials as corrosion inhibitors. *Br. Corros. J.*, **16**, 221.
112. Muller, B. (2002) Corrosion inhibition of aluminium and zinc pigments by saccharides. *Corros. Sci.*, **44** (7), 1583–1591.
113. Bouyanzer, A. and Hammouti, B. (2004) Naturally occurring ginger as corrosion inhibitor for steel in molar hydrochloric acid at 353 K. *Bull. Electrochem.*, **20** (2), 63–66.
114. Chetouani, A., Hammouti, B., and Benkaddour, M. (2004) Corrosion inhibition of iron in hydrochloric acid solution by jojoba oil. *Pigm. Resin Technol.*, **33** (1), 26–31.
115. Chaieb, E.A., Bouyanzer, A., Hammouti, B., and Benkaddour, M. (2005) Inhibition of the corrosion of steel in 1~M HCl by eugenol derivatives. *Appl. Surf. Sci.*, **246** (1–3), 199–206.
116. Bouyanzer, A. and Hammouti, B. (2004) A study of anti-corrosive effects of Artemisia oil on steel. *Pigm. Resin Technol.*, **33** (5), 287–292.
117. Benabdellah, M., Benkaddour, M., Hammouti, B., Bendahhou, M., and Aouniti, A. (2006) Inhibition of steel corrosion in 2 M H_3PO_4 by artemisia oil. *Appl. Surf. Sci.*, **252** (18), 6212–6217.
118. Bouyanzer, A., Hammouti, B., and Majidi, L. (2006) Pennyroyal oil from *Mentha pulegium* as corrosion inhibitor for steel in 1~M HCl. *Mater. Lett.*, **60** (23) 2840–2843.
119. Ananda, R., Sathiyanathan, L., Maruthamuthu, S., Selvanayagam, M., Mohanan, S., and Palaniswamy, N. (2005) Corrosion inhibition of mild steel by ethanolic extracts of Ricinus

communis leaves. *Indian J. Chem. Technol.*, **12** (3), 356–360.

120. Minhaj, A., Saini, P.A., Quarishi, M.A., and Farooqi, I.H. (1999) A study of natural compounds as corrosion inhibitors for industrial cooling systems. *Corros. Prev. Control*, **46** (2), 32–38.
121. Zucchi, F. and Omar, I. (1985) Plant extracts as corrosion inhibitors of mild steel in HCl solutions. *Surf. Tech.*, **24** (4), 391–399.
122. Cabrera, G., Ramos, E., Perez, J., Santhomas, J., and Azucar, C. (1977 (Patent), April–June 1977, Abstract 13–20.
123. El-Etre, A.Y. (1998) Natural honey as corrosion inhibitor for metals and alloys. i. copper in neutral aqueous solution. *Corros. Sci.*, **40** (11), 1845–1850.
124. El-Etre, A.Y. and Abdallah, M. (2000) Natural honey as corrosion inhibitor for metals and alloys. II. C-steel in high saline water. *Corros. Sci.*, **42** (4), 731–738.
125. Parikh, K.S. and Joshi, K.J. (2004) Natural compounds onion (Allium Cepa), Garlic (Allium Sativum) and bitter gourd (Momordica Charantia) as corrosion inhibitors for mild steel in hydrochloric acid. *Trans. SAEST*, **39** (1–2), 29–35.
126. Chetouani, A. and Hammouti, B. (2003) Corrosion inhibition of iron in hydrochloric acid solutions by naturally Henna. *Bull. Electrochem.*, **19** (1), 23–26.
127. Chetouani, A. (2003) Etude numérique de problèmes non linéaire et application aux problèmes de dynamique de populations, PhD thesis. Faculty of Sciences, Oujda, Morocco.
128. Hammouti, B., Kertit, S., and Mellhaoui, M. (1995) BGUGAINE: a natural pyrrolidine alkaloid product as corrosion inhibitor of iron in acid chloride solution. *Bull. Electrochem.*, **11** (12), 553–555.
129. Hammouti, B., Kertit, S., and Mellhaoui, M. (1997) Electrochemical behaviour of bgugaine as a corrosion inhibitor of iron in 1 M HCl. *Bull. Electrochem.*, **13** (3), 97–98.
130. Kertit, S., Hammouti, B., and Mellhaoui, M. (1995) Morroccan Patent No. 23,910.
131. Saleh, R.M. and El-Hosaray, A.A. (1972) Proceedings of the 13th Seminar on Electrochemistry, CECRI, Karaikudi.
132. El-Hosary, A., Gowish, M.M., and Saleh, R.M. (1980) Proceedings of the 2nd International Symposium and Oriented Basic Electrochemistry, SAEST, IIT, Madras, Technical Session, Vol. VII, Paper 6.24.
133. Ramesh, S.P., Vinod Kumar, K.P., and Sethuraman, M.G. (2001) Extract of Andrographis paniculata as corrosion inhibitor of mild steel in acid medium. *Bull. Electrochem.*, **17** (3), 141–144.
134. Sethuraman, M.G., Vadivel, P., and Vinod Kumar, K.P. (2001) Tea wastes as corrosion inhibitor for mild steel in acid medium. *J. Electrochem. Soc. India*, **50** (3), 143–147.
135. Kliskic, M., Radosevic, J., Gudic, S., and Katalinic, V. (2000) Aqueous extract of Rosmarinus officinalis L. as inhibitor of Al–Mg alloy corrosion in chloride solution. *J. Appl. Electrochem.*, **30** (7), 823–830.
136. Sanghvi, M.J., Shukla, S.K., Misra, A.N., and Padh, M.R. (1995) Mehta, 5th National Congress on Corrosion Control, New Delhi, p. 46.
137. Sanghvi, M.J., Shukla, S.K., Misra, A.N., Padh, M.R., and Mehta, G.N. (1996) Corrosion inhibition of mild steel in hydrochloric acid by acid extracts of Sapindus trifolianus, Acacia concian and Trifla. *Trans. MFAI*, **5**, 143–147.
138. Patel, N.S., Jauhari, S., and Mehta, G.N. (2009) Inhibitive effect by acid extracts of Malachra capitata leaves on the sulphuric acid corrosion of mild steel. *Asian J. Res. Chem.*, **2** (4), 427–431.
139. Smita, V. and Mehta, G.N. (1997) Acid corrosion of MS and its inhibition by acid extracts of Lawsonia inermis. *Trans. SAEST*, **32** (2–3), 36–39.
140. Smita, V. and Mehta, G.N. (1997) Effect of acid extracts of powdered seeds of Eugenia jambolans on corrosion of

MS in HCI –Study of DC polarization techniques. *Trans. SAEST*, **32** (4), 20–23.

141. Sakthivel, P., Nirmala, P.V., Umamaheswari, S., Arul Antony, A., Kalignan, G.P.P., Gopalan, A., and Vasudevan, T. (1999) Corrosion inhibition of mild steel by extracts of Pongamia glabra and Annona squamosa in acidic media. *Bull. Electrochem.*, **15** (2), 83–86.
142. Verma, S. and Mehta, G.N. (1999) Effect of acid extracts of ACACIA ARABICA on acid corrosion of mild steel. *Bull. Electrochem.*, **15** (2), 67–70.
143. Ebenso, E.E. and Ekpe, U.J. (1996) Kinetic study of corrosion and corrosion inhibition of mild steel in H_2SO_4 using *Carica papaya* leaves extract. *West Afr. J. Biol. Appl. Chem.*, **41**, 21–27.
144. Ekpe, U.J., Ebenso, E.E., and Ibok, U.J. (1994) Inhibitory action of *Azadirachta* leaves extract on corrosion of mild steel in tetraoxosulphate (VI) acid. *J. West Afr. Assoc.*, **37**, 13–30.
145. Loto, C.A. (1998) The effect of *Vernonia amydalina* (bitter leaf) extracts on corrosion of mild steel in 0.5 M HCl and H_2SO_4 solutions. *Niger. Corr. J.*, **1** (1), 19–20.
146. Avwiri, O. and Igho, F.O. (2003) Inhibitive action of *Vernonia amygdalina* on the corrosion of aluminium alloys in acidic media. *Mater. Lett.*, **57** (22–23), 3705–3711.
147. Chauhan, L.R. and Gunasekaran, G. (2007) Corrosion inhibition of mild steel by plant extract in dilute HCl medium. *Corros. Sci.*, **49** (3), 1143–1161.
148. Orubite, K.O. and Oforka, N.C. (2004) Inhibition of the corrosion of mild steel in hydrochloric acid solutions by the extracts of leaves of *Nypa fruticans* Wurmb. *Mater. Lett.*, **58** (11), 1768–1772.
149. Abdallah, M. (2004) Guar gum as corrosion inhibitor for carbon steel in sulfuric acid solutions. *Port. Electrochem. Acta*, **22** (2), 161–175.
150. Martinez, S. and Stern, I. (2001) Inhibitory mechanism of low-carbon steel corrosion by mimosa tannin in sulphuric acid solutions. *J. Appl. Electrochem.*, **31** (9), 973–978.
151. Oguzie, E. (2005) Inhibition of acid corrosion of mild steel by Telfaria occidentalis. *Pigm. Resin Technol.*, **34** (6), 321–326.
152. Patel, N.S., Jauhari, S., and Mehta, G.N. (2009) Inhibitor for the corrosion of mild steel in H_2SO_4. *S. Afr. J. Chem.*, **62**, 200–204.
153. Oguzie, E. (2006) Studies on the inhibitive effect of *Occimum viridis* extract on the acid corrosion of mild steel. *Mater. Chem. Phys.*, **99** (2–3), 441–446.
154. Oguzie, E. (2007) Corrosion inhibition of aluminium in acidic and alkaline media by *Sansevieria trifasciata* extract. *Corros. Sci.*, **49** (3), 1527–1539.
155. El-Etre, A.Y. (2005) Khillah extract as inhibitor for acid corrosion of SX 316 steel. *Appl. Surf. Sci.*, **252** (24), 8521–8525.
156. El-Etre, A.Y., Abdallah, M., and El-Tantawy, Z.E. (2005) Corrosion inhibition of some metals using lawsonia extract. *Corros. Sci.*, **47** (2), 385–395.
157. El-Etre, A.Y. (2003) Inhibition of aluminum corrosion using *Opuntia* extract. *Corros. Sci.*, **45** (11), 2485–2495.
158. El-Etre, A.Y. (2001) Inhibition of acid corrosion of aluminum using vanillin. *Corros. Sci.*, **43** (6), 1031–1039.
159. Li, Y., Zhao, P., Liang, Q., and Hou, B. (2005) Berberine as a natural source inhibitor for mild steel in 1 M H_2SO_4. *Appl. Surf. Sci.*, **252** (5), 1245–1253.
160. Raja, P.B. and Sethuraman, M.G. (2008) Natural products as corrosion inhibitor for metals in corrosive media – A review. *Mater. Lett.*, **62** (1), 113–116.
161. Verma, S. and Mehta, G.N. (1999) Effect of acid extracts of calotropis giganta latex on sulphuric acid corrosion of mild steel–study by DC polarisation techniques. *Chem. Environ. Res.*, **8**, 131–136.
162. Chauhan, L.R. and Gunasekaran, G. (2004) Eco friendly inhibitor for corrosion inhibition of mild steel in phosphoric acid medium. *Electrochim. Acta*, **49** (25), 4387–4395.

163. Raja, P.B. and Sethuraman, M.G. (2008) Inhibitive effect of black pepper extract on the sulphuric acid corrosion of mild steel. *Mater. Lett.*, **62** (17–18), 2977–2979.

164. De Souza, F.S. and Spinelli, A. (2009) Caffeic acid as a green corrosion inhibitor for mild steel. *Corros. Sci.*, **51** (3), 642–649.

165. Okafor, C., Ikpi, M.E., Uwah, I.E., Ebenso, E.E., Ekpe, U.J., and Umoren, S.A. (2008) Inhibitory action of *Phyllanthus amarus* extracts on the corrosion of mild steel in acidic media. *Corros. Sci.*, **50** (8), 2310–2317.

166. Eddy, N. (2009) Inhibitive and adsorption properties of ethanol extract of *Colocasia esculenta* leaves for the corrosion of mild steel in H_2SO_4. *Int. J. Phys. Sci.*, **4** (4), 165–171.

167. Cheng, S., Chen, S., Liu, T., Chang, X., and Yin, Y. (2007) Carboxymethylchitosan as an ecofriendly inhibitor for mild steel in 1 M HCl. *Mater. Lett.*, **61** (14–15), 3276–3280.

6
Green Corrosion Inhibitors: Status in Developing Countries

Sanjay K. Sharma and Alka Sharma

6.1
Introduction

The phenomenon of corrosion has been known ever since the discovery of metals. Corrosion is a natural process associated with the behavior of a metal in its environment. It is the chemical or electrochemical reaction between materials, usually a metal and its environment, resulting in the deterioration of the metal and its properties, which can lead to many serious problems. The most important chemical property of a metal or alloy is its reaction with atmospheric air when left unprotected. This is because metals have a natural tendency to revert back to their combined states. The process by which the metal has a tendency to revert to its natural stable state as an ore is called *corrosion* [1–4].

The primary driving force of corrosion is based upon the transformation of metal (e.g., iron) from its natural state to steel. The refining of iron ore into steel, which is essentially an unstable state of iron, requires the addition of energy. This energy is the driving force of corrosion. Hence, the process of iron returning to its natural state is the corrosion. The process may involve a second step in which an oxide, hydroxide, or carbonate of the metal may form and deposit at the corrosion site. Thus, corrosion takes place when a metal dissolves in, or is disintegrated by, some corrosive agents, may be acids, alkalies, salts, or atmosphere. In certain cases, the second-step corrosion products or deposits formed may be protective against further corrosion.

The surfaces of all metals (except for a noble metal like gold) in air are covered with oxide films. The basic phenomenal concepts of corrosion can be very briefly elucidated as follows: when such a metal comes in contact with an aqueous solution, the oxide film tends to dissolve. In acidic solutions, the oxide film may dissolve completely, leaving a bare metal surface, which is said to be in the active state, whereas in near-neutral solutions, since the solubility of the oxide will be much lower than in acid solution, the extent of dissolution tends to be lesser. The underlying metal may then become exposed initially only at localized points where, owing to some discontinuity in the metal, for example, the presence of an inclusion or a grain boundary, the oxide film may be thinner or more prone to

Green Corrosion Chemistry and Engineering: Opportunities and Challenges, First Edition.
Edited by Sanjay K. Sharma.

dissolution than elsewhere. If the near-neutral solution contains inhibiting anions, this dissolution of the oxide film may be suppressed and the oxide film stabilized to form a passivating oxide film. The formation of a passivating oxide film on metal surfaces is an important aspect of corrosion protection. However, under certain circumstances, the passivating film may not be protective, and so corrosion may occur in the passive state.

The scientific investigation on corrosion started at the beginning of the nineteenth century when Nicholson and Caryle discovered the electrolytic decomposition of water by the electric current supplied by a galvanic battery [5]. Humphrey Davy established a relationship between the production of electricity and the oxidation of zinc, in which one of the two metals was copper and the other acted as the generator of electricity. In 1830, the Genevan chemist Auguste de la Rive developed the basis of the electrochemical theory of corrosion. At the beginning of the twentieth century, this theory was taken up by Whitney, and completed by Hoar and Evans at the end of the 1920s [6]. Academician Aleksandra Naumovich Frumkin, one of the founders of present day twentieth-century electrochemistry, made a substantial contribution to the corrosion theory. Studies of the adsorption effect of organic substances on the electrocapillary curve of a mercury electrode led Frumkin to experimentally check the Gibbs adsorption equation and to the discovery of the famous isotherm that became known as the *Frumkin isotherm*. The Frumkin isotherm combined with the model of two parallel capacitors, also put forward by him, has played an important role in the study of the adsorption processes on electrodes [7–10].

Corrosion, in an aqueous environment and in an atmospheric environment (which also involves thin aqueous layers), is an electrochemical process because it involves the transfer of electrons between a metal surface and an aqueous electrolyte solution. Fortunately, most metals react with the environment to form more or less protective films of corrosion reaction products that prevent the metal from turning into solution as ions. Most metals corrode on contact with water (and moisture in the air), acids, bases, salts, oils, aggressive metal polishes, and other solid and liquid chemicals. Metals will also corrode when exposed to gaseous materials like acid vapors, formaldehyde gas, and sulfur-containing gases. They even form compounds such as sulfides, carbonates, sulfates, and so on, depending upon the presence of impurities. Such compounds have less energy than pure metals.

Corrosion is recognized as one of the most serious problems in our modern societies, and the resulting losses each year are in hundreds of billions of dollars. Studies regarding the cost of corrosion have been undertaken by several countries including the United States, United Kingdom, Japan, Australia, Kuwait, Germany, Finland, Sweden, India, Italy, China, and some of the African countries as well. The studies have ranged from formal and extensive efforts to informal and modest efforts. The common finding of these studies was that the annual corrosion costs ranged from approximately 1 to 5% of the gross national product (GNP) of each nation [3, 4, 11–14]. A survey conducted in Italy in the past 20 years shows that the impact of corrosion on the domestic economy accounted for more than 3%

of the gross domestic product (GDP) [11–13]. In the United States alone, this amounts to over US$ 340 billion per year (i.e., 3.1% of the GDP) (as estimated by the National Association of Corrosion Engineers (NACE)) [3, 4, 14]. In developing countries, such as India also, as per a survey conducted by the scientists of CECRI, Karaikudi, approximately US$ 10 billion per year is lost solely due to corrosion. These figures are perhaps less intimidating considering that corrosion occurs with varying degrees and types of degradation, whenever metallics are used. Corrosion not only causes economic loss to any country, but also to safety, technology, health, and cultural losses to the civilization on the whole. Structurally, the loss of thickness of the cross section due to corrosion attack leads to a smaller resistant area, resulting in a decrease in the structural performance in terms of strength, stiffness, and ductility, and even the local failure of a component or joint could affect the stability of the whole structure. The corrosion phenomenon can produce a significant reduction in the fatigue strength, mainly in zones with high-stress concentrations. Moreover, corrosion contributes to depletion of our natural resources also. While corrosion is inevitable, it can be predicted and controlled and, globally, at least 25% of this enormous economic loss could easily be saved by the better use of existing knowledge in corrosion protection and promoting relevant research in this domain.

More than this enormous economic loss, failure of structural metals such as steel due to corrosion in certain situations may entail even loss of life [1–4, 14–18]. Corrosion products of some metals such as copper and lead may contaminate food and cause food poisoning. With the advancement in medical sciences, the use of metal prosthetic devices in the body has amplified. New alloys and better techniques of implantation have been developed, but corrosion continues to create problems due to the failure of these implanted metallic products.

Corrosion loss not only includes the cost of replacing corroded structures and machineries but also includes the cost of using corrosion-resistant metals and alloys. It also causes loss in efficiency due to diminished heat transfers through accumulated corrosion products, loss of valuable products through leakages, and effects on safety and reliability. Owing to corrosion, metal loses its strength, ductility, and finally it may fail suddenly to exhibit its metallic characteristics.

Currently, the study of metal corrosion phenomena has become an important industrial and academic topic, especially in aggressive media. This is due to the increased industrial applications of acid solutions. The most important fields of applications being acid pickling, industrial cleaning, acid descaling, oil-well acid in oil recovery, and the petrochemical processes. Aqueous solutions of acids are among the most corrosive media. Therefore, the rate of corrosion at which metals are destroyed in acidic media is very high, especially when soluble corrosion products are formed.

Chemistry has an important role to play in achieving a sustainable civilization on Earth. Currently, there is an increasing demand for extensive applied research, education, information, transfer of knowledge, and technology, and technical development in the protection of metal corrosion. Research, where considerable emphasis has been placed on the connection between practical problems and basic scientific principles, is considered to be of vital importance. It is further believed that the research carried out in this domain will undoubtedly be of technological and economic benefit, aiming to achieve the prevention and control of metallic corrosion.

Owing to the tremendous economic loss it can cause, corrosion has and continues to be the subject of extensive study especially with a view to its minimization at acceptable expenses in terms of the three Es: economy, environment, and energy.

Pure metal is extracted from its ore with enormous use of energy, and a large amount of energy is lost during their decay via the corrosion process. Combating metal corrosion is energy saving without corrosion management. Corrosion-prevention methods (based on technology and management) can be considered as a probable tool for conservation of energy and resource.

This chapter presents the corrosion prevention of metals and their alloys, employing *Green Corrosion Inhibitors* and the status of their usage in developing countries. Before revealing about their usage, a brief account of inhibitor mechanism is also given to get an overview of the inhibitory action toward combating corrosion.

6.2 Protection against Corrosion

Some metals are more chemically active than others. Some react more readily to form ions or compounds. These reactions are linked to a metal's readiness to release electrons in the formation of compounds or ions. For example, some metals combine with oxygen or hydrogen quite easily; others are comparatively inert. Potassium, at one extreme, readily combines with oxygen. Gold, at the other extreme, is extremely stable. When aluminum or zinc are exposed to air, oxidation occurs. However, the oxide that forms adheres to the metal underneath. In this way, the oxide acts as a protective coating to prevent further contact between the bare metal and air. In contrast, when iron is oxidized, the rust that forms may flake off almost as rapidly as it occurs. The reason for this is that the oxidation of iron usually results in a rather porous material. In some cases, this deposit flakes off or washes away. In doing so, it continually exposes more metal. Even when iron oxide does not separate from the metal surface, it does not form a sufficiently solid coating to prevent continued corrosion. To control this corrosion of iron, coatings of other materials are usually applied to the surface.

The basic requirements of any metal used in industry or at home are durability and ability to give good services continuously. It has been estimated that failure of metals due to mechanical defects is only 40% as compared to 70% due to corrosion. Corrosion control and prevention methods for metals are a major aspect of any scheme of metal care. Corrosion mitigation can be accomplished by several ways [2, 3, 15–18], such as the following:

- correct material selection;
- correct design;

- change of metal electrode potential;
- alteration of environment;
- use of inhibitors;
- protective coatings (metallic, chemical conversion, and organic coatings).

The concept of placing a protective barrier between materials and their environment is so ancient that its origin is lost in the mist of history. The best way to combat metal corrosion is by using *inhibitors*.

As defined by NACE, *inhibitor* is a substance that decreases the corrosion rate, when added in a small quantity to a corrosive environment containing a metal or an alloy [19]. The concentration of a given inhibitor needed to protect a metal will depend on a number of factors such as composition of the environment, temperature, velocity of the liquid, the presence or absence in the metal of internal or external stresses, composition of the metal, and the presence of any other metal contact. The type and quantity of inhibitors required for a given metal can be determined by experiments. These experiments should include measurements on the location and degree of the corrosive attack as a function of the inhibitor type and concentration. Corrosive attacks may occur even if experimentation shows that adequate levels and the correct type of inhibitor are used. Inhibitors function in various ways, and the basic concept of the inhibitor functioning has been briefly outlined in Section 6.3.

It is well established that synthetic chemicals are most effective for inhibiting metal corrosion. Soluble hydroxides, chromates, phosphates, silicates, molybdates, and carbonates are added to decrease the corrosion rate of carbon steel and alloys in various corrosive environments [20–26]. But, unfortunately, most of these compounds are not only expensive but also toxic to living beings; as a result, their use as inhibitors has come under severe criticism. These inhibitors may cause reversible (temporary) or irreversible (permanent) damage to the organ systems (kidneys or liver) or they may disturb a biochemical process or disturb an enzyme system at some site in the body. The toxicity may manifest either during the synthesis of the compound or during its application. Chromium, especially chromates (hexavalant chromium), has been found to cause irritation of the respiratory tract, cause ulcers and perforations of the nasal septum, and cause lung cancer in workers employed in chromium-manufacturing plants in West Germany and the United States [27].

Consequently, search for effective as well as environmentally acceptable corrosion inhibitors as alternatives to these toxic inhibitors is considered to be important for prevention of corrosion of metal and its alloys in an aggressive medium. In the past few years, attempts have been made to use plant extracts as *green inhibitors* for combating corrosion of metals and their alloys in aggressive media. The extracts of natural products contain a large variety of organic compounds rich in π-electrons, donor atoms, namely O–, N–, S–, P–, and so on, with which they get adsorbed on the metal surface and form a compact barrier film. Thus, such investigations are found to be very fruitful and encouraging in saving not only metals and energy but also the environment.

6.3 Inhibitors

According to Norman E. Hammer [19], the definition of inhibitor favored by the NACE is *a substance that retards corrosion when added to an environment in small concentrations*. It is well established that inhibitors function in one or more ways to control corrosion: by adsorption of a thin film onto the surface of a corroding material, by inducing the formation of a thick corrosion product, or by changing the characteristics of the environment resulting in reduced aggressiveness [28–31]. These mechanisms cover most of the observed effects and form the basis for experimental work leading to research, development, and engineering applications of inhibitors.

6.3.1 Mechanism of Inhibition

Inhibition is achieved by one or more of several mechanisms. Some inhibitors retard corrosion by adsorption to form an invisibly thin film only a few molecules thick; others form visible bulky precipitates that coat the metal and protect it from attack. Another common mechanism consists of causing the metal to corrode in such a way that a combination of adsorption and corrosion products forms a passive layer. As per definition, substances which, when added to an environment, retard corrosion but do not interact directly with the metal surface are inhibitors. These types of inhibitors create conditions in the environment that are more favorable for the formation of protective precipitates, or they remove an aggressive constituent from the environment.

Inhibition is unlikely to be a simple phenomenon, but rather a sequence of processes that is more or less well identified. Inhibitors can act on the medium by modifying its properties, thus reducing its aggressiveness toward the metal. Their action is probably mainly within the volume of the Helmholtz double layer. They can act on the metal by modifying its state, or can modify the anodic or cathodic reactions, or even both at the same time. They form a monomolecular protective layer on the metal and thus create a barrier to the environment. Thus, transition of the metal/solution interface from a state of active dissolution to the passive state is attributed to the adsorption of the inhibitor molecules at the metal/solution interface, forming a protective film. The rate of adsorption is usually rapid and, hence, the reactive metal surface is shielded from the aggressive environment [32].

In order to be effective, inhibitors need to be adsorbed at the metal surface. This is the first stage of the mechanism of inhibitors. The adsorption process can occur through the replacement of solvent molecules from the metal surface by ions and molecules accumulated in the vicinity of the metal/solution interface. Ions can accumulate at the metal/solution interface in excess of those required to balance the charge on the metal at the operating potential. These ions replace solvent molecules from the metal surface and their centers reside at the inner Helmholtz plane. This phenomenon is termed *specific adsorption, contact adsorption*. The anions are adsorbed when the metal surface has an excess positive charge in an amount greater than that required to balance the charge corresponding to the applied potential. Aromatic compounds (which contain the benzene ring) undergo particularly strong adsorption on many electrode surfaces. The bonding can occur between metal surface atoms and the aromatic ring of the adsorbate molecules or ligand substituent groups. The exact nature of the interactions between a metal surface and an aromatic molecule depends on the relative coordinating strength toward the given metal of the particular groups present [33].

Two types of adsorption can be distinguished as follows:

- Physical adsorption (physisorption): this can either be of the Van der Waals type, involving weak attraction forces and spreading out over the whole surface, or of the electrostatic type if the inhibitor is dissociated in the solution. The cations resulting from this dissociation fix on cathodic sites. Such an inhibitor is called a *cathodic inhibitor*;
- Chemical adsorption (chemisorption): this is formed by a covalent bonding using electrons common to the inhibitor and the metal. This type of adsorption is irreversible and normally leads to inhibition of the anodic reaction.

Most of the inhibitors function as chemically or physically adsorbed films, which either alter the electrochemical characteristics of the metal or serve as mechanical barriers to the normal corrosion process. The choice of an inhibitor to prevent or minimize the corrosion of a metal depends chiefly on the nature of metal and the corrosive environment.

The inhibitor may function as depicted below:

1) adsorbed as a film on the surface of a corroding metal;
2) induce the formation of a thick corrosion product;
3) change the characteristics of the environment either by producing precipitate or inactivating an aggressive constituent so that it does not corrode the metal.

An inhibitor that forms a protective thin film and at the same time retards aggressiveness of the corrosive medium is classified as types (1) and (3) and is thus considered as an excellent inhibitor technically.

6.3.2
Choice of Inhibitors

The choice of inhibitor must conform to applicable standards and regulations concerning toxicity and environmental protection. While choosing an inhibitor, besides its effectiveness, the legal acceptability must also be considered. This limits the use of certain inhibitors, in particular, sodium and potassium chromate, which were widely used because they are the most effective inhibitors for aluminum. The same applies to the volatile inhibitors. Their replacement by other inorganic salts such as vanadates, molybdates, phosphates, dichromates, and arsenates, and so on, has been investigated for several years [20–26]. But environmental restrictions imposed on the use of heavy-metal-based corrosion inhibitors, especially chromates due to their biotoxicity, oriented scientific researches toward exploring nontoxic and environmentally friendly corrosion inhibitors.

However, there are several considerations when choosing an inhibitor:

- The cost of the inhibitor can sometimes be very high when the material involved is expensive or when the amount needed is large.
- The toxicity of the inhibitor can cause jeopardizing effects on human beings and other living species.
- The availability of the inhibitor will determine its selection and, if availability is low, the inhibitor often becomes expensive.
- Environment friendliness.

The use of corrosion inhibitors has become one of the foremost methods of combating corrosion. To use them effectively, one must identify problems that can be solved by the use of corrosion inhibitors. Second, the economics involved must be considered, namely, whether or not the loss due to corrosion exceeds the cost of the inhibitor and maintenance and operation of the attendant injection system. Third, the compatibility of inhibitors with the process to avoid adverse effects such as foaming, decreases in catalytic activity, degradation of another material, loss of heat transfer, and so on, must also be considered. Finally, the inhibitor must be applied under conditions that produce maximum effect.

Thus, corrosion inhibitors are better understood keeping these four points of view in mind:

1) their effects on the corrosion process;
2) their interactions with various aggressive environments;
3) properties of the inhibitors themselves;
4) possible effects of inhibitors on unit operations.

The corrosion process can be influenced by inhibitors. Corrosion inhibitors are used in acid treatment solutions to significantly reduce the overall and local pickling attack and the hydrogen absorption of steel. It has been speculated that organic inhibitors are more effective with mild steel; also, in particular, polar organic compounds containing sulfur and nitrogen are good corrosion inhibitors for the acidic dissolution of metals [34–40]. The high electron density on the –S and –N atoms in these organic molecules helps them to get chemisorbed on the metal surface [41].

The interactions of corrosion inhibitors with aggressive environments and the influence of their properties themselves can be well understood by reviewing a few studies that have been carried out by various workers on various corrosion inhibitors on different metals/alloys in an aggressive medium. One such study was carried out by A. M. Al-Turkustani, S. T. Arab, and R. H. Al-Dahiri on *Aloe*. They observed that the inhibition efficiency of the *Aloe* extract increased with an increase in its concentration. The aqueous extract of the *Aloe* plant investigated in this study contains anthraquinones, coumarins, and reducing substances, which are mostly composed of oxygen-containing compounds [42]. Hence, it may be suggestive that the inhibition due to *Aloe* extract is because of the adsorption of its constituents through their oxygen active centers. Moreover, these compounds may also form complexes with metallic cations. The complexes can cause blocking of the microanodes and/or the microcathodes that are generated on the aluminum surface when in contact with accordant (HCl), and hence can retard the dissolution of the aluminum [43]. In hydrochloric acid, *Aloe* compounds form its hydrochloride or, more specifically, it exists as a moiety of protonated *Aloe* or *Aloe* compound's ions and chloride ion [44].

The inhibiting mechanism of *Aloe* can be elucidated by knowing the adsorption processes going on at the metal surface in the acidic medium. When the aluminum electrode is immersed in a dilute solution of hydrochloric acid containing *Aloe*, three kinds of species can be adsorbed on its surface, which are as follows [45]: (i) if the metal surface is positively charged (with respect to potential of zero charge (PZC)), the chloride ions will first be adsorbed on the metal surface, which in turn will attract the *Aloe* compound's ions and protonated water molecules. Therefore, a close-packed triple layer will form on the metal surface and inhibit aluminum ions to enter the solution. Hence, with an increase in the positive charge on the metal surface, the adsorption of *Aloe* would increase and the *Aloe* content of the solution would decrease; (ii) if the metal surface is negatively charged with respect to PZC, the protonated water molecules and the *Aloe* compound's ions would be directly adsorbed on the metal surface. With an increase in the negative charge on the metal surface, adsorption of *Aloe* increases and its concentration in solution decreases; and lastly (iii) when the metal surface attains the potential at which the surface charge becomes zero, none of the ions (neither cations nor anions) adsorb on the surface through their centers. A few *Aloe* compounds molecules may, however, be physically adsorbed through their planar π-orbitals on the metal surface (with vacant d-orbitals). Because of this possibility, the concentration of *Aloe* in solution at PZC also decreases instead of remaining the same. However, the decrease is small due to the fact that, in the case of planar adsorption, the molecular density of *Aloe* on the aluminum surface becomes smaller to a lesser degree than in the case of later-ionic adsorption. The amount of *Aloe* adsorbed is smaller in the former case than in the latter.

The properties of the inhibitors themselves also play a greater role in corrosion-inhibition process. Recently, several papers have suggested that some amino acids (AAs) can function as corrosion inhibitors for metals such as copper, iron, and steel [46–48]. K. Barouni *et al.* studied the inhibition effect of five AAs on the corrosion of copper in molar nitric solution. Here, it was found that valine and glycine accelerated the corrosion process, but arginine, lysine, and cysteine inhibit the corrosion phenomenon. Cysteine is the best inhibitor. On the contrary, when Hluchan *et al.*, examined the behavior of 22 AAs as potential corrosion inhibitors for iron in a strong acidic medium (1 M HCl), they observed that the corrosion-inhibition property is more in case of AAs with a longer hydrocarbon chain or greater number of additional amino groups. It was found that AAs with shorter chains, such as glycine, aspartic acid, and glutamic acid, showed only 50% reduction in the corrosion rate.

Moreover, the change in pH of the test solution was also found to affect the inhibiting tendency of the inhibitors. Polyaspartic acid showed greater corrosion inhibition than aspartic acid in the pH range between 8 and 10. Further, polyaspartic acid was found to act as a corrosion accelerator below pH 7, but as a corrosion inhibitor above pH 10 for iron and steel; which could be explained as the ability of polyaspartic acid to complex with the iron as a function of pH. At lower pH values, the iron–polyaspartate complexes were stable, whereas at higher pH no such complexes could be predicted by the author [38].

Thus, the behavior of an inhibitor can be affected by certain factors, such as the following:

- temperature: adsorption decreases with increase in temperature;
- pH: chromates are effective at a pH above 3;
- the surface/volume ratio;
- nature of metal used;
- the inhibitor's concentration.

Temperature increases affect the rate of corrosion, because an increase in temperature increases the rate of the electrochemical corrosion reaction. As a rule, an increase in temperature results in a stepping up of the thermal agitation of the molecules of the medium and, hence, the corrosive action of the medium. Studies show that the corrosion of steel may be stepped up three to four times the normal rate when the temperature of water is increased from 288 to 333.15 K. Over 333.15 K, the rate may double with every 20° increase in temperature.

Although many synthetic compounds show good anticorrosive activity, most of them are highly toxic to both human beings and the environment. The safety and environment issues of corrosion inhibitors that have arisen in industries have always been a global concern. These toxic effects have led to the use of natural products as anticorrosion agents, which are eco-friendly, harmless, as well as cost effective. In recent years, many alternative eco-friendly corrosion inhibitors have been developed ranging from rare earth elements [49–55] to organic compounds [56–68].

6.4 Natural Products as Green Corrosion Inhibitors

Several aromatic plants are popular for domestic and commercial uses. Collectively they are called medicinal and aromatic plants (MAPs). About 12.5% of the 422 000 plant species documented worldwide are reported to have medicinal values [69, 70]. In an Asian country such as India, of the 17 000 species of higher plants, 7500 are known for medicinal uses [71]. This proportion of medicinal plants is the highest proportion of plants known for their medical purposes in any country for the existing flora of that respective country. Currently, approximately 25% of drugs are derived from plants, and many others are synthetic analogs built on prototype compounds isolated from plant species in modern pharmacopoeia [70–72].

Worldwide there are 12 megabiodiversity centers, which are considered to be rich in plant wealth and medicinal plant heritage. To unveil regarding the inhibitory action of the natural products, it is necessary to know their chemical constituents and their effectiveness. The prospects of a search for new and potential corrosion inhibitors in the plant kingdom is promising and is an accepted practice, as they are rich sources of a variety of alkaloids, flavonoids, tannins, vitamins, AAs, and steroids as well as volatile oils as a major chemical component [73, 74]. These substances, for their use in the inhibition of the acid cleaning process, have continued to receive attention as replacements for synthesized organic inhibitors. Investigation into the use of some plants as modes for corrosion inhibitors has proposed parts of economic plants as pickling inhibitors [75–90]. Not much has actually been achieved using natural products, compared to the extensive research on organic inhibitors especially N- and S-containing organic compounds.

Corrosion inhibition occurs via adsorption of their molecules on the corroding metal surface, and the efficiency of inhibition depends on the mechanical, structural, and chemical characteristics of the adsorption layers formed under a particular condition. In order to mitigate metal corrosion, the main strategy is to effectively isolate the metal from the corrosive agents by the use of corrosion inhibitors. Inorganic substances such as phosphates, vanadates, chromates [20–26], dichromates, and arsenates have been found to be effective as inhibitors of aluminum corrosion, but being highly toxic their use as inhibitors has come under severe criticism. Among the alternative inhibitors, organic compounds containing polar functional groups with N, O, and/or S atoms in a conjugated system have been found to possess good inhibiting properties. But, unfortunately, most of these compounds are not only expensive but also toxic to living beings. Hence, the significance of plant extracts as corrosion inhibitors becomes more as they are eco-friendly, environmentally acceptable, cheap, easily available, and safe inhibitors of corrosion, that is, *green corrosion inhibitors*. Moreover, these inhibitors are biodegradable and do not contain heavy metals or other toxic compounds.

Plant extracts have become important as environmentally acceptable, readily available, and renewable sources for a wide range of inhibitors. They are rich sources of ingredients, which have very high inhibition efficiency. In 1930, plant extracts (dried stem, leaves, and seeds) of Celandine (*Chelidonium majus*) and other plants were used in H_2SO_4 pickling baths. In fact, the first patented corrosion inhibitors used were either natural products, such as flour, yeast, and so on, or by-products of food industries for restraining iron corrosion in acid media [91]. The anticorrosion activity of *Acacia nilotica, Lawsonia inermis, Garcinia kola*, onion, garlic, bitter gourd, *Zenthoxylum alatum* extract, *Nypa fructicans wurmb, Occium viridis* extract, and so on, for mild steel and other metals in acidic media has been studied by a large number of workers [75–83]. Likewise, the inhibitory action of extracts of various plants/plant parts (such as, *Osccimum bascilicum, Pachylobus edulis* exudate gum, *Gongronema latifolium, Vernonia Amygdalina, Sansevieria trifasciata* extract, *Delonix regia*, and *Opuntia* extract on the corrosion of aluminum and its alloys in aggressive environment has been investigated [84–90]. Muller [92] investigated the effect of *saccharides* (reducing sugars – fructose and mannose) on the corrosion of aluminum and zinc in alkaline medium.

The leaf extracts of *Chromolaena odorata* L. (LECO) has been studied as a possible source of green inhibitor for corrosion of aluminum in 2 M HCl using gasometric and thermometric techniques at 30 and 60 °C [93]. The inhibition efficiency increases with extract concentration but decreases with temperature. The adsorption of LECO on the aluminum surface is in accord with the Langmuir adsorption isotherm. Both kinetic and thermodynamic parameters governing the adsorption process were calculated and discussed. From the experimental results obtained, it can be concluded that LECO, which is biodegradable, environmentally benign, and is obtained from a renewable resource with minimal health and safety concerns, has the potential to be a cost-effective alternative to synthetic corrosion inhibitors. Results obtained showed that the LECO functioned as an excellent corrosion inhibitor for aluminum in the acidic environment and could find possible applications in metal surface anodizing and surface coating in industries.

Gum Arabic (GA) is the oldest and best known of all natural gums. It is water soluble and a branched, neutral, or slightly acidic complex mixture of arabinogalactan, oligosaccharides, polysaccharides, and glucoproteins and contains calcium, magnesium, and potassium salts. It has excellent emulsifying properties. The hydrophobic polypeptide backbone strongly adsorbs at the oil–water interface, while the attached carbohydrate units stabilize the emulsion by steric and electrostatic repulsion. Because it is rich in organic constituents with adsorptive properties, it has been studied for combating aluminum and mild steel corrosion in an acidic medium [94] and it was found to act as a very significant inhibitor in an aggressive medium for aluminum.

Similarly, the inhibition efficacy of *Punica granatum* extract on the corrosion of brass in 1N HCl has been studied by mass loss measurements at various times and temperatures by P. Deepa Rani and S. Selvaraj [95]. The inhibition efficiency is observed to be markedly higher in an HCl environment with addition of *P. granatum* extract compared with those in the inhibitor-free solution. The maximum inhibition efficiency of 94.52% is achieved at 1000 ppm of bioinhibitor concentration. The inhibition efficiency increases with an increase in the concentration of the inhibitor but decreases with rise in temperature and time. On the basis of the values of activation energy, free energy of adsorption, and variation of inhibition efficiency with temperature, a physical adsorption mechanism is proposed for the adsorption of *P. granatum* on the surface of brass.

Hitherto saps of certain plant leaves such as *Annona squamosa, Accacia Arabica, Accacia conicianna, Azadirachta Indica, Carica papaya, Ricimus communis, Eugenia jambolans, Embilica officianilis, Telferia occidentalis, Terminalia chebula, Terminalia belivia, Sapindus trifolianus, Swertia angustifolia, Pongamia glabra,* and *Eucalyptus* sp., have been studied for the corrosion inhibition of mild steel in acid media in addition to the use of some herbs such as coriander, hibiscus, anis, black cumin, and garden cress as new types of green inhibitors for acid corrosion of steel [96–103].

Some of the fruits such as Nicotiana, *Ricimus communis* seed oil, acacia gum, and lignin along with *Papaia, Poinciana pulcherrima, Cassia occidentalis, Datura metel,* and *Datura stramonium* have also been used as efficient corrosion inhibitors for steel [104–108]. Oil extracts of ginger, jojoba, eugenol, acetyl-eugenol, artemisia oil, and *Mentha pulegium* are used for corrosion inhibition of steel in acid media [109, 110]. Saps of certain plants are very useful corrosion inhibitors. *Calotropis procera, Azydracta indica, Andrographis paniculata,* and *Auforpio turkiale* sap are useful as acid corrosion inhibitors. Quinine has been studied for its anticorrosive effect on carbon steel in 1 M HCl. The inhibition effect of *Zenthoxylum alatum* extract on the corrosion of mild steel in aqueous *ortho*-phosphonic acid was investigated [111–113].

Various techniques can be employed for the surface morphological studies for the verification of the adsorption of the inhibitor molecules on the metal surface, that is, justifying geometric blocking effects, for example, the alcoholic extract of the bioinhibitor (*P. granatum*) and the corrosion product (with inhibitor) is analyzed by UV, IR, and X-Ray diffraction (XRD) studies, which leads to the bioinhibitor as an adsorption inhibitor [95].

The corrosion-inhibition activity in many of these plant extracts could be due to the presence of heterocyclic constituents such as AAs, vitamins, alkaloids, flavonoids, and so on. Even the presence of tannins, cellulose, and polycyclic compounds normally enhances film formation over the metal surface, thus aiding corrosion. It is also likely that a mixture of constituents present may potentiate the inhibitive effect of one particular constituent. But it is still amazing to see reports wherein inhibition to the extent of 99% efficiency is achieved [114–128]. It is certain that natural compounds will emerge as effective inhibitors of corrosion in the coming years due to their biodegradability, easy availability, and nontoxic nature. Careful perusal of the literature clearly reveals that the era of green inhibitors has already begun.

6.5 Green Corrosion Inhibition: Research and Progress

In recent years, especially in the developing countries, the research in corrosion is oriented to the development of "*green corrosion inhibitors,*" compounds with good inhibition efficiency but low risk of environmental pollution [115]. The successful use of naturally occurring substances to inhibit the corrosion of metals in acidic environment has been reported by several authors [75–90]. On surveying literature, it has been observed that the locally available plant extracts have been successfully used as inhibitors, with very promising efficiency [93–128]. Efforts geared toward finding naturally occurring organic substances or biodegradable organic materials as a suitable replacement of organic/inorganic corrosion inhibitors has been intensified in our research. As a contribution to the current interest in environmentally friendly *green* corrosion inhibitors, the present work is a review on the globally used *green inhibitors* for combating metal corrosion.

A literature survey reveals that depending on the chemical constituents of the plant extract the protective propensity may differ by varying the test conditions. Use of *Lawsonia alba Lam.* syn. *Lawsonia inermis, Piper nigrum* L., *Murraya koenigii, Mangifera Indica, Trigonella foenum gracum, Psidium guajava, Acacia nilotica, Ocimum tenuiflorum* syn. *Ocimum sanctum Linn.*, and so on for combating acid corrosion of aluminum (grades 1100 and 6063) resulted in almost 50–99% of inhibition efficiency [121–128]. Reports shows that these two grades of aluminum, AA 1100 (Si + Fe 0.95%, Cu 0.05%, aluminum minimum 99.00%) and AA 6063 (Si 0.20–0.60%, Fe 0.35%, Cu 0.10%, Mn 0.10%, Mn 0.45–0.90%, Cr 0.10%, Zn 0.10%, Ti 0.10%, and remainder is aluminum), have excellent resistance toward atmospheric corrosion in marine, urban, and industrial environments [129–132].

Piper nigrum L. seeds (*PnL*Ss) contain 5–9% of the alkaloids Piperine, Piperidine, Piperittine, and 1–2.5% of volatile oil, the major constituents of which are α- and β-pinene, limonene, and phellandrene [133–135]. In one study, black pepper (*Piperaceae*) was found to comprise 33.7% β-caryophyllene, besides, chavicine, thiamine, riboflavin, nicotinic acid, resins, and metals like Ca, P, and Fe [136] (Figure 6.1). Most among these organic compounds are antibacterial, fungicide, and antioxidants; moreover, compounds like piperine, piperidine, thiamin, riboflavin, nicotinic acid, and so on contain nitrogen and oxygen, which strengthen their adsorptive property over metal surface and hence the anticorrosive behavior [137–140]. On adding the ethanolic extract of *PnLS* to the aggressive solution, there was appreciable weight loss compared to that with the aggressive solution alone, depicting the inhibitory propensity of *PnLS*. Similarly, with increasing concentration of *PnLS*, a comparable decrease in the corrosion rate of aluminum and hence an increase in the inhibition efficiency (IE%) of *PnLS* as a function of its concentration was observed (Figures 6.2 and 6.3). The maximum inhibition efficiency (IE) (85.69%) with a corresponding minimum corrosion rate (ρ_{corr} = 0.4906 mmpy) was observed at a higher concentration (0.4575 g/l) of *PnLS*.

6.5.1
The Proposed Mechanism for the Inhibitory Behavior of the Extracts

The mechanism of inhibition can be understood by knowing the mode of interaction of the inhibitor molecules with the electrode. Inhibitors function by adsorption and/or hydrogen bonding to the metal [141]. This in turn depends on the chemical composition, structure of the inhibitor, the nature of the metal surface, and the properties of the medium. Structural and electronic parameters, such as the type of the functional group, steric, and electronic effects, are generally responsible for the inhibition efficiency of any inhibitor, that is, the adsorption mechanism. Since the compound has to block the active corrosion sites present on the metal surface, the adsorption occurs by the inhibitor's free electrons linking with the metal. The plant extracts constitute organic compounds containing the following [73, 74]:

- lone pair of electrons present on a hetero atom (e.g., N, S, P, and O);

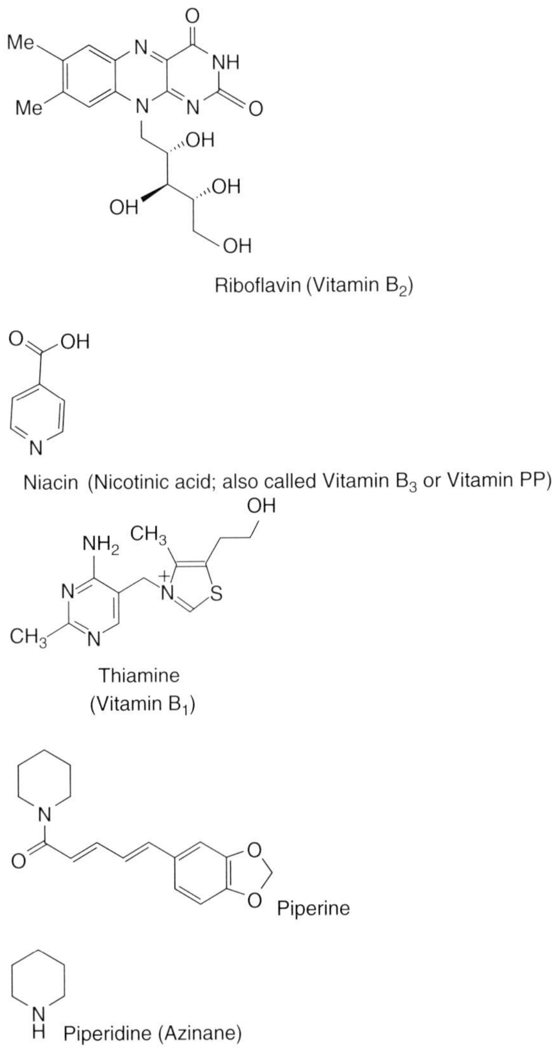

Riboflavin (Vitamin B_2)

Niacin (Nicotinic acid; also called Vitamin B_3 or Vitamin PP)

Thiamine (Vitamin B_1)

Piperine

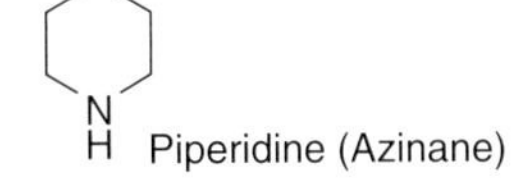

Figure 6.1 Structures of some of the main chemical constituents of *Piper nigrum* L. seeds (*PnLS*).

- π-bond (these molecules are the very effective but often harmful to the environment);
- triple bond (e.g., cyano groups);
- heterocyclic compounds such as pyridine ring pyrrole, imidazole, and so on.

The plant extracts comprise tannins, alkaloids, polyphenols, gallic acids, volatile oil, and vitamins such as thiamine, riboflavin, nicotinic acid, resins, aromatic oils, as well as metals such as Ca, P, and Fe. Many among these organic compounds

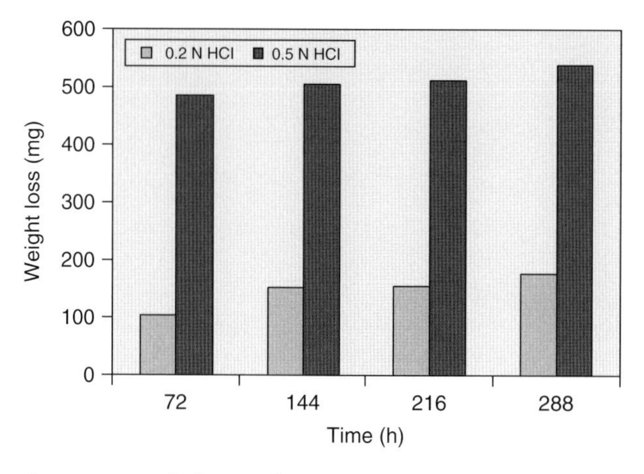

Figure 6.2 Weight loss (milligrams) versus immersion time period (hours) for various concentrations of ethanolic extracts of *PnLS*.

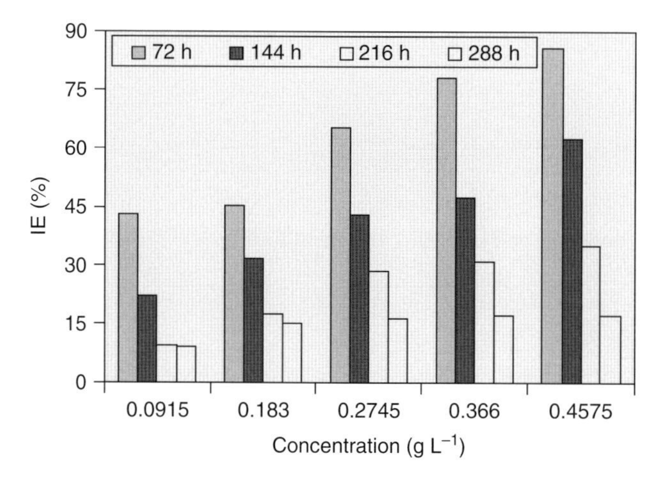

Figure 6.3 Inhibition efficiency (IE %) versus concentration of the extract (grams per liter) for various immersion time periods (hours).

are antibacterial, fungicide, and antioxidants; moreover, compounds such as AAs and vitamins, such as thiamin, riboflavin, nicotinic acid, and so on, contain hetero atoms such as O– and N–, which strengthen their adsorptive property over the metal surface and hence the anticorrosive behavior [94–128, 142].

As a rule of thumb it holds that N-containing compounds exert their best efficiencies in HCl [142], and S-containing compounds are best in H_2SO_4; therefore, it can be suggested that the high IE (%) of green inhibitors is due to their active organic constituent containing O–, N–, and S–. The inhibiting influence of these molecules is attributed to their adsorption through the –NH, C=O, OH, COOH, and so on groups and also may be due to the presence of more π-electrons in

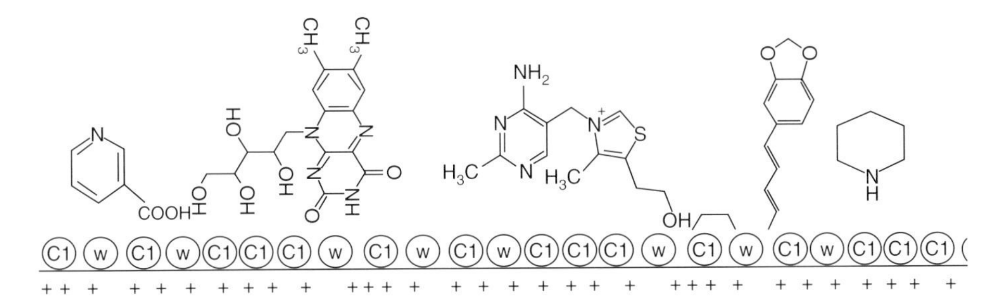

Figure 6.4 A sketch representing one of the possible orientations of the active organic constituent(s) of plant extract during the adsorption phenomenon over the metal surface.

the rings [142]. The high efficiency may also be because the active constitutes act together (synergistically) for their better protective performance toward acid corrosion of metal. These organic molecules get physisorbed/chemisorbed on the metal surface forming a protective film.

The active organic molecules consisting of O- and N-containing groups orient themselves over the metal surface so as to cover it horizontally either via weak bonding (physisorbed) or strong bonding (chemisorbed). The adsorbed organic molecules may interact with each other as well as with the electrode surface. In the process, water molecules must be displaced from the metal surface. The possible mode of adsorption of these compounds on the metal surface is depicted pictorially (Figure 6.4).

6.6 Green Corrosion Inhibition in Developing Countries

6.6.1 Usage of Metals and Present Corrosion Management: Practice and Prevention

It is a well known and proven fact and can also be justified by surveying through the literature that no country can survive without the use of metals; metals are an essential part of our present civilization. Because it is a metallic characteristic to react with the atmosphere and deteriorate due to corrosion, this leads to weakening of the metal or metallic structures. Hence, it becomes very significant to save the metal from corrosion to be able to save the economy, energy, as well as the environment of the country using it.

There are various methods used for the protection of metal against corrosion, but the best method is to protect it by using inhibitors. In the past, chemical inhibitors were widely used for protecting metals from getting corroded; but later the researchers discovered that chemical inhibitors lead to toxicity of the environment either during their synthesis processes or during application. With the safety of the global environment as the top priority, researchers are inclined to discover nontoxic, safe to use, easily available, and cheaper inhibitors for combating metal corrosion. An extensive survey of literature clearly depicts the growing use of the extracts of various plant parts (leaves, seeds, fruits, etc.) for the protection of many metals, namely steel, copper, aluminum, zinc, brass, tin, and so on, which are generally used in industries or for domestic purposes against corrosion. Categorically, in developing countries and cities, such as Nigeria, Egypt, Baltimore, Morocco, Columbia, Brazil, French Guiana, Algeria, Johannesburg, Jeddah, India, Iran, China, France, Italy, Atlanta, Mexico, Rio de Janeiro, Bucharest, Poland, Czech Republic, Romania, Somalia, Argentina, and so on [143–159], the usage of locally available plants and the extract of their parts for combating metal corrosion was observed to be significant. The researchers got very promising and encouraging results from their investigations regarding the use of natural resources as corrosion inhibitors.

On using the locally and abundantly available plants as green inhibitors, significant protection of the metals was observed. It is not possible to summarize all the findings of the investigations here; hence, only very recent observations of the researches going on around the globe have been tabulated in Table 6.1.

6.6.2 Summary of Researchers' Work to Develop Green Inhibition Science

For the safety of the environment as well as economic concerns, it is now necessary to develop corrosion protection science, employing green inhibitors. Moreover, focusing on and strengthening of these three areas to reduce the losses due to corrosion is also recommended:

1) In terms of information dissemination: establishment of a corrosion-prevention service center is desired with technical experts from almost all the industry-based cities and countries to increase communication among academic institutions and the industry and to build modules for getting more numbers of trained engineers.
2) In terms of education: it is essential to spread awareness from the elementary school to the university level for conservation of material resources and the environment.
3) In terms of research and development: it is necessary to have appropriate monitoring and inspection of the equipment and machines that are being used for corrosion-prevention control, along with greater awareness of corrosion-protection techniques among technicians, academicians, as well as researchers. Besides this, the usage of green inhibitors for corrosion protection outside the laboratories should be correctly implemented for all practical purposes.

It may be concluded that if inhibitors from natural products, which are less toxic, eco-friendly, with good anticorrosive activity, easily available, and cost effective are used to combat corrosion, then the chemical scientist may not be required to compromise with hygiene factors and heavy losses due to corrosion. To address

Table 6.1 Metal protection employing green inhibitors: as practiced globally.

Country (s)	Plant as inhibitor	Findings for the metal protection	Reference(s)
Nigeria, Jeddah	*Aloe vera* extract	Zinc in 2 M HCl; aluminum in 0.5 M HCl (84% IE)	[143, 144]
French Guiana	*Annona squamosa* extract	Mixed-type inhibitor for C38 steel in 1 M HCl	[145]
Sharjah (UAE), Nigeria	*Azadirachta indica (leaves, seeds, fruits)* extract	C steel in 1 M HCl (87% IE); mild steel in 2.0 M H_2SO_4 (82% IE)	[146, 147]
Malaysia	*Uncaria gambir* extract	Mixed-type inhibitor for mild steel in aqueous (~70% IE)	[148]
Malaysia	*Vegetal tannins*	Iron and steel in acidic and near-neutral media	[149]
India	*Vitis vinifera* seed and skin extract	Pure copper and one of its alloy (Cu-27Zn brass) in a natural seawater environment (92.3% IE)	[150]
Iran	*Lawsonia inermis* extract	Mixed type of inhibitor for mild steel in 1 M HCl (92% IE)	[151]
Morocco, France	*Rosmarinus officinalis* oil	C38 steel in 2.0 M H_2SO_4 (~60% IE)	[152]
French Guiana	*Palicourea guianensis* extract	Mixed-type inhibitor for C38 steel in 1 M HCl (89% IE)	[153]
Somalia and Italy	*Papaia, Poinciana pulcherrima, Cassia occidentalis* and *Datura stramonium* seeds and *Papaia, Calotropis procera B, Azydracta indica* and *Auforpio turkiale* sap	Mild steel in 1 N HCl (88–96% IE)	[154]
Columbia	Tobacco extracts	Aluminum and steel in acidic medium	[155]
China	*Jasminum nudiflorum* Lindl. leaves extract	Mixed type of inhibitor for cold rolled steel (CRS) in 1.0 M HCl	[156]
Latin America	*Zygophyllum album*	X52 mild steel in 1M H_2SO_4	[157]
Mexico	Opuntia-Ficus-Indica (Nopal) mucilage	Steel in a concretelike environment (alkaline solutions)	[158]
Mmabatho (SA)	*Solanum melongena* leaves extract	Mild steel in HCl	[159]

this crucial issue, research is going on for combating corrosion by employing *green corrosion inhibitors* in order to have a cleaner, greener, and safer environment as well as effectual prevention of corrosion.

Acknowledgments

We are thankful to authorities of Rajasthan University, Jaipur (India), and JECRC Foundation, Jaipur (India), for the support and encouragement.

References

1. Moore, J.J. (1994) *Corrosion of Metals. A Text Book of Chemical Metallurgy*, Butterworth-Heinemann Ltd, Boston, pp. 351–393.
2. Jones, D.A. (1996) *Principles and Prevention of Corrosion*, 2nd edn, Prentice Hall, Upper Saddle River, NJ.
3. Roberge, P.R. (2007) *Corrosion Inspection and Monitoring*, John Wiley & Sons, Inc., New York.
4. Shaw, B.A. and Kelly, R.G. (2006) *Electrochem. Soc. Interface*, 24–26.
5. Lynes, W. (1951) *J. Electrochem. Soc.*, **98**, 3c–10c.
6. Hoar, T.P. and Evan, U.R. (1932) *Proc. R. Soc. Lond.*, **137**, 343.
7. Frumkin, A. (1925) *Z. Phys. Chem.*, **116**, 498.
8. Frumkin, A. (1926) *Z. Phys. Chem.*, **35**, 792.
9. Frumkin, A. (1932) *Z. Phys. Chem.*, **160**, 11.
10. Frumkin, A. and Kolotyrkin, J. (1941) *Acta Physicochim. URSS*, **14**, 469.
11. Landolfo, R., Cascini, L., and Portioli, F. (2010) *Sustainability*, **2**, 2163–2175.
12. CC Technologies Laboratories & NACE International (2001) *Corrosion Costs and Preventive Strategies in the United States*, Publication No. FHWA-RD-01-156, U.S. Federal Highway Administration, Washington, DC.
13. Landolfo, R., Di Lorenzo, G., and Guerrieri, M.R. (2005) Proceedings of the Italian National Conference on Corrosion and Protection, Senigallia, Ancona, Italy, 25 June–1 July.
14. Miksic, B., Boyle, R., and Wuertz, B. (2004) F.N. Speller Award Lecture at CORROSION, New Orleans, LA, March 2004; Miksic, B., Boyle, R., and Wuertz, B. (2004) *Corrosion*, **60** (6) 515–522.
15. Evans, U.R. (1960) *The Corrosion and Oxidation of Metals*, E. Arnold, London, (1960, following previous editions), with 1st supplementary volume (1968), 2nd Supplementary Volume (1976).
16. Narayan, R. (1983) *An Introduction to Metallic Corrosion and its Prevention*, 1st edn, Oxford and IBH, India.
17. Fontana, M.G. and Greene, N.D. (1986) *Corrosion Engineering*, 3rd edn, McGraw-Hill, Singapore.
18. Uhlig, H.H. (1971) *Corrosion and Corrosion Control*, 2nd edn, John Wiley & Sons, Inc., New York.
19. NACE Glossary of Corrosion Terms (1965) *Mater. Prot.*, **4** (1) 79.
20. Stranick, M.A. (1984) *Corrosion*, **40** (3), 296–301.
21. Vukasovich, M.S. and Farr, J.P.G. (1986) *Mater. Perform.*, **25** (5), 9–18.
22. Shams El Din, A.M., Mohammed, R.A., and Haggag, H.H. (1997) *Desalination*, **114** (1), 85–95.
23. Frankel, G.S. and McCreery, R.L. (2001) *Electrochem. Soc. Interface*, **10**, 34–38.
24. Buchheit, R.G., Guan, H., Mahajanam, S., and Wong, F. (2003) *Prog. Org. Coat.*, **47**, 174–182.
25. Bastos, A.C., Ferreira, M.G., and Simoes, A.M. (2006) *Corros. Sci.*, **48** (6), 1500–1512.
26. Iannuzzi, M. and Frankel, G.S. (2007) *Corros. Sci.*, **49**, 2371–2391.

27. Foster, T., Blenkinsop, G.N., Blattler, P., and Szandorowski, M. (1991) *J. Coat. Technol.*, **63** (801), 91–99.
28. Heckerman, N. (1965) Fundamentals of Inhibitors in *NACE Basic Corrosion Course*, NACE, Houston, TX.
29. Riggs, O.L. and Nathan, C.C. (eds) (1973) Theoretical Aspects of Corrosion Inhibitors and Inhibition, in *Corrosion Inhibitors*, NACE, Houston, pp. 7–27.
30. Roebuck, A. H. and Nathan, C. C. (eds) (1973) Inhibition of Aluminium, in *Corrosion Inhibitors*, NACE, Houston, pp. 240–244.
31. Sastri, V.S. (1998) *Corrosion Inhibitors: Principles and Applications*, John Wiley & Sons, Ltd, Chichester.
32. Ritchie, I.M., Bailey, S., and Woods, R. (1999) *Adv. Colloid. Interface. Sci.*, **80**, 183.
33. Trabenelli, G. and Mansfeld, F. (1987) *Corrosion Mechanisms*, Marcel Dekker, New York, p. 109.
34. Hackerman, N. and Hurd, R.M. (1961) *Corrosion*, **116**, 166.
35. Chatterjee, P., Benerjee, M.K., and Makherjee, K.P. (1991) *Indian J. Technol.*, **29**, 191.
36. Elachouri, M., Hajji, M.S., Kertit, S., Essassi, E.M., Salem, M., and Courert, R. (1995) *Corros. Sci.*, **37**, 381.
37. Mernari, B., Elattari, H., Traisnel, M., and Bentsis, F. (1998) *Corros. Sci.*, **40**, 391.
38. Bentiss, F., Traisnee, M., and Lagrenee, M. (2000) *Corros. Sci.*, **42**, 127.
39. Elkadi, L., Mernari, B., Traisnel, M., Benliss, F., and Lagrenee, M. (2000) *Corros. Sci.*, **42**, 703.
40. Rajalakshmi, R. and Subhashini, S. (2010) *E-J. Chem.*, **7** (1), 325–330.
41. Iita, B. and Offiong, O.E. (1999) *Mater. Chem. Phys.*, **60**, 79.
42. Al-Turkustani, A.M., Arab, S.T., and Al-Dahiri, R.H. (2010) *Mod. Appl. Sci.*, **4** (5), 105.
43. Quraishi, M., Farooqi, I., and Saini, P. (1999) *Corrosion*, **55** (5), 493.
44. Banerjee, G. and Malhotra, S.N. (1992) *Corrosion*, **48** (1), 10.
45. Hunt, G.B. and Holiday, A.K. (1980) *Organic Chemistry*, Butterworth, London, p. 229.
46. Barouni, K., Bazzi, L., Salghi, R., Mihit, M., Hammouti, B., Albourine, A., and El Issami, S. (2008) *Mater. Lett.*, **62** (19), 3325–3327.
47. Silverman, D.C., Kalota, D.J., and Stover, F.S. (1995) *Corrosion*, **51** (11), 818–825.
48. Hluchan, V., Wheeler, B.L., and Hackerman, N. (1988) *Werkst. Korros.*, **39**, 512.
49. Bernal, S., Botana, F.J., Calvino, J.J., Marcos, M., Pérez-Omil, J.A., and Vidal, H. (1995) *J. Alloys Comp.*, **225** (1-2), 638–641.
50. Bethencourt, M., Botana, F.J., Calvino, J.J., Marcos, M., and Rodriguez-Chacon, M.A. (1998) *Corros. Sci.*, **40** (11), 1803–1819.
51. Arenas, M.A., Conde, A., and De Damborenea, J.J. (2002) *Corros. Sci.*, **44**, 511–520.
52. Tang, L., Mu, G., Zhang, J., and Liu, G. (2003) *Zhongguo Xitu Xuebao*, **21** (3), 272–277.
53. Mishra, A.K., Balasubramaniam, R., and Tiwari, S. (2007) *Anti-Corros. Methods Mater.*, **54** (1), 37–46.
54. Koszegi, S., Pasternak, A., Felhosi, I., and Kalman, E. (2007) *Korrozios Figyelo*, **47** (5), 133–140.
55. Yasakau, K.A., Zheludkevich, M.L., and Ferreira, M.G.S. (2008) *J. Electrochem. Soc.*, **155** (5), C169–C177.
56. Meena; S.L. (2002) Electrochemistry of Some Organic Compounds and their role as Corrosion Inhibitors, PhD Thesis. Department of Chemistry, University of Rajasthan, Jaipur, India.
57. Salih Al-Juaid, S. (2003) *Orient. J. Chem.*, **19** (3), 555–558.
58. Ashassi-Sorkhabi, H., Ghasemi, Z., and Seifzadeh, D. (2005) *Appl. Surf. Sci.*, **249**, 408–418.
59. Yurt, A., Ulutas, S., and Dal, H. (2006) *Appl. Surf. Sci.*, **253** (2), 919–925.
60. Awad, M.I. (2006) *J. Appl. Electrochem.*, **36**, 1163–1168.
61. Scendo, M. (2007) *Corros. Sci.*, **49**, 3953–3968.
62. Ismail Khaled, M. (2007) *Electrochim. Acta*, **52**, 7811–7819.
63. Ju, H. and Li, Y. (2007) *Corros. Sci.*, **49**, 4185–4201.

64. Fuchs-Godec, R. (2007) *Electrochim. Acta*, **52** (15), 4974–4981.
65. Helal, N.H., El-Rabiee, M.M., Abd El-Hafez, G.M., and Badawy, W.A. (2008) *J. Alloys Comp.*, **456**, 372–378.
66. Ergun, U., Yuezer, D., and Emreguel, K.C. (2008) *Mater. Chem. Phys.*, **109** (2–3) 492–499.
67. Achary, G., Naik, A.Y., Kumar, S.V., Venkatesha, T.V., and Sherigara, B.S. (2008) *Appl. Surf. Sci.*, **254** (17), 5569–5573.
68. Yazdzad, A.R., Shahrabi, T., and Hosseini, M.G. (2008) *Mater. Chem. Phys.*, **109** (2–3), 199–205.
69. Zoebelein, H. and Böllert, V. (2001) *Dictionary of Renewable Resources*, Wiley-VCH Verlag GmbH, New York.
70. Rao, M.R., Palada, M.C., and Becker, B.N. (2004) *Agrofor. Syst.*, **61–62** (1–3), 107–122.
71. Kala, C.P., Dhyani, P.P., and Sajwan, B.S. (2006) *J. Ethnobiol. Ethnomed.*, **2**, 32.
72. Prajapati, N.D., Purohit, S.S., Sharma, A.K., and Kumar, T. (2003) *A Handbook of Medicinal Plants*, Agrobios (India), Jodhpur.
73. Harborne, J.B. (1973) *Phytochemical Methods: A Guide to Modern Techniques of Plant Analysis*, Chapman & Hall, London.
74. Pullaiah, T. (2002) *Medicinal Plants in India*, vol. 2, Regency Publisher, New Delhi.
75. Parikh, K.S. and Joshi, K.J. (2004) *Chem. Eng. World*, **39** (10), 64–68.
76. Orubite, K.O. and Oforka, N.C. (2004) *Mater. Lett.*, **58**, 1768–1772.
77. Nagarajan, P. and Sulochana, N. (2006) *J. Ind. Counc. Chem.*, **23** (2), 51–55.
78. Abdel-Gaber, A.M., Abd-El-Nabey, B.A., Sidahmed, I.M., El-Zayady, A.M., and Saadawy, M. (2006) *Corros. Sci.*, **48** (9), 2765–2779.
79. Ebenso, E.E. and Okafor, P.C. (2007) *Pigm. Resin Technol.*, **36** (3), 134–140.
80. Okafor, P.C., Osabor, V.I., and Ebenso, E.E. (2007) *Pigm. Resin Technol.*, **36** (5), 299–305.
81. Chauhan, L.R. and Gunasekaran, G. (2007) *Corros. Sci.*, **49** (3), 1143–1161.
82. El-Etre, A.Y. (2008) *Mater. Chem. Phys.*, **108** (2–3), 278–282.
83. Radojcic, I., Berkovic, K., Kovac, S., and Vorkapic-Furac, J. (2008) *Corros. Sci.*, **50** (5), 1498–1504.
84. El-Etre, A.Y. (2003) *Corros. Sci.*, **45**, 2485–2495.
85. Lakshmi, S.P., Chitra, A., Rajendran, S., and Anuradha, K. (2005) *Surf. Eng.*, **21** (3), 229–231.
86. Oguzie, E.E., Onuchukwu, A.I., Okafor, P.C., and Ebenso, E.E. (2006) *Pigm. Resin Technol.*, **35** (2), 63–70.
87. Abiola, O.K., Oforka, N.C., Ebenso, E.E., and Nwinuka, N.M. (2007) *Anti-Corros. Meth. Mater.*, **54** (4), 219–224.
88. Oguzie, E.E. (2007) *Corros. Sci.*, **49**, 1527–1539.
89. Oguzie, E.E., Onuoha, G.N., and Ejike, E.N. (2007) *Pigm. Resin Technol.*, **36** (1), 44–49.
90. Umoren, S.A., Obot, I.B., and Ebenso, E.E. (2008) *E-J. Chem.*, **5** (2), 355–364.
91. (a) Putilova, I.N., Balezin, S.A., and Barannik, V.P. (1960) *Metallic Corrosion Inhibitors*, Pergamon Press, Oxford, London; (b) Baldwin, J. (1895) Details – First Patent for an Inhibitor, issued to British Patent No. 2,327.
92. Muller, B. (2002) *Corros. Sci.*, **44**, 1583–1591.
93. Obot, I.B. and Obi-Egbedi, N.O. (2010) *J. Appl. Electrochem.*, **40** (11), 1977–1984.
94. Umoren, S.A. (2008) *Cellulose*, **15**, 751–761.
95. Rani, P.D. and Selvaraj, S. (2010) *J. Phytol.*, **2** (11), 58–64.
96. Noor, E.A. (2008) *J. Eng. Appl. Sci.*, **3**, 23.
97. Buchweishaija, J. and Mhinzi, G.S. (2008) *Port. Electrochim. Acta*, **26**, 257.
98. Oguzie, E.E. (2008) *Corros. Sci.*, **50**, 2993.
99. Okafor, P.C., Ikpi, M.E., Uwaha, I.E., Ebenso, E.E., Ekpe, U.J., and Umoren, S.A. (2008) *Corros. Sci.*, **50**, 2310.
100. Valek, L. and Martinez, S. (2007) *Mater. Lett.*, **61**, 148.
101. Singh, A., Singh, V.K., and Quraishi, M.A. (2010) *J. Mater. Environ. Sci.*, **1** (3), 162–174.
102. Quraishi, M.A., Singh, A., Singh, V.K., Yadav, K.D., and Singh, A.K. (2010) *Mater. Chem. Phy.*, **122**, 114.

103. Quraishi, M.A., Yadav, D.K., and Ishtiaque, A. (2009) *Open Corros. J.*, **2**, 56.
104. Noor, E.A. (2009) *J. Appl. Electrochem.*, **39**, 1465.
105. De Souza, F.S. and Spinelli, A. (2009) *Corros. Sci.*, **51**, 642.
106. Raja, P.B. and Sethuraman, M.G. (2008) *Mater. Lett.*, **62**, 1602.
107. El-Etre, A.Y. (2003) *Corros. Sci.*, **45**, 2485.
108. Badiea, A.M. and Mohana, K.N. (2009) *J. Mater. Eng. Perform.*, **18**, 1264.
109. Chauhan, L.R. and Gunasekaran, G. (2007) *Corros. Sci.*, **49**, 1143.
110. El-Etre, A.Y., Abdallah, M., and El-Tantawy, Z.E. (2005) *Corros. Sci.*, **47**, 385.
111. Orubite, K.O. and Oforka, N.C. (2004) *Mater. Lett.*, **58**, 1768.
112. Evic Grassino, A.N., Grabaric, Z., Pezzani, A., Fasanaro, G., and Voi, A.L. (2009) *Food Chem. Toxicol.*, **47**, 1556.
113. Torres-Acosta, A.A. (2007) *J. Appl. Electrochem.*, **37**, 835.
114. Pandian, B.R. and Mathur, G.S. (2008) *Mater. Lett.*, **62** (1), 113–116.
115. Trabanelli, G. and Mansfeld, F. (eds) (1987) *Corrosion Mechanisms*, Marcel Dekker, Inc., New York, p. 119.
116. Sharma, S.K., Mudhoo, A., Jain, G., and Sharma, J. (2010) *Green Chem. Lett. Rev* ., **3** (1), 7–10.
117. Sharma, S.K., Mudhoo, A., Jain, G., and Khamis, E. (2009) *Green Chem. Lett. Rev.*, **2** (1), 47–51.
118. Sharma, S.K., Mudhoo, A., and Khamis, E. (2009) *J. Corros. Sci. Eng.*, **11** (14), 1–25.
119. Sharma, S.K., Jain, G., Sharma, J., and Mudhoo, A. (2009) *Int. J. Appl. Chem.*, **6** (1), 83–94.
120. Sharma, S.K., Jain, G., Mudhoo, A., and Sharma, J. (2009) *Rasayan J. Chem.*, **2** (2), 332–339.
121. Nair, R.N., Sharma, S., Sharma, I.K., Verma, P.S., and Sharma, A. (2010) *Rasayan J. Chem.*, **3** (4), 783–795.
122. Sharma, A., Verma, P.S., Sharma, I.K., and Nair, R.N. (2009) *Proc. World Acad. Sci. Eng. Technol. (PWASET)*, **39**, 745–753 (ISSN: 2070-3740).
123. Nair, R.N., Kharia, N., Sharma, I.K., Verma, P.S., and Sharma, A. (2007) *J. Electrochem. Soc. India*, **56** (1/2), 41–47.
124. Nair, R.N., Sanjay, Sharma, I.K., Verma, P.S., and Sharma, A. (2007) CRSI Sponsored 9th National Symposium in Chemistry (NSC-9), Department of Chemistry, University of Delhi, Delhi, February 1–4, 2007.
125. Nair, R.N., Sharma, I.K., Verma, P.S., and Sharma, A. (2009) National Seminar on Recent Trends in Chemistry (NSRTC 09), January 21–22, 2009, organized by Department of Chemistry, Punjabi University, Patiala.
126. Yadav, S., Sharma, I.K., Verma, P.S., and Sharma, A. (2010) National Convention of Electrochemists (NCE-15), February 18–19, 2010, at VIT University, Vellore, organized by SAEST in association with CECRI, Karaikudi and VIT University, Vellore.
127. Sharma, S., Nair, R.N., Verma, P.S., and Sharma, A. (2010) 9th International Symposium on Advances in Electrochemical Science and Technology (ISAEST-9) Dec 2–4, 2010, organized by SAEST and CECRI, Karaikudi held at Chennai.
128. Sharma, S., Sharma, I.K., Verma, P.S., and Sharma, A. (2010) 12th International Conference on Applied Physical Sciences (CONIAPS XII), December 22–24, 2010, organized by the University of Rajasthan, Jaipur.
129. Hatch, J. E. (ed.) (1984) *Aluminium: Properties and Physical Metallurgy*, Aluminium Association Inc., ASM International, Metals Park.
130. Davis, J.R. (1993) *ASM Specialty Handbook: Aluminum and Aluminum Alloys*, ASM International, Materials Park, p. 31.
131. Kammer, C. (1999) *Aluminum Handbook*, 1st edn, vol. 1, Aluminum Verlag, Düsseldorf, p. 125.
132. Vargel, C. (2004) *Corrosion of Aluminum*, Elsevier, Oxford.
133. Spath, E. and Englaender, G. (1935) Occurrence of piperidine in black pepper. *Ber. Dtsch. Chem. Ges. [Abt.] B: Abh.*, **68B**, 2218–2221.

134. Evans, W.C. (1996) *Trease and Evans' Pharmacognosy*, WB Saunders Company Ltd, London.
135. Samuelsson, G. (1999) *Drugs of Natural Origin: A Textbook of Pharmacognosy*, Swedish Pharmaceutical Press, Stockholm.
136. Tewtrakul, S. *et al.* (2000) *J. Essent. Oil Res.*, **12**, 603–608.
137. Singh, G., Marimuthu, P., Catalan, C., and de Lampasona, M.P. (2004) *J. Sci. Food Agric.*, **84** (14), 1878–1884.
138. Muralidharan, S., Chandrasekar, R., and Iyer, S.V.K. (2000) *Proc. Ind. Acad. Sci. (Chem. Sci.)*, **112** (2), 127–136.
139. Khaled, K.F., Babic'-Samardzija, K., and Hackerman, N. (2004) *J. Appl. Electrochem.*, **34**, 697–704.
140. Pandian, B.R. and Mathur, G.S. (2008) *Mater. Lett.*, **62** (17–18), 2977–2979.
141. Aramaki, K. and Hackerman, N. (1968) *J. Electrochem. Soc.*, **115**, 1007.
142. Noor, E.A. (2007) *Int. J. Electrochem. Sci.*, **2**, 996–1017.
143. Abiola Olusegun, K. and James, A.O. (2010) *Corros. Sci.*, **52** (2), 661–664.
144. Al-Turkustani, A.M., Arab, S.T., and Al-Dahiri, R.H. (2010) *Mod. Appl. Sci.*, **4** (5), 105–124.
145. Lebrini, M., Robert, F., and Roos, C. (2010) *Int. J. Electrochem. Sci.*, **5**, 1698–1712.
146. Ayssar, N., Ideisan, A-A., Ibrahim, A.-R., and Maysoon, A.-K. (2010) *Int. J. Corros.*, 2010, 9. (Article ID: 460154). doi: 10.1155/2010/460154
147. Okafor Peter, C., Ebenso Eno, E., and Ekpe Udofot, J. (2010) *Int. J. Electrochem. Sci.*, **5**, 978–993.
148. Hazwan, H.M. and Jain, K.M. (2010) *J. Phys. Sci.*, **21** (1), 1–13.
149. Rahim, A.A. and Kassim, J. (2008) *Recent Pat. Mater. Sci.*, **1**, 223–231.
150. Rani, P.D. and Selvaraj, S. (2010) *Rasayan J. Chem.*, **3** (3) 473–482.
151. Ostovari, A., Hoseinieh, S.M., Peikari, M., Shadizadeh, S.R., and Hashemi, S.J. (2009) *Corros. Sci.*, **51** (9), 1935–1949.
152. Ouariachi, E.E., Paolini, J., Bouklah, M., Elidrissi, A., Bouyanzer, A., Hammouti, B., Desjobert, J.-M., and Costa, J. (2010) *Acta Metall. Sin. (Engl. Lett.)*, **23** (1), 13–20.
153. Lebrini, M., Robert, F., and Roos, C. (2011) *Int. J. Electrochem. Sci.*, **6**, 847–859.
154. Fabrizio, Z. and Hashi, O.I. (1985) *Surf. Technol.*, **24** (4), 391–399.
155. Davis, G.D., von Fraunhofer, J.A., Krebs, L.A., and Dacres, C.M. (2001) Corrosion 2001, Paper 1558.
156. Li, X.-H., Deng, S.-D., and Fu, H. (2010) *J. Appl. Electrochem.*, **40**, 1641–1649. doi: 10.1007/s10800-010-0151-5
157. Gherraf, N., Namoussa, T.Y., Ladjel, S., Ouahrani, M.R., Salhi, R., Belmnine, A., Hameurlain, S., and Labed, B. (2009) *Am.-Eurasian J. Sustain. Agric.*, **3** (4), 781–783.
158. Torres-Acosta, A.A. (2007) *J. Appl. Electrochem.*, **37**, 835–841. doi: 10.1007/s10800-007-9319-z
159. Eddy, N.O., Awe, F., and Ebenso, E.E. (2010) *Int. J. Electrochem. Sci.*, **5**, 1996–2011.

7
Innovative Silanes-Based Pretreatment to Improve the Adhesion of Organic Coatings

Michele Fedel, Flavio Deflorian, and Stefano Rossi

7.1
Introduction

Commonly, metal surfaces are pretreated prior the application of an organic coating, thus the overall properties of the complete protection system improve remarkably [1]. For many decades, chromium conversion treatments, involving the use of chromic-acid-containing Cr^{6+} species, have been used for their corrosion protection and adhesion promotion performances [2]. Their success was because of the formation of a few decades of nanometers of complex mixture of chromium compounds as a consequence of Cr^{6+} reduction; the reduction of either water, hydrogen ion, or dissolved oxygen to form hydroxyl ions at the metal surface; and the presence of oxidizing Cr^{6+} species, responsible for the self-healing potential in the presence of defects [3]. Moreover, the chromate conversion coatings promote a very good initial adhesion with paint, lacquer, and organic finishes [4]. However, in the early 1990s, the restrictions on the use of Cr^{6+}-based conversion treatments became more and more stringent owing to the health and environmental issues related to the use of these kinds of finishing [5]. In fact, it was proved that hexavalent chromium can cause lung cancer in humans as a result of inhalational exposure in certain occupational settings. Hexavalent chromium was recognized as a known carcinogen by the World Health Organization (WHO) and by the U.S. Environmental Protection Agency (EPA), but it is also toxic and can lead to allergic contact dermatitis as a result of direct skin exposure to powders or liquids containing hexavalent chromium [6, 7]. In addition, Cr^{6+} has a high oxidizing potential, and thus, it is also dangerous for the environment. Aiming at the replacement of chromate conversion treatment in the metal finishing industry, in the last two decades, several different *environmentally friendly* pretreatments were studied. Among them, a lot of studies deal with:

- pretreatments with protecting mechanism similar to Cr^{6+}, acting as passivating agents, such as molybdates, permanganates, vanadates, and tungstates [8, 9];
- rare earth based conversion treatment, mainly based on cerium and lanthanum salts [10–12];

Green Corrosion Chemistry and Engineering: Opportunities and Challenges, First Edition.
Edited by Sanjay K. Sharma.

- Cr^{3+} conversion layers that do not contain oxidizing Cr^{6+} species responsible for the self-healing of flaws [13];
- fluotitanate and fluozirconate conversion treatment obtained using acidic compounds of titanium and zirconium oxides [14];
- phosphate conversion treatment that leads to the formation of phosphate compound on metals surface [15, 16];
- metal alkoxides and/or organically modified metal alkoxides applied onto the metal substrates via sol–gel route [17, 18].

In this panorama, in the last few decades, thin sol–gel films from hybrid silicon alkoxide precursors are of growing interest as efficient pretreatments to promote the adhesion between organic coatings and metal surfaces addressing the environmental issues [19]. In particular, in the very last few years, the development of water based silicon alkoxides sol–gel pretreatments promoted a great diffusion of these conversion layers as a "green" technology [20]. The first part of this chapter deals with the basic chemistry of silicon alkoxide molecules reported, as well as their main physical properties and the effect of different parameters on the kinetics of the hydrolysis and condensation reactions among the different. Moreover, the formation mechanism of sol–gel coatings obtained from silicon alkoxide precursors is analyzed, with particular interest to the dip coating technique. Subsequently, two different sections are devoted to the interaction between metal surfaces and silicon alkoxide molecules and the interaction between organic polymeric material and silicon alkoxides, which are described suggesting the possible bonding mechanisms. Eventually, the corrosion protection properties of the silicon alkoxide sol–gel films as protective layer on different metals are discussed. In the last section, the performances of experimental sol–gel films applied as coupling agent are explained, reporting a few real cases.

7.2
Hybrid Silane Sol–Gel Coatings

In the last two decades, the potential of organofunctional silane molecules for the replacement of chrome conversion treatments has been the topic of a huge number of studies [21–28]. Silanes have been widely studied as coupling agents between inorganic and organic materials since the first works of Plueddemann [29]. Concerning corrosion science, a lot of efforts have been made to apply these materials as adhesion promoters between metallic substrates and the organic coatings used for protection against the corrosion phenomena [30–32]. The chemistry of silicon alkoxides and the mechanism of interaction of these molecules with a metallic substrate and an organic coating have been widely investigated [33–35]. In the next section, a brief summary of the most important chemical properties of the silicon alkoxide precursors used for corrosion protection purposes is reported.

7.2.1
Basic Chemistry of the Silicon Alkoxides and Organofunctional Silicon Alkoxides

Silane molecules are commonly applied onto metal surfaces by means of the hydrolysis and condensation of silicon alkoxides. Starting from the molecular precursors, it is possible to obtain solid networks by means of the *sol–gel* process [36]. The sol–gel process involves hydrolysis reactions of silicon alkoxide and further condensation reactions among the different hydrolyzed molecules. The hydrolytic decomposition of the alkoxide leads to the formation of hydrated metal oxides, which are then subject to dehydration [37]. This is the main principle of the sol–gel technology of inorganic materials. Silicon alkoxides easily undergo hydrolysis, forming hydrated oxides not containing any extra ions (in contrast to precipitation from the aqueous solutions of inorganic salts).

Schematically, the formation of an inorganic network starting from silicon alkoxides can be expressed as

$$Si(OR)_n \xrightarrow[nR(OH)]{+H_2O} Si(OH)_n \xrightarrow[H_2O]{Drying} Si_xO_y$$

Being R an organic group such as methyl ($-CH_3$) and ethyl ($-CH_2-CH_3$).

The deposition of hybrid silane sol–gel coatings generally occurs in four stages [38]:

1) hydrolysis of the silicon alkoxide molecules;
2) condensation reactions and polymerization of monomers to form chains and particles;
3) growth of the particles (or chains);
4) agglomeration of the polymer structures followed by the formation of a network that entraps the remaining solution (gelation).

After gelation, aging (increase in cross-linking associated with shrinkage), drying (loss of volatile compounds from the pore of the structure), and densification (collapse of the open structure leading to the formation of a dense structure) [39] can be promoted.

The hydrolysis occurs in presence of water. This process is the basis for the sol–gel method of preparation of coatings. With respect to other metal alkoxides, such as Sn,Ti,Zr,Ce$(OR)_4$, the hydrolysis of silicon alkoxides is not very quick because of their low reactivity. The chemical reactivity of the metal alkoxides toward nucleophilic reactions depends on the strength of the nucleophile, the electrophilic character of the metal atom, and its ability to increase its coordination number. In this sense, the electronegativity of Si is rather high (1.74, therefore the electrophilicity is low), its partial charge is rather small (0.32), and it remains fourfold coordinated in the precursor as well as in the oxide [40]. Therefore, with respect to the other metal alkoxides, silicon alkoxides are weak Lewis acids [41]. Thus, the hydrolysis rate (as well as the condensation rate) of silicon alkoxides is low and must be increased by adding catalysts. Commonly, the rate of the hydrolysis (and condensation) reactions is increased by acid or base catalysis.

Under acidic catalyzed conditions, inorganic or organic acids protonate the negatively charged OR groups (bonded to the silicon atom by Si–OR bonds), leading to the formation of pentacoordinate intermediates by the interaction with water. In fact, after protonation, the electron density is withdrawn from the central silicon atom, making it more electrophilic and thus more susceptible to the attack by water, leading to an increase in the rate of hydrolysis. Acid catalysts lead to the prolongation of alkoxy groups that become better leaving groups: a nucleophilic substitution of OR groups by OH groups occurs [42]. A schematic representation of the hydrolysis reactions under acidic conditions is reported in Figure 7.1.

Under alkaline catalyzed conditions, the silicon atom is attacked directly by hydroxide, leading to the formation of pentacoordinate intermediates by the interaction with water. The reaction proceeds by the nucleophilic attack of an OH^- ion on the silicon atom. A schematic representation of the hydrolysis reactions under basic conditions is reported in Figure 7.2.

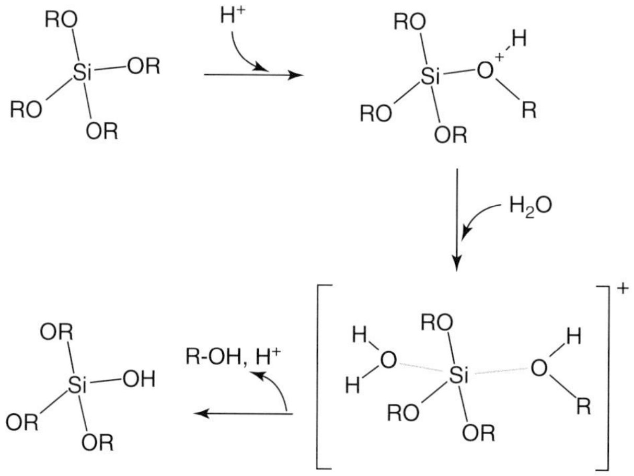

Figure 7.1 Hydrolysis reaction in acidic conditions.

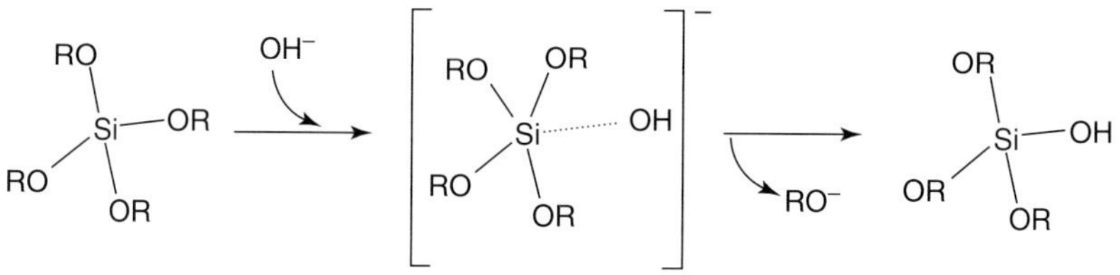

Figure 7.2 Hydrolysis reaction in alkaline conditions.

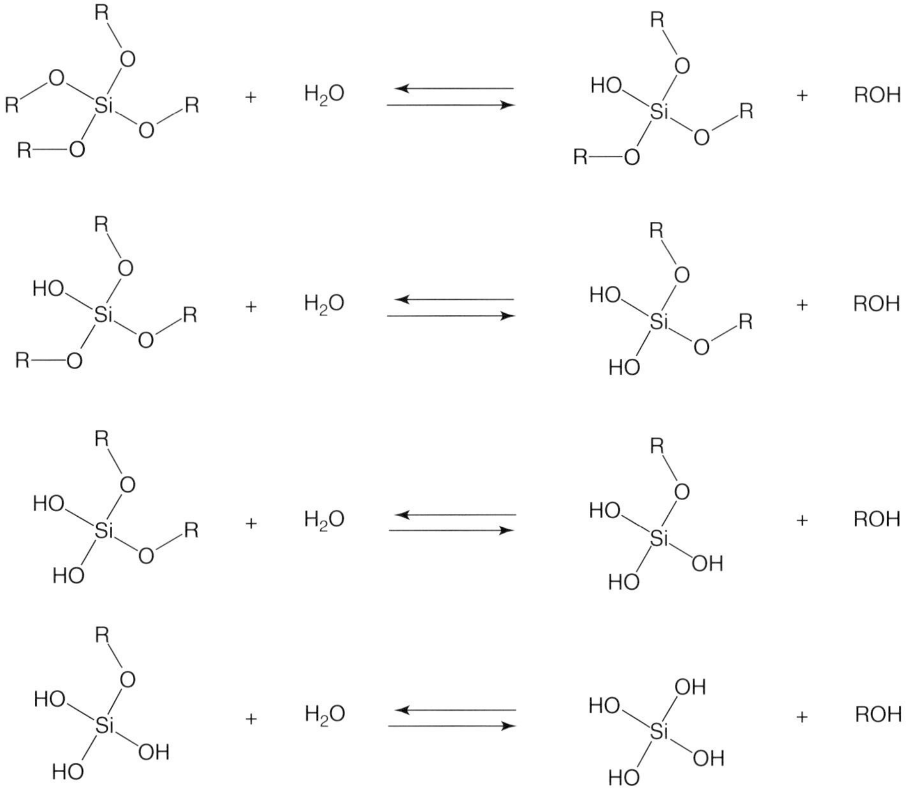

Figure 7.3 Hydrolysis reactions for a silicon tetralkoxide: different steps.

The different catalysts not only affect the hydrolysis and condensation reaction but also influence the shapes of the polymeric structure being formed. In fact, if an acidic catalysis is carried out, linear molecules with a low degree of cross-linking tend to form. On the other hand, if a basic catalysis is carried out, branched clusters of hydrolyzed silicon alkoxide molecules tend to form.

Regardless of the catalysis conditions, tetraalkoxysilane molecules displace Si–OR groups with Si–OH groups (where R is an organic group) following approximately the steps schematically reported in Figure 7.3 [43].

Notice that alcohols are byproducts obtained by the hydrolysis reaction. Instead of the complete hydrolysis to give silicic acid $Si(OH)_4$, the condensation between two silanol groups may occur to form a Si–O–Si network with the elimination of water [44]. Although this is the most common and faster mechanism to form a polymerized network, other condensation mechanisms (which are reported in Figure 7.4) are possible [45].

In particular, it has been established that in principle, condensation at higher temperatures may take place by elimination of an ether (R_2O), following the third

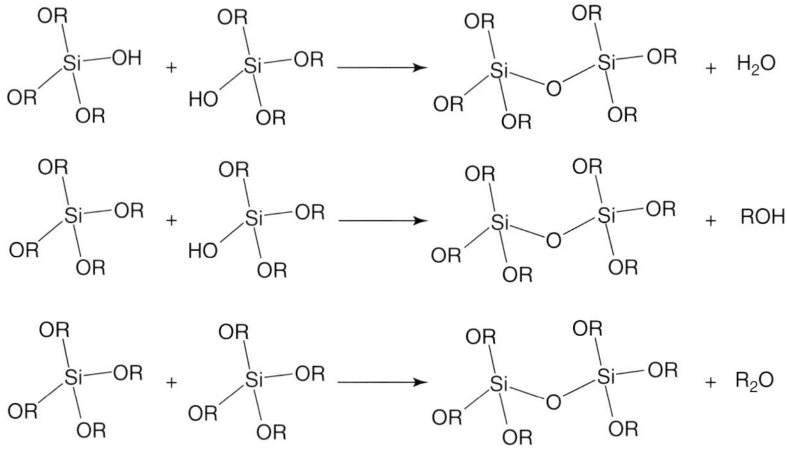

Figure 7.4 Condensation, alcolation, and oxolation reactions.

reaction in Figure 7.4 [46]. The second reaction showed in Figure 7.4 consists of an alcolation: one silanol group and one silicon alkoxide form an oxygen bridge between the silicon atoms (Si–O–Si bond) with elimination of an alcohol.

The three different mechanisms lead to an increase in the connectivity of the network produced by condensation reactions. Actually, as previously anticipated, the main mechanism through which the polymerization takes place is condensation (first reaction depicted in Figure 7.4), through which the silanols form a Si–O–Si network with the elimination of water. As well as for the hydrolysis reactions, the rate of the condensation reactions can be increased by means of acidic or basic catalysis. A schematic representation of the condensation reactions under acidic conditions is reported in Figure 7.5.

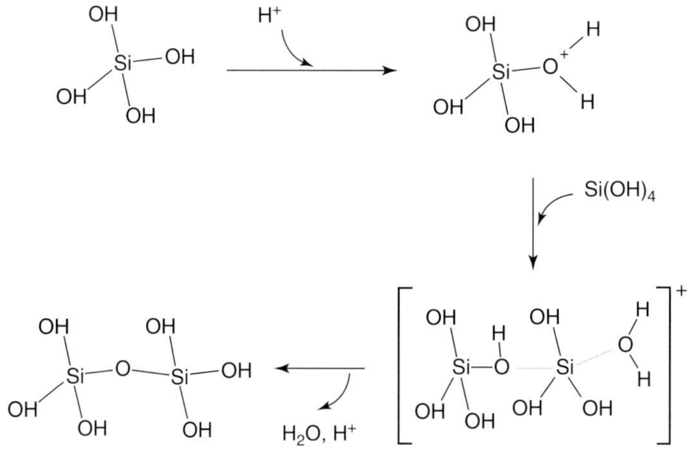

Figure 7.5 Condensation reactions with acid catalyst.

If the condensation is base catalyzed, the condensation reactions follow the steps schematically reported in Figure 7.6.

It is worthy to highlight that hydrolysis and condensation reactions are affected by a huge number of parameters, which can affect and modify the reaction kinetics as well as the morphology of the structure being formed. The main parameters are [47]

- pH of the mixture containing the silicon alkoxides;
- nature of the alkoxides;
- alkoxides/water ratio;
- nature of the solvent.

As explained before, the pH of the solution containing the silicon alkoxides dramatically affects the rates of the hydrolysis and condensation reactions. Figure 7.7

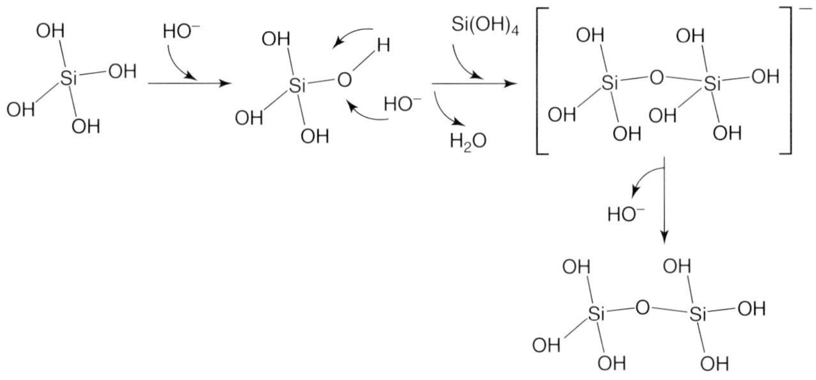

Figure 7.6 Condensation reactions with base catalyst.

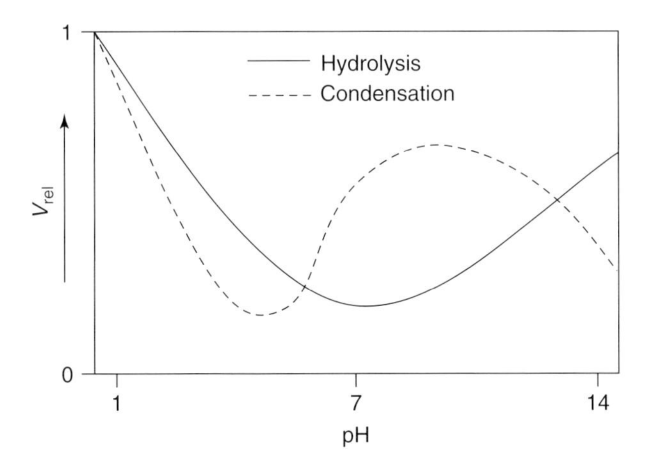

Figure 7.7 Hydrolysis and condensation with the pH of the solution.

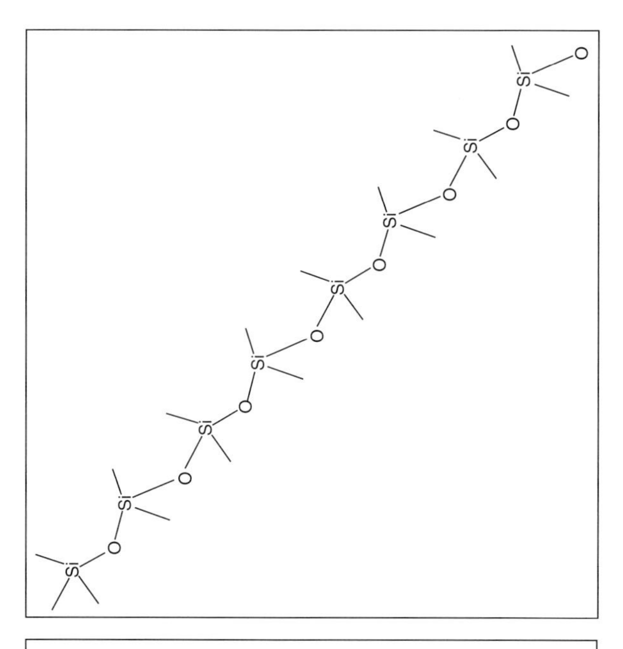

Figure 7.8 Linear structures: acid catalyst.

[47] shows, for example, the effect of pH on the rates of the hydrolysis and condensation reactions for a tetraalkoxysilane.

Considering the hydrolysis reaction, the minimum reaction rate is at pH 7, while concerning the condensation rate, the minimum is between pH 4 and 5. As far as the condensation reaction is concerned, the minimum of the reaction rate corresponds to the isoelectric point of silica. Looking at the graph depicted in Figure 7.7, it is possible to appreciate that at low pH, the rate of hydrolysis is faster than the rate of condensation. For neutral and alkaline pH, hydrolysis is the slower process, except for highly alkaline solutions (approaching the pH 13–14). The different kinetics of hydrolysis and condensation processes in the presence of the different catalysts affect the structure of the polymeric structure that is forming. In fact, under acidic conditions, hydrolysis is faster than condensation, and thus, long polymeric chains with small branches are formed. On the other hand, if condensation is faster than hydrolysis, hydrolyzed species are quickly consumed, leading to the formation of highly branched structures.

The chemical nature of the organic substituent (R) affects the rate of the hydrolysis reactions. In particular, the steric hindrance of the organic substituent (R) significantly modifies the hydrolysis reaction rate. Thus, the higher the length of the polymeric chain R, the slower the rate of hydrolysis, even in presence of a catalyst. Similarly, increasing the degree of branched structure of the R substituent decreases the reaction rate of the hydrolysis process. In addition, the hydrolysis rate is affected by the electron density on the silicon atom. In fact, the efficiency of the nucleophilic or electrophilic catalyst is strongly influenced by the electron density on the silicon atom. The electron density on the silicon atom is lower when Si is in the condensed dimeric form of ≡Si–O–Si≡ and is higher when the silicon atom is bonded to R or OR groups (i.e., ≡Si–R, ≡Si–OR). In this sense, the hydrolysis and condensation reaction rates increase with a decrease in electron density.

To complete the hydrolysis of a silicon tetraalkoxide, four equivalents of water is needed (in the absence of condensation reactions). In general, the water/alkoxy group ratio influences the hydrolysis and condensation kinetics. By increasing the water content, the hydrolysis reactions are favored with respect to the condensation reactions. In fact, in principle, the condensation reactions are reversible, and therefore, high amounts of water (i.e., high water/alkoxy group ratio) promote the formation of silanol species with respect to the condensated species. The water/alkoxy group ratio affects also the morphology of the forming network. In fact, this parameter affects the shape of the polymers that are forming as the kind of catalyst, leading to the formation of linear structures or highly branched structures. Linear polymeric structures (Figure 7.8) are formed when an acid catalyst is used and/or in presence of a high water/alkoxy group ratio.

On the other hand, basic catalysts and/or low water/alkoxy group ratio promotes the formation of branched structures (Figure 7.9).

As far as organofunctional alkoxysilanes are concerned, the mechanisms through which hydrolysis and condensation reactions occur are approximately the same as those previously described for tetraethoxysilanes. Organofunctional alkoxysilanes are hybrid molecules containing one or more nonhydrolyzable substituent bonded

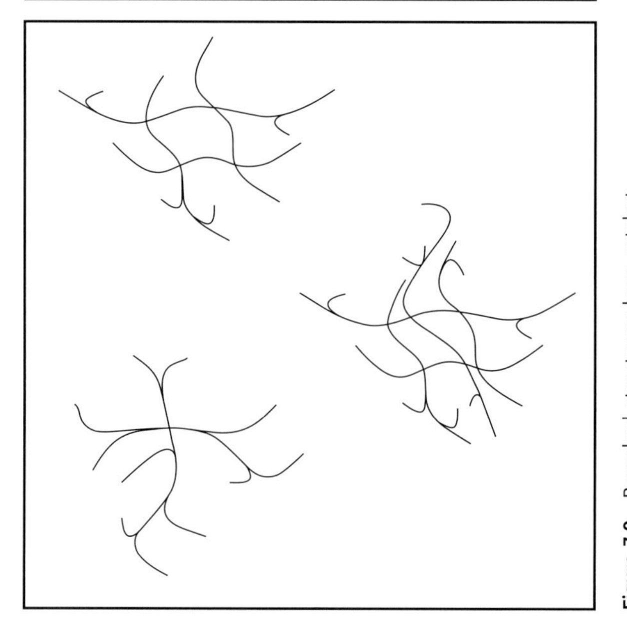

Figure 7.9 Branched structures: base catalyst.

to the silicon atom. The structure of this kind of molecules is approximately

$$(\mathrm{R-O})_{4-n}\,\mathrm{SiR}_n^*$$

where 4−*n* alkoxy groups (R–O) and *n* nonhydrolyzable substituents (R*) are present [48]. The R* group is the carbon chain containing a functional group (amino, urethanic, epoxy, etc.). The silicon/carbon bond is not hydrolyzable [49], and therefore, it is stable in water at room temperature. Examples of the most common organofunctional silicon alkoxides used for corrosion protection purposes are reported in Figure 7.10.

Figure 7.11 shows, for example, the corresponding hydrolysis reaction for an organofunctional silane molecule characterized by one organofunctional group R*.

Also, in this case, the presence of a basic or acid catalyst increases the hydrolysis and condensation reaction rates. It was demonstrated [50] that also, in this case, the slower rate of hydrolysis corresponds to approximately pH 7 and that an acidic or alkaline condition promote acceleration of both the hydrolysis and condensation reactions.

Organofunctional silane molecules usually contain one organically modified group (stable in water), which ensures a second step of polymerization if either thermal or photochemical curing is provided [51]. Moreover, the organic functionalization of the alkoxysilane molecules with organic groups allows the sol–gel film to chemically interact with organic molecules, such as polymers. A sol–gel film made up of organofunctional molecules contains organic chemical groups able to interact with an organic coating. In this sense, it is possible to design a sol–gel film with organically modified silane molecules, which can act as a coupling agent between metallic inorganic substrates and an organic coating [52]. In fact, organofunctional

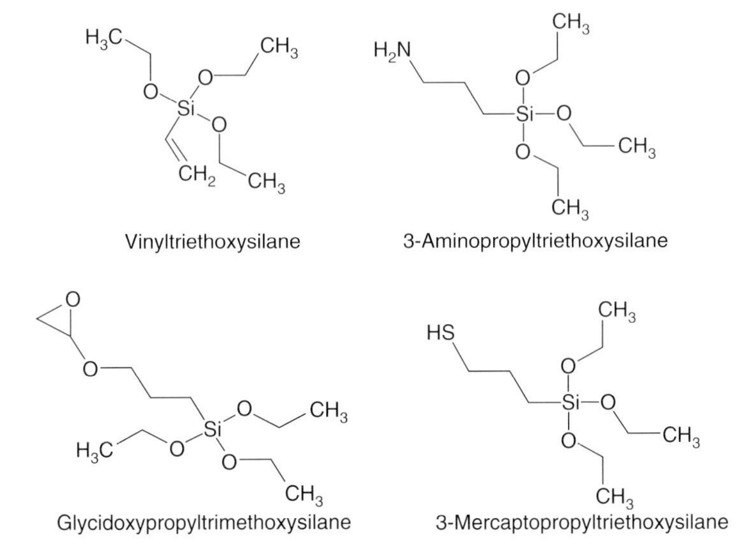

Figure 7.10 Examples of organofunctional silanes.

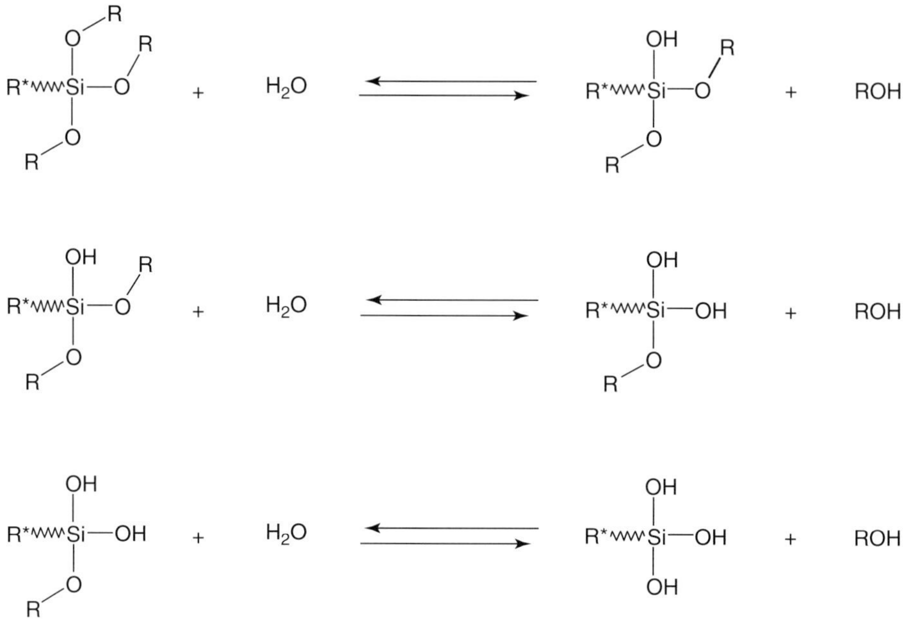

$$R^*\sim Si(OR)_3 + H_2O \rightleftharpoons R^*\sim Si(OH)(OR)_2 + ROH$$

$$R^*\sim Si(OH)(OR)_2 + H_2O \rightleftharpoons R^*\sim Si(OH)_2(OR) + ROH$$

$$R^*\sim Si(OH)_2(OR) + H_2O \rightleftharpoons R^*\sim Si(OH)_3 + ROH$$

Figure 7.11 Hydrolysis reactions for an organically modified alkoxide: different steps.

silane molecules contain both alkoxide groups, which hydrolyze on reacting with water; generate the sol–gel network; and bond to the metal substrate and organic functionalities that are embedded in the silane film. The organic functionalities can polymerize and chemically interact with an organic coating. The latter mechanism of interaction is discussed in detail in Section 7.2.4.

7.2.2 Dip Coating

Dip coating is a widespread technique used to apply sol–gel films onto metallic substrates. By dip coating, it is possible to obtain a thin sol film on metal substrates. As far as application of metal substrate for adhesion promotion purposes is concerned, the percentage of silicon alkoxides in the solution is very low, around 1–10%. In fact, usually, thin conversion films are needed, and thus, much diluted silicon alkoxide solutions are prepared [53].

The dip coating technique consists of three different steps.

1) **Immersion** of the sample in a homogeneous solution of hydrolyzed silicon alkoxides and **maintenance** in this solution for a certain time.
2) **Withdrawal** of the sample from the solution at a certain velocity: in this phase, a film starts to form on the surface of the sample.

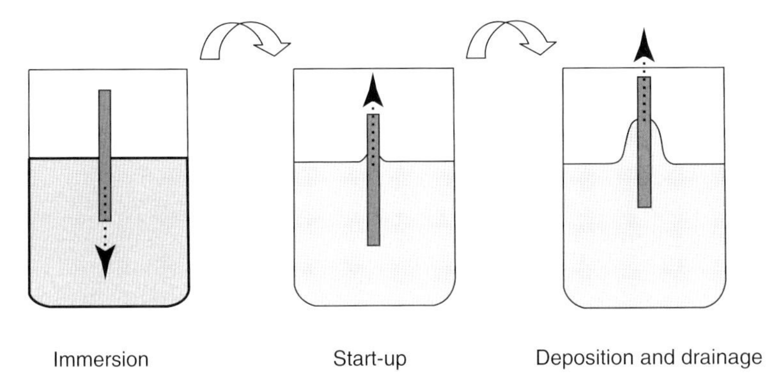

Figure 7.12 Schematic representation of the first stages of the dipping procedure.

3) **Aging** or **curing** of the film obtained on the surface of the sample: cross-linking occurs during shrinkage, and eventually, the open structure collapses, leading to the formation of a dense material.

Actually, the real film deposition is performed by withdrawing a metal substrate from a bath containing a homogeneous solution of hydrolyzed silicon alkoxides, which would finally form the desired structure. Figure 7.12 schematically represents the first stage of the dipping procedure [54].

Since solution containing low percentage (1–10 wt%) of silicon alkoxides can be considered as Newtonian fluids and when both withdrawal speed and viscosity are relatively low, the coating thickness (t) obtained by dipping the metal sample in a homogeneous solution of hydrolyzed silicon alkoxides can be calculated by the equation developed by Landau and Levich [55] and Derjaguin (Eq. (7.1)).

$$t = 0.94 \times \frac{(\eta \times \nu)^{\frac{2}{3}}}{\gamma^{\frac{1}{6}} \times \sqrt{\rho \times g}} \tag{7.1}$$

which means that the coating thickness (t) can be calculated using the solution viscosity (η) and density (ρ), the withdrawing speed (ν), the surface tension of the metal (γ), and the acceleration due to gravity (g).

At small withdrawing speed, the static meniscus is hardly deformed by the entrained film. In particular, only the top of the static meniscus is affected by the deposition [56].

Therefore, the rheological properties of the solution (related to the physical properties of the fluid, i.e., fluid density and viscosity), the surface tension of the metal, and the withdrawal speed influence the thickness of the final sol–gel coating [57] of the metal substrate. In particular, considering Eq. (7.1), the higher the viscosity and the withdrawal speed, the higher the thickness of the final sol–gel coating [58]. Figure 7.13 [59] shows what happens during the extraction of a sample from a homogeneous solution of hydrolyzed silicon alkoxides.

The evaporation of the solvent and gelation may occur during the extraction from the flask containing a homogeneous solution of hydrolyzed silicon alkoxides. Thus,

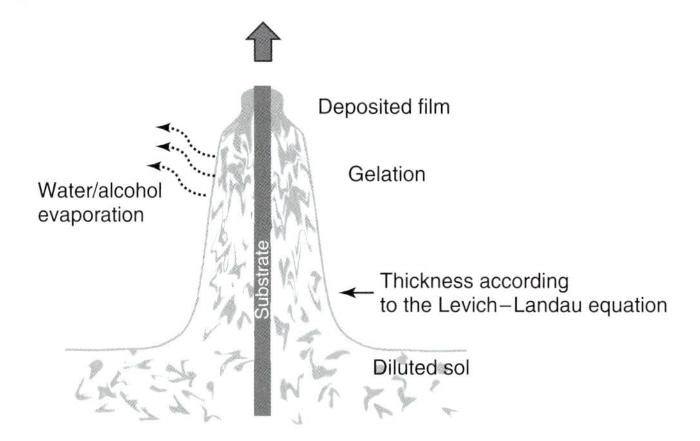

Figure 7.13 Schematic representation of evaporation and gelation [59].

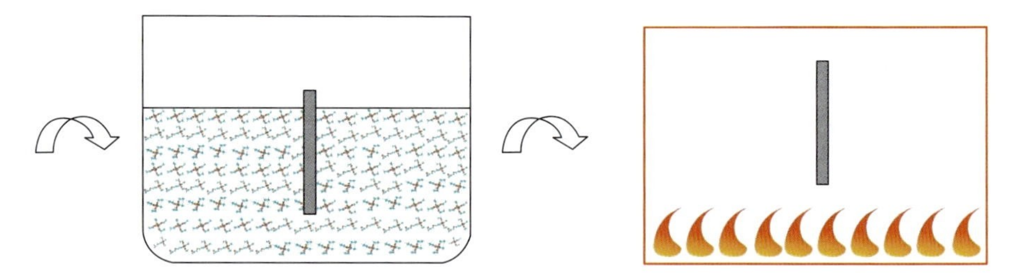

Figure 7.14 Last stage of the sol–gel film production by dipping: heat treatment.

it is clear that the physical properties of the solvent also affect the properties of the final sol–gel coating applied to the metal substrate.

Commonly, the metal samples are maintained in the homogeneous solution of hydrolyzed silicon alkoxide molecules for a lapse of time that varies from tens of seconds to a few minutes. During this lapse of time, the formation of weak hydrogen bonds between the metal hydroxides in equilibrium on the surface of the metal takes place, and the molecules self-organize themselves onto the surface of the metal.

After the withdrawal, commonly, a heat treatment is performed in order to provide enough thermal energy to promote the condensation reactions. In fact, by supplying heat to the sol–gel film, it is possible to increase the solvent evaporation rate and, at the same time, increase the kinetics of the condensation reactions. Figure 7.14 schematically depicts the last stages of the deposition of the sol–gel layer (heat treatment).

7.2.3 Interaction between Silicon Alkoxides and Metallic Substrates

Considering the interaction between a metal substrate and hydrolyzed silicon alkoxide molecules, actually we refer to the bonding mechanism between the silanols and the oxide of the metal substrate because of the metal to protect is always covered by its natural oxide and hydroxide. Moreover, according to the study of Plueddemann [29], oxide surfaces with a high density of hydroxyl groups promote the interaction between silanols and the inorganic substrate. As far as the application of silane sol–gel films onto metal is concerned, the hydrolysis of the alkoxysilanes has a dramatic effect on the final properties of the sol–gel film. In fact, it is necessary to have sufficient number of silanol (Si–OH) groups to interact with the metal substrate [60], because only the hydrolyzed species are able to interact with the metal hydroxides of the substrate, leading to the formation of weak hydrogen bonds, in a first step, and strong covalent bonds in a second step. Therefore, the number of silanols in the solution strongly affects the corrosion protection performances of the final sol–gel film. In fact, the competition between the hydrolysis and condensation reactions implies that after a certain degree of hydrolysis, the number of silanols in the solution begins to decrease as a result of the condensation reactions. If the condensation reactions take place in the solution, rather than on the surface of the metal to protect, the number of silanol (Si–OH) groups to interact with the metal hydroxides (Me–OH) decreases [58]. In addition, when the molecules grows due to condensation reactions the average distance between the metal–oxygen–silicon bonds on metal surface increases. Figure 7.15a

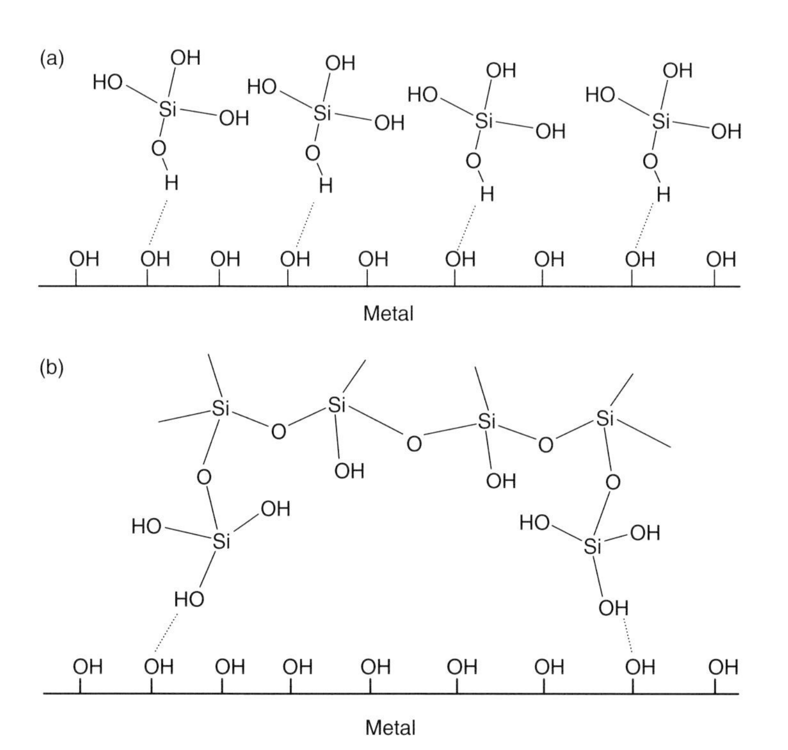

Figure 7.15 (a) Interaction between hydrolyzed molecules and a metal substrate, (b) and interaction between branched condensed molecules and a metal substrate.

shows schematically the average distance between the metal–oxygen–silicon bonds during the initial stage of interaction on the metal surface when only hydrolyzed molecules interact with the substrate.

On the other hand, Figure 7.15b depicts the average distance between the metal–oxygen–silicon bonds during the initial stage of the interaction on the metal surface when branched condensed molecules interact with the substrate.

The correlation between the amount of silanol in the solution and the corrosion protection performances of the final sol–gel coating was investigated by van Ooij [58]. Working on Al substrate with bistriethoxysilylpropyltetrasulfide (BTSE), the author demonstrated that after fixing the other parameters, there is an optimal time of hydrolysis to obtain the maximum value of silanols in the solution and, consequently, the best corrosion protection properties of the final sol–gel coating (Figure 7.16), evaluated by means of electrochemical impedance spectroscopy (EIS) measurements, DC polarization, and immersion tests.

After the formation of weak hydrogen bonds, the condensation reactions among the different silanol groups and between the metal hydroxides and the silanols take place [61].

In fact, the silane solution of hydrolyzed molecules is in contact with the hydroxyl groups that are in equilibrium with the oxide on the surface of the metal. Thus, the silanol groups can interact not only among themselves but also with the hydroxyl groups in equilibrium on the metal substrate. This is a critical point for the application of silane sol–gel films onto a metal substrate: the adhesion between the metal oxide and the sol–gel film dramatically depends on the interaction of the silanol groups with the metal hydroxides on the surface of the inorganic substrate. The metal surface has to be rich in hydroxyl groups to permit the adsorption of the silanols to occur, according to the mechanism depicted in Figure 7.17 [62–64].

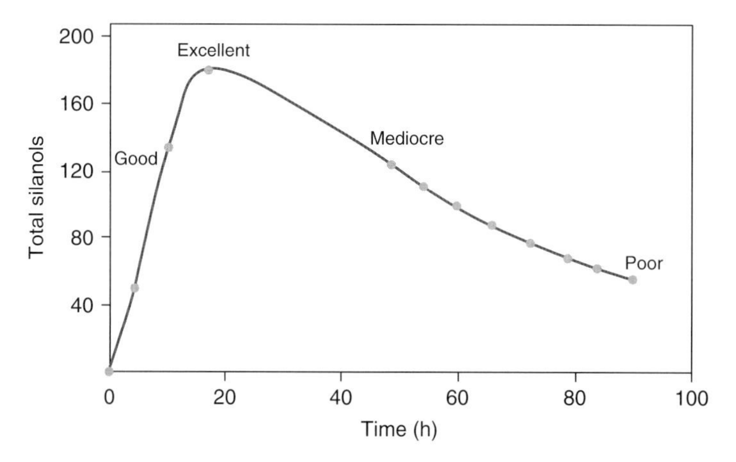

Figure 7.16 Effect of the time of hydrolysis on the corrosion protection properties of the sol–gel coating for BTSE [58].

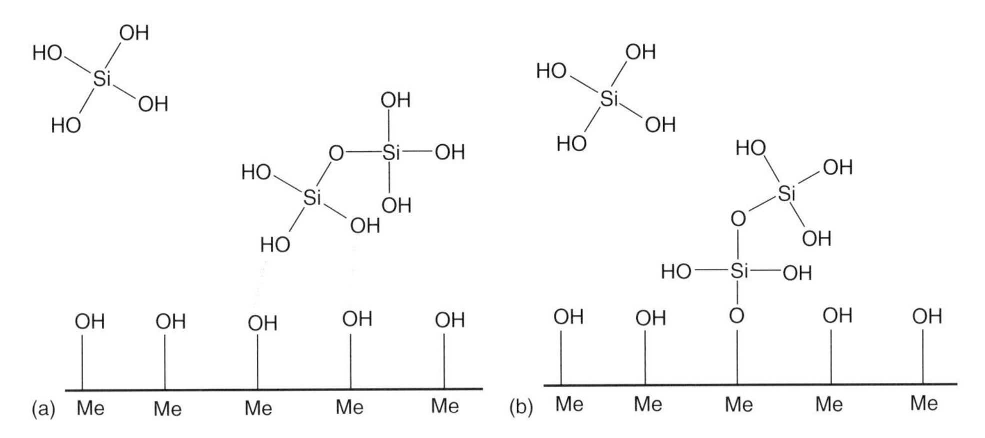

Figure 7.17 Schematic representation of (a) hydrolysis and (b) condensation reactions onto metal surface (Me).

Considering the interaction between silanols and metal hydroxides, first of all, the formation of weak hydrogen bonds between the hydroxyl groups in equilibrium on the metal surface and the silanol groups takes place (Figure 7.17a). On drying, condensation reactions occur between the Si–OH groups of the silanol and the M–OH hydroxides of the metal, leading to the formation of covalent metal–siloxane bonds (M–O–Si), according to the mechanism depicted in Figure 7.17b and in the following reaction (silanization reaction) [65].

$$\mathrm{Si{-}OH_{(solution)} + M{-}OH_{(surface)} \rightleftharpoons Si{-}O{-}M_{(interface)} + H_2O}$$

If the rate of hydrolysis is sufficiently high compared to that of the condensation between silanol groups, many Si–O–M bonds can be formed between small-sized silanols and the metal. On the other hand, if the rate of hydrolysis is relatively low compared to that of the condensation between silanol groups, few Si–O–M bonds are formed between large-sized hindered silanols and the substrate [57, 63].

In fact, the silanization reaction occurs in competition with the condensation reaction of the silanol groups, which interact among themselves, according to the following reaction

$$\mathrm{Si{-}OH_{(solution)} + Si{-}OH_{(solution)} \rightleftharpoons Si{-}O{-}Si_{(solution\ or\ film)} + H_2O}$$

As a result of the hydrolysis and, consequently, the condensation reactions of the alkoxy groups linked to the silicon atom, a sol–gel film is formed [51]. The structure of the film is strongly related to the chemistry of the silane molecules involved, the deposition condition, and the properties of the homogeneous solution of hydrolyzed silicon alkoxides.

However, the final structure, obtained after a convenient curing of the film, can be schematically represented as depicted in Figure 7.18, where M represents the metal substrate.

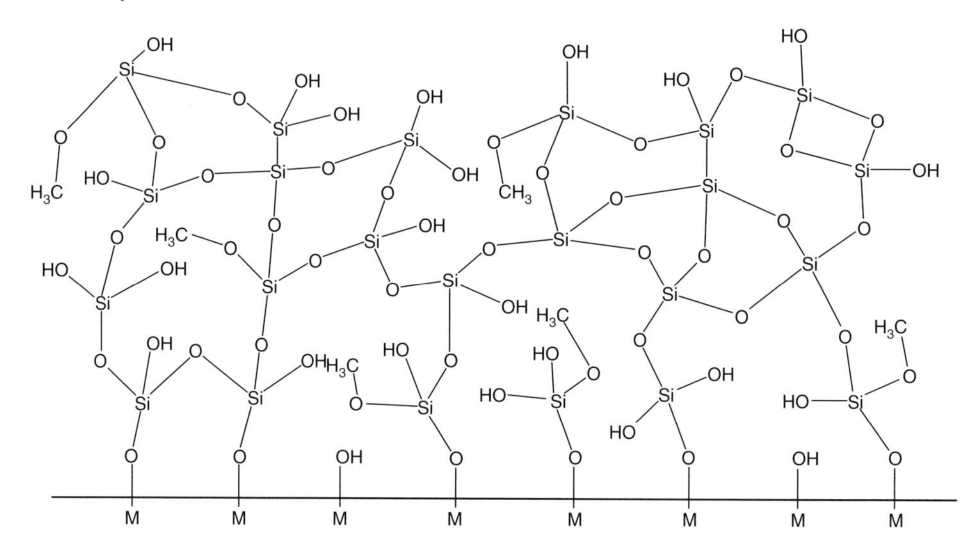

Figure 7.18 Two-dimensional schematic representation of the structure of a silane sol–gel film bonded onto a metal (M) substrate.

The Si–O–Si network is formed when condensation reactions occur among the different silanols.

Considering Figure 7.18, notice that the number of hydroxyl groups depends on the curing process: higher the temperature and/or longer the heat treatment, lower the number of uncondensed Si–OH bonds. As a result of an uncompleted hydrolysis of silicon alkoxides, residual not-hydrolyzed organic groups (such as $-OCH_3$ and $-OCH_2CH_3$) might be embedded onto the Si–O–Si network.

The final thickness of the sol–gel film depends on different parameters, such as

- the concentration of the silicon alkoxides in the starting solution and the chemistry of the molecules;
- the time and temperature of curing;
- in case of dip coating, the previously described parameters of the Levich–Landau–Derjaguin equation.

The effect of the rheological and physical properties of a homogeneous solution of hydrolyzed silicon alkoxide molecules on the thickness of a sol–gel coating was previously described. It was reported that solution viscosity (η) and density (ρ), the withdrawing speed (ν), and the surface tension of the metal (γ) affect the final thickness of the sol–gel layer in a way described by the Levich–Landau–Derjaguin equation. The concentration of the silicon alkoxides in the starting solution and the chemistry of the molecules affect the viscosity and density of the solution. Therefore, these parameters are implicitly considered in the Levich–Landau–Derjaguin equation. In particular, it was demonstrated [66] that there is a basically linear relationship between the film thickness and the concentration of the silicon alkoxides in the solution. Thus, the higher the concentration of the silicon alkoxides in the solution, the thicker the final sol–gel film. An increase in the concentration of the silicon alkoxides in the solution will lead to a linear increase in the thickness of the final sol–gel film.

In general, the thermal curing dramatically influences the thickness of the sol–gel coating. It was demonstrated [67] that as the duration of the thermal curing process increases, the thickness of the final sol–gel film decreases. Commonly, after a certain lapse of time of thermal treatment, a plateau value of thickness is reached, and a further decrease of the sol–gel film thickness is no more appreciable. In fact, the degree of condensation reactions among the Si–OH groups increases continuously during the curing treatment, till the complete conversion of the silanols in a Si–O–Si continuous network. The decrease in thickness observed after increasing the thermal curing time is due to the densification of the Si–O–Si network promoted by the thermal energy provided to the sol–gel film by heating. This densification, related to a decrease in the mean distance among the different silanol chains, leads to a shrinkage of the film, which involves a decrease in the film thickness.

For the same reasons, an increase in the curing temperature leads to a decrease in the thickness of the sol–gel coating [52]. In fact, the condensation rates increase by increasing the amount of thermal energy provided to the sol–gel coating.

7.2.4
Interaction between Silicon Alkoxides and an Organic Polymeric Material

This mechanism of interaction is schematically represented in Figure 7.19.

The organic group linked to the silicon atom can be properly designed to chemically interact with the organic coating to form covalent bonds between the hybrid organically modified sol–gel film and the polymeric chain of the organic material. Obviously, it is necessary to use organically modified silicon alkoxides to ensure a potential chemical interaction between the sol–gel coating and the polymeric material. In fact, the interaction between the organic functionalizations bonded to the silicon atom of the silanols and the active substituent of the organic polymeric material determines the extent of adhesion. However, the adhesion is affected not only by the formation of strong covalent bonds between the organofunctional groups of the sol–gel coating and the organic polymeric matrix. In fact, according to Plueddemann [21], if the oligomeric siloxanol network is compatible in the liquid matrix of the polymeric material, it is possible to obtain a copolymer at the sol–gel coating–resin interface during resin cure (due to the formation of the previously described covalent bonds). Alternatively, if the oligomeric siloxanol network has only partial solution compatibility with the liquid matrix of the organic polymeric material, it is possible to obtain an interpenetrating polymer network (IPN, schematically reported in Figure 7.20) during resin cure.

Commonly, the formation of an IPN is characterized by a merely partial (or completely absent) copolymerization between the organofunctional molecules and the organic polymeric material.

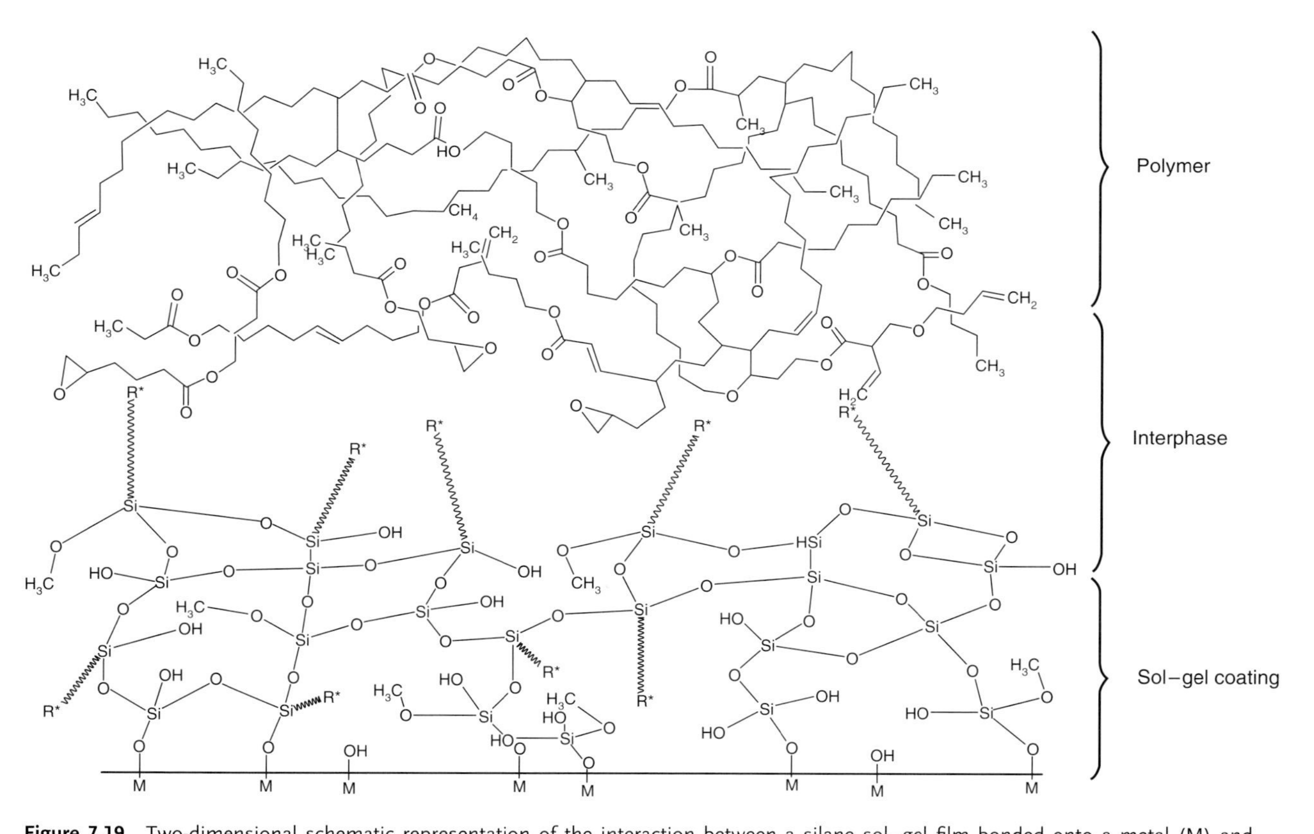

Figure 7.19 Two-dimensional schematic representation of the interaction between a silane sol–gel film bonded onto a metal (M) and an organic coating.

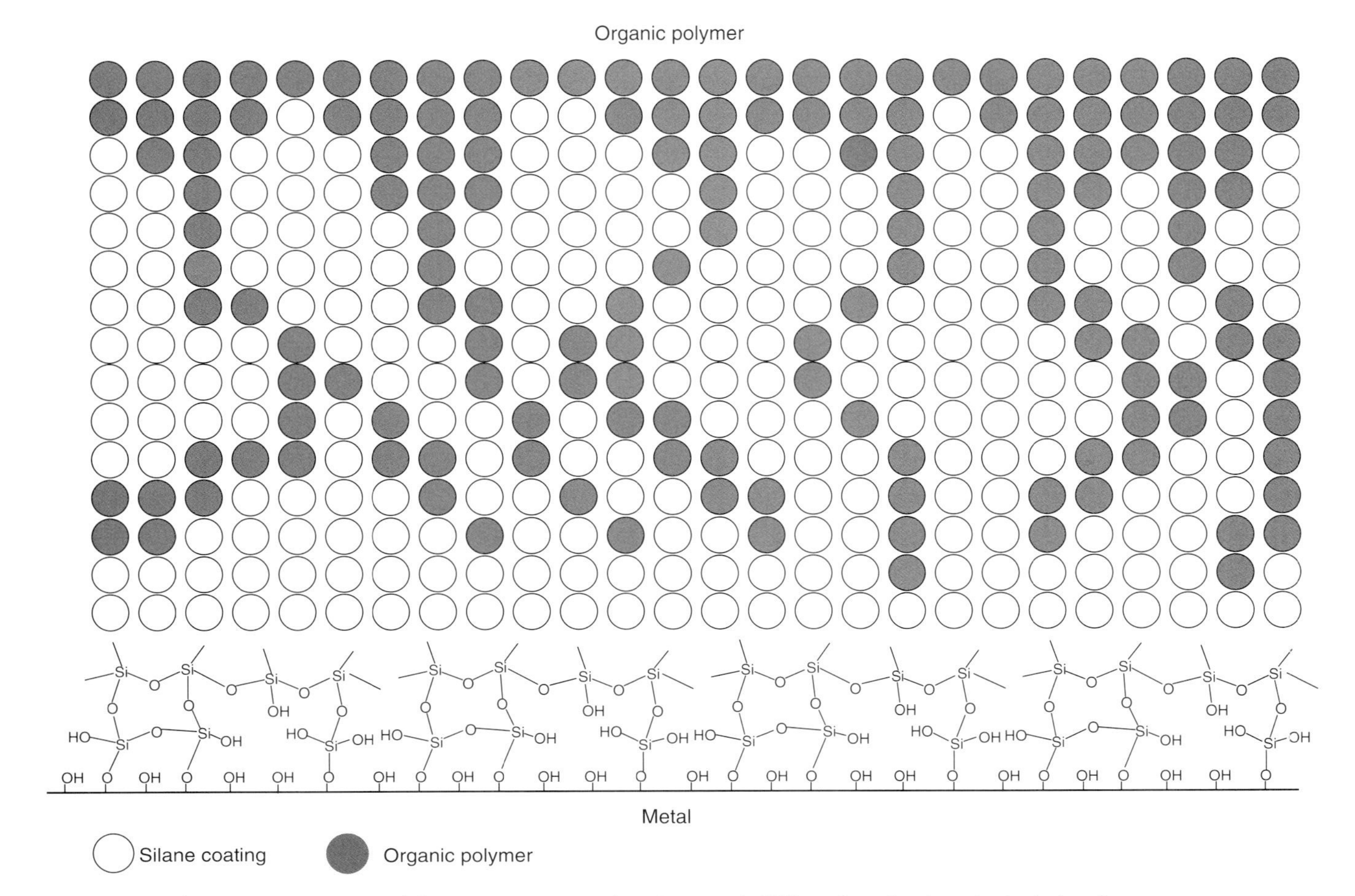

Figure 7.20 Schematic representation of the interpenetrating polymeric network (IPN) at the sol–gel coating/resin interface.

7.3
Corrosion Protection by Sol–Gel Coatings

Commonly, metal surfaces are protected with an organic coating to improve the durability or the esthetic properties.

Cleaning pretreatments or chemical conversion coatings are commonly applied on many metal surfaces, such as steel, galvanized steel, aluminum, and aluminum alloys, to improve the adhesion between the metallic substrate and an organic coating and enhance the protective or appearance properties of the product or structure. The increasing attention to environmental concerns led to the development of new pretreatments more compatible with the environment [68]. To address these environmentally friendly issues, in the early 1990s, the use of silane hybrid molecules containing functional groups was introduced as an alternative for the conversion treatments containing chromium VI (which were discussed in Section 7.1) for corrosion protection of metal and adhesion promotion between metals and organic coatings. In this sense, the surface conversion treatment of silicon hybrid molecules forms a sort of interphase that imparts resistance against degradation of the substrate and improves adhesion with the organic polymeric coating. In comparison to traditional chemical conversion coatings, sol–gel coatings obtained from silicon alkoxide precursors have many advantages: easy application (mainly by dipping but also by spray deposition) even on multimetal systems [69], low environmental impact, high adhesion stability in humid environments (wet adhesion) [70] because of the high stability of the Si–O–Si chemical bond [71], and relatively good corrosion protection properties in proportion to their thickness (commonly only a few hundreds of nanometers).

7.3.1
Corrosion Protection Properties of Organofunctional Sol–Gel Coatings

Sol–gel coatings obtained from silicon alkoxide precursors not only ensure the adhesion between metal substrates and organic coatings (described in Section 7.2.3) but also provide a thin, but efficient, barrier against oxygen diffusion to the metal [72]. As far as corrosion protection properties are concerned, the permeability of a coating to water and oxygen has a dramatic effect on the effectiveness of the coating itself. In fact, when water and oxygen reach the surface of the metal, the cathodic reduction of oxygen may occur, as well as the anodic dissolution of the metal substrate, following the well-known reactions

$$\mathrm{M} \longrightarrow \mathrm{M}^{n+} + n\mathrm{e}^-$$

$$\mathrm{H_2O} + \frac{1}{2}\mathrm{O_2} + 2\mathrm{e}^- \longrightarrow 2\mathrm{OH}^-$$

where M stands for metal.

Therefore, the barrier against oxygen diffusion provided by the sol–gel coating obtained from silicon alkoxide precursors affects the kinetics of the oxygen reduction reaction, decreasing the rate of the cathodic reaction. In addition, the sol–gel coatings obtained from silicon alkoxide precursors are also able to limit the access of water and aggressive species (ions) to the surface of the metal. In fact, if the sol–gel coating covers the entire metal surface and covalent M–O–Si (metal–oxygen–silicon) bonds are present at the metal oxide/sol–gel coating interface, the optimal corrosion protection level is ensured.

Sol–gel coatings obtained from silicon alkoxide precursors can also be used as a final coating, if a short-term protection of the substrate is enough for a specific application.

Compared to the traditional chromate conversion treatments, the only drawback of silane sol–gel coatings is that they do not provide an active protection to the metallic substrate [73–75]. In fact, when water and aggressive ions reach the surface of the metal, silane layers are not able to ensure an active inhibition of the corrosion process and chromate compounds [76].

For this reason, and in general to improve the protection properties of the silane layers, several attempts were made by adding corrosion inhibitors, such as cerium and lanthanum salts [77–79] or CeO_2 and LaO_2 nanopowders [80], to the silane sol–gel films. In addition, the performances of silane layers filled with silica particles were also evaluated [81–84], aiming at an improvement of the barrier properties of the sol–gel film itself.

7.3.2 Experimental Methods of Investigation of the Properties of the Silicon Alkoxide Sol–Gel Coatings as Coupling Agents

Sol–gel coatings obtained from silicon alkoxide precursors are applied as thin film on metals as a pretreatment for the further application of an organic polymeric coating.

Considering the neat sol–gel coating prior the application of an organic polymer, the properties of the sol–gel coatings obtained from silicon alkoxide precursors can be investigated using traditional analytical chemical techniques such as Fourier transform infrared spectroscopy (FT-IR), Raman spectroscopy, differential scanning calorimetry (DSC), thermogravimetric analysis (TGA), NMR (nuclear magnetic resonance), and ssNMR (solid-state nuclear magnetic resonance). Concerning the physical/chemical properties (such as thickness, composition, surface roughness, morphology) of the sol–gel coatings applied onto a metal surface, the analytical techniques used for the characterization are time-of-flight secondary ion mass spectroscopy (ToF-SIMS), X-ray photoelectron spectroscopy (XPS), glow discharge optical emission spectroscopy (GDOES), ellipsometry, and microscopic techniques, such as transmission electron microscopy (TEM), scanning electron spectroscopy (SEM), and atomic force microscopy (AFM).

As far as the corrosion protection properties of the sol–gel coatings obtained from silicon alkoxide precursors are concerned, electrochemical techniques are commonly used to characterize the different materials. In particular, EIS measurements are able to provide very interesting data about the dielectric and resistance properties of the sol–gel coatings [85]. By means of EIS, it is possible to gain information about the degradation kinetics, water absorption, and ion barrier properties of the sol–gel coatings during immersion time in an aggressive solution. Anodic and cathodic polarization measurements provide information [86] about the extent of surface covered by the sol–gel coating and about the barrier properties of the coating against water and oxygen diffusion to the metal interface. Localized electrochemical techniques such as scanning vibrating electrode technique (SVET) [87] and localized electrochemical impedance spectroscopy (LEIS) [88] are also applied to study the corrosion protection properties of the sol–gel coatings obtained from silicon alkoxide precursors. These techniques are commonly use to investigate the evolution of localized corrosion phenomena in the presence of a small defect in the sol–gel coating, in particular, if a corrosion inhibitor pigment (organic or inorganic) is embedded onto the coating. Eventually, a few authors [89] also used the scanning Kelvin probe (SKP) technique to characterize the metal oxide/sol–gel coating interface.

Generally, the techniques used to investigate the corrosion protection properties of a complete protection system for metals consisting of a sol–gel coating obtained from silicon alkoxide precursors coated with an organic coating (paint) are the same as those used for the characterization of traditional pretreated and painted metals. Therefore, exposition in the salt spray chamber of intact and scribed samples (according to ISO 9227 and ASTM B117) for the determination of the wet resistance and the susceptibility to blisters formation, adhesion measurements both in dry and wet conditions (according to ISO 4624), cathodic disbonding test (according to ISO 15711) for the investigation of the resistance against the advance of the cathodic front, and EIS measurements carried out both on intact and scribed samples are the experimental measurements that are commonly used to investigate the effectiveness of the sol–gel coating obtained from silicon alkoxide precursors as a pretreatment to improve the corrosion protection properties of a painting system.

7.3.3 Practical Examples of Corrosion Protection by Silicon-Based Sol–Gel Coatings

In this section, a few real examples of the applications of sol–gel coatings obtained from silicon alkoxide precursors applied onto different metals are reported.

The first example, reported from [52], deals with the properties of an experimental mixture of silicon alkoxides used to obtain the formation of an organofunctionalized sol–gel coating on galvanized steel surfaces by the dip coating application technique. The silane solution consisted of an experimental mix of three different silane molecules: γ-glycidoxypropyltrimethoxysilane (γ-GPS), tetraethoxysilane (TEOS), and methyltriethoxysilane (MTES). γ-GPS is an organofunctional silane (Figure 7.10) that shows a short carbon backbone with an epoxy functionalized tail and a Si atom substituted with three (–O–CH_3) groups. Owing to the presence of the epoxy group, this molecule ensures a potential interaction with an organic

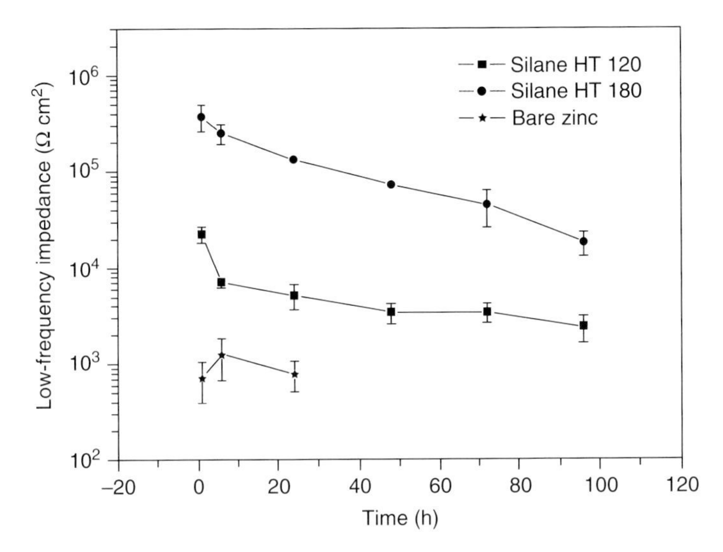

Figure 7.21 Evolution of the low-frequency impedance values during immersion in sodium chloride solution.

polymeric material. The effect of two different curing treatments was investigated: 20 min at 120 and 180 °C. The two samples were named Silane HT120 and Silane HT180, respectively. The corrosion protection performances of the uncoated sol–gel coatings were investigated by means of EIS in a 0.1 M NaCl solution. The evolution of the impedance modulus at 0.01 Hz ($|Z|_{0.01}$) during immersion time is reported in Figure 7.21.

The sample Silane HT180 maintains a high value of the total impedance even for about 120 h of continuous immersion in the solution. The barrier properties of this film seem to be good and probably correlated with the dense and stable network of Si–O–Si bonds formed by curing the sol–gel coating at a high temperature. On the other hand, the sample treated with the same silane solution, but cured at a lower temperature, shows always lower values of the $|Z|_{0.01}$ impedance modulus during immersion time than the sample cured at higher temperatures. The curing treatment led to a densification of the Si–O–Si network (the thickness of the samples cured at 120 °C is about 300 nm with respect to the 100 nm thickness of the samples cured at 180 °C), improving the barrier properties of the sol–gel coating and, therefore, the total impedance of the system.

Considering the previous practical example, a 60 μm thick epoxy-polyester powder coating was applied on the sol–gel pretreated samples, and this complete protection system was investigated. In particular, Table 7.1 reports the linear detachment (measured in millimeters) from the artificial scribe for the different samples after the cathodic disbonding test (conditioning of the scribed samples: −1.6 V vs. Ag/AgCl during 30 min, repeated five times).

Sample Silane HT180 shows the best performance during cathodic polarization. Notice the correlation between the good corrosion protection properties of the

Table 7.1 Linear extent of detachment for the different samples.

	Linear extent of the detachment (mm)
Untreated HDG	5.0–8.0
Silane HT120	4.0–5.0
Silane HT180	1.0–3.0

samples Silane HT180 highlighted by the EIS measurements and the very low scribe creep evidenced by the same sample (coated with the organic polymer) during the cathodic disbonding measurements. The treatment with the experimental mixtures of silicon alkoxide molecules cured at 180 °C seems to ensure good adhesion between the metallic substrate and the organic coating.

The second example, reported from [90], deals with the properties of the same experimental mixtures of silicon alkoxide molecules used in the previous example, which was applied on galvanized steel sheets hot dip galvanized (HDG) by the dip coating technique and used as a pretreatment for the electrophoretic application of an epoxy cataphoretic polymer. Also, in this case, the effect of two different curing treatments was investigated: 20 min at 120 and 180 °C.

In this case, the wet adhesion and the resistance against scribe creep were roughly estimated by means of salt spray chamber exposure of silane pretreated samples after the application of the epoxy cataphoretic polymer. Scribed samples were exposed in the salt spray chamber for about 500 h. The comparison between the experimental results after the accelerated aging is reported in Table 7.2.

Notice that the samples pretreated with the silicon alkoxide sol–gel film show a smaller delaminated area, related to the limited extent of the delamination process, with respect to the untreated substrate. In this case, the good barrier properties of a sol–gel coating obtained from silicon alkoxide precursors treated at 180 °C are not able to ensure the best corrosion protection properties probably because the high barrier properties of this film lead to a noticeable degradation during the electrodeposition of the epoxy cataphoretic polymer, likely due to the hydrogen production and bubbling at the cathode. In fact, for this specific application, the sol–gel coating obtained from silicon alkoxide precursors treated at lower

Table 7.2 Linear extent of delamination for the different samples.

	Linear extent of delamination (mm)
A 120	1.5
A 180	1.9
Untreated HDG	3.6

temperatures (120 °C) and, therefore, characterized by lower barrier properties shows a better compatibility with the epoxy cataphoretic polymer deposition, which leads to a lower scribe creep (i.e., higher resistance) after the exposure in the salt spray chamber. This fact is probably due to a lower deterioration of the silane sol–gel film cured at 120 °C during the electrodeposition of the epoxy cataphoretic polymer, which probably leads to the formation of a homogeneous and defect-free cataphoretic film.

These two examples show the versatility of the sol–gel coatings obtained from silicon alkoxide precursors, which can be used in different conditions by optimizing the physical/chemical properties of the sol–gel film itself as a function of the specific needs.

Therefore, an optimal compromise can be found by balancing the thickness, the barrier properties, the curing temperature, and the parameters of dip coating the silane sol–gel layer depending on the specific application.

References

1. Deflorian, F., Rossi, S., and Fedrizzi, L. (2006) *Electrochim. Acta*, **51**, 6097.
2. Lunder, O., Walmsley, J.C., Mack, P., and Nisancioglu, K. (2005) *Corros. Sci.*, **47**, 1604.
3. Bellezze, T., Roventi, G., and Fratesi, R. (2002) *Surf. Coat. Technol.*, **155**, 221.
4. Eppensteiner, F.W. and Jennkind, M.R. (2007) *Met. Finish.*, **105**, 413.
5. Caruana, C.M. (2006) *Met. Finish.*, **104**, 44.
6. U.S. Department of Health and Human Services, Public Health Service, Agency for Toxic Substances and Disease Registry (2008) Toxicological Profile for Chromium, *www.atsdr.cdc.gov*. (accessed 2010).
7. (2008) *Focus Pigm.*, **2**, 6, *www.osha.org*. (accessed 2010).
8. Almeida, E., Fedrizzi, L., and Diamantinio, T.C. (1998) *Surf. Coat. Technol.*, **105**, 97.
9. Almeida, E., Diamantino, T.C., Figueiredo, M.O., and Sà, C. (1998) *Surf. Coat. Technol.*, **106**, 8.
10. Lingjie, L., Jingle, L., Shenghai, Y., Yujing, T., Qiquan, J., and Fusheng, P. (2008) *J. Rare Earths*, **26**, 383.
11. Scholes, F.H., Soste, C., Hughes, A.E., Hardin, S.G., and Curtis, P.R. (2006) *Appl. Surf. Sci.*, **253**, 1770.
12. Campestrini, P., Terryn, H., Hovestad, A., and de Wit, J.H.W. (2004) *Surf. Coat. Technol.*, **176**, 365.
13. Zhang, X., van den Bos, C., Sloof, W.G., Hovestad, A., Terryn, H., and de Wit, J.H.W. (2005) *Surf. Coat. Technol.*, **199**, 92.
14. Nordliena, J.H., Walmsley, J.C., østerberg, H., and Nisancioglu, K. (2002) *Surf. Coat. Technol.*, **153**, 72.
15. Song, Y., Shan, D., Chen, R., Zhang, F., and Han, E.-H. (2009) *Surf. Coat. Technol.*, **203**, 1107.
16. Van Roy, I., Terryn, H., and Goeminne, G. (1998) *Coll. Surf.*, **136**, 89.
17. Sathyanarayana, M.N. and Yaseen, M. (1995) *Prog. Org. Coat.*, **26**, 275.
18. Twite, R.L. and Bierwagen, G.P. (1998) *Prog. Org. Coat.*, **33**, 91.
19. Deflorian, F., Rossi, S., Fedrizzi, L., and Fedel, M. (2008) *Prog. Org. Coat.*, **63**, 338.
20. Neuder, H., Sizemore, C., Kolody, M., Chiang, R., and Lin, C.-T. (2003) *Prog. Org. Coat.*, **47**, 225.
21. Plueddeman, E.P. (ed.) (1992) Reminiscing on silanes coupling agents, in *Silanes and Other Coupling Agents*, VSP, Utrecht, pp. 3–19.
22. Mittal, K.L. (ed.) (2000) *Silanes and Other Coupling Agents*, vol. 2, VSP, Utrecht.
23. Mittal, K.L. (ed.) (2004) *Silanes and Other Coupling Agents*, vol. **3**, VSP/Brill.

24. Mittal, K.L. (ed.) (2007) *Silanes and Other Coupling Agents*, vol. **4**, VSP/Brill, Leiden.
25. Mittal, K.L. (ed.) (2009) *Silanes and Other Coupling Agents*, vol. **5**, VSP/Brill, Leiden.
26. Jayseelan, S.K. and van Ooij, W.J. (2001) *J. Adhes. Sci. Technol.*, **15**, 967.
27. Deflorian, F., Rossi, S., and Fedrizzi, L. (2006) *Electrochim. Acta*, **51**, 6097.
28. Ferreira, M.G.S., Duarte, R.G., Montemor, M.F., and Simoes, A.M.P. (2004) *Electrochim. Acta*, **49**, 2927.
29. Plueddemann, E.P. (1970) *Composites*, **1**, 321.
30. Zhu, D. and van Ooij, W.J. (2003) *Corros. Sci.*, **45**, 2177.
31. van Ooij, W.J., Zhu, D., Stacy, M., Seth, A., Mugada, T., Gandhi, J., and Puomi, P. (2005) *Tsinghua Sci. Technol.*, **10**, 639.
32. Zhu, D. and van Ooij, W.J. (2004) *Prog. Org. Coat.*, **49**, 42.
33. Plueddemann, E.P. (1990) *Silane Coupling Agents*, 2nd edn, Plenum Press, New York.
34. Pu, Z., van Ooij, W.J., and Mark, J.E. (1997) *J. Adhes. Sci. Technol.*, **11**, 29.
35. Montemor, M.F., Simães, A.M., Ferreira, M.G.S., Williams, B., and Edwards, H. (2000) *Prog. Org. Coat.*, **38**, 17.
36. Livage, J. (1986) *J. Solid State Chem.*, **64**, 322.
37. Chrusciel, J. and Olusarski, L. (2003) *Mat. Sci.*, **21**, 461.
38. Wang, D. and Bierwagen, G.P. (2009) *Prog. Org. Coat.*, **64**, 327.
39. Tracton, A.A. (2006) *Coatings Materials and Surface Coatings*, Chapter 1, CRC, pp. 1–14.
40. Goedecker, S., Deutsch, T., and Billard, L. (2002) *Phys. Rev. Lett.*, **88**, 1.
41. Kessler, V.G. and Seisenbaeva, G.A. (2008) New insight into mechanisms of sol-gel process and new materials and opportunities for bioencapsulation and biodelivery, in *Sol-gel Methods for Material Processing* (eds. P. Innocenzi, Y.L. Zub, and V.G. Kessler), Springer, Dordrecht, pp. 141–153.
42. Livage, J., Sanchez, C., Henry, M., and Doeuff, S. (1989) *Solid State Ionics*, **32/33**, 633.
43. Abel, M.L., Joannic, R., Fayos, M., Lafontaine, E., Shaw, S.J., and Watts, J.F. (2006) *Int. J. Adhes.*, **26**, 16.
44. Corriu, R.J.P. and Douglas, W.E. Organosilicate oligomers and nanostructured materials, (2001) in *Organosilicate Oligomers and Nanostructured Materials, in Silicon-containing Polymers*, Chapter 25 (eds. R.G. Jones, W. Ando, and J. Chojnowski), Kluwer Academic Publisher, pp. 667–695.
45. Dislich, H. (1985) *J. Non-Cryst. Solids*, **73**, 599.
46. Schmidt, H., Scholze, H., and Kaiser, A. (1984) *J. Non-Cryst. Solids*, **63**, 1.
47. Schubert, U. and Hüsing, N. (2005) *Synthesis of Inorganic Materials*, Chapter 4, 2nd edn, Wiley-VCH Verlag gmbH, pp. 200–214.
48. Wu, K.H., Chao, C.M., Yeh, T.F., and Chang, T.C. (2007) *Surf. Coat. Technol.*, **201**, 5782.
49. Donley, M.S., Mantz, R.A., Khramov, A.N., Balbyshev, V.N., Kasten, L.S., and Gaspar, D.J. (2003) *Prog. Org. Coat.*, **47**, 401.
50. van Ooij, W.J., Zhu, D., Stacy, M., Seth, A., Mugada, T., Gandhi, J., and Puomi, P. (2005) *Tsinghua Sci. Technol.*, **10**, 639.
51. Sheffer, M., Groysman, A., and Mandler, D. (2005) *Corros. Sci.*, **45**, 2893.
52. Fedel, M., Olivier, M., Poelman, M., Deflorian, F., Rossi, S., and Druart, M.-E. (2009) *Prog. Org. Coat.*, **66**, 118.
53. De Graeve, I., Tourwé, E., Biesemans, M., Willem, R., and Terryn, H. (2008) *Prog. Org. Coat.*, **63**, 38.
54. Brinker, C.J. and Scherer, G.W. (1990) *Sol-Gel Science: The Physics and Chemistry of Sol-Gel Processing*, Academic Press Inc., New York.
55. Landau, L.D. and Levich, B.G. (1942) *Acta Physiochim*, **17**, 42–54.
56. Quere, D. (2007) Three-Phases capillarity, in *Thin Films of Soft Matter* (eds. S. Kalliadasis and U. Thiele), Springer, pp. 115–136.
57. Krechetnikov, R. and Homsy, G.M. (2006) *J. Fluid Mech.*, **559**, 429.
58. van Ooij, W.J. (2004) Proceedings ICEPAM 2004, Oslo, Norway,

16–18/2004, *http://www.sintef.no.* (accessed 2010)

59. Brinker, C.J., Hurd, A.J., and Ward, K.J. (1988) Fundamentals of sol-gel thin-film formations, in *Ultrastructure Processing of Advanced Ceramics* (eds J.D. Mackenzie and D.R. Ulrich), John Wiley & Sons, Inc., New York.
60. Cecchetto, L., Denoyelle, A., Delabouglise, D., and Petit, J.-P. (2008) *Appl. Surf. Sci.*, **254**, 1736.
61. Kim, H.-J., Zhang, J., Yoon, R.-H., and Gandour, R. (2004) *Surf. Coat. Technol.*, **188-189**, 762.
62. Boerio, F.J. and Williams, J. (1981) *Appl. Surf. Sci.*, **7**, 19.
63. Li, G., Wang, X., Li, A., Wang, W., and Zheng, L. (2007) *Surf. Coat. Technol.*, **201**, 9571.
64. Song, J. and van Ooij, W.J. (2003) *J. Adhes. Sci. Technol.*, **17**, 2191.
65. Chovelon, J.M., Aarch, L.E.L., Charbonnier, M., and Romand, M. (1995) *J. Adhes.*, **50**, 43.
66. Metroke, T.L., Gandhi, J.S., and Apblett, A. (2004) *Prog. Org. Coat.*, **50**, 231.
67. Franquet, A., Terryn, H., and Vereecken, J. (2003) *Appl. Surf. Sci.*, **211**, 259.
68. Fedrizzi, L., Deflorian, F., Boni, G., Bonora, P.L., and Pasini, E. (1996) *Prog. Org. Coat.*, **29**, 89.
69. Deflorian, F., Rossi, S., and Fedrizzi, L. (2006) *Electrochim. Acta*, **27**, 6097.
70. Deflorian, F., Rossi, S., Fedrizzi, L., and Fedel, M. (2008) *Prog. Org. Coat.*, **63**, 338.
71. Cabral, A.M., Duarte, R.G., Montemor, M.F., and Ferreira, M.G.S. *Prog. Org. Coat.* **54**, 322.
72. Suegama, P.H., de Melo, H.G., Recco, A.A.C., Tschiptschin, A.P., and Aoki, I.V. (2008) *Surf. Coat. Technol.*, **202**, 2850.
73. Cabral, A.M., Duarte, R.G., Montemor, M.F., Zheludkevich, M.L., and Ferreira, M.G.S. (2005) *Corros. Sci.*, **47**, 896.
74. Wang, H. and Akid, R. (2007) *Corros. Sci.*, **49**, 4491.
75. Wang, H. and Akid, R. (2008) *Corros. Sci.*, **50**, 1142.
76. Montemor, M.F. and Ferreira, M.G.S. (2008) *Prog. Org. Coat.*, **63**, 330.
77. Montemor, M.F. and Ferreira, M.G.S. (2007) *Electrochim. Acta*, **52**, 6976.
78. Palomino, L.M., Suegama, P.H., Aoki, I.V., Montemor, M.F., and De Melo, H.G. (2009) *Corros. Sci.*, **51**, 1238.
79. Rosero-Navarro, N.C., Pellice, S.A., Durán, A.A., and Aparicio, M. (2008) *Corros. Sci.*, **50**, 1283–1291.
80. Montemor, M.F., Trabelsi, W., Lamaka, S.V., Yasakau, K.A., Zheludkevich, M.L., Bastos, A.C., and Ferreira, M.G.S. (2008) *Electrochim. Acta*, **53**, 5913.
81. Suegama, P.H., De Melo, H.G., Recco, A.A.C., Tschiptschin, A.P., and Aoki, I.V. (2008) *Surf. Coat. Technol.*, **202**, 2850.
82. Castro, Y., Ferrari, B., Moreno, R., and Durán, A. (2004) *Surf. Coat. Technol.*, **182**, 199–203.
83. Castro, Y., Ferrari, B., Moreno, R., and Durán, A. (2005) *Surf. Coat. Technol.*, **191**, 228–235.
84. Montemor, M.F., Cabral, A.M., Zheludkevich, M.L., and Ferreira, M.G.S. (2006) *Surf. Coat. Technol.*, **200**, 2875.
85. Amirudin, A. and Thierry, D. (1995) *Prog. Org. Coat.*, **26**, 1.
86. Deflorian, F., Rossi, S., Fedel, M., and Motte, C. (2010) *Prog. Org. Coat.*, doi: 10.1016/j.porgcoat.2010.04.007
87. Raps, D., Hack, T., Wehr, J., Zheludkevich, M.L., Bastos, A.C., Ferreira, M.G.S., and Nuyken, O. (2009) *Corros. Sci.*, **51**, 1012.
88. Barranco, V., Carmona, N., Galván, J.C., Grobelny, M., Kwiatkowski, L., and Villegas, M.A. (2010) *Prog. Org. Coat.*, **68**, 347.
89. Rossi, S., Fedel, M., Deflorian, F., and Vadillo, Md.C. (2008) *C. R. Chim*, **11**, 984–994.
90. Fedel, M., Druart, M.-E., Olivier, M., Poelman, M., Deflorian, F., and Rossi, S. (2010) *Prog. Org. Coat.*, doi: 10.1016/j.porgcoat.2010.04.003

8
Corrosion of Austenitic Stainless Steels and Nickel-Base Alloys in Supercritical Water and Novel Control Methods

Lizhen Tan, Todd R. Allen, and Ying Yang

8.1
Introduction

8.1.1
Supercritical Water and Its Applications

A pressure–temperature phase diagram of water is shown in Figure 8.1. A triple point exists at ~0 °C and ~6 × 10^{-4} MPa referring to a three-phase equilibrium of solid, liquid, and gas. Supercritical water (SCW) exists in a regime above the critical point at ~374 °C and ~22.05 MPa.

Significant thermal property changes have been observed as water goes into the supercritical regime. Figure 8.2 shows examples of the property variations of water under isobaric (25 MPa) and isothermal (500 °C) conditions, which were adapted from Refs. [1, 2]. Comparing the isobaric plot to the isothermal one, properties show more dramatic changes at the critical temperature under the isobaric condition than those at the critical pressure under the isothermal condition. As shown in the two plots, thermophysical properties, such as thermal conductivity (k), density (ρ), viscosity (μ), and dielectric constant (ε), generally decrease with increasing temperature and decreasing pressure. In contrast, thermodynamic properties, such as enthalpy (H) and entropy (S), follow the opposite trend, that is, they increase with increasing temperature and decreasing pressure. One exception is heat capacity (C_P) variation at the critical temperature under the isobaric condition, which approaches infinity and behaves differently from the other properties.

Owing to the special properties, SCW has become a hot research topic in exploring its potential applications since the 1990s as suggested by the yearly number of published papers as shown in Figure 8.3. SCW has been primarily used for waste treatment, for hydrothermal syntheses of materials, and as coolants in power production.

As a result of the significant property changes, high-temperature water behaves like many organic solvents, leading to high solubilities of organic compounds in near-critical water and complete miscibility with SCW. This feature promotes the applications of SCW to more *green* or environmentally benign chemical processes.

Green Corrosion Chemistry and Engineering: Opportunities and Challenges, First Edition.
Edited by Sanjay K. Sharma.

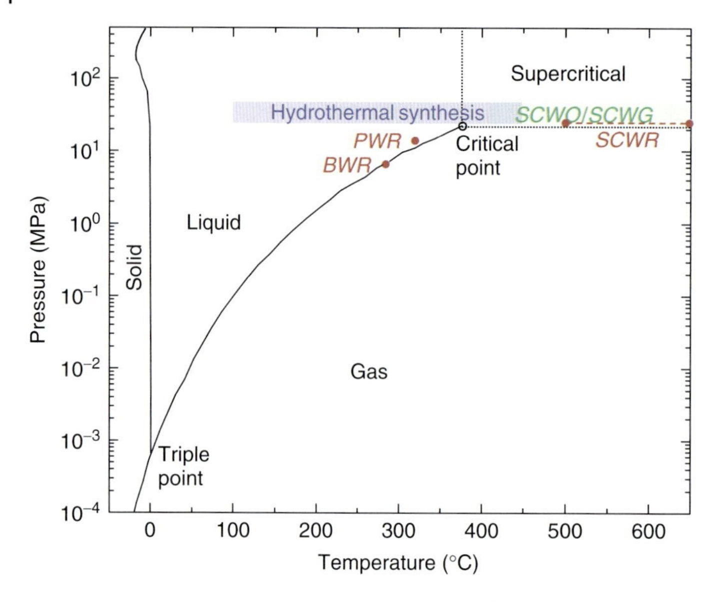

Figure 8.1 Pressure–temperature phase diagram and typical applications of water. (PWR, pressurized water reactor; BWR, boiling water reactor; SCWR, supercritical-water-cooled reactor; SCWO/SCWG, supercritical water oxidation/supercritical water gasification).

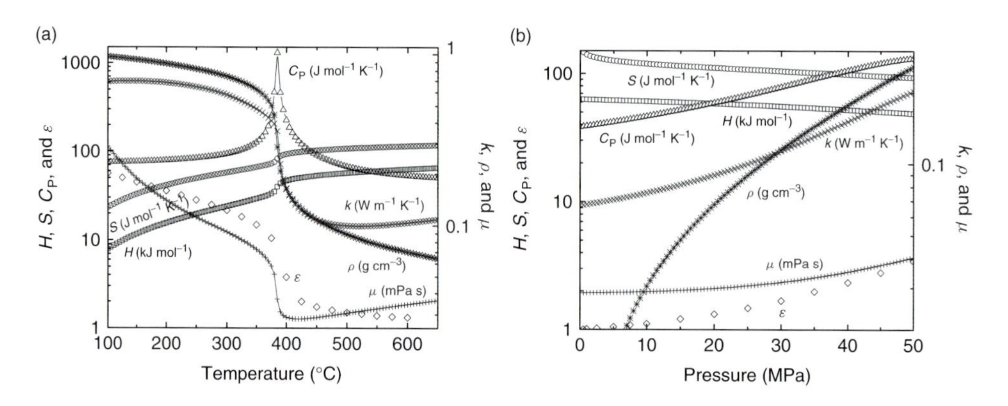

Figure 8.2 Isothermal (a) 500 °C and isobaric (b) 25 MPa properties of water: enthalpy (H), entropy (S), heat capacity (C_P), thermal conductivity (k), density (ρ), viscosity (μ) [1], and dielectric constant (ε) [2].

For example, supercritical water oxidation (SCWO) has been developed to destruct organic compounds and toxic wastes. Hydrogen peroxide (H_2O_2) is usually added as an oxidant to increase destruction efficiency. Moreover, gases are also miscible in SCW, which motivates the recent development of supercritical water gasification (SCWG). Compared to other thermochemical methods, SCWG of biomass can directly deal with wet biomass without drying and has higher yield of H_2 in addition to other common gaseous products such as CO, CO_2, and CH_4. SCWO/SCWG

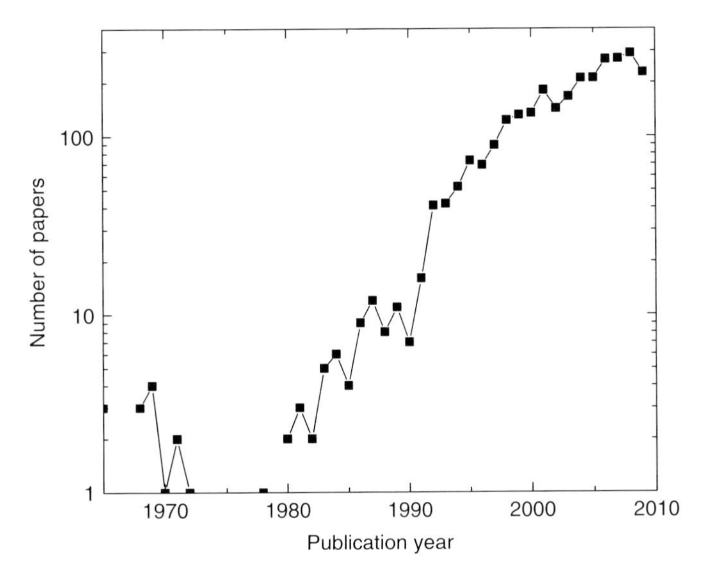

Figure 8.3 Yearly publicated number of papers about SCW included in the Science Citation Index (*http://thomsonreuters.com/products_services/science/science_products/a-z/science_citation_index*).

is usually operated at 400–650 °C and 23–50 MPa, as schematically illustrated in Figure 8.1. The temperature and pressure can be controlled to obtain expected products. The major species produced during SCWO of hydrocarbons are CO_2 and H_2O, which is a key advantage over incineration since no nitrogen oxide (NO_x) forms in this process. If halogens are present in the waste, corresponding acids will be produced during reactions.

Water in both subcritical and supercritical conditions are suited to materials syntheses from respective precursors, for example, metal oxide powders from metal salt precursors, because of their corrosive and dehydration properties. The metal/oxygen ratios in the final products can vary according to the choice of starting material(s) in the reaction and/or the control of the reducing/oxidizing atmospheres by introducing appropriate reductants/oxidants such as H_2, CO, or O_2. Hydrothermal syntheses have been normally operated at 100–450 °C and 23–50 MPa, as schematically illustrated in Figure 8.1. The size distribution and morphology of specific oxide powder products can be controlled by relatively small changes in reaction times, temperature, pressure, and concentration of the metal ions in solution. A systematic discussion of SCWO/SCWG and hydrothermal synthesis can be found in Refs. [3–6].

Another application of SCW is as coolants in power production owing to its unique properties such as low density, high enthalpy, and single phase. To enhance power generation efficiency, the proposed operational temperatures and pressures have been greatly increased in the Generation IV nuclear reactor concepts [7]. For example, the outlet temperatures are expected to be raised up to ~500 °C/~25 MPa for the supercritical water-cooled reactor (SCWR) from ~285 °C/~7 MPa of the boiling water reactor (BWR) and ~320°C/~15 MPa of the pressurized water reactor (PWR). The CANDU[1)]-SCWR concept envisaged an even higher temperature to 650 °C at ~25 MPa. Such increase in the operational conditions is expected to increase the thermal efficiency to ~44% compared to ~33% of the current light water reactors (LWRs) [7].

8.1.2 Austenitic Stainless Steels and Ni-Base Alloys and Their General Corrosion Behavior

8.1.2.1 Austenitic Stainless Steels and Ni-Base Alloys

Austenitic stainless steels and Ni-base alloys are an important category of structural materials for many applications. In addition to excellent mechanical properties, good corrosion resistance is a critical performance requirement for the materials used in aggressive environments. The development history of some representative materials, from standard austenitic stainless steels (SS) such as 304, 316L, and D9 to superaustenitic stainless steel such as alloys 800H and NF709 and Ni-base alloys such as alloys 617, 625, 690, and 718, is shown in Figure 8.4. More details on the history and applications of these materials can be obtained from Ref. [8]. The typical chemical compositions in weight percentage (wt%) of the representative materials are listed in Table 8.1. According to the effect of the alloying elements on phase stability, Mo, Nb, Al, Ti, Si, and V favor ferrite formation, while Co, Mn, Cu, C, and N favor austenite formation. These alloying elements play a significant role in the mechanical properties and/or corrosion resistance.

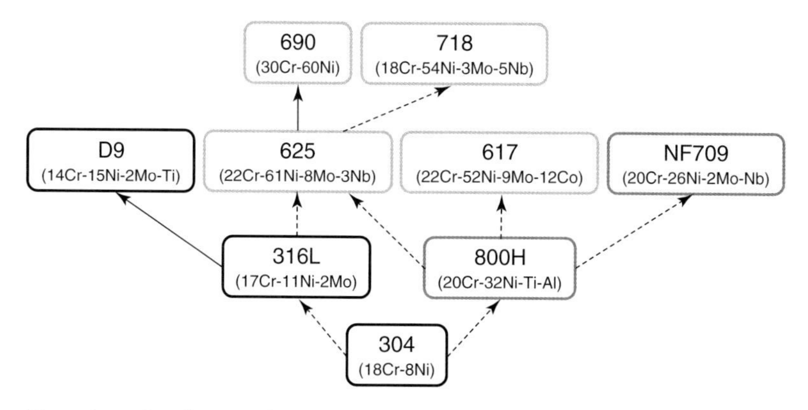

Figure 8.4 The alloy tree showing the distant (dashed arrows) and close (solid arrows) relationships between the austenitic stainless steels and Ni-base alloys. The black, gray, and light gray borders of the alloys denote the three categories of alloys: austenitic stainless steels (304, 316L, D9), superaustenitic stainless steels (800H, NF709), and Ni-base alloys (617, 625, 690, 718), respectively.

1) CANDU – (Canada Deuterium Uranium) is a registered trademark of the Atomic Energy of Canada Limited.

Table 8.1 Typical chemical compositions (wt%) of representative alloys.

Alloys	Fe	Cr	Ni	Mo	Nb	Al	Ti	Si	V	Co	Mn	Cu	C	N
304	70.51	18.30	8.50	0.37	–	–	–	0.65	–	–	1.38	–	0.035	0.068
316L	67.11	16.62	11.20	2.06	–	–	–	0.65	–	0.05	1.86	0.24	0.022	0.020
D9	65.49	13.70	15.80	1.65	–	–	0.34	0.80	0.010	–	2.03	–	0.039	0.004
NF709	49.93	20.41	26.00	1.50	0.34	–	–	0.47	–	–	0.99	–	0.078	0.102
800H	45.26	20.42	31.59	–	–	0.50	0.57	0.13	–	–	0.76	0.42	0.069	–
718	18.04	18.11	53.85	2.96	5.03	0.57	1.05	0.10	0.017	0.04	0.08	0.03	0.031	–
690	10.23	29.58	59.32	0.01	–	0.20	0.35	0.03	0.013	0.01	0.20	0.01	0.032	–
625	4.39	21.90	61.22	8.43	3.17	0.20	0.21	0.11	0.014	0.05	0.08	0.19	0.010	–
617	1.47	22.05	52.32	9.35	–	1.07	0.38	0.15	0.009	12.69	0.07	0.11	0.015	–

As shown in Figure 8.4, the materials were developed based on austenitic SS 304 (Unified Numbering System, UNS S30400). SS 316L (UNS S31603) was developed from SS 304 by reducing C and adding Mo to improve general corrosion and pitting resistance from a corrosion point of view. SS 316L was further developed by adding Ti, creating alloy D9 (UNS S38660). The Ti combines with excess carbon, leading to reduced risk of intergranular corrosion.

Developed from SS 304, alloy 800H (UNS N08810) possesses a composition swinging on the border of SS and Ni-base alloys because neither Fe nor Ni is present at more than 50% concentration, although this alloy is designated as "N" (nickel-base) in the UNS system.[2] The recently developed alloy NF709 (20Cr-25Ni-1.5Mo) has not been assigned a UNS designation.

Experienced a series of intermediate austenitic stainless steels or Ni-base alloys (not shown here), Ni-base Inconel alloy 625 (UNS N06625) was developed from SS 316L or Incoloy alloy 800H. Alloy 625 was further developed into Inconel alloys 690 (UNS N06690) and 718 (UNS N07718). Inconel alloy 617 (UNS N06617) was developed from alloy 800H.

8.1.2.2 General Corrosion Behavior

Corrosion behavior of austenitic stainless steels and Ni-base alloys is dependent on environments to which they are exposed. For example, the classic pitting corrosion of austenitic stainless steels was found to be triggered by the presence of chloride in the environment, which reacts with sulfide inclusions such as MnS in stainless steels [9]. Thus, a material may exhibit a different corrosion behavior corresponding to the respective environments.

2) UNS, jointly sponsored by the American Society for Testing and Materials (ASTM) and the Society of Automotive Engineers (SAE), is operated in accordance with ASTM E527/SAE J1086, Recommended Practice for Numbering Metals and Alloys.

This chapter focuses on corrosion behavior of materials used in the SCWR. When water becomes supercritical, any impurities dissolved in it would be deposited on reactor materials. When such impurities are deposited within the reactor core, they could have harmful consequences for the thermal conductivity and structural stability of the core materials, as well as serious implications for neutronics and for reactor stability. Additionally, they may increase corrosive stress outside the core. Therefore, the intake water in the SCWR should be as pure as possible, which is also an essential requirement in the BWR.

Water vapor or steam oxidation of austenitic stainless steels and Ni-base alloys provides a reference for their corrosion behavior in SCW. Oxidation has been observed as the predominant corrosion phenomenon of austenitic and super-austenitic stainless steels during steam exposures. Multilayer oxide scales usually form, which are primarily composed of a Fe- or (Fe,Cr)-enriched oxide near the outer surface and a Cr-enriched oxide near the oxide–metal interface [10, 11]. A parabolic rate law is usually followed for the growth of the oxide scales [11]. Water vapor corrosion of SS 304 at 600 °C showed a relatively thin layer of $(Cr,Fe)_2O_3$ plus large oxide islands consisting mainly of Fe_2O_3. Conversion to $CrO_2(OH)_2$ (g) from $(Cr,Fe)_2O_3$ occurred during the exposure, resulting in breakaway oxidation [12]. Steam oxidation of alloy 800H at 700 °C showed a uniform duplex thin oxide scale on the alloy, with the outer oxide layer being composed of Fe_3O_4 with some Fe_2O_3 at the surface and the inner oxide layer primarily consisting of $(Fe,Cr)_3O_4$ [13].

Ni-base alloys showed corrosion behavior different from austenitic and super-austenitic stainless steels. Steam oxidation studies of alloy 617 indicated that Cr_2O_3 and $MnCr_2O_4$ formed on surface within a short period of time (e.g., ~1000 h) and additional NiO formed after a longer exposure at 800 °C and 4 MPa [14]. Atmospheric steam exposure of alloy 625 at 700 °C showed the formation of a protective Cr_2O_3 layer [13]. Primary Cr_2O_3 with some oxide exfoliation was also observed on alloy 690 exposed to steam at 600–800 °C [15]. Severe pitting was observed on alloy 718 samples that were subjected to subcritical water at 320 °C [16].

8.2 Thermodynamics of Alloy Oxidation

The ion product, or dissociation constant, for water as it approaches the critical point, is about three orders of magnitude higher than that for ambient liquid water. Accordingly, it can boast a higher H^+ and OH^- ion concentration than liquid water under certain conditions, which results in acidic and basic environments. As one exceeds the critical point, however, the dissociation constant decreases dramatically. For example, the dissociation constant is about nine orders of magnitude lower at 600 °C and ~25 MPa than it is at ambient conditions, resulting in a poor medium for ionic chemistry [17]. Thus, reactions are expected to primarily occur between alloys and O_2 in SCW. Furthermore, iron hydroxide and nickel hydroxide are not stable in SCW [18, 19], for example, iron hydroxide can only be present above ~5 GPa at temperatures above the critical point of water [18].

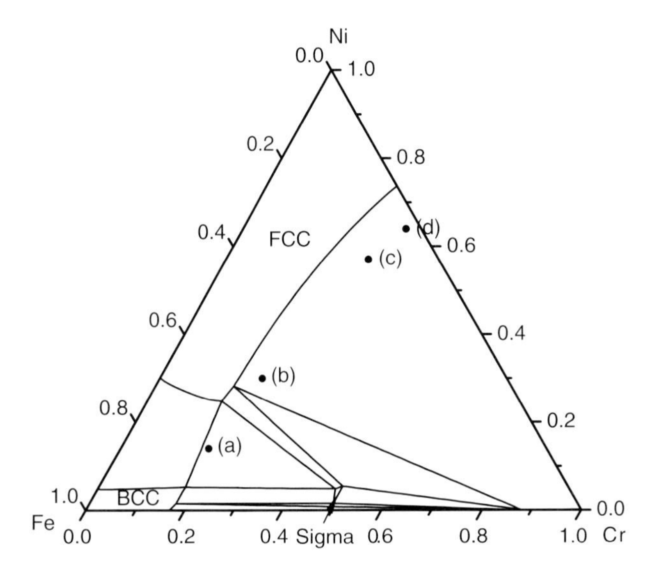

Figure 8.5 Isothermal Fe-Cr-Ni ternary phase diagram at 500 °C (wt%): (a) 68Fe-18Cr-14Ni, (b) 49Fe-21Cr-30Ni, (c) 14Fe-29Cr-57Ni, and (d) 3Fe-33Cr-64Ni.

Since the major components of austenitic stainless steels and Ni-base alloys are Fe, Cr, and Ni, thermodynamic calculations of the Fe-Cr-Ni-O system are expected to give a good first-order approximation of the oxidation thermodynamics of the materials. A thermodynamic database was modified from the work of Luoma [20], and the calculation was performed using Pandat.[3] Although the SCW in the SCWR is designed to be operated at a pressure of 25 MPa, the calculations were performed at 0.1 MPa because of the absence of thermodynamic data at higher pressures. The high pressure is expected to make a difference on the fugacity of oxygen (f_{O_2}), but it will not affect the fugacity of solid phases. The dependence of the fugacity coefficient (ϕ) of gases on temperature and pressure, summarized by Gamson and Watson [21], indicates that the fugacity coefficient of oxygen (ϕ_{O_2}) is <1.2 at 500 °C and 25 MPa. The calculated diagrams corrected by the f_{O_2} ($f_{O_2} = \phi_{O_2} P_{O_2}$) did not show any noticeable difference from those without the correction. Therefore, the following thermodynamic analysis, based on calculations at 0.1 MPa, can reasonably describe the oxidation thermodynamics at 25 MPa.

The 500 °C isothermal Fe-Cr-Ni ternary phase diagram is shown in Figure 8.5, labeled with four model alloys. Considering the stabilization effect of the alloying elements on either ferrite or austenite, the model alloys (a) 68Fe-18Cr-14Ni, (b) 49Fe-21Cr-30Ni, (c) 14Fe-29Cr-57Ni, and (d) 3Fe-33Cr-64Ni approximately represent the standard austenitic stainless steels (304, 316L, and D9), superaustenitic stainless steels (NF709 and 800H), Ni-base alloys 718 and 690, and Ni-base alloys

3) A thermodynamic calculation software developed by CompuTherm LLC (http://www.computherm.com).

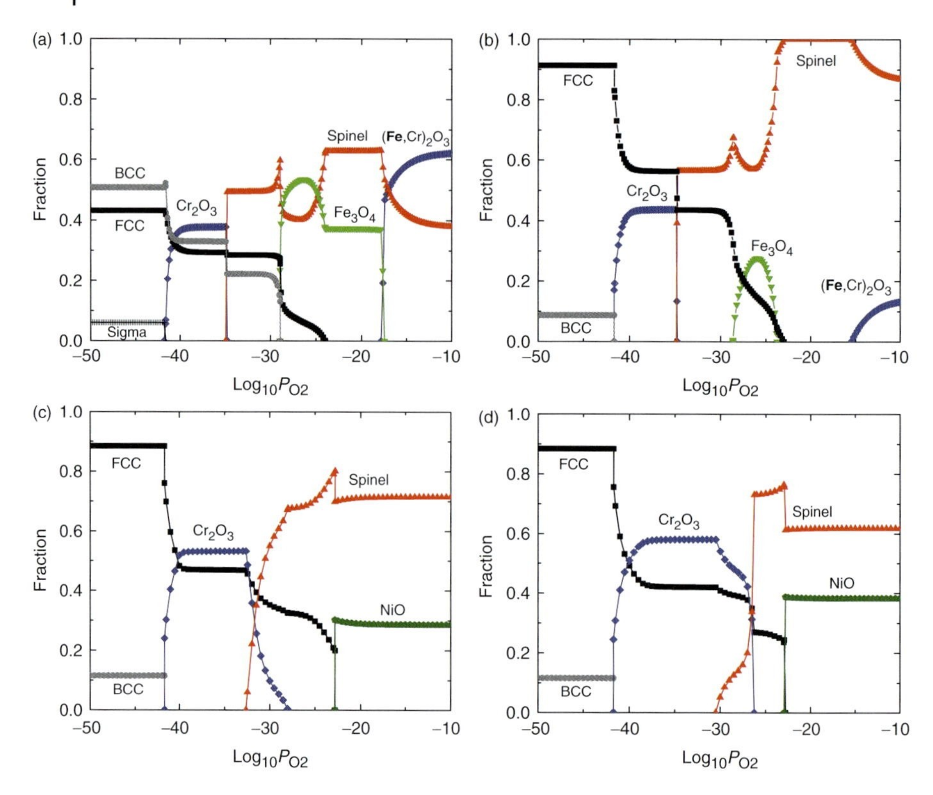

Figure 8.6 Calculated phase mole fraction as a function of oxygen partial pressure of alloys (a) 68Fe-18Cr-14Ni, (b) 49Fe-21Cr-30Ni, (c) 14Fe-29Cr-57Ni, and (d) 3Fe-33Cr-64Ni at 500 °C.

625 and 617, respectively. The equilibrium microstructures of these commercial materials are primarily composed of a face-centered cubic (FCC) phase in addition to a body-centered cubic (BCC) phase at 500 °C. The amount of BCC phase ranges from ~51% in mole fractions for Fe-rich alloy (a) to ~10% for Cr-rich alloys (b–d).

Figure 8.6 shows the calculated 500 °C isothermal equilibrium diagrams of the four model alloys reacting with oxygen. The increase of the oxygen partial pressure (P_{O_2}) schematically represents oxide scales formed on the four model alloys from metallic matrix to scale surface. The oxide scales are primarily composed of corundum (i.e., Cr_2O_3 and $(Fe,Cr)_2O_3$), spinel, and magnetite (i.e., Fe_3O_4) on the stainless steels (a and b), but only Cr_2O_3, spinel, and an additional halite (i.e., NiO) on the Ni-base alloys (c and d). An example of the chemical constituents of the phases in the alloy (a) at 500 °C is shown in Figure 8.7. The corundum is Cr_2O_3 at the metal–oxide interface but $(Fe,Cr)_2O_3$ at surface. Spinel exists over a large range of P_{O_2}, with a composition changing from $FeCr_2O_4$ at low P_{O_2} to $(Fe,Cr,Ni)_3O_4$ at high P_{O_2}. Magnetite shown with symbols is included in Figure 8.7 (b), which exists over a small range of P_{O_2}.

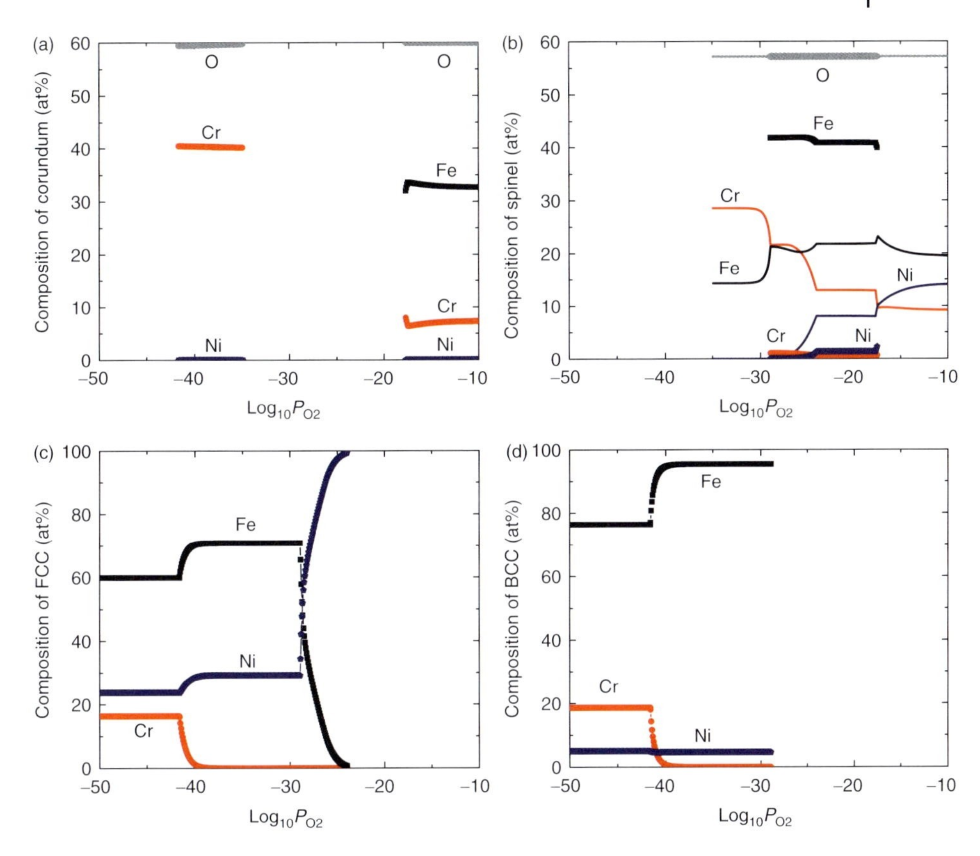

Figure 8.7 (a)–(d) Calculated constituents of the phases in atomic percentage (at%) in alloy 68Fe-18Cr-14Ni at 500 °C under different oxygen partial pressures.

Summarizing the calculations as shown in Figures 8.6 and 8.7, a cartoon has been schematically drawn in Figure 8.8, representing the constituents of the oxide layer structures formed on the respective alloys. The horizontal axis schematically corresponds to the increase in P_{O_2} from left to right. The width of the layers does not correlate with the thickness of the oxide scales. The dash-dot lines are expected to be the interfaces dividing the oxide scales into the outer layer and the inner layer. Calculations indicate that Ni is enriched at the interfaces. Comparing the oxide scales, the austenitic steels (a and b) and the Ni-base alloys (c and d) exhibit significant differences in oxide components and layer structures. The austenitic stainless steels (a) and the superaustenitic steels (b) have similar oxide scales except for the greater amounts of Fe_3O_4 and $(Fe,Cr)_2O_3$ on the austenitic stainless steels (a). Despite similar oxide scales on the Ni-base alloys, the significantly lower Fe contents in alloys 625 and 617 reduce the formation of $FeCr_2O_4$. The calulations indicate that the decreasing amount of Fe leads to the reduced amount of Fe_3O_4 as well as $(Fe,Cr)_2O_3$. The increasing amounts of Cr and Ni lead to the increased amount of Cr_2O_3 and NiO, respectively.

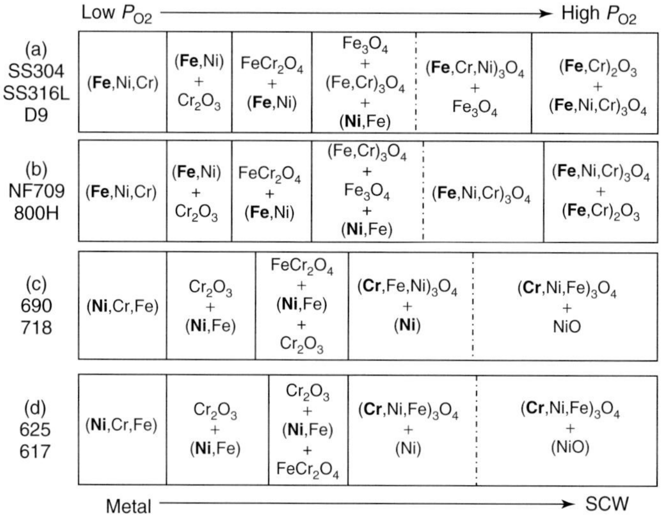

Low P_{O2} ⟶ High P_{O2}

(a) SS304 SS316L D9	(**Fe**,Ni,Cr)	(**Fe**,Ni) + Cr_2O_3	$FeCr_2O_4$ + (**Fe**,Ni)	Fe_3O_4 + $(Fe,Cr)_3O_4$ + (**Ni**,Fe)	$($**Fe**$,Cr,Ni)_3O_4$ + Fe_3O_4	$($**Fe**$,Cr)_2O_3$ + $($**Fe**$,Ni,Cr)_3O_4$
(b) NF709 800H	(**Fe**,Ni,Cr)	(**Fe**,Ni) + Cr_2O_3	$FeCr_2O_4$ + (**Fe**,Ni)	$(Fe,Cr)_3O_4$ + Fe_3O_4 + (**Ni**,Fe)	$($**Fe**$,Ni,Cr)_3O_4$	$($**Fe**$,Ni,Cr)_3O_4$ + $($**Fe**$,Cr)_2O_3$
(c) 690 718	(**Ni**,Cr,Fe)	Cr_2O_3 + (**Ni**,Fe)	$FeCr_2O_4$ + (**Ni**,Fe) + Cr_2O_3	$($**Cr**$,Fe,Ni)_3O_4$ + (**Ni**)	$($**Cr**$,Ni,Fe)_3O_4$ + NiO	
(d) 625 617	(**Ni**,Cr,Fe)	Cr_2O_3 + (**Ni**,Fe)	Cr_2O_3 + (**Ni**,Fe) + $FeCr_2O_4$	$($**Cr**$,Ni,Fe)_3O_4$ + (Ni)	$($**Cr**$,Ni,Fe)_3O_4$ + (NiO)	

Metal ⟶ SCW

Figure 8.8 Schematic comparison of the thermodynamically predicted oxide layer structures formed on the representative alloys (on the left) at 500 °C. The enriched species are denoted with bold letters and the sequence in the parentheses with the richer one(s) listed at the front. The dash-dot lines denote the interface, where metallic phase stops.

The calculations of the alloy oxidation at higher temperatures (e.g., 600 °C) did not show a significant difference from those at 500 °C. The topologies of the diagrams, as shown in Figure 8.6, are shifted to the right side with a higher P_{O_2}, which is consistent with the Ellingham diagram where the equilibrium P_{O_2} of the oxides increases with the increasing temperature [22]. The mole fraction ratio of FCC/BCC phases in the metallic matrix increases with the increasing temperature as a result of the increased FCC phase region at higher temperatures (Figure 8.5).

8.3 Corrosion of Austenitic Stainless Steels and Ni-Base Alloys in SCW

Samples of the austenitic stainless steels and Ni-base alloys listed in Table 8.1 were tested in deionized water with a flow rate of ~1 m s^{-1} at a constant pressure of ~25 MPa and a variety of temperatures. Any galvanic interaction between the samples was eliminated using alumina washers as insulators between the samples and the sample holder. Deionized water with a conductivity of ~0.1 μS cm^{-1} was used for the tests. The outlet conductivity was measured to be ~0.8 μS cm^{-1}. Water chemistry analysis indicates that the outlet conductivity increase was attributed to the metallic ions released from the samples during the tests, such as Fe, Mn, and Ni, in a descending order of content.

8.3.1
Weight Change

The weight change (Δw) of the samples tested at 500 °C for up to 3000 h is shown in Figure 8.9 in a logarithmic scale. Most of the samples showed weight gains except for several samples of the Ni-base alloys 718, 625, and 617 showing weight loss. The weight loss data, denoted with respective triangle symbols, are schematically shown below the x-axis because of the logarithmic scale. Basically, the stainless steels (304, 316L, and D9) showed the greatest weight gains with a slight fluctuation with the exposure time. The superaustenitic stainless steels (NF709 and 800H) showed weight gains slightly lower than those of the stainless steels with a larger fluctuation with the exposure time, especially for alloy 800H. The Ni-base alloys (718, 690, 625, and 617) showed the lowest weight gains and the occurrence of weight loss with a large fluctuation with the exposure time.

The weight changes of these three categories of alloys (316L, D9, NF709, 800H, and 625) at higher (600 °C) and lower (360 °C) temperatures are shown in Figure 8.10a,b, respectively. The weight gain trend of the three categories of alloys at 600 °C is similar to that at 500 °C. In contrast, a distinct weight gain trend was not observed for these alloys at 360 °C (i.e., subcritical condition) because of the limited data and the occurrence of weight loss.

Figures 8.9 and 8.10 indicate that the exposure temperature played a significant effect on the weight change of the austenitic and superaustenitic stainless steels

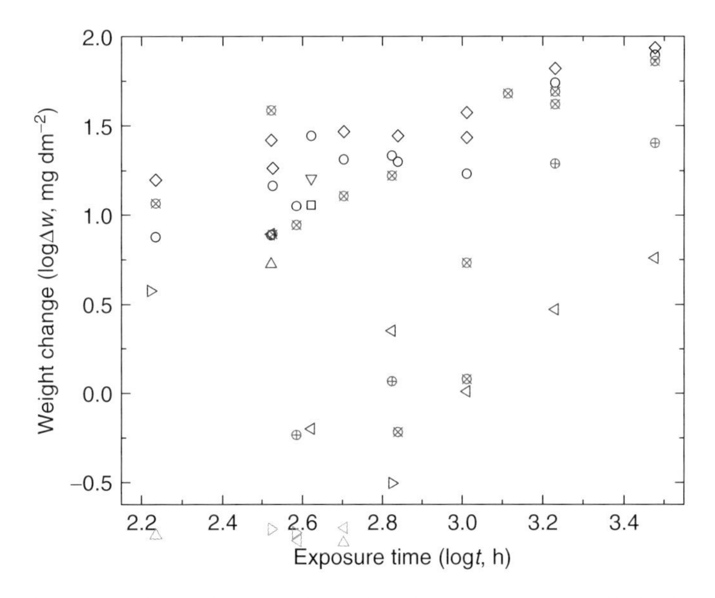

Figure 8.9 Weight change (Δw) as a function of exposure time (t) for the austenitic stainless steels and Ni-base alloys tested at 500 °C: □, 304 [23]; ○, 316L; ◇, D9; ⊕, NF709; ⊗, 800H; △, 718; ▽, 690 [23]; ◁, 625; and ▷, 617. The symbols beneath the x-axis denote weight loss (i.e., negative weight change).

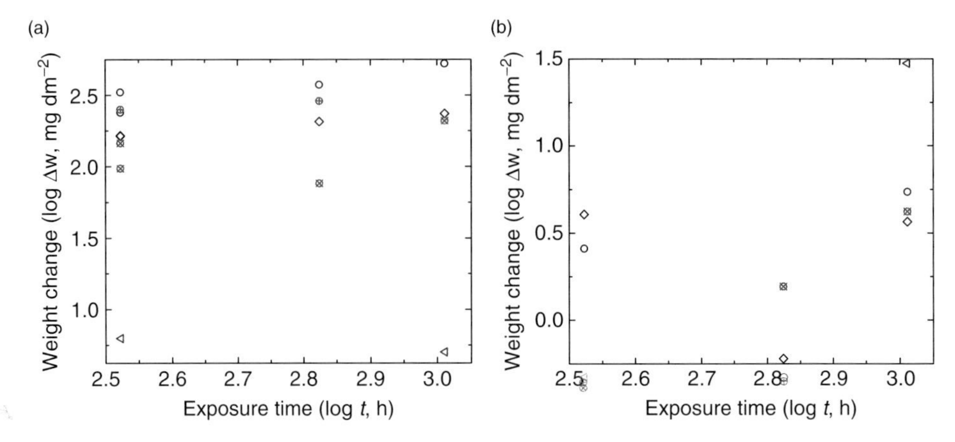

Figure 8.10 Weight change (Δw) as a function of exposure time (t) for the austenitic stainless steels and Ni-base alloys tested at (a) 600 °C and (b) 360 °C, namely, 316L, D9, NF709, 800H, and 625, denoted with the same symbols used in Figure 8.9. The symbols beneath the x-axis denote weight loss (i.e., negative weight change).

but not on the Ni-base alloys. The Ni-base alloys showed both weight gain and loss. The large variation in their weight changes seems independent of the exposure temperatures. Furthermore, the weight gains decreased with increasing chromium and nickel contents in alloys, which is consistent with the literature reports on the effect of chromium and nickel on oxidation [24, 25].

The oxidation kinetics of these three category alloys were then studied by fitting the weight gain data using the following equation

$$\Delta w = k \times t^n \tag{8.1}$$

where k (mg dm^{-2}) and n are the preexponential constant and time exponent, respectively. A linear function is produced by applying logarithm on both sides of Eq. (8.1):

$$\log \Delta w = \log k + n \log t \tag{8.2}$$

The data at 360 °C were not used for fitting because of their large scattering. The parameters for fitting the data at 500 and 600 °C are summarized in Table 8.2. The parameter R^2 is used for measuring the significance of the slope (n) of the fitted line differing from zero [26]. The consistency between the data and the fitted line suggests the quality of the fitting. All the data of the austenitic SS 316L and D9 can be fitted with the equation, although the fitting qualities are not good for the SS 316L. The values of the n parameter in Table 8.2 suggest that the growth kinetics of the SS 316L samples roughly followed a linear rate law ($n = 1$) at 500 °C, which changed to a parabolic rate law ($n = 0.5$) at 600 °C. The D9 samples approximately followed a parabolic rate law at 500 °C but a cubic rate law ($n = 1/3$) at 600 °C. A good fitting was obtained for the limited data of the NF709 samples at 500 °C with an abnormal large growth rate ($n \approx 2$), which suggests that the weight gain of NF709 is expected to surpass that of alloy 800H within a short period of time

Table 8.2 Fitting parameters of the weight change data in Figures 8.9 and 8.10. Some data points with enormous scattering were excluded in the fitting to show clear tendencies of the predominant data.

Alloy	Temperature (°C)	k (mg dm^{-2})	n	R^2	Comments
304	–	–	–	–	Only 1 data point
316L	500	0.118	0.799	0.77	All the 11 data (roughly linear)
	600	13.002	0.528	0.71	All the 4 data (parabolic)
D9	500	0.742	0.576	0.85	All the 9 data (parabolic)
	600	24.980	0.324	0.99	All the 4 data (cubic)
NF709	500	2.9×10^{-6}	2.037	0.93	All the 4 data (high rate)
	600	–	–	–	Only 2 data
800H	500	0.016	1.071	0.97	8 data excluded 5 data (linear)
	600	6.366	0.504	0.44	3 data excluded 1 data (parabolic)
718	–	–	–	–	1 weight gain and 2 weight loss data
690					Only 1 data
625	500	0.002	0.977	0.66	5 data excluded 1 weight gain and 2 weight loss data (linear)
	600				Only 2 data
617	–	–	–	–	2 weight gain and 2 weight loss data

even though the initial weight gain of NF709 is much less than that of 800H. Disregarding the significantly scattered data, the 800H samples approximately followed a linear rate law at 500 °C and a possible parabolic rate law at 600 °C. Reliable growth kinetics could not be obtained for the Ni-base alloys because of the limited and scattered data.

8.3.2 Surface Morphology

Typical surface morphologies of the samples exposed to the SCW at 500 °C are shown in Figure 8.11. A compact oxide scale was observed on the austenitic stainless steels (e.g., D9 in Figure 8.11a). Some oxide exfoliation was observed on the superaustenitic steel NF709, with significant spallation on the alloy 800H (Figure 8.11b). The alloys 617 and 690 showed good integrity of the surface with limited oxide exfoliation (e.g., alloy 617 in Figure 8.11c). Significant pitting was observed on the alloys 625 and 718 (e.g., alloy 718 in Figure 8.11d) because of the presence of the Nb- and/or Ti-rich precipitate (γ''-phase) in these alloys [27].

Typical surface morphologies of the samples exposed to the SCW at 360 and 600 °C were also studied. Figure 8.12 shows two examples of 800H at 360 °C and D9 at 600 °C. Sparsely distributed oxide particles from nanometers to a few micrometers were observed on the samples at 360 °C (e.g., alloy 800H in

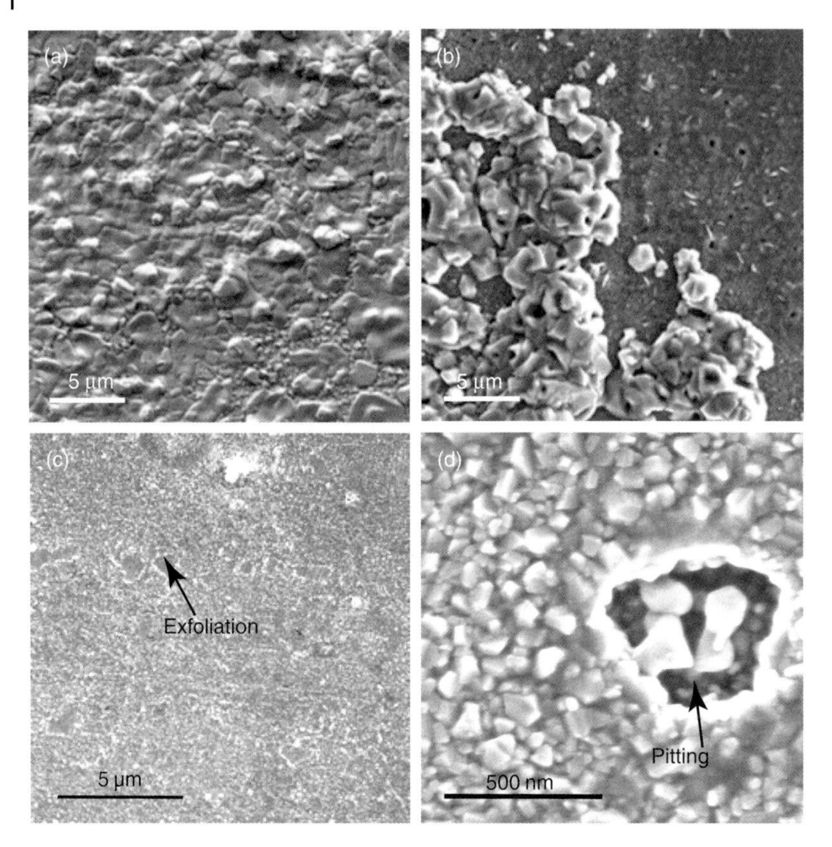

Figure 8.11 Surface morphologies of the oxide formed on the samples of the alloys (a) D9 (1026 h), (b) 800H (1026 h), (c) 617 (667 h), and (d) 718 (505 h) at 500 °C.

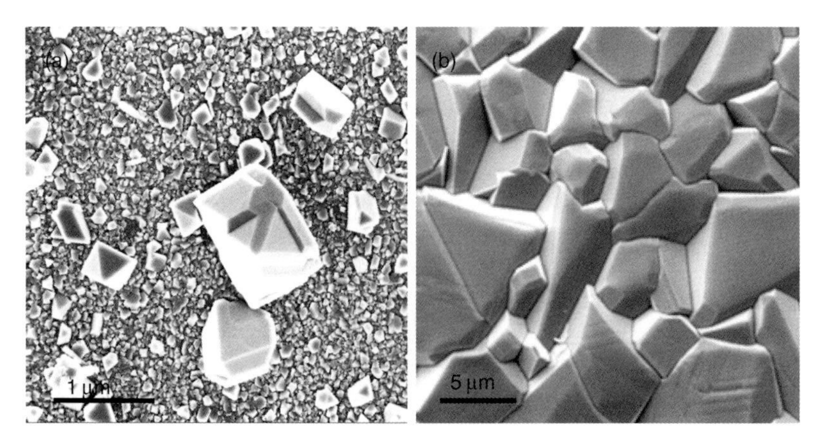

Figure 8.12 Surface morphologies of the oxide formed on the samples of alloys (a) 800H at 360 °C for 1026 h and (b) D9 at 600 °C for 1000 h.

Figure 8.12a). As temperature increases, the oxide grains increased and the oxide scale became continuous (e.g., alloy D9 in Figure 8.12b). Oxide exfoliation at different levels was also observed on some of the samples at 600 °C, which is not desired because it may cause blockage inside the superheater/reheater tubes or a severe erosion damage at the turbine blades [28].

8.3.3 Oxide Layer Structure

Figure 8.13 shows a typical oxide layer structure formed on each of the three categories of alloys: austenitic stainless steels 304 (a), 316L (b), and D9 (c); superaustenitic steel 800H (d); and Ni-base alloys 690 (e) and 617 (f). A dual-layer oxide scale was observed in the first two categories of steels and some Ni-base alloys. The interface of the outer and inner oxide layers was found to be the original surface of the samples before the exposures. Irregular surface contours of the outer layers were observed on the austenitic stainless steels. The irregularity became diminished as the Cr and Ni contents increase in the alloys. Furthermore, high

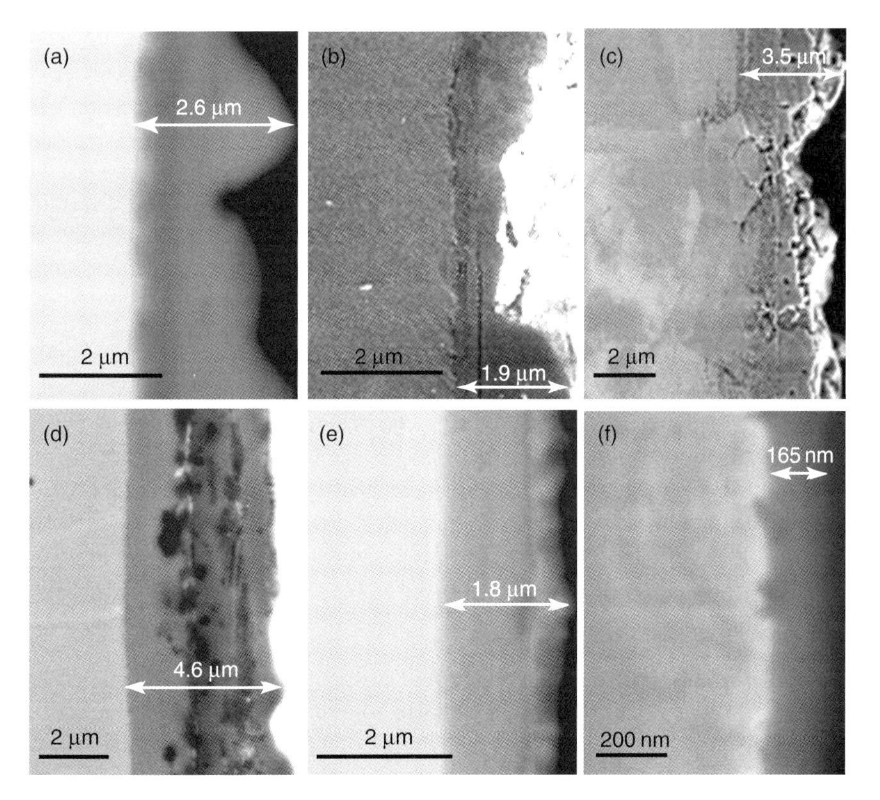

Figure 8.13 Oxide layer structure formed on the alloy samples exposed to the SCW at 500 °C: (a) 304 for 419 h [23], (b) 316L for 1026 h [29], (c) D9 for 505 h [29], (d) 800H for 505 h, (e) 690 for 419 h [23], and (f) 617 for 667 h.

Cr and Ni contents benefited the thickness reduction of oxide scales, for example, <200 nm for the alloy 617. The oxide scale on alloy 617 was too thin to distinguish its morphology.

Energy dispersive X-ray spectroscopy (EDS) and X-ray diffraction (XRD) analyses provided further composition and phase information on the oxide scales. Dual-layer oxide scales, including inner spinels ($Ni(Fe,Cr)_2O_4$ and $FeCr_2O_4$) and outer magnetite (Fe_3O_4), were reported on stainless steels 304 and 316L exposed at 400–550 °C [23]. However, only spinel $(Fe,Cr)_3O_4$ with localized metallic Ni enrichment was observed in the inner layer together with outer Fe_3O_4 on alloy D9 [29]. The presence of Ni-rich austenite phase (Ni,Fe) in the inner oxide scale on alloy D9 was also confirmed by transmission electron microscopic (TEM) analysis [30]. The oxide constituents without $Ni(Fe,Cr)_2O_4$ in the inner layer on alloy D9 are consistent with the thermodynamic prediction as presented in Section 8.2. A higher exposure temperature of 600 °C did not change the constituent phases of the scales but changed Cr_2O_3 from a discrete morphology at 500 °C into a continuous layer at some regions between the oxide scale and the substrate steel [31]. The elevated temperature as well as small grain size of alloy D9 (primarily 1–20 μm) promoted the formation of Cr_2O_3.

Compared to the oxide scale on alloy D9 at 500 °C, an additional thin layer of hematite (Fe_2O_3) developed at the surface of alloy 800H [32]. Oxide exfoliation was mainly observed at the interface of inner spinel and outer magnetite and some at the interface of magnetite and hematite [32–34]. Ni-enriched regions became pronounced at higher temperatures in addition to the occurrence of mushroom-shaped internal oxidation at 600 °C. Oxide scale and its exfoliation behavior on NF709 were similar to that on alloy 800H [29]. The constituents of oxide scales on the austenitic and superaustenitic stainless steels are consistent with the thermodynamic predictions except that the thermodynamic predicted $(Fe,Ni,Cr)_3O_4$ (or $Ni(Fe,Cr)_2O_4$) spinel in the outer layer was not observed. This is due to kinetic reasons discussed in detail later.

An inner layer of Ni, Cr_2O_3, and $Ni(Fe,Cr)_2O_4$ together with an outer layer of NiO was observed on alloy 690 exposed at 400–550 °C [23]. NiO predominated at lower temperatures, but Cr_2O_3 and spinel became more prevalent with increasing temperatures. Compared to the oxide scale on alloy 690, an additional $(Cr,Fe)_2O_3$ was revealed on the surface of alloy 718 at 500 °C by Auger electron spectroscopy (AES) coupled with XRD analyses [27, 29, 35]. Although Cr-rich $(\mathbf{Cr},Fe)_2O_3$ is not expected to form on both alloys 690 and 718 according to Figures 8.6c and 8.8c, the higher Fe content in alloy 718 may have favored the formation of $(\mathbf{Cr},Fe)_2O_3$.

Oxides of Cr_2O_3, $Ni(Cr,Fe)_2O_4$, $(Cr,Fe)_2O_3$, and NiO were observed on alloy 625 at 500 °C [27, 29]. The NiO dispersed at the surface, and its amount diminished with increased exposure time, suggesting its nonprotective nature. The presence of $(Cr,Fe)_2O_3$ on alloy 625 is not consistent with the thermodynamic predictions as shown in Figures 8.6d and 8.8d. Detailed microstructure characterization of the thin oxide scales is necessary to elucidate the discrepancy. Compared to alloy 625, oxide scales on alloy 617 at 500 °C were primarily composed of Cr_2O_3 and $(Ni,Co)Cr_2O_4$. The significant amount of Co addition (substituting for Ni) and the small amount of Fe favored the formation of (Ni,Co)Cr_2O_4 sharing the same crystal structure as Ni(Cr,Fe)$_2$$O_4$ NiO was not observed on alloy 617, which may have exfoliated during the exposure.

8.4 Novel Corrosion Control Methods

8.4.1 Microstructural Optimization

Grain boundaries are common defects existing in crystalline materials, and they play a major role in determining the physical, mechanical, electrical, and chemical properties of crystalline materials [36]. Grain boundary engineering (GBE), usually implemented by means of thermomechanical processing, has shown a remarkable improvement in the properties of FCC polycrystalline metals and alloys with low stacking fault energies [37]. The improved properties have been primarily attributed to the increased fraction of low-Σ coincidence site lattice boundaries (CSLBs), with $\Sigma \leq 29$ in the materials, where Σ is the reciprocal density of coincident sites at the grain boundary between two adjoining grains. With the coincidence site lattice (CSL) model, grain boundaries can be classified into two categories, that is, low-Σ CSLBs and general boundaries including high-Σ CSLBs ($\Sigma > 29$) and random boundaries.

One of the applications of GBE is to improve oxidation resistance. Successful examples of applying GBE to the mitigation of oxide exfoliation and the reduction of oxide growth kinetics have been reported [32–35]. Figure 8.14 shows surface morphologies of alloys 800H and 617 exposed to the SCW at 600 and 500 °C, respectively. The GBE-treated alloy 800H with ~75% low-Σ CSLBs showed a continuous compact oxide scale without noticeable exfoliation. In contrast, significant exfoliation occurred on the as-received alloy 800H sample with ~51% low-Σ CSLBs. Oxide exfoliation also occurred on the as-received alloy 617 sample with ~53% low-Σ CSLBs. But its exfoliation was far less than the as-received alloy 800H sample because of the high Ni and Cr contents in alloy 617. The GBE-treated alloy 617 sample with ~79% low-Σ CSLBs also showed improved resistance to oxide exfoliation compared to the as-received sample. X-ray photoelectron spectroscopy (XPS) analysis revealed a single Cr_2O_3 layer on the GBE-treated alloy 617 sample instead of additional (Ni,Co)Cr_2O_4 on the as-received sample [34].

Similar oxide exfoliation was observed on the as-received alloy 800H but not on the GBE-treated alloy 800H exposed to SCW at 500 °C. Cross-sectional regions with well-retained oxide scales are shown in Figure 8.15 for the as-received and GBE-treated alloy 800H samples. The as-received sample had a slightly thicker (~0.5 μm) oxide scale than the GBE-treated one. Both samples displayed dual-layer oxide scales. The inner layer contains spinel (S: (Fe,Cr)$_3$$O_4$) interwoven with metallic phase (Fe,Ni) as characterized by TEM [38], which resulted from the inward diffusion of oxygen. Some Cr-rich islands were dispersed in the inner oxide layer, as shown in the dark regions of Figure 8.15. The darker region corresponds to

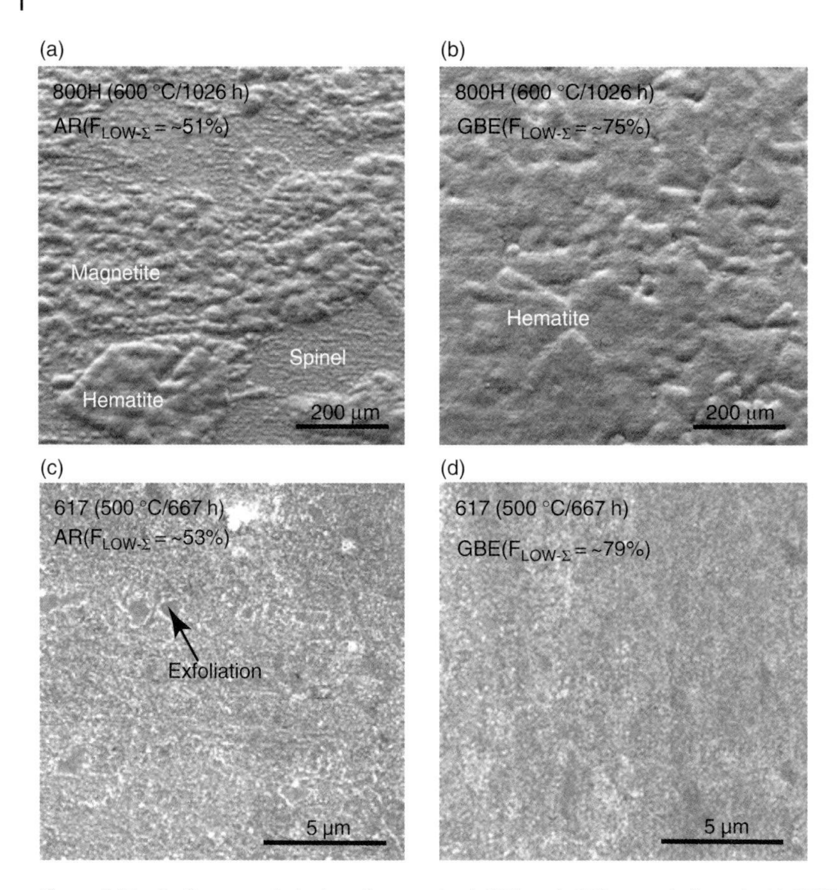

Figure 8.14 Surface morphologies of as-received (AR) and GBE-treated alloys (a, b) 800H and (c, d) 617 exposed to SCW ($F_{Low-\Sigma}$, fraction of low-Σ CSLBs).

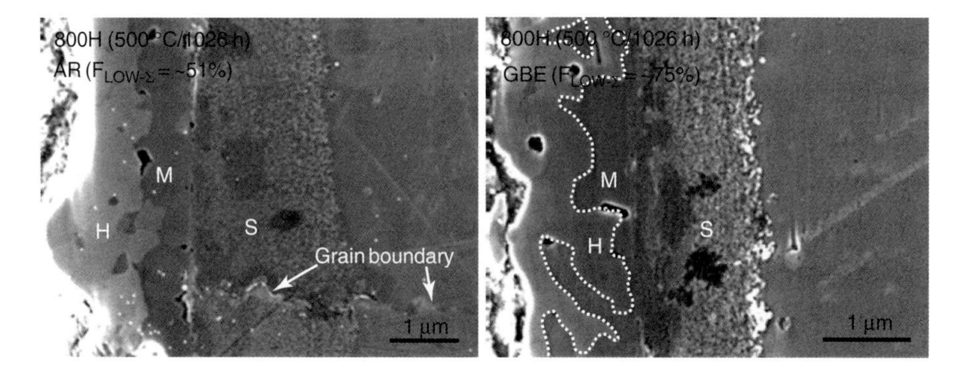

Figure 8.15 Cross sections of oxide scales formed on the (a) as-received (AR) and (b) GBE-treated alloy 800H exposed to SCW at 500 °C for 1026 h. S, M, and H denote spinel, magnetite, and hematite, respectively.

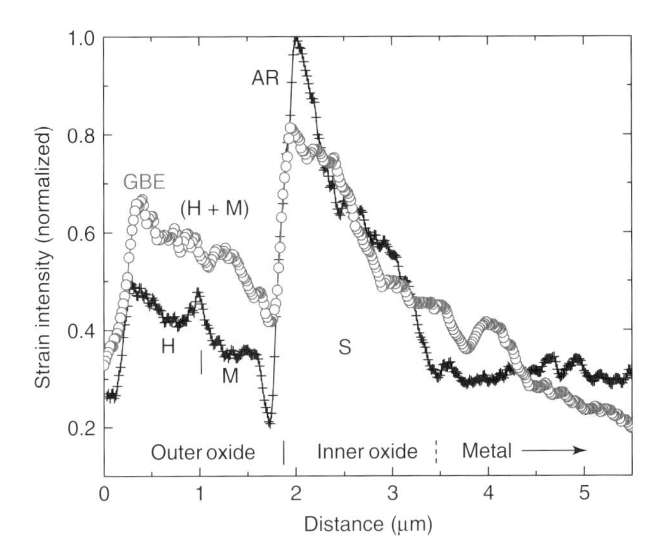

Figure 8.16 Normalized strain intensity across the oxide scales formed on the as-received (AR) and GBE-treated alloy 800H samples exposed to SCW at 500 °C [34].

higher Cr content. The outward diffusion of Fe developed the outer layer composed of a layer of hematite (H: Fe_2O_3) at the surface followed by a layer of magnetite (M: Fe_3O_4) in sequence. Approximately parallel interfaces of S–M and M–H were developed on the as-received sample. Unlike the relatively smooth M–H interface on the as-received sample, a serrated M–H interface on the GBE-treated sample is highlighted with a white dotted line in Figure 8.15. It is also noticed that the volume ratio of H/M on the GBE-treated sample is greater than that on the as-received one. Figure 8.15 also shows oxide growth along a general boundary on the as-received sample, which promoted the formation of a nodule at the surface and further inward growth into the bulk, leading to ~1.5 μm thicker oxide scale compared to its adjacent regions.

Strain intensity accumulated in the oxide scales was characterized by electron backscatter diffraction (EBSD). Figure 8.16 shows normalized strain intensity across the oxide scales formed on the as-received and GBE-treated alloy 800H samples exposed to SCW at 500 °C. Strain largely accumulated in the spinel's inner oxide layer and approached the maximum close to the interface of the outer-inner oxide layers. When crossing the interface from the inner layer to the outer layer, a large drop in strain intensity occurred. This strain intensity change in the oxide scale on the as-received sample is about twice of that on the GBE-treated sample. The strain intensity changes could be attributable to the temperature-induced variations in the volume thermal expansion coefficients (α_v) of the species. Figure 8.17 shows the effect of temperature on the α_v of Fe_2O_3, Fe_3O_4, $FeCr_2O_4$, Cr_2O_3, Ni, and 800H [39–41]. The presence of a Ni-rich metallic phase and trace amount of Cr_2O_3 in the spinel layer would not develop large strains because of their similar α_v. In contrast, the remarkable differences between Fe_3O_4 and spinel in the values of

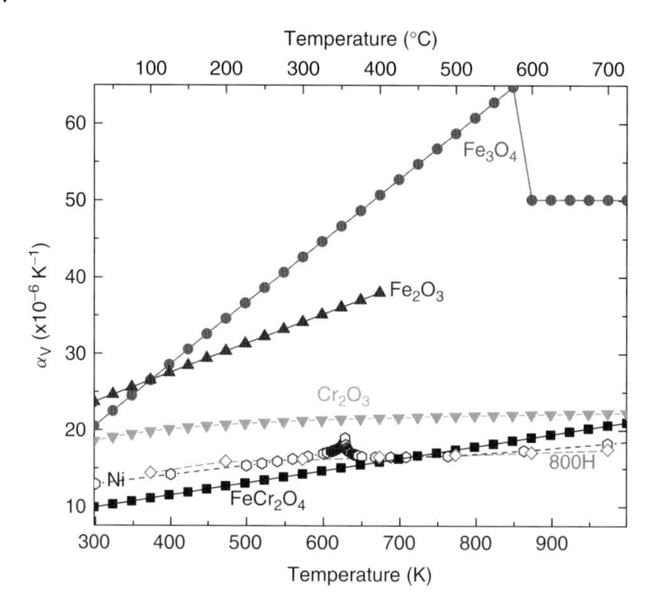

Figure 8.17 Volume thermal expansion coefficients (α_v) of magnetite (Fe_3O_4), hematite (Fe_2O_3), chromia (Cr_2O_3), spinel ($FeCr_2O_4$), Ni, and alloy 800H.

α_v and their variations as a function of temperature would develop a large strain change on the as-received sample. The $\alpha_{vFe_2O_3}$, however, is smaller than $\alpha_{vFe_3O_4}$ at elevated temperatures and follows a function of temperature similar to that of spinel. Therefore, the spinel phase has a more compatible volume expansion with hematite than with magnetite, which leads to less strain change at the S–H interface. The serrated M–H interface and the increased amount of hematite in the outer layer on the GBE-treated sample increase the probability of a larger S–H interface and thus mitigate the changes in strain intensities.

Figure 8.18 shows oxide scales formed on the as-received and GBE-treated alloy 800H samples exposed to SCW at 600 °C for 1026 h. A thin Ni layer was formed at the interface of the outer-inner layers. Mushroom-shaped oxide grew inward on the as-received samples. In contrast, the inward growth of the oxide was more irregular on the GBE-treated samples. Image analysis on Figure 8.18a indicated that the average thickness of the oxide scale on the GBE-treated sample is about twice of that on the as-received sample. The ratio of the average to maximum thickness (T_{ave}/T_{max}) of the outer layer of the GBE-treated sample approaches one, suggesting good integrity of the scale. In contrast, the T_{ave}/T_{max} ratio of the outer layer of the as-received sample is much smaller, which suggests poor integrity of the scale with significant exfoliation. The T_{ave}/T_{max} ratio of the inner layer of the GBE-treated sample is slightly larger than that of the as-received sample [38], suggesting greater oxide coverage. The morphologies of the inner layers, for example, mushroom shaped, were affected by the microstructure of the samples, which is discussed later.

Figure 8.18b shows the detailed oxide scale formed on the GBE-treated sample. The presence of the phases were identified with EDS, AES, and EBSD [38]. A

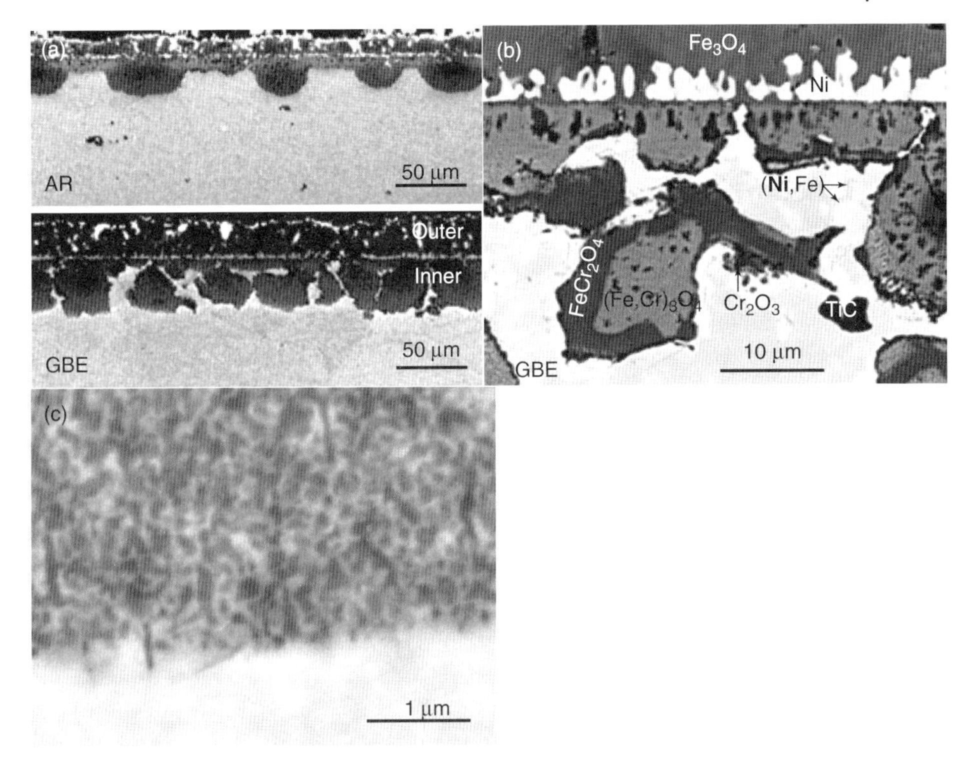

Figure 8.18 Backscattered electron images (BEIs) of cross-sectional oxide scales formed on the as-received (AR) and GBE-treated alloy 800H samples exposed to SCW at 600 °C for 1026 h. (a) low magnification images showing the overall oxide layer structures; (b) high magnification image of the GBE-treated sample showing the detailed microstructure of the inner oxide layer; (c) high magnification image of the AR sample showing the front microstructure of the inward oxidation.

metallic Ni-rich regime was observed at the outer-inner oxide and the oxide–metal interfaces in the form of Ni and (**Ni**,Fe), respectively. The outer layer is composed of magnetite (Fe_3O_4). The inner layer is predominantly composed of two types of spinels ($(Fe,Cr)_3O_4$ in light gray and $FeCr_2O_4$ in gray) with a trace amount of Cr_2O_3 (in dark gray). The $(Fe,Cr)_3O_4$ is not only surrounded by $FeCr_2O_4$ but also dispersively decorated with $FeCr_2O_4$ islands inside. Figure 8.18c shows an example of the front microstructure of inward oxidation with $(Fe,Cr)_3O_4$ interweaving with Ni-rich metallic phase.

8.4.2
Grain Size Refinement

Shot peening is a cold working process using high-speed beads peening on the surface of a workpiece, generating dimples. Consequently, compressive stresses are introduced on the stretched surface. Furthermore, shot peening generally refines surface grains into submicrometer to nanometer sized ones [42]. The grain refinement mechanism involves dislocation activities resulting in subgrain boundaries and thus evolving into nanometer and submicrometer grains [43]. It has been reported that shot peening increases hardness and durability, fatigue life [44], stress corrosion cracking resistance [45], and closing of porosity [46]. Shot peening may also mitigate oxide exfoliation and reduce oxide growth kinetics due to compressive stresses and refined grain size at surface, which may alter the microstructure of the oxide scale.

Figure 8.19 shows the surface morphologies and cross-sectional oxide scales formed on the as-received and shot-peened alloy 800H exposed to SCW at 500 °C for 3000 and 667 h, respectively. Compared to the predominant large Fe_3O_4 grains (in micrometers) on the as-received sample, only a few submicrometer Fe_3O_4 grains were discretely distributed close to the cracks usually located at ridges or edges of dimples on the shot-peened sample. The surface of the shot-peened sample was primarily covered with a fine compact oxide scale as shown in Figure 8.19b. An ~70 μm thick deformation zone was developed on the shot-peened sample, which consisted of an ultrafine grain (15–21 nm) region at the surface, analyzed by XRD, and a sequential transition region with many slip bands in the subsurface [47]. The overlapped edges of dimples generated many cracks of ~1 μm depth. Figure 8.19c shows the oxide scale formed close to a crack marked with a black arrow. Oxidation occurred along the crack and resulted in an oxide scale thicker (~1–2 μm) than the layer (~0.5 μm) at the surface. EDS and grazing-incidence XRD analyses indicated that the oxide scale was primarily composed of an inner Cr_2O_3 (gray) and an outer $FeCr_2O_4$ (dark gray in Figure 8.19c) in addition to a few discrete submicrometer

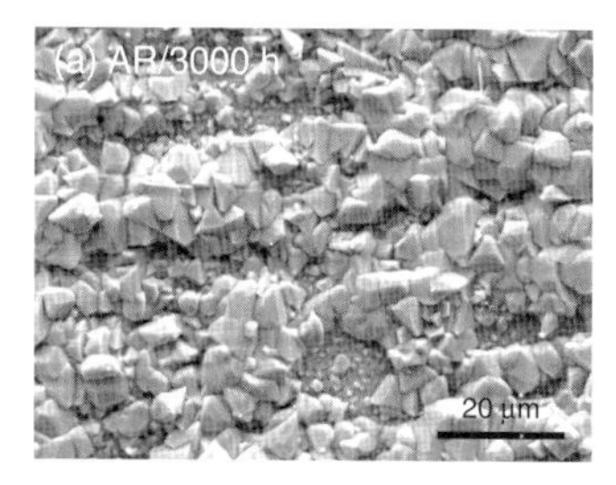

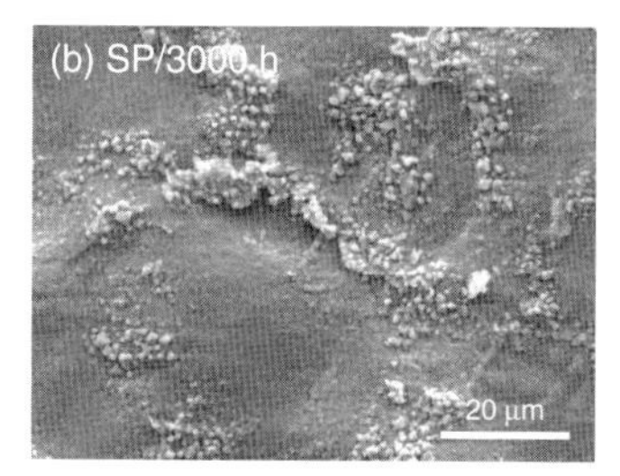

Figure 8.19 Surface morphologies of the (a) as-received and (b) shot-peened (SP) samples, and (c) cross-sectional oxide scale (left side) of an SP sample exposed to SCW at 500 °C [47].

Fe_3O_4 grains at the surface (Figure 8.19b). A slightly Ni-enriched metallic phase (light gray in Figure 8.19c) was observed beneath the oxide scale. Noticeable grains and twins (e.g., marked with white arrows) in nanometer and submicrometer size were observed close to the surface after the exposure. The cracks induced by the overlapped dimples did not deteriorate the overall oxidation performance of the shot-peened samples within the period of SCW exposures.

8.4.3 Performance Comparison of the Corrosion Control Methods

Figure 8.20 shows the weight change (Δw) as a function of exposure time (t), in a logarithmic scale, of alloys 800H and 617 in the as-received, GBE-treated, and shot-peened conditions exposed to the SCW at 500 °C. The data of the GBE-treated and shot-peened samples showed significantly smaller scattering than the as-received samples. Using a process similar to that used in Table 8.2, the kinetic parameters to describe the data in Figure 8.20 were obtained and summarized in Table 8.3. The results show that the as-received alloy 800H samples would follow a linear ($n = 1$) rate for the experimental data excluding the five data points with enormous scattering. The data of the GBE-treated alloy 800H samples can be well represented with $n = 0.84$, suggesting a mixed oxidation kinetics of a linear ($n = 1$) and a parabolic rate law ($n = 0.5$). The data of the shot-peened alloy 800H samples were described with $n = 1.20$. The k parameter had a much

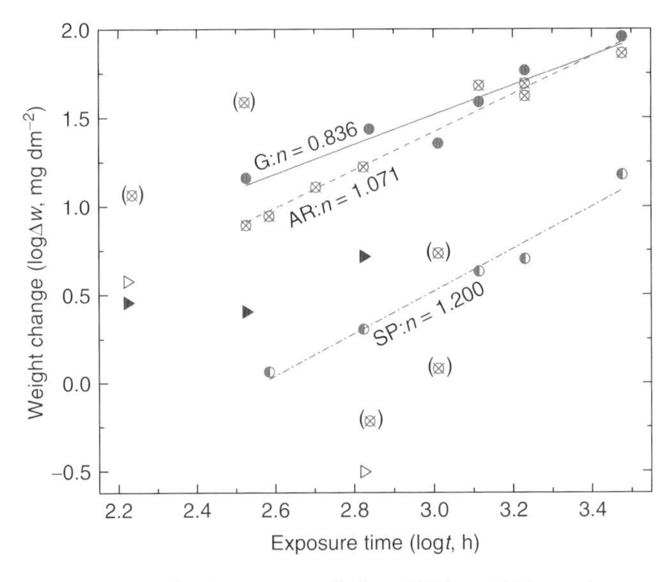

Figure 8.20 Weight change (Δw) of alloys 800H and 617 samples exposed to SCW at 500 °C. Open circle with a cross sign, solid circle, and half filled circle denote the as-received, GBE-treated, and shot-peened alloy 800H samples, respectively. Open and solid triangles denote the as-received and GBE-treated alloy 617 samples, respectively. Fitted lines are for 800H.

Table 8.3 Fitting parameters of the weight change data in Figure 8.20.

Alloy	Condition	k (mg dm^{-2})	n	R^2	Comments
800H	AR	0.016	1.071	0.97	8 data excluded 5 data in ( )
	G	0.102	0.836	0.88	All the 6 data
	SP	8.3×10^{-4}	1.200	0.97	All the 5 data
617	AR	–	–	–	2 weight gain and 2 weight loss data
	G	0.277	0.429	0.20	All the 3 data

smaller value for the shot-peened samples (8.3×10^{-4} mg dm^{-3}) than for the GBE-treated samples (0.102 mg dm^{-3}). Therefore, although the oxidation of the shot-peened samples had a higher rate (n), the significantly smaller constant k can keep the weight gain of the shot-peened samples in an extremely small regime. Assuming the oxidation growth kinetics does not change over time, it would take 5.5×10^5 h (or ~62.8 years) for the shot-peened samples to surpass the weight gain of the GBE-treated samples. In contrast, it would only take about 2650 h for the as-received samples to surpass the weight gain of the GBE-treated samples, although the as-received samples showed smaller weight gain than the GBE-treated samples in a short period of exposure. Thus, during the practical service life, the oxidation resistance of alloy 800H would be the best in the shot-peened condition and better in the GBE-treated condition compared to the as-received condition. The limited data of alloy 617 samples could not produce predictive trends. However, Figure 8.20 indicates that the GBE-treated alloy 617 samples had a stable oxide growth compared to the large scattering data of the as-received samples with two weight loss data points as shown in Figure 8.9.

8.5 Factors Influencing Corrosion

8.5.1 Test Conditions

As shown in Figure 8.2a, dramatic changes on the isobaric thermophysical properties of water, such as heat capacity, thermal conductivity, and viscosity of water, occur when temperature crosses the critical point at 374 °C. These factors have the potential to play a significant role in the corrosion behavior of materials.

First of all, the sharp peak of heat capacity and large increase of thermoconductivity when temperature decreases from 450 to 350 °C (i.e., the vicinity of critical temperature) is beneficial for adsorbing more heat in water at lower temperature and thus mitigate the heat rise and delay the onset of the oxidation of the steels [48]. Second, oxygen diffusivity in water ($D_{O_{2-}}$ H_2O) is affected by temperature (T)

and the viscosity of water (μH), which can be simply expressed as

$$D_{O_{2-}}\ H_2O = \frac{k \times T}{\mu H_2O} \tag{8.3}$$

where k is a constant related to the properties of H_2O and O_2, for example, the association parameter for the solvent water, molecular weight of water, and molar volume of oxygen. A detailed description of the $D_{O_{2-}}\ H_2O$ can be found in Ref. [49]. $D_{O_{2-}}\ H_2O$ is proportional to T in degrees Kelvin and inversely proportional to μH_2O. μH_2O is also a function of T, which has significantly greater values at temperatures below the critical point of water as shown in Figure 8.2a. Thus, a small decrease of T across the critical point of water results in a remarkable increase in μH_2O that plays a significant effect in reducing $D_{O_{2-}}\ H_2O$. The significantly reduced $D_{O_{2-}}\ H_2O$ at temperatures below the critical point hindered oxygen diffusion into metals [50], leading to less oxidation and weight gains.

The isothermal thermophysical properties of water as a function of pressure, as shown in Figure 8.2b, do not show a dramatic change at the critical point as those as a function of temperature in the isobaric case. However, pressure has shown some effects on oxidation. Studies on steam exposures of austenitic stainless steels suggested that the scale integrity is worsened by increasing pressure, subsequently leading to an enhanced oxidation rate [51]. Voids and cavities were found in the outer scale layer under higher steam pressures, which may be attributable to the local cessation of the scale growth, the different diffusion rates in the outer and inner scale layers, and/or a weakening of the oxide grain boundaries [52]. Similar pressure effects on oxidation are expected to occur during SCW exposure.

Another test condition that affects oxidation and scale exfoliation is the oxygen content dissolved in the water. Compared to the exposures with deoxygenated SCW (e.g., $<\sim$25 ppb), severe exfoliation of the outer oxide scale with hematite at the surface was observed on alloy D9 exposed to SCW at 500 °C with a high oxygen content ($\sim$2 ppm) [31]. In contrast, alloy 800H showed a more compact oxide scale with less exfoliation in high-oxygen SCW ($\sim$2 ppm) compared to that in deoxygenated SCW [29, 38].

8.5.2 Effect of Thermodynamics and Kinetics

Experimental observations exhibit good consistency with thermodynamic predictions for the most part, except for some discrepancies primarily in the outer layer of the oxide scales. It should be mentioned that the thermodynamic equilibrium calculations in Section 8.2 were based on an assumption of constant alloy compositions at every P_{O_2}. However, alloy composition usually does not follow the assumption in a practical oxidation situation because of the different diffusivities of the species. Thus, the observed experimental results are not completely consistent with the predicted results as shown in Figures 8.6 and 8.8.

Ni was reported to have the smallest diffusion coefficient (D) in austenitic Fe-Cr-Ni stainless steels with $D_{Cr} > D_{Fe} > D_{Ni}$ [53]. The preferential formation of Cr_2O_3, favored by thermodynamics and kinetics, could restrain the outward diffusion of Ni due to the limited solubility of Ni in Cr_2O_3. In contrast, the diffusion of Fe is less constrained because of the large solubility of Fe in Cr_2O_3 [54]. At lower temperatures (e.g., 500 °C), the reaction of $2Fe + 2Cr_2O_3 + O_2 = 2FeCr_2O_4$ [55] is favorable, leading to only a limited amount of Cr_2O_3 that can hardly be detected. At higher temperatures (e.g., 600 °C), the detectable amount of Cr_2O_3 increased because of the increased diffusivity of Cr at elevated temperatures. The *unconstrained* outward diffusion of Fe favored the preferential formation of Fe_3O_4 at the surface of the exposed stainless steels, leaving $(Fe,Cr)_3O_4$ beneath it in the inner layer. Owing to the limited solubility of Ni in iron oxides such as Fe_3O_4, $FeCr_2O_4$, and $(Fe,Cr)_3O_4$ [54], Ni enrichment was observed forming equilibrium with $(Fe,Cr)_3O_4$ (Figure 8.18c) and at the interfaces of the metal–$FeCr_2O_4$ and the inner–outer (Fe_3O_4) layers (Figure 8.18). The semicontinuous Ni-enrichment layer at the interface of the inner–outer layers on alloy 800H exposed at 600 °C was characterized as *pure* Ni (>97.5 at% Ni) with AES, which is consistent with the thermodynamic calculation as shown in Figure 8.7c. Such pure Ni was also observed at the interface of the inner–outer layers at 500 °C, but its amount was far less than that at 600 °C. The $Ni(Fe,Cr)_2O_4$ predicted by thermodynamic calculation was not observed in the outer layer at both temperatures, which is probably due to the blocked outward diffusion of Ni by the Fe_3O_4 layer. The pure Ni layer could have acted as a diffusion barrier layer at 600 °C, which restrained the outward diffusion of Cr but did not affect Fe because of the larger diffusivity of Fe in Ni ($D_{Fe} \approx 4D_{Cr}$ in Ni) [56]. However, the pure Ni layer was not an effective barrier to the inward diffusion of oxygen because $D_O \approx 10^5 D_{Fe}$ in Ni [57]. Although corundum $(Fe,Cr)_2O_3$ is not expected to form on the surface of alloy 800H according to the P_{O_2} of the SCW with a low oxygen content (~25 ppb) (i.e., $\log P_{O_2} = -16$ [58]), it was observed at the surface of the exposed samples, which may have been favored by the kinetics in the alloy oxidation system.

8.5.3
Effect of Microstructure

With alloy 800H as an example, the GBE-treatment primarily changed the microstructure in three aspects: (i) significantly increased population of low-Σ CSLBs, (ii) significantly reduced tangling dislocation network clusters and *free* dislocation regions, and (iii) introduced a large amount of nanoprecipitates in the matrix [59]. The effect of the nanoprecipitates on corrosion is not clear currently, but they somewhat enhanced the strength of the GBE-treated samples [34, 60]. Shortcut diffusion via grain boundary networks and dislocation network clusters is expected to have played a predominant role, compared to lattice or volume diffusion, at the exposure temperatures used in these studies. These shortcut diffusion paths, especially for the dislocation clusters due to their extremely larger population compared to grain boundaries, promoted the formation of the mushroom-shaped inner oxide layer in the as-received samples (Figure 8.18). The inward diffusion of oxygen was preferentially transported through the shortcut paths, leading to the formation of Cr_2O_3, most of which eventually transformed to Cr-rich spinel such as $FeCr_2O_4$ as shown with the dark gray islands in the inner layer in Figure 8.18b. The continuous supply of oxygen through these paths (dislocation clusters) resulted in radial expansion into the metal, leading to the formation of a mushroom-shaped inner layer. On the other side, the outward diffusion of Ni was retarded by the $FeCr_2O_4$ formed on these shortcut paths and resulted in the formation of few pure Ni regions at the adjacent interface of the inner–outer layers but a relatively continuous pure Ni layer at the regions without mushroom-shaped inner oxidation. The formation mechanism of the mushroom-shaped inner oxidation is consistent with the experimental observations. Contrary to the results in the as-received samples, such mushroom-shaped inner oxidation was not observed in the GBE-treated samples because of the presence of dispersive dislocations rather than tangling dislocation clusters in the as-received samples [59]. The dispersive dislocations resulted in not only a relatively uniform inward growth (or larger oxide coverage) but also deeper penetration of the inner oxide layer in the GBE-treated samples.

8.5.4 Effect of Grain Size

Analyses indicated that the oxide scale on the shot-peened 800H samples was different from that on the as-received and GBE-treated samples after the SCW exposures. Cr-rich oxides such as Cr_2O_3 and $FeCr_2O_4$ predominated in the shot-peened samples in contrast to Fe-rich oxides such as Fe_3O_4 and $(Fe,Cr)_3O_4$ in the as-received and GBE-treated samples. The formation of Cr-rich oxides on the shot-peened samples mainly resulted from the refined grain size in the surface region. The samples with ultrafine grain size provided remarkably increased population of grain boundaries as shortcut diffusion paths for Cr flux since Cr has the highest diffusion coefficient in the Fe-Cr-Ni system [53].

According to the equations of Cr flux via lattice (j_L) and grain boundaries (j_B) for a semi-infinite geometry assuming negligible transfer to and from the grain [61, 62]:

$$j_L \approx \frac{d^2 D_L (C_B - C_1)}{(\pi D_L t)^{1/2}} \tag{8.4}$$

and

$$j_B \approx \frac{2dw D_B (C_B - C_1)}{(\pi D_B t)^{1/2}} \tag{8.5}$$

the flux ratio of Cr via grain boundaries versus lattice will be

$$\frac{j_B}{j_L} = \frac{2w}{d} \left(\frac{D_B}{D_L} \right)^{1/2} \tag{8.6}$$

where d is grain size; w is grain boundary width, which is usually ~0.5 nm [63]; D_L and D_B are the lattice and grain boundary diffusion coefficients, respectively; C_B and C_1 are the bulk concentration and fixed surface concentration, respectively; and t is the period of oxidation. Although D_L and D_B of Cr in alloy 800H are not available, the data in alloy 800 [64] should be applicable for this calculation because of the same chemical compositions between them. Thus

$$D_{L,Cr} = 3.24 \times 10^{-4} \exp\left(\frac{-287.4 \text{ KJ mol}^{-1}}{RT}\right) \text{m}^2 \text{ s}^{-1}, (T = 1060\text{–}1510 \text{ K}) \tag{8.7}$$

$$D_{B,Cr} = 5.80 \times 10^{-5} \exp\left(\frac{-184.2 \text{ KJ mol}^{-1}}{RT}\right) \text{m}^2 \text{ s}^{-1}, (T = 775\text{–}1170 \text{ K}) \tag{8.8}$$

Thereby, the ratio of Cr flux via grain boundaries versus lattice (j_B/j_L) as a function of grain size (d) can be calculated and is plotted in Figure 8.21. The fraction of Cr flux via grain boundaries [$j_B/(j_B+j_L)$] as a function of grain size (d) is also plotted in this figure. The shot-peened samples, with significantly decreased grain size (~20 nm) in the surface region, had a Cr flux ratio (j_B/j_L) of ~5 in contrast to that of ~0.001 in the as-received samples with a general grain size of ~100 μm. Therefore, the shot-peened samples had ~80% Cr transported via grain boundaries in contrast to only ~0.1% Cr in the as-received samples transported via grain boundaries. The significantly increased Cr flux via grain boundaries due to grain size refinement greatly promoted the formation of the predominant Cr-rich oxides in the shot-peened samples.

The formation of Cr-rich oxides, such as Cr_2O_3 and $FeCr_2O_4$, improved not only oxidation resistance but also oxide exfoliation resistance because of the small

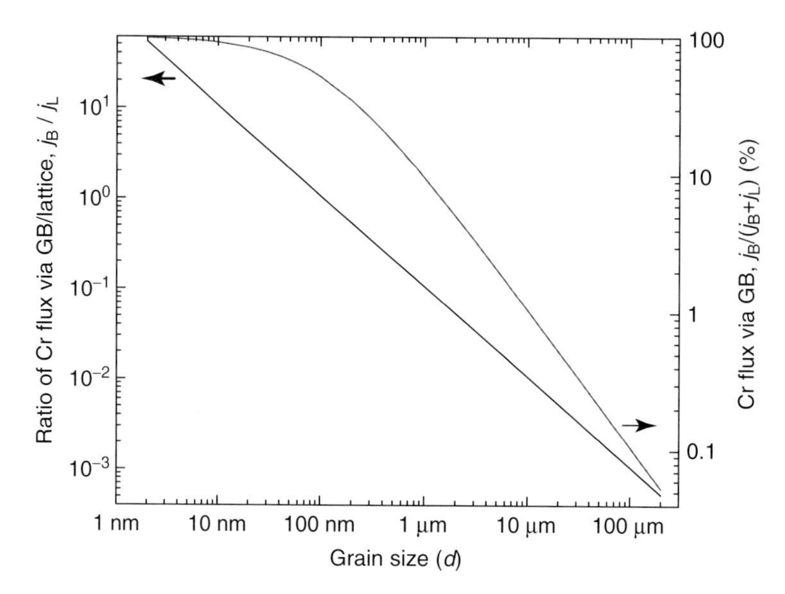

Figure 8.21 Flux ratio of Cr via grain boundary (GB) versus lattice (j_B/j_L) and Cr flux via GB [$j_B/(j_B + j_L)$] as a function of grain size (d) [47].

difference in volume thermal expansion coefficients between the oxides and the metal, as shown in Figure 8.17. Furthermore, the much smaller grain size of oxides in the shot-peened samples than the as-received samples is expected to be beneficial for mitigating thermal shock-induced oxide exfoliation. This is because the thermal conductivity of ultrafine-grained surface layer is less than that of coarse-grained samples, which is due to the larger volume fraction of interfaces in ultrafine-grained layer [65]. Such grain refinement benefit was also observed in ferritic-martensitic steels ($>\sim 9$ wt% Cr) that had been subjected to severe plastic deformation by means of shot peening or equal channel angular pressing (ECAP), showing improved oxidation resistance with less oxide exfoliation after SCW exposures [66].

8.6 Summary

SCW has become an important superfluid with a variety of applications because of its unique properties. One of its proposed new applications is as a coolant in the SCWR, with increased pressure and temperature to enhance thermal efficiency. Representative austenitic stainless steels and Ni-base alloys are proposed for use in this system, and their performance in SCW is reviewed. Oxidation was observed as the major corrosion phenomenon in these materials. Stainless steels (304, 316L, and D9), superaustenitic steels (NF709 and 800H), and Ni-base alloys (718, 690, 625, and 617) showed weight gains in a descending order with different levels of oxide exfoliation. Pitting was observed on alloys 718 and 625 because of the presence of Nb- and/or Ti-rich precipitates (γ''-phase). Alloy oxidation thermodynamics provided rational predictions and explanations to the formation of oxide scales on the respective materials.

Examples of applying novel corrosion control methods such as GBE and shot peening to alloys 800H and 617 showed remarkable improvement in oxidation resistance and mitigation in oxide exfoliation. Both GBE and shot peening altered the microstructure of oxide scales. GBE significantly changed the characteristics of grain boundary and dislocation networks of the bulk materials, which act as shortcut diffusion paths for preferential diffusion species. Consequently, straight interface and mushroom-shaped oxide were eliminated in the outer and inner oxide layers, respectively, on the GBE-treated alloy 800H samples. Shot peening significantly reduced grain size to the nanoscale in the surface region of alloy 800H. Although the shot-peened 800H samples showed a greater oxidation rate than the as-received and GBE-treated samples because of its greater grain boundary area, the predominant protective Cr-rich oxides (Cr_2O_3 and $FeCr_2O_4$) helped mitigating oxide exfoliation and is expected to maintain long-term low weight gains, compared to the as-received and GBE-treated samples with predominant Fe-rich oxides (Fe_2O_3, Fe_3O_4, and $(Fe,Cr)_3O_4$). The effects of test conditions (e.g., water properties, temperature, pressure, and dissolved oxygen content) and kinetics associated with microstructure (dislocation networks and grain size) on oxidation were discussed.

References

1. Lemmon, E.W., McLinden, M.O., and Friend, D.G. (2009) Thermophysical properties of fluid systems, in *NIST Chemistry WebBook* (eds P.J. Linstrom and W.G. Mallard), NIST Standard Reference Database Number 69, National Institute of Standards and Technology, Gaithersburg, MD, *http://webbook.nist.gov* (accessed 27 July 2009).
2. Uematsu, M. and Franck, E.U. (1980) Static dielectric constant of water and steam. *J. Phys. Chem. Ref. Data*, **9**, 1291–1306.
3. Noyori, R. (1999) Supercritical fluids. *Chem. Rev.*, **99**, 353–634.
4. Weingartner, H. and Franck, E.U. (2005) Supercritical water as a solvent. *Angew. Chem. Int. Ed.*, **44**, 2672–2692.
5. Sun, Y.P. (ed.) (2002) *Supercritical Fluid Technology in Materials Science and Engineering – Syntheses, Properties, and Applications*, Marcel Dekker, Inc., New York.
6. Brunner, G.H. (ed.) (2004) *Supercritical Fluids as Solvents and Reaction Media*, Elsevier B.V.
7. U.S. DOE Nuclear Energy Research Advisory Committee and the Generation IV International Forum (2002) A Technology Roadmap for Generation IV Nuclear Energy Systems, (GIF-002-00), December 2002.
8. Farrar, J.C.M. (2004) *The Alloy Tree – A Guide to Low-Alloy Steels, Stainless Steels and Nickel-base Alloys*, Woodhead Publishing Limited and CRC Press LLC.
9. Ryan, M.P., Williams, D.E., Charter, R.J., Hutton, B.M., and McPhail, D.S. (2002) Why stainless steel corrodes. *Nature*, **415**, 770–774.
10. Robertson, J. (1989) The mechanism of high-temperature aqueous corrosion of steel. *Corros. Sci.*, **29**, 1275–1291.
11. Fry, A., Osgerby, S., and Wright, M. (2002) Oxidation of Alloys in Steam Environments – A Review NPL Report MATC(A) 90, Crown (ISSN 1473-2734).
12. Asteman, H., Svensson, J.-E., Norell, M., and Johansson, L.-G. (2000) Influence of water vapor and flow rate on the high-temperature oxidation of 304L; effect of chromium oxide hydroxide evaporation. *Oxid. Met.*, **54**, 11–26.
13. Otsuka, N. and Fujikawa, H. (1991) Scaling of austenitic stainless-steels and nickel-base alloys in high-temperature steam at 973 K. *Corrosion*, **47**, 240–248.
14. Abe, F. and Yoshida, H. (1953) Corrosion behaviours of heat resisting alloys in steam at 800 °C and 400 atm pressure. *Z. Metallkunde*, **76**, 219–225.
15. Hussain, N. (2000) Oxidation behaviour of superalloys at elevated temperatures under different oxidizing atmospheres. PhD Thesis. University of The Punjab, Pakistan.
16. Zhang, L., Han, E., Zhang, Z., Guan, H., and Ke, W. (2003) Corrosion of stainless steel and nickel base alloys in subcritical water condition. *Jinshu Xuebao*, **39**, 649–654.
17. Savage, P.E. (1999) Organic chemical reactions in supercritical water. *Chem. Rev.*, **99**, 603–621.
18. Gleason, A.E., Jeanloz, R., and Kunz, M. (2008) Pressure-temperature stability studies of FeOOH using X-ray diffraction. *Am. Mineral.*, **93**, 1882–1885.
19. Palmer, D.A., Bénézeth, P.D., and Wesolowski, J. (2004) Solubility of nickel oxide and hydroxide in water. 14th International Conference on the Properties of Water and Steam – Water, Steam and Aqueous Solutions for Electric-Power Advances in Science and Technology, August 29 –September 3, 2004, Kyoto, Japan, pp. 264–269.
20. Luoma, R. (2002) *A Thermodynamic Analysis of the System Fe-Cr-Ni-C-O*, Acta Polytechnica Scandinavica, Chemical Technology Series, Vol. 292, Finnish Academies of Technology, Helsinki, pp. 1–91.
21. Gamson, B.W. and Watson, K.M. (1944) National Petroleum News 36. Tech. Sect. R623.
22. Gaskell, D.R. (2003) *Introduction to the Thermodynamics of Materials*, 4th edn, Taylor & Francis Books, Inc.
23. Was, G.S., Teysseyre, S., and Jiao, Z. (2006) Corrosion of austenitic alloys in supercritical water. *Corrosion*, **62**, 989–1005.

24. Young, D.J. (2008) *High Temperature Oxidation and Corrosion of Metals*, Elsevier, Oxford.
25. Croll, J.E. and Wallwork, G.R. (1972) The high-temperature oxidation of iron-chromium-nickel alloys containing 0-30% chromium. *Oxid. Met.*, **4**, 121–140.
26. Fonticella, R. (1998) *The Usefulness of the R^2 Statistic*, Casualty Actuarial Society Forum, Maryland, p. 55.
27. Tan, L., Ren, X., Sridharan, K., and Allen, T.R. (2008) Corrosion behavior of Ni-base alloys for advanced high temperature water-cooled nuclear plants. *Corros. Sci.*, **50**, 3056–3062.
28. Armitt, J., Holmes, D.R., Manning, M.I., Meadowcroft, D.B., and Metcalfe, E. (1978) Electric Power Research Institute, Palo, Alto, CA. The spalling of steam-grown oxide from superheater and reheater tube steels. EPRI Report FP 686. Technical Planning Study 76–655
29. Allen, T.R. and Was, G.S. (2008) Candidate materials evaluation for supercritical water-cooled reactor, Technical Report, DE-FC07-05ID14664 V.1. doi: 10.2172/944040
30. Chen, Y., Sridharan, K., and Allen, T.R. (2007) Corrosion of candidate austenitic stainless steels for supercritical water reactors. Corrosion Nashville, TN, paper number 07408.
31. Chen, Y., Sridharan, K., and Allen, T.R. (2005) Corrosion behavior of NF616 and D9 as candidate alloys for supercritical water reactors. Corrosion Houston, TX, paper number 05391.
32. Tan, L., Sridharan, K., and Allen, T.R. (2006) The effect of grain boundary engineering on the oxidation behavior of Incoloy alloy 800H in supercritical water. *J. Nucl. Mater.*, **348**, 263–271.
33. Tan, L., Sridharan, K., and Allen, T.R. (2008) Altering corrosion response via grain boundary engineering. *Mater. Sci. Forum*, **595–598**, 409–418.
34. Tan, L., Sridharan, K., Allen, T.R., Nanstad, R.K., and McClintock, D.A. (2008) Microstructure tailoring for property improvements by grain boundary engineering. *J. Nucl. Mater.*, **374**, 270–280.
35. Ren, X., Sridharan, K., and Allen, T.R. (2007) Corrosion behavior of alloys 625 and 718 in supercritical water. *Corrosion*, **63**, 603–612.
36. Flewitt, P.E.J. and Wild, R.K. (2001) *Grain Boundaries – Their Microstructure and Chemistry*, John Wiley & Sons, Ltd.
37. Tan, L., Allen, T.R., and Busby, J.T. (2010) Grain boundary engineering for structure materials of nuclear reactors. *J. Nucl. Mater.*, submitted.
38. Tan, L., Allen, T.R., and Yang, Y. (2011) Corrosion behavior of Incoloy alloy 800H (Fe-21Cr-32Ni) in supercritical water. *Corros. Sci.*, **53**, 703–711.
39. Skinner, B.J. (1966) in Thermal expansion. *Handbook of Physical Constants* (ed. S.P. Clark Jr.), Geological Society of America Memoir, p. 75.
40. Kollie, T.G. (1977) Measurement of the thermal-expansion coefficient of nickel from 300 to 1000 K and determination of the power-law constants near the Curie temperature. *Phys. Rev. B*, **16**, 4872–4881.
41. Incoloy alloy 800H & 800HT (2004) Special Metals Publications: SMC-047, *http://www.specialmetals.com.*
42. Valentine, K.B. (1948) Recrystallization as a measurement of relative shot peening intensities. *Trans. Am. Soc. Metal.*, **40**, 420–434.
43. Tao, N.R., Wang, Z.B., Tong, W.P., Sui, M.L., Lu, J., and Lu, K. (2002) An investigation of surface nanocrystallization mechanism in Fe induced by surface mechanical attrition treatment. *Acta Materialia*, **50**, 4603–4616.
44. Haga, S., Harada, Y., and Tsubakino, H. (2006) Fatigue life prolongation of carburized steel by means of shot-peening. *Mater. Sci. Forum*, **505–507**, 775–780.
45. Zoeller, H.W. and Cohen, B. (1966) Shot peening for resistance to stress-corrosion cracking. *Met. Eng. Q.*, **6**, 16.
46. Scholtes, B. and Vohringer, O. (1993) Origin, determination and assessment of near-surface microstructural alterations due to shot peening processes. *Materialwiss. Werkstofftech.*, **24**, 421–431.
47. Tan, L., Ren, X., Sridharan, K., and Allen, T.R. (2008) Effect of shot-peening

on the oxidation of alloy 800H exposed to supercritical water and cyclic oxidation. *Corros. Sci.*, **50**, 2040–2046.

48. Castaldi, M.J., LaPierre, R., Lyubovski, M., Pfefferle, W., and Roychoudhury, S. (2005) Effect of water on performance and sizing of fuel-processing reactors. *Catal. Today*, **99**, 339–346.
49. Wilke, C.R. and Chang, P. (1955) Correlation of diffusion coefficients in dilute solutions. *AIChE J.*, **1**, 264–270.
50. Davis, J.R. (ed.) (1997) *Heat-Resistant Materials, ASM Specialty Handbook*, ASM International, p. 147.
51. Mongomery, M. and Karlsson, A. (1995) Survey of oxidation in steamside conditions. *VGB Kraftwerkstechnik*, **75**, 235–240.
52. Otoguro, Y., Sakakibara, M., Saito, T., Ito, H., and Inoue, Y. (1988) Oxidation behavior of austenitic heat-resisting steels in a high temperature and high pressure steam environment. *Trans. Iron Steel Inst. Jpn.*, **28**, 761–768.
53. Rothman, S.R., Nowicki, L.J., and Murch, G.E. (1980) Self-diffusion in austenitic Fe-Cr-Ni alloys. *J. Phys. F: Met. Phys*, **10**, 383–398.
54. Young, D.J. (2008) *High Temperature Oxidation and Corrosion of Metals*, Elsevier, Oxford.
55. McIntosh, M.S. (2005) Mechanisms and factors affecting chromium oxide particle reduction in iron-chromium honeycombs. PhD Thesis. Georgia Institute of Technology.
56. Neumann, G. and Tuijn, C. (2009) *Self-Diffusion and Impurity Diffusion in Pure Metals: Handbook of Experimental Data*, Elsevier Ltd.
57. Park, J.W. and Altstetter, C.J. (1987) The diffusion and solubility of oxygen in solid nickel. *Metall. Mater. Trans. A*, **18**, 43–50.
58. Tan, L., Yang, Y., and Allen, T.R. (2006) Oxidation behavior of iron-based alloy HCM12A exposed in supercritical water. *Corros. Sci.*, **48**, 3123–3138.
59. Tan, L., Rakotojaona, L., Allen, T.R., Nanstad, R.K., and Busby, J.T. (2011) Microstructure optimization of Incoloy alloy 800H (Fe-21Cr-32Ni). *Acta Mater.*, **528**, 2755–2761.
60. Nanstad, R.K., McClintock, D.A., Hoelzer, D.T., Tan, L., and Allen, T.R. (2009) High temperature irradiation effects in selected Generation IV structural alloys. *J. Nucl. Mater.*, **392**, 331–340.
61. Lobb, R.C. and Evans, H.E. (1981) Formation of protective oxide film on chromium-depleted stainless steel. *Met. Sci.*, **15**, 267–274.
62. Crank, J. (1975) *The Mathematics of Diffusion*, 2nd edn, Oxford University Press, p. 32.
63. Fisher, J.C. (1951) Calculation of diffusion penetration curves for surface and grain boundary diffusion. *J. Appl. Phys.*, **22**, 74–82.
64. Paul, A.R., Kaimal, K.N.G., Naik, M.C., and Dharwadkar, S.R. (1994) Lattice and grain boundary diffusion of chromium in superalloy Incoloy-800. *J. Nucl. Mater.*, **217**, 75–81.
65. Guo, F.A., Trannoy, N., and Lu, J. (2006) Characterization of the thermal properties by scanning thermal microscopy in ultrafine-grained iron surface layer produced by ultrasonic shot peening. *Mater. Chem. Phys.*, **96**, 59–65.
66. Ren, X., Sridharan, K., and Allen, T.R. (2009) Effect of grain refinement on corrosion of ferritic-martensitic steels in supercritical water environment. *Mater. Corros.*, **60** (9999), doi: 10.1002/maco.200905446

9
Metal–Phosphonate Anticorrosion Coatings

Konstantinos D. Demadis, Maria Papadaki, and Dimitrios Varouchas

9.1
Introduction

"Corrosion" is a broad scientific/technological area that is truly an interdisciplinary mixture of several other sciences, Figure 9.1. Hence, its definition depends on the particular discipline of science considered. All definitions, albeit diverse, have a significant common theme: the change of the mechanical properties of metals in an undesirable way. According to ISO 8044, corrosion is a "physico-chemical interaction, that is usually of an electrochemical nature, between a metal and its environment which results in changes in the properties of the metal and which may often lead to impairment of the function of the metal, the environment, or the technical system of which these form a part" [1]. Apart from technological aspects, the "economics" of corrosion is a significant factor in the marketplace. The cost of corrosion has been estimated (based on several reports) to be in the order of an astonishing 1–5% of the national Gross Domestic Product (GDP) [2]. The worldwide cost of corrosion for the production of all grades of pulp is about US$3 billion/year. These figures exclude the cost of production losses, unexpected shutdowns to make repairs to corroded equipment, and lost working time. Corrosion cost in the US electric power industry reaches US $10 billion/year, according to the Electric Power Research Institute (EPRI) [3]. Also, it has been reported by EPRI that corrosion-related phenomena are responsible for at least 55% of all unplanned outages. They also add at least 10% to the average annual household electricity bill. The impact of corrosion on all industrial sectors in almost all countries can be easily seen. For example, in 1993 it was estimated that 60% of all maintenance costs for North Sea oil production platforms were related to corrosion either directly or indirectly. A report on inspection results of several offshore production plants showed that corrosion was a factor in 35% of structures, 33% of process systems, and 25% of pipelines. The importance of preventing corrosion issues is also exemplified in the recent unfortunate incident of 267 000 gal crude oil spillage in Prudhoe Bay, Alaska. BP was accused of faulty maintenance of a corrosion-ladden pipeline [4].

Green Corrosion Chemistry and Engineering: Opportunities and Challenges, First Edition.
Edited by Sanjay K. Sharma.

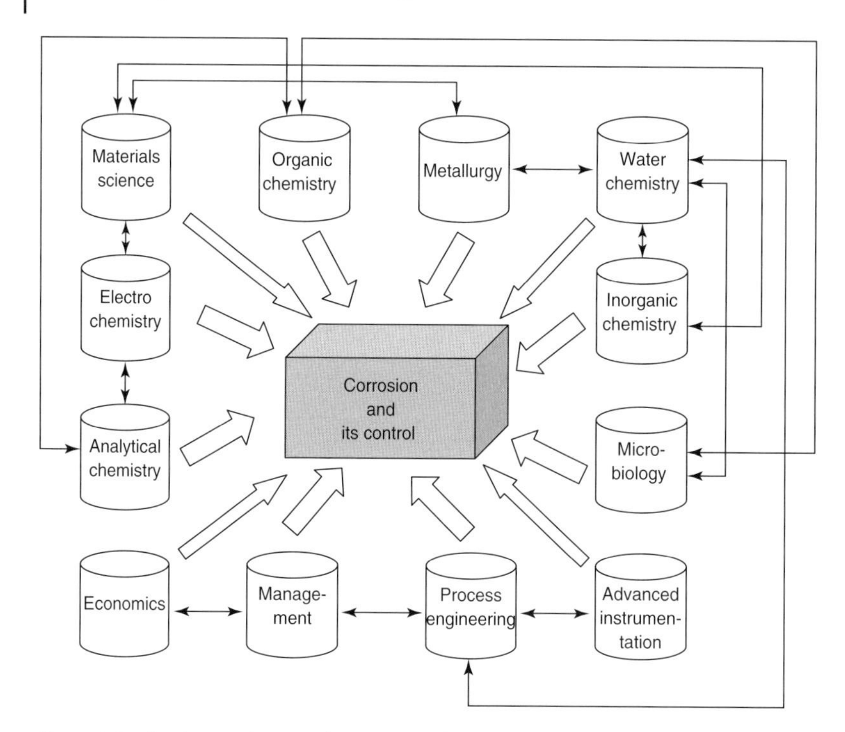

Figure 9.1 A scheme showing the true interdisciplinary nature of corrosion, and how it creates synergies between diverse scientific disciplines.

Corrosion control and management can be based on several strategies, one of which is by chemicals. Herein, the present review focuses on water-related corrosion and water-soluble corrosion inhibitors. Thus, chemical additives are added to the process stream that is in contact with the metallic surface to be protected. These are called *corrosion inhibitors*. These are chemical additives that delay or (ideally) stop metallic corrosion [5].

There exists a diverse list of effective corrosion inhibitors that are able to decrease the metallic corrosion in a wide range of pH conditions. However, within the framework of the present chapter, we limit our discussion to the class of inhibitors that operates by complexation with metal ions. Kuznetsov has named these "corrosion inhibitors of the complexing type" [6]. In our discussion, we consider a particular type of corrosion inhibitors, namely (poly)phosphonic acids and their synergy with metal ions present in the water streams.

At appropriate pH regions, phosphonic acids exist in their (partially or completely) deprotonated form. There are several reports on the deprotonation chemistry of phosphonic acids [7]. Thus, in the presence of metal cations (commonly alkaline earth metals), they form sparingly soluble compounds, which eventually precipitate on to the metallic surface to form an ideally two-dimensional (2D) protective thin film. There are several such phosphonate inhibitors for various water-related technologies. Some representative examples are shown in Figure 9.2.

9.2 The Scope of Green Chemistry and Corrosion Control

"Green Chemistry" has become a topic of intense discussion and debate during the last decade [8–11]. The concept of green chemistry has infiltrated to all aspects of chemistry and related technological fields. One important area where green chemistry can potentially find several applications is the use of environmentally acceptable additives for water treatment.

Chemical additives are used to condition water so that the following problems do not occur, or are minimized [12]:

- scale formation–deposition of sparingly soluble salts;
- corrosion of metal surfaces;
- development of biofouling.

Naturally, chemicals that are added to condition water have various purposes, and thus widely different physicochemical properties. For example, organophosphonates are used to combat calcium/barium/strontium salt formation and deposition [13]. Control of other types of scales, for example, silica requires a more thoughtful and at times "exotic" approach [14]. Colloidal silica presents a problem that has been poorly solved and still represents an area of development for inhibitor chemistries.

An important definition of a "green chemical" has been given by Anastas and Warner [8]. They have given a broad definition of green chemistry based on 12 principles that relate to several steps, from chemical synthesis to chemical usage. A green chemical should be synthesized in a safe and energy efficient manner and its toxicity should be minimal, whereas its biodegradation should be optimal. Lastly, its impact on the environment should be as low as possible.

The OSPAR Commission (Oslo and Paris Commission [15]) is the international body responsible for harmonization of the strategies and legislation in the North-East Atlantic Region. The Commission has stated that every effort should be made to combat eutrophication and achieve a healthy marine environment where eutrophication does not occur by the year 2010. Chemicals are classified differently depending on the particulars of the geographical area. The guidelines set by OSPAR are as follows:

1) biodegradability ($>60\%$ in 28 days; if $<20\%$, the chemical is a candidate for substitution);
2) toxicity (LC_{50} or $EC_{50} > 1$ mg/l for inorganic species, LC_{50} or $EC_{50} > 10$ mg/l for organic species);
3) bioaccumulation (Log pow < 3, pow $=$ partition in octanol/water).

When a chemical fulfils two out of three requirements and its biodegradability is superior to 20% in 28 days, it is eligible to be listed on the PLONOR list (Pose Little Or No Risk). This emphasizes the biodegradability factor and influences usage of water additives.

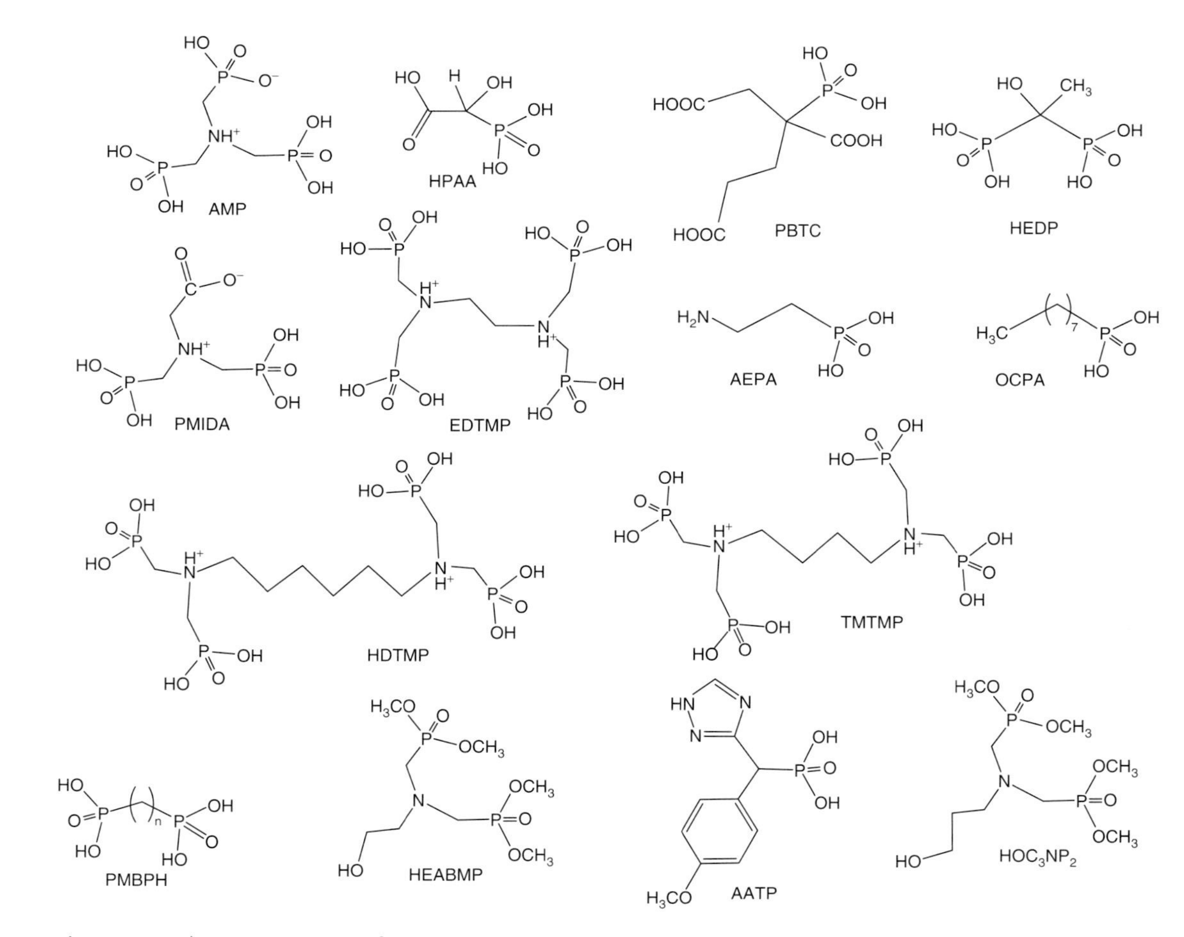

Figure 9.2 Schematic structures of several phosphonate-based corrosion inhibitors.

9.3
Metal–Phosphonate Materials: Structural Chemistry

The emerging fields of supramolecular chemistry, crystal engineering, and materials chemistry have made long strides forward with impressive growth during the last decades. The chemistry of phosphonate ligands has undoubtedly played an important role in widening these areas of research [16–25]. Phosphonate ligands have attracted considerable attention in the context of fundamental research, but they have also been extensively used in several other technologically/industrially significant areas, such as water treatment [26–29], oilfield drilling [30, 31], minerals processing [32], corrosion control [33–35], metal complexation and sequestration [36], dental materials [37], enzyme inhibition [38–40], bone targeting [41–43], cancer treatment [44], and so on. There exists a plethora of metal–phosphonate materials whose crystal structures exhibit attractive features that depend on several variables, such as nature of M^{n+} (metal oxidation state, ionic radius, and coordination number in particular), number of phosphonate groups on the ligand backbone, presence of other functional moieties (carboxylate, sulfonate, amine, hydroxyl), and, naturally, on process variables (reactant ratio and concentration, temperature, pressure, etc.). The synthesis of metal–phosphonates is usually carried out in aqueous solutions (or in mixtures of water and a polar organic solvent such as alcohols, DMF, acetone), so it is not surprising that water is commonly found in their lattice, participating in extensive hydrogen bonding, which is predominant in these architectures resulting in 1-, 2-, and 3D supramolecular networks [45]. Lastly, it is worth noting that the vast majority of metal–phosphonates are coordination polymers, although there are reports of metal–phosphonate complexes.

The structural literature of metal–phosphonate materials is fairly large. However, herein we present a selective overview, focusing on metal–phosphonate materials that contain metal ions that are found in natural waters (Mg, Ca, Sr, Ba, Cu, and Zn) and phosphonates that have been used as corrosion inhibitors (see Figure 9.2). It is materials with the above-mentioned metal ions and various phosphonates that act synergistically as anticorrosion protective films.

9.3.1
Phosphonobutane-1,2,4-Tricarboxylic Acid (PBTC)

Phosphonobutane-1,2,4-tricarboxylic acid (PBTC) (Figure 9.2) was first launched by the research laboratories of Bayer Co. (commercial name Bayhibit) [46] and has survived intense use in the water-treatment field for several decades, owing to its ease of preparation on a large scale and robustness to decomposition by oxidizing biocides [27]. Up until 2005 there was no published structural information on PBTC or its metal complexes. PBTC acid contains one phosphonate and three carboxylate groups. It was crystallized from very concentrated aqueous solutions. The presence of protonated phosphonate and carboxylate groups as well as the water molecule creates an intricate network of hydrogen bonding. The complexity of the structure can be seen in Figure 9.3.

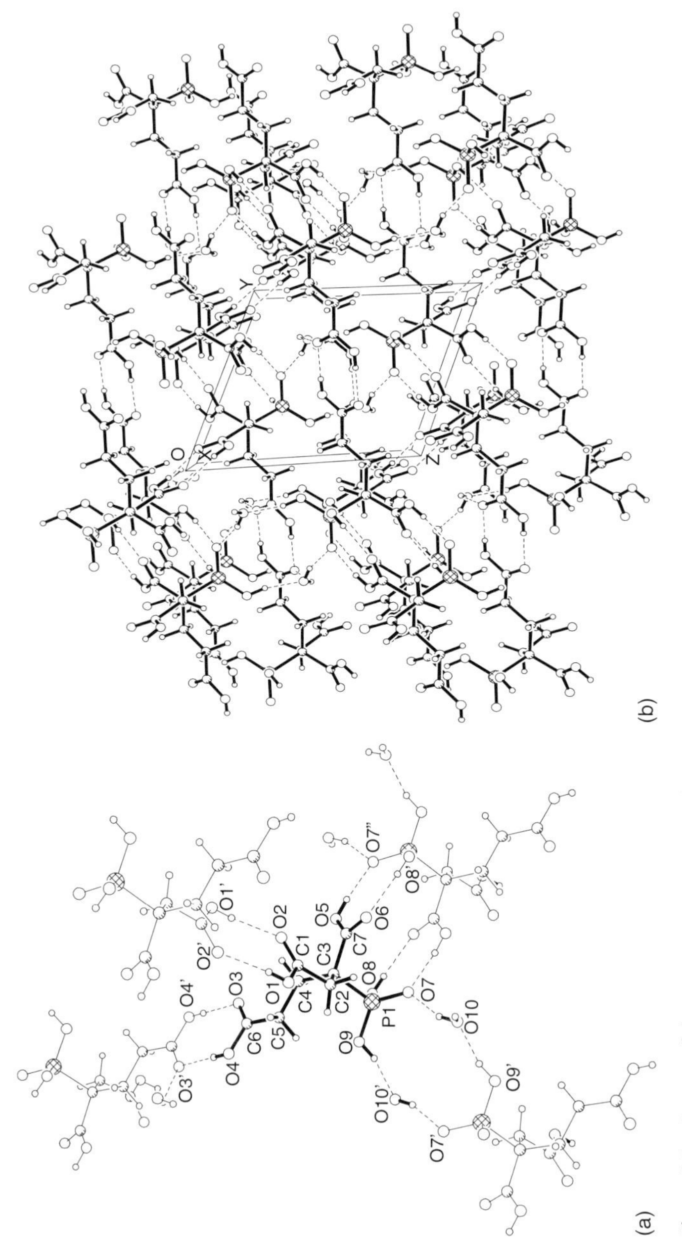

Figure 9.3 A view of the PBTC·H_2O structure showing all hydrogen bonds (a) and packing diagram of the PBTC·H_2O structure (b). Reproduced from Ref. [27].

The $-PO_3H_2$ group acts as both donor and acceptor and forms two sets, a total of four, H- bonds. The first set forms between the $-P{=}O$ (from one molecule), the $-P{-}O(H)$ (from a neighboring molecule) and two water molecules of crystallization. This H-bonding mode forms an eight-member ring (not counting the H atoms) and is locally centrosymmetric. The second set consists of two hydrogen bonds. These are formed between the $P{=}O$ portion of the $-PO_3H_2$ group and the $-OH$ group of the carboxylate of a neighboring molecule, and between the second $-P{-}O(H)$ group and the carbonyl $-C{=}O$ portion of the same carboxylate. The $-COOH$ group at the 2′ position participates in hydrogen-bonding interactions with the $-PO_3H_2$ group as described above. The $-COOH$ group at the 4′ position forms a commonly seen "dicarboxylate dimer" with a neighboring carboxylate also at the 4′ position. This "dimer" sits on a "local" inversion center. The $-COOH$ group at the 1′ position forms the aforementioned "dicarboxylate dimer" with a neighboring carboxylate (also at the 1′ position) forming a six-member ring. The H-bonding distance is 2.7674(15) Å. All three carboxylate and the phosphonate groups in PBTC are protonated. The $P{=}O$ double bond length is 1.4928(10) Å, whereas the P–O single bonds are 1.5294(10) and 1.5578(10) Å. The P–C bond length is 1.8465(12) Å and it falls in the normal range (1.8–1.9 Å) for such bonds [47].

9.3.2
Ethylenediamine-Tetrakis(Methylenephosphonic Acid) (EDTMP)

Ethylenediamine-*tetrakis*(methylenephosphonic acid) (EDTMP) is a tetraphosphonate with four methylenephosphonate groups attached to two different N atoms. It is an excellent scale inhibitor for sparingly soluble salts [48]. Its molecular structure is shown in Figure 9.4. EDTMP is a zwitter-ion in low pH regions. Both N atoms are protonated because of their high basicity, whereas two phosphonate groups (one per N atom) are monodeprotonated. The remaining two phosphonate moieties are fully protonated. EDTMP can be easily synthesized from ethylenediamine, hypophosphorus acid, and formaldehyde in a Manich-type reaction. Several aminomethylenephosphonates can be synthesized in a similar manner [49].

9.3.3
Dimethylaminomethylene-Bis(Phosphonic Acid) (DMABP)

Dimethylaminomethylene-bis(phosphonic acid) (DMABP) is a diphosphosphonate that belongs to the "gem-bisphosphonate" family of phosphonates [50]. Both phosphonate groups are linked to the same C atom. The molecular structure of DMABP is shown in Figure 9.5. As expected, the N atom is protonated and one of the two phosphonate groups is monodeprotonated. DMABP can be prepared from dimethylformamide, phosphorus acid, and phosphorus trichloride [51]. Complexation studies have been reported with Indium(III), Gallium(III), Iron(III), Gadolinium(III), and Neodymium(III) ions [51]. Linear coordination polymers with tungstate centers have been reported in which the negatively charged chains $[(O_3PC(H)N(CH_3)_2PO_3)W_2O_6]^{4-}$ are the repeat units [52].

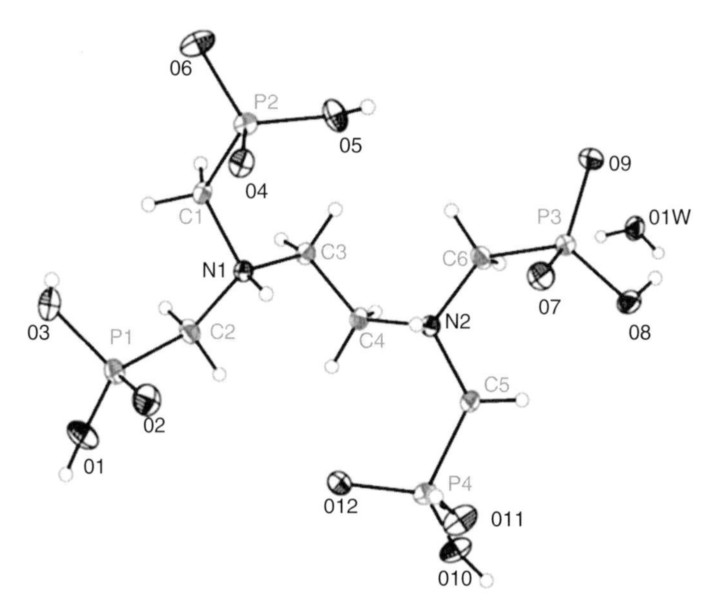

Figure 9.4 Molecular structure of EDTMP (as its diammonium salt, hydrate) with the numbering scheme. The two NH_4^+ cations have been omitted.

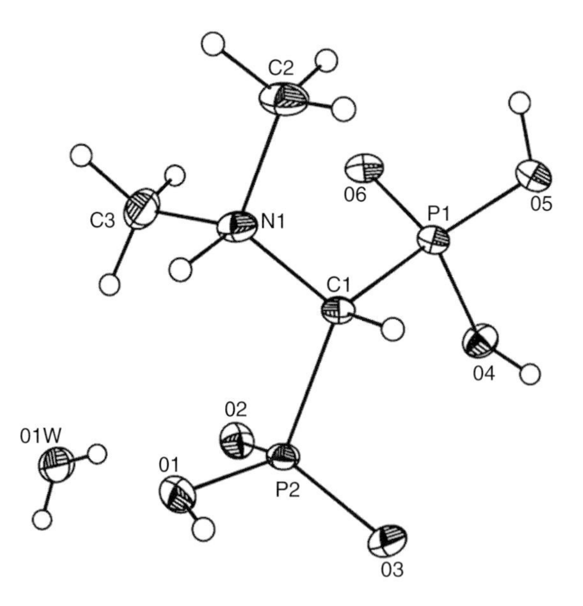

Figure 9.5 Molecular structure of DMABP hydrate, with the numbering scheme.

9.3.4
Magnesium (or Zinc)-(Amino-Tris-(Methylenephosphonate)), Mg(or Zn)-AMP

These metal–AMP (amino-tris-(methylenephosphonate)) organic–inorganic hybrids are isostructural [53, 54]; therefore, only the crystal structure of the Zn analog is discussed herein. In the structure of Zn-AMP, each phosphonate group is singly deprotonated, whereas the N atom is protonated. Therefore, AMP maintains its "zwitter ion" character in the crystal lattice. Zn^{2+} is coordinated by three phosphonate Os and three H_2O molecules. Notably, there are no lattice H_2O molecules. The asymmetric unit is shown in Figure 9.6 (upper). AMP forms an eight-member chelate ring with Zn^{2+}. Zn–O(P) bond distances range from 2.0459(13) to 2.1218(13) Å. Bond angles point to a slightly distorted octahedral

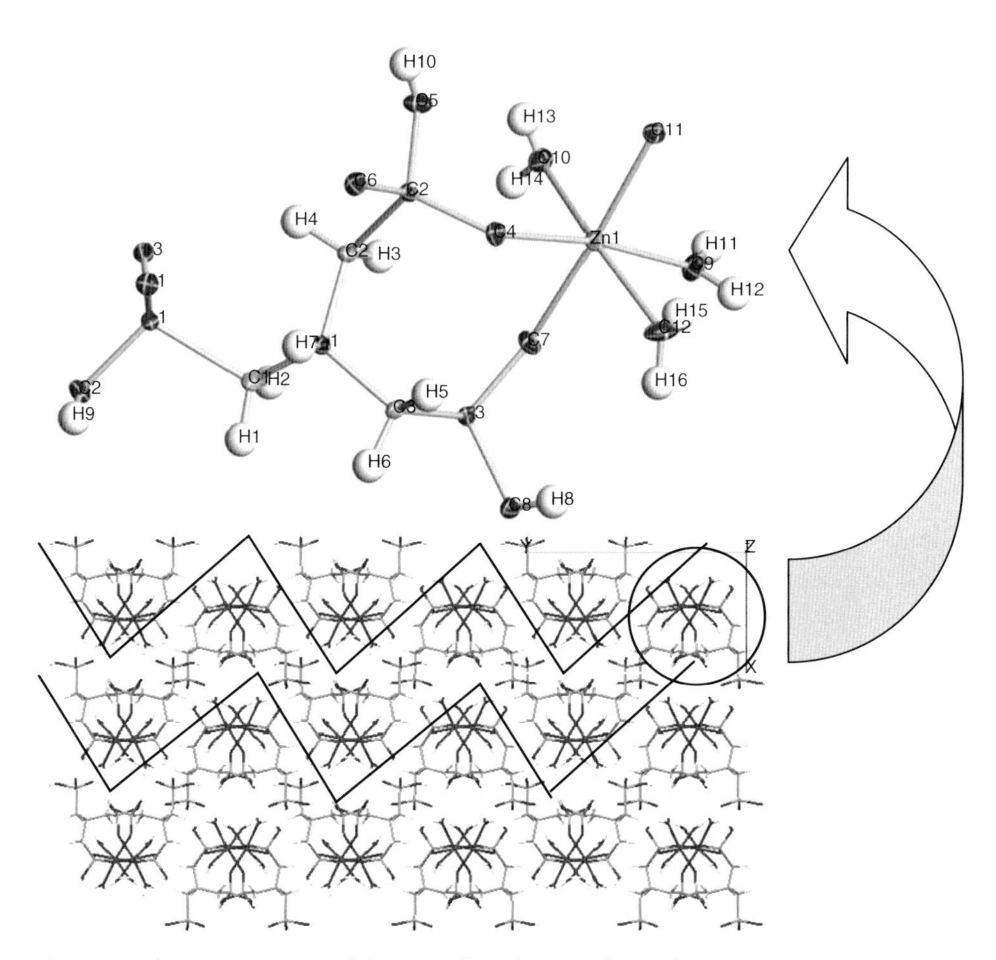

Figure 9.6 The asymmetric unit of the $Zn[HN(CH_2PO_3H)_3(H_2O)_3]_x$ polymer (upper). Packing diagram of the Zn-AMP lattice showing the corrugated structure and an isolated zigzag chain (lower).

geometry, with the largest deviation being 166.90(6)° for the O12-Zn-O10 angle. The third phosphonate arm is surprisingly *not* coordinated to Zn^{2+}, but is exclusively involved in H-bonding through O1, O2, and O3 (*vide infra*).

A zigzag chain parallel to the *c*-axis is formed by Zn^{2+}, Figure 9.6 (lower). The Zn^{2+} centers are located at the corners of the zigzag chain, whereas the "linear" portion of the zigzag is made of the noncoordinated, hydrogen-bonded phosphonate groups. Besides, the three metal-bonded phosphonate oxygens (O4, O7, and O11), three additional oxygens (O5, O8, and O2) are protonated, and the remaining three O atoms serve as hydrogen bond acceptors.

The H_2O molecule (O9) located trans to a coordinated phosphonate (P2) participates in two H bonds with O6 (1.961 Å) of a Zn-bound phosphonate (P2) and O1 (1.937 Å) of a free phosphonate (P1). The three H_2O molecules form their hydrogen bonds, mostly in the *a*-axis direction. The H bonds create a 3D network of H-bonded linear chains. The overall effect is the formation of 2D corrugated sheets that nest within each other running along the *ab* diagonal. However, these sheets are made up of individual chains where the noncoordinated phosphonate groups overlap. An isostructural series of $M[HN(CH_2PO_3H)_3(H_2O)_3]_x$ (M = Cd, Ni, Co, Mn) compounds has been prepared by Clearfield *et al.* [55].

9.3.5 Calcium-(Amino-Tris-(Methylenephosphonate)), Ca-AMP

The complexity of the polymeric structure can be seen in Figure 9.7 [26]. There are no discrete molecular units of the Ca-AMP complex. Instead, the methylenephosphonate "arms" participate in an intricate network of intermolecular and intramolecular interactions involving Ca atoms and hydrogen bonds. The result is a complex polymeric 3D structure caused mainly by multiple bridging of the AMP molecules. Each phosphonate group is monodeprotonated. The protonated O atom (–P–O–H) remains noncoordinated. The remaining two P–O groups bridge two neighboring Ca atoms in a Ca–O–P–O–Ca arrangement. There are four Ca-AMP "units" per unit cell. The overall Ca:AMP molar ratio is 1 : 1. Electroneutrality is achieved by charge balance between the divalent Ca and the triply deprotonated/monoprotonated AMP ligand. There are also 3½ water molecules in the unit cell. One is coordinated to Ca. Water molecules of crystallization serve as "space fillers" and also participate in extensive hydrogen-bonding superstructures.

The Ca is surrounded by six oxygens, five from phosphonate groups and one from water (Figure 9.7, left). Ca–O(P) distances range from 2.2924(14) to 2.3356(14) Å. The Ca–O(H_2O) distance is 2.3693(17) Å, somewhat longer than Ca–O(P) distances. The Ca atom is situated in a slightly distorted octahedral environment, as judged by the O–Ca–O angles, which show slight deviations from idealized octahedral geometry. Ca–O(P) bond lengths can be compared to similar bonds found in the literature.

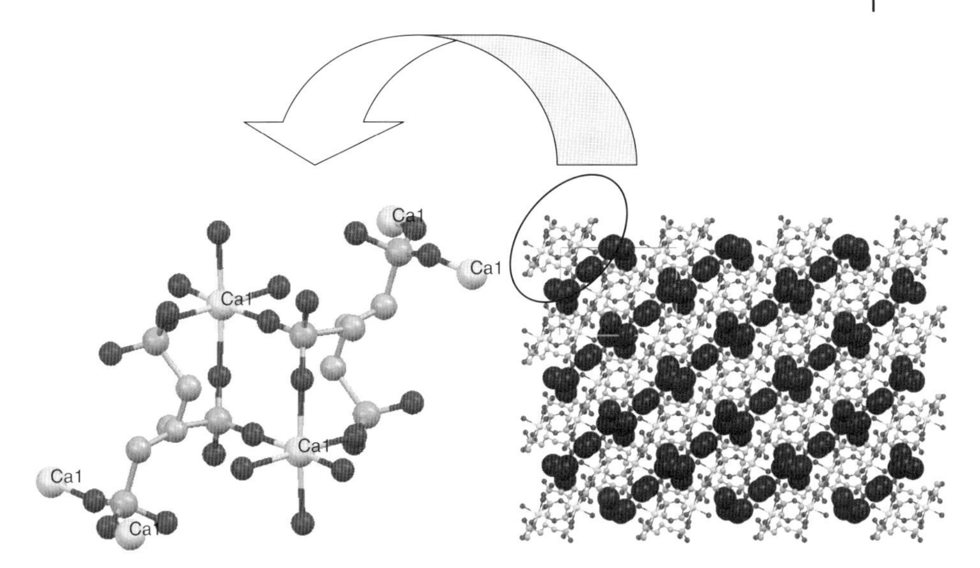

Figure 9.7 Fragment of the structure of $\{Ca[HN(CH_2PO_3H)_3(H_2O)(2.5H_2O)]\}_x$ polymer, showing the octahedral coordination of Ca (left). Layers of the structure showing waters of crystallization (exaggerated spheres) between the 2D sheets (right).

One AMP ligand per Ca acts as a bidentate chelate, forming an eight-member ring. Each Ca center is coordinated by four AMP phosphonate oxygens in a monodentate fashion. Each of these methylenephosphonate groups is simultaneously coordinated to a neighboring Ca atom. A water molecule completes the octahedron.

All three phosphonate groups in AMP are monodeprotonated. This formally separates the P–O bonds into three groups: P–O–H (protonated), P=O (phosphoryl), and $P–O^-$ (deprotonated). The P–O(H) bond lengths are 1.5684(15), 1.5703(16), and 1.5802(14) Å. On the other hand, P=O and $P=O^-$ bond lengths are crystallographically indistinguishable and are found in the 1.4931(15)–1.5102(14) Å range. This observation, coupled with the fact that all Ca–O(P) distances are very similar, points to the conclusion that the negative charge on each $-PO_3H^-$ is delocalized over the O–P–O moiety. It is worth noting that only the deprotonated P–O groups coordinate to the Ca atoms, whereas the protonated P–OH's remain noncoordinated. P–C bond lengths are unexceptional, 1.8382(20), 1.8347(20), and 1.8330(20) Å. Figure 9.7 (right) clearly shows that this structure is a 2D layered structure with water molecules filling the space between the layers.

9.3.6
Strontium-(Amino-Tris-(Methylenephosphonate)), Sr-AMP

Strontium-AMP is a 3D polymer consisting of $[Sr(AMP)]_n$ [54]. The Sr atoms are seven-coordinate, with five monohapto and one chelating AMP ligands and Sr–O bond lengths ranging from 2.4426(17) to 2.9060(17) Å. The structure can be viewed as Sr "dimers" connected by AMP ligands (Figure 9.8). The Sr centers in these

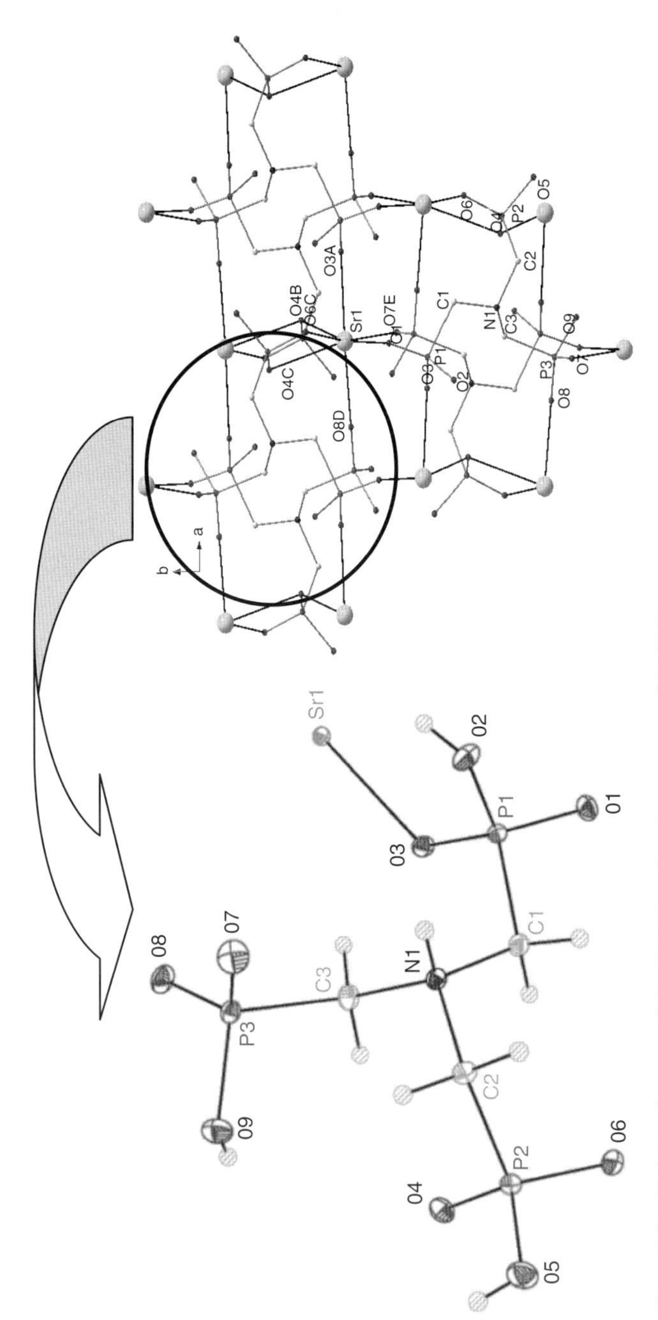

Figure 9.8 Chelating function of the AMP ligand in the structure of Sr-AMP.

"dimers" are bridged by a phosphonate oxygen. This bridging oxygen, together with a second oxygen atom of the same phosphonate group, coordinates in a chelating manner to the same Sr center. The Sr atoms form layers separated by AMP ligands, with each AMP bridging four Sr atoms in one layer and two in the next one.

Distorted hexagons are formed by Sr^{2+} centers. In this Sr-based, distorted hexagonal honeycomb lattice, the short distances are 4.445(1) Å and the long distances are 6.150(28) Å.

9.3.7 Barium-(Amino-Tris-(Methylenephosphonate)), Ba-AMP

Barium-AMP has a 3D polymeric structure with the formula $\{Ba(AMP)(H_2O)\}_n$ (Figure 9.9) [54]. AMP maintains its "zwitter ion" character in the crystal lattice of Ba-AMP. The coordination modes of both symmetry-independent AMP^{2-} octadentate ligands are identical. Metric features of the Ba-coordinated AMP ligand show insignificant variations from those in "free" AMP [56].

Ba(1) is nine-coordinated, bound by nine O atoms, eight originating from phosphonate oxygens and one from H_2O, whereas Ba(2) is ten-coordinated, linked by nine phosphonate and one H_2O oxygens, Figure 9.9. The geometry at Ba^{2+} does not approximate either of the idealized polyhedra, the bicapped square antiprism or the bicapped dodecahedron; this observation is not surprising in view of the steric requirements of the triphosphonate ligand in the structure. There are no lattice H_2O molecules of crystallization. The closest Ba···Ba contact is 4.3691(10) Å. This Ba···Ba close proximity has its origin in the triple Ba(1)-μ-O-Ba(2) bridge by O atoms from three different AMP ligands. The Ba–O(H_2O) bond distances are Ba(1)–O(51) 2.841(10) Å and Ba(2)–O(52) 2.956(12) Å. Note that the longer Ba–O(H_2O) bond distance is associated with the 10-coordinate Ba^{2+} center. All bridging O atoms belong to phosphonate moieties that act as chelates for one Ba^{2+} and form four-membered rings. This bridging motif has been observed in the structure of Ba-glyphosate, in which Ba is eight-coordinate [57].

The 3D structure of $\{Ba(AMP)(H_2O)\}_n$ can be seen as a layer of Ba atoms lying in the *bc* plane interconnected via AMP ligands in the *a* direction. Ba···Ba distances between the triply μ-O bridged Ba atoms are 4.637(1) and 4.369(1) Å, while between doubly or singly μ-(O–P–O) bridged Ba atoms they are 7.163(1) or 7.687(1) Å apart (Figure 9.9).

9.3.8 Zinc-Hexamethylene-Diamine-Tetrakis(Methylenephosphonate), Zn-HDTMP

The crystal structure of Zn-HDTMP (hexamethylene-diamine-*tetrakis*(methylene phosphonate)) shows that it is a 3D coordination polymer [58]. The Zn–O distances are unexceptional and consistent with other structurally characterized

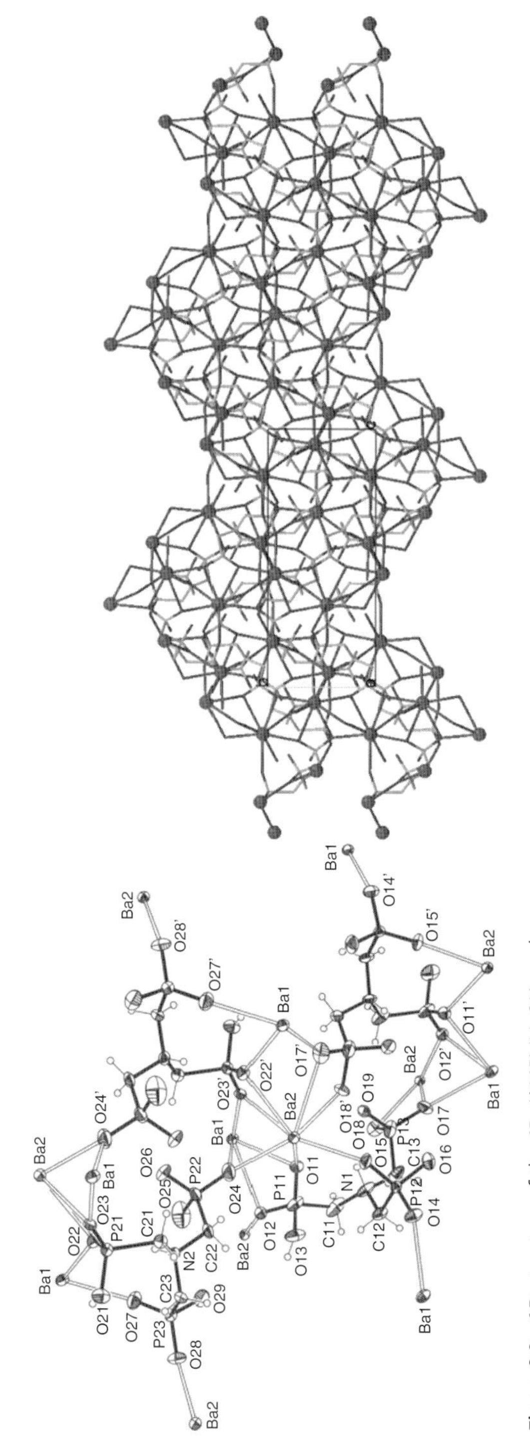

Figure 9.9 3D structure of the $\{Ba(AMP)(H_2O)\}_n$ polymer.

Zn-phosphonates [59]. Zn^{2+} is found in a distorted octahedral environment (Figure 9.10) formed exclusively by phosphonate oxygens. An interesting feature is that the sixth oxygen ligand for Zn^{2+} originates from a protonated phosphonate oxygen, O(9), and forms a long interaction (2.622(3) Å) with Zn^{2+}. Apparently, this interaction offers local stabilization because of a strong hydrogen bond, O(9)–H(9)···O(3), 1.879 Å. Two Zn^{2+} centers and the aminomethylene-bis-phosphonate portions of HDTMP form an 18-membered ring (Figure 9.10), while there is a concentric 8-membered ring formed by the same Zn^{2+} centers and the protonated methylenephosphonate arm involved in the long Zn···O(9) interaction. The lattice water interacts weakly with O5 (2.700 Å) and O(2) (2.964 Å). The absence of chelate rings is noteworthy, in contrast to several metal animomethylene–phosphonate structures [60]. HDTMP's four phosphonate groups are coordinated to six different Zn^{2+} centers. O1 (from P(1)) and O4 (from P(2)) act as unidentate ligands to Zn^{2+}. O(10) and O(12) (both from P(4)) bridge two Zn^{2+} centers that are 4.395 Å apart. O(7) and O(9) (both from P(3)) also bridge two Zn^{2+} centers but because of the long O(9) ··· Zn interaction (2.622 Å), their distance is much longer, 5.092 Å.

The structure of Zn-HDTMP can be compared to that of $Ca[(HO_3PCH_2)_2N(H)CH_2C_6H_4CH_2N(H)(CH_2PO_3H)_2]\cdot 2H_2O$ possessing a flexible cyclohexane ring linker [61]. Major structural differences between the two include the bidentate chelation of the tetraphosphonate to the metal center. These are absent in Zn-HDTMP.

9.3.9 Strontium or Barium-Hexamethylene-Diamine-Tetrakis(Methylenephosphonate), Sr or Ba-HDTMP

HDTMP reacts with alkaline-earth metal salts to give polymeric materials as products. The Sr and Ba-HDTMP materials have been prepared in high yields and are structurally characterized. They are isostructural, but their structure is notably different from that of Zn-HDTMP discussed above.

The metal centers are eight-coordinated (Figure 9.11). Two of the ligands are phosphonate oxygens from two neighboring HDTMP ligands and the remaining six ligands are water molecules. It is important to note that two phosphonate moieties of HDTMP (one per "side") are monodeprotonated, but not coordinated to a metal ion. They are hydrogen bonded to a neighboring water of crystallization. The structure of Sr/Ba-HDTMP could be seen as 2D sheetlike topology made up by zigzag chains that form a corrugated sheet (Figure 9.11). The structure of Sr/Ba-HDTMP can be compared to that of a similar material, Co-HDTMP [62]. The latter is a linear structure, not polymeric because of the trans configuration of the two, Co-coordinated phosphonate groups that are bonded to an octahedral Co center.

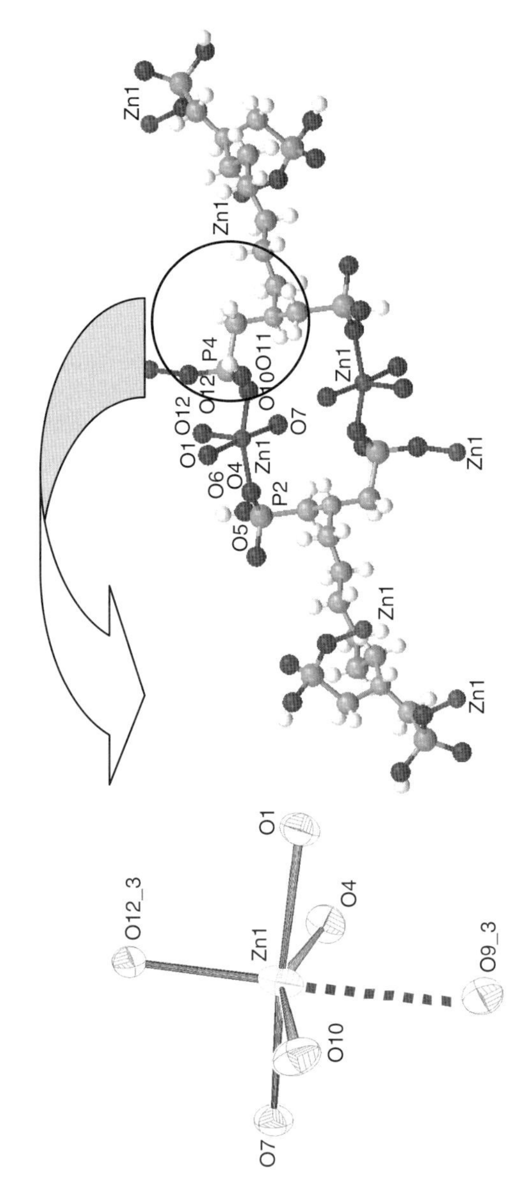

Figure 9.10 Coordination environment of the Zn^{2+} center (left) and the asymmetric unit (right).

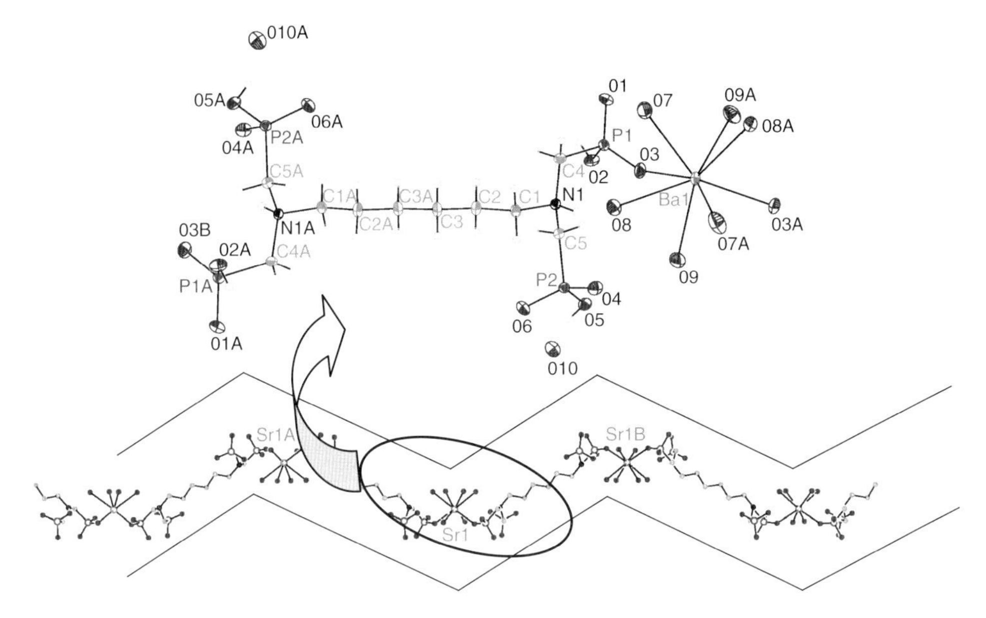

Figure 9.11 Fragment of the Ba-HDTMP structure. The eight-coordinated Ba center and the two noncoordinated phosphonate moieties can be clearly seen (upper). Zigzag chains in the structure of Sr-HDTMP (lower). Partially reproduced with permission from Ref. [45c]. © American Chemical Society.

9.3.10
Zinc-Tetramethylene-Diamine-Tetrakis(Methylenephosphonate), Zn-TDTMP

The reaction of Zn^{2+} with HDTMP afforded a Zn-HDTMP inorganic–organic hybrid coordination polymer. However, a reaction under the same conditions between Zn^{2+} and tetramethylene-diamine-*tetrakis*(methylenephosphonate) (TDTMP) gave a dramatically different material in which Zn^{2+} was not coordinated to the tetraphosphonate, but was found in an octahedral hexa-aqua coordination environment (Figure 9.12) [63].

It appears that water coordinates more strongly to Zn^{2+} than TDTMP. The presence of phosphonates and metal ions that are not coordinated to them is rarely encountered in the literature [64].

9.3.11
Strontium and Calcium-Ethylene-Diamine-Tetrakis(Methylene Phosphonate), Sr-EDTMP and Ca-EDTMP

In contrast to Zn^{2+}, EDTMP reacts with soluble Sr^{2+} salts to give 1D coordination polymers (Figure 9.13) [65]. In the structure of Sr-EDTMP, the tetraphosphonate acts as a chelate for a Sr center with two of its phosphonate groups (originating from different N atoms), whereas it bridges two different Sr centers with two phosphonate moieties (originating from the same N atom). The Sr centers are

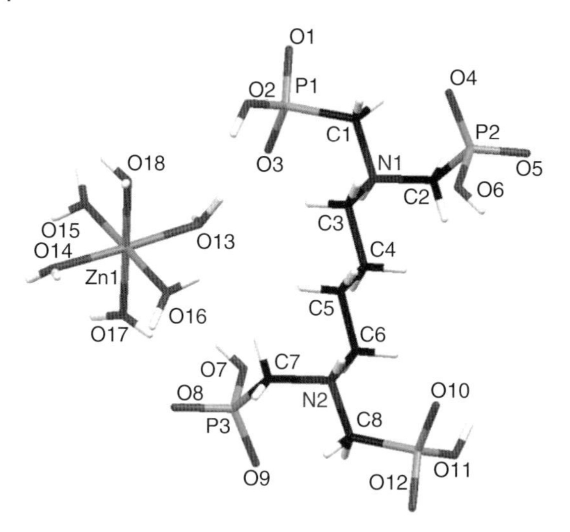

Figure 9.12 The crystal structure of the Zn-TDTMP material, showing the "free" phosphonate and the $Zn(H_2O)_6^{2+}$ cation.

octahedral with phosphonate oxygens occupying the basal positions and water oxygens completing the octahedron in the axial positions. The result of Sr chelation is a rare, 11-membered ring. Bridging creates 1D "rods" (Figure 9.13). The Ca-EDTMP material is isostructural to Sr-EDTMP.

9.3.12
Barium-Phosphonomethylene-Imino-Diacetate, Ba-PMIDA

Phosphonomethylene-imino-diacetate (PMIDA) is a "mixed" phosphonate/carboxylate that possesses two carboxylate and one aminomethylenephosphonate groups. Its reaction with soluble Ba salts at pH ~ 6 affords a 2D coordination polymer, Ba_2-PMIDA (Figure 9.14). The ligand is completely deprotonated with a "4-" charge that coordinates two crystallographically independent Ba ions. Both Ba centers are eight-coordinated. Ba–O bond distances are in the range 2.739(3)–3.056(3) Å for Ba(1) and 2.655(3)–2.968(3) Å for Ba(2). Ba(1) and Ba(2) are 4.798 Å apart.

The structure of Ba_2-PMIDA can be viewed as 1D linear rods that run along the *b*-axis (Figure 9.14). These 1D "polymers" are held together via hydrogen bonding mediated by waters of crystallization positioned in the space between the rods. Metal–PMIDA materials have been reported in the literature [66].

9.3.13
Tetrasodium-Hydroxyethyl-Amino-Bis(Methylenephosphonate), Na_4-HEABMP

The crystal structure of Na_4-HEABMP (hydroxyethyl-amino-bis(methylene phosphonate)) could be described as a 2D polymeric layered structure

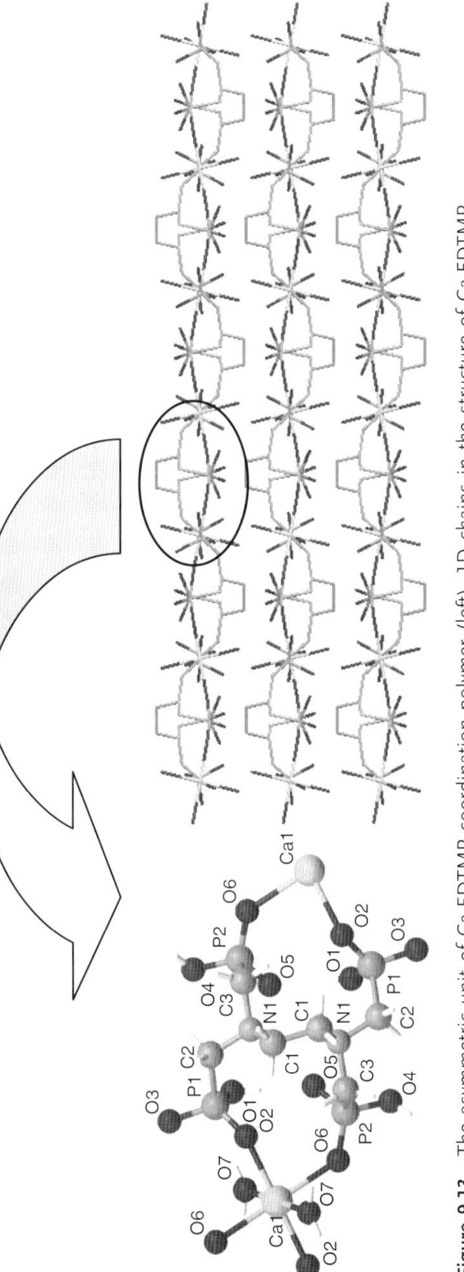

Figure 9.13 The asymmetric unit of Ca-EDTMP coordination polymer (left). 1D chains in the structure of Ca-EDTMP.

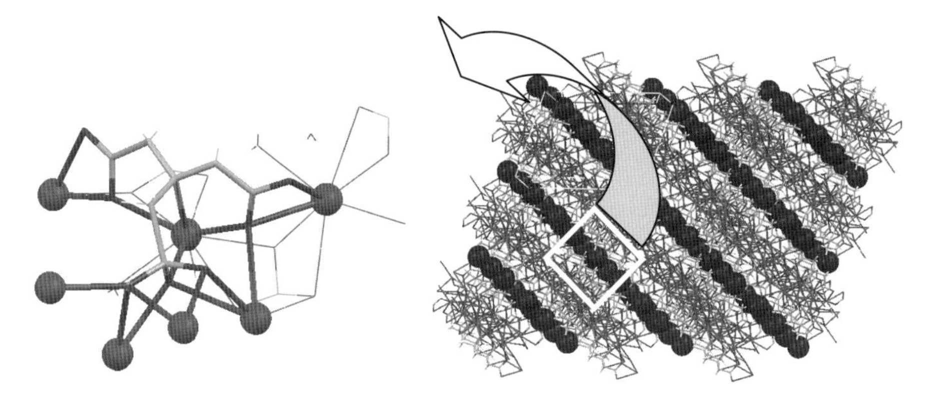

Figure 9.14 The coordination environment of PMIDA in the structure of Ba-PMIDA (left). A view of the 2D layered structure. Interlayer water molecules are shown as exaggerated spheres (right).

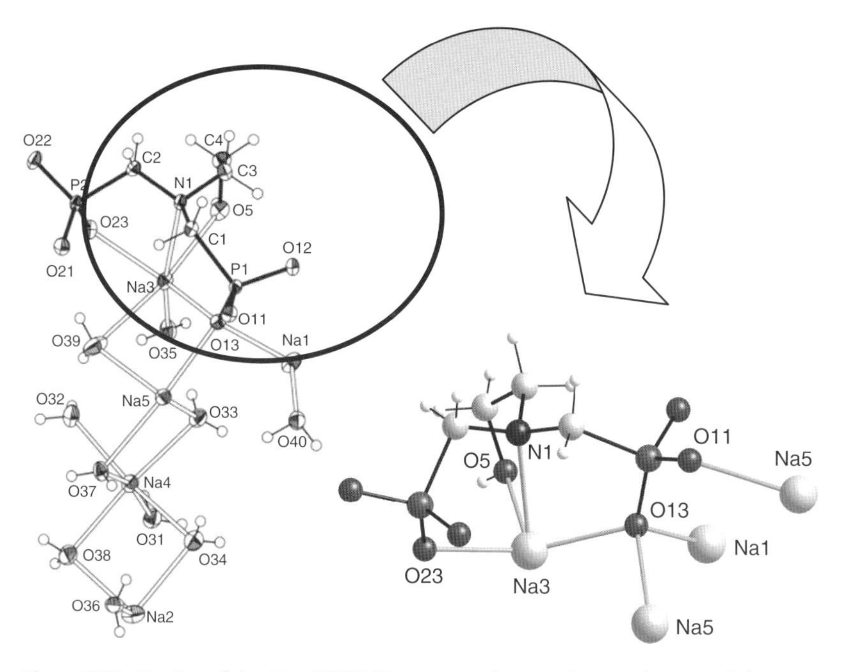

Figure 9.15 Portion of the Na_4-HEABMP structure showing the coordination of the HEABMP ligand and the hydrated Na ions (left). "Ball and stick" representation showing the "cavity" created by the HEABMP tetraanionic ligand, thus "trapping" a Na ion (right).

hydrogen bonded into a 3D supramolecular polymeric network [67]. The symmetry-independent part of Na_4–HEABMP and the coordination mode of the HEABMP tetraanion are shown in Figure 9.15.

Its structure consists of a "three-arm" backbone stemming from the N atom. Two "arms" are fully deprotonated methylene phosphonate ($-CH_2PO_3^{2-}$) moieties and the third is a hydroxyethyl ($-CH_2CH_2OH$) moiety. One of the methylene phosphonate arms uses only one oxygen donor atom (O23) to coordinate terminally the Na3 atom. The other arm uses two O donors (O11 and O13) to coordinate four Na cations. The donor O11 acts as a monodentate and terminally coordinates Na5 from the adjacent formula unit, while O13 is triply bridging Na1, Na3, and Na5 with very similar Na–O distances. The O5 atom of the hydroxyethyl arm and the N1 atom are involved in the coordination of Na3.

Na5 is "nested" in an octahedral environment formed by four H_2O lattice molecules and two O atoms from PO_3 groups, coming from adjacent molecules. Na–O(H_2O) interactions (all of them of bridging origin) are in the range of 2.3049(11)–2.5773(15) Å.

Both phosphonate groups in HEABMP are fully deprotonated. P–O bond lengths are nearly equivalent in both groups showing rather minor differences, and range from 1.5141(8) to 1.5368(8) Å. The P–O bond length equivalency implies even distribution of the negative charge over all three oxygens per $-PO_3$ group. P–C bond lengths fall in the normal range (1.8–1.9 Å) and are 1.8375(11) Å and 1.8279(12) Å. The N atom is not protonated as expected owing to the high pH of crystal preparation.

9.3.14
Calcium-Phosphonobutane-1,2,4-Tricarboxylate, Ca-PBTC

Crystalline Ca-PBTC, $Ca(PBTC)(H_2O)_2{\cdot}2H_2O$, is obtained by reacting $CaCl_2{\cdot}2H_2O$ and PBTC in a 1 : 1 molar ratio. It can also be prepared in high yields from CaO or $Ca(OH)_2$ and PBTC in heterogeneous aqueous medium. Its crystal structure reveals a polymeric material with PBTC acting as a tetradentate chelate [68], Figure 9.16.

The Ca^{2+} center is seven-coordinated in a capped octahedral environment, bound by two phosphonate oxygens, three carboxylate oxygens, and two water molecules. The phosphonate oxygens act as bridges between two neighboring Ca^{2+} centers located 6.781 Å apart. The phosphonate group and the carboxylate group oxygen atoms at the 2-position form a six-membered chelate with the Ca^{2+} center. The phosphonate group is doubly deprotonated. The –O–P–O– moiety bridges two Ca^{2+} centers. On the basis of the similar Ca–O(phosphonate) bond distances of 2.378(2) and 2.385(2) Å, the negative charge is delocalized over the entire O–P–O moiety. Ca–O_{water} distances, Ca(1)–O(11) 2.352(3) and Ca(1)–O(10) 2.445(3) Å, are consistent with those reported in the literature. There are numerous hydrogen-bonding interactions in the structure of $Ca(H_3PBTC)(H_2O)_2{\cdot}2H_2O$. Twelve out of thirteen oxygens in the structure (except O(3)) participate in an intricate network of hydrogen bonds. The shortest O···O interactions are $O_{carboxylate}(2) \cdots O(8)_{phosphonate} = 2.518$ Å, $O_{carboxylate}(4) \cdots O(9)_{phosphonate} =$ 2.510 Å, and $O_{carboxylate}(6) \cdots O(7)_{phosphonate} = 2.652$ Å.

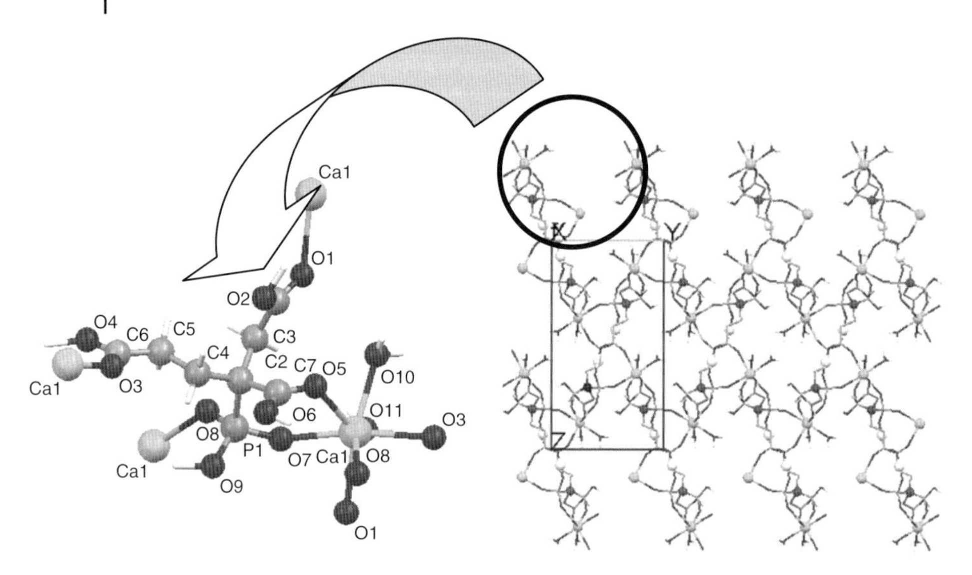

Figure 9.16 Fragment of the $[Ca(H_3PBTC)(H_2O)_2{\cdot}2H_2O]_n$ coordination polymer, showing the coordination environment of the seven-coordinated Ca^{2+} and the tetradentate chelation mode of H_3PBTC^{2-} to four Ca^{2+} centers (left). Packing of the structure along the *a*-axis (right).

9.3.15
Calcium-Hexamethylene-Diamine-Tetrakis(Methylenephosphonate), Ca-HDTMP

The layered structure of $CaH_6DTMP.2H_2O$ (Figure 9.17) is composed of a six-coordinated Ca^{2+} center surrounded exclusively by phosphonate oxygens from the H_6DTMP^{2-} ligand (embedded *within* the layers) with one crystallographically independent lattice water between the hybrid layers [69]. The basic building block for these phosphonate materials is given in Figure 9.17, where the eight-membered ring chelate and the organic ligand are shown. The phosphonate groups bridge the calcium polyhedra giving infinite chains along the *a*-axis, which is a common structural feature of all these crystalline hybrids. It must be highlighted that the layers are only held together by H bonding and that the organic ligand is located within the layers in contrast to a large number of pillared metal–phosphonates, where the phosphonate ligand connects neighboring inorganic layers.

9.3.16
Calcium-Hydroxyphosphonoacetate, Ca-HPAA

When water-soluble Ca^{2+} salts are reacted with a racemic mixture of *R*- and *S*-hydroxyphosphonoacetic acid (HPAA) at circumneutral pH, a highly crystalline product is quantitatively obtained. Its molecular structure from single crystal X-ray diffraction data revealed a linear trimer, whose Ca atoms are bridged by two tris-deprotonated HPAA ligands (Figure 9.18). There are some noteworthy structural features. The central Ca is six-coordinated, whereas the two Cas at the

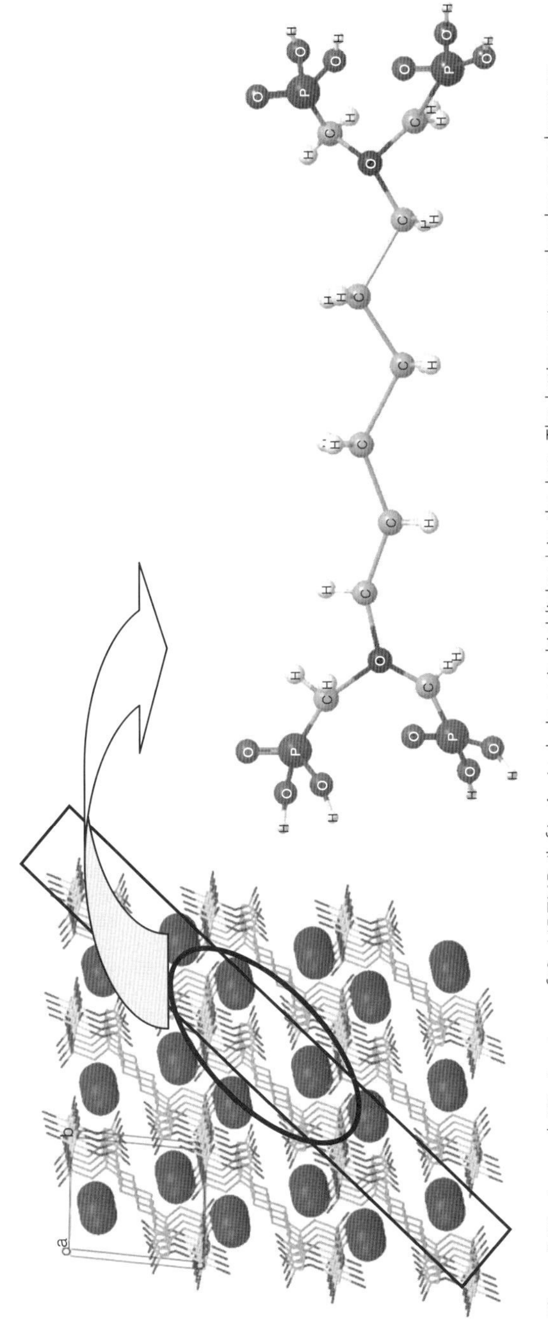

Figure 9.17 Layered 2D structure of Ca-HDTMP (left). A single layer is highlighted in the box. The lattice water molecules are shown as exaggerated spheres. The basic tetraphosphonate structural building block (right).

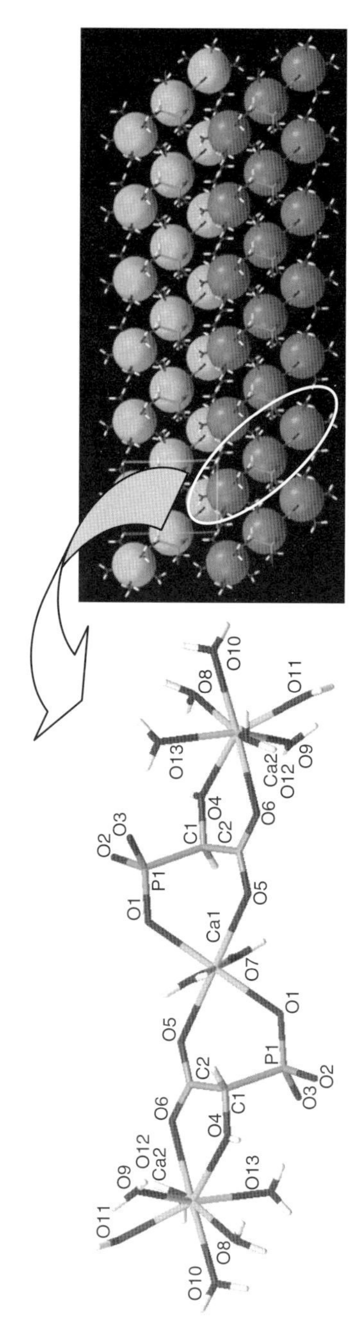

Figure 9.18 The molecular structure of the Ca-HPAA trimer (left) and the packing (along the *c*-axis) of two neighboring layers of trimers (right).

end are eight-coordinated. The central Ca is surrounded by two chelating $HPAA^{3-}$ ligands. Each $HPAA^{3-}$ ligand binds to the central Ca through a phosphonate and carboxyl Os. The slightly distorted octahedron is completed with two water molecules trans to each other. The Cas at the end sit in a coordination environment that resembles that of a bicapped trigonal prism. They are coordinated by a carboxyl O and the –OH group and six water molecules. Both *R* and *S* isomers of HPAA are incorporated into the trinuclear complex. The phosphonate moiety is completely deprotonated ($-PO_3^{2-}$); however, it is found to act as a monodentate ligand for the central Ca.

This is particularly surprising, as possible substitution of at least one of the labile water molecules (at the end Cas) would have occurred, resulting in a hypothetical 2D structure. Stabilization by strong hydrogen bonds could explain why the $-PO_3^{2-}$ moiety is found deprotonated, yet noncoordinating. Indeed, both noncoordinating phosphonate Os participate in a total of seven H bonds. Specifically, O(2) interacts with three Ca_{end}-bound waters from neighboring trimers (contacts 2.710–2.900 Å) and the –OH group of HPAA. O(3) is H bonded to three Ca-bound water molecules (two inter- and one intramolecular interactions; contacts 2.740–2.774 Å).

9.3.17 Strontium and Barium-Hydroxyphosphonoacetate, Sr and Ba-HPAA

9.3.17.1 $Sr[(HPAA)(H_2O)_3]\cdot H_2O$

This is a 2D, layered polymer consisting of "$Sr(HPAA)(H_2O)_3$" units that are connected through a Sr–O(carboxylate) linker [70]. The Sr atoms are eight-coordinated. Sr–O bond distances range from 2.5030(17) to 2.6916(18) Å. Consequently, each $HPAA^{2-}$ ligand is coordinated to three symmetry related Sr^{2+} centers (Figure 9.19). On the basis of the virtually equal C–O bond lengths (C(2)-O(5) 1.258(3) Å and C(2)-O(6) 1.260(3) Å), the negative charge is delocalized over the entire O–C–O moiety. The $Sr-O_{carboxylate}$ bond distances are 2.6055(18) Å (monodentate), 2.6178(17) Å (bridging), and 2.6916(18) Å (bridging). The monodeprotonated $-PO_3H^-$ group is coordinated to only one Sr, with a Sr–O(4) bond distance of 2.5030(17) Å. The P–O bond lengths (P–O(4) 1.5000(17) Å, P–O(2) 1.516(2) Å, and $P-O_H(3)$ 1.5680(17) Å) point to delocalization of the negative charge over the O(4)–P–O(2) moiety. The hydroxyl group remains protonated with its oxygen atom found at a 2.5786(17) Å from Sr. The coordination environment of Sr could best be described as a *bicapped octahedron*. There is one H_2O of crystallization per asymmetric unit. O–Sr–O angles range from 70.13(5)° to 98.96(7)°. Each HPAA acts as a double chelate bridge between two Sr ions (a carboxyl and the hydroxyl oxygens chelate one Sr, while the other carboxyl and a phosphono-oxygen chelate another Sr). This type of HPAA bridging induces sufficient "strain" to create a zigzag chain with alternating "Sr" and "HPAA" units with a Sr–Sr–Sr dihedral angle of 136.27°. These zigzag chains are connected through Sr–O(carboxyl) linkages found on every other Sr, thus creating a corrugated layer that runs parallel to the *bc* plane. One could envision this architecture as a "double inorganic layer," the bridging mode of the carboxylate oxygen, O(5), being responsible for it. The closest Sr ··· Sr intralayer contact is 4.231

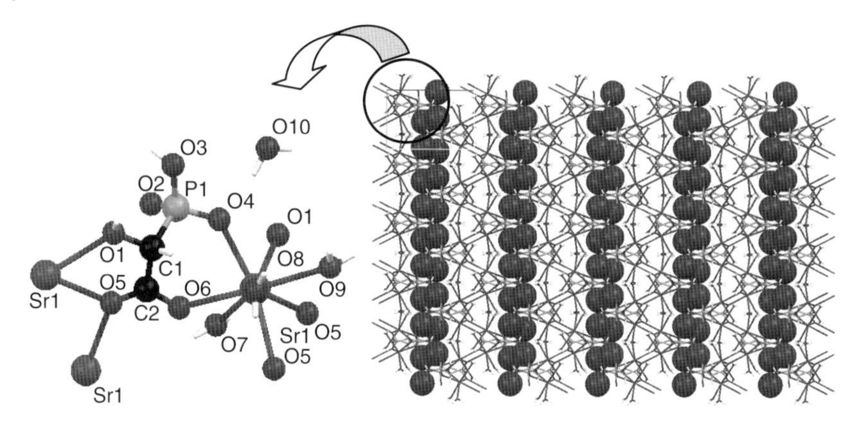

Figure 9.19 Coordination environment of Sr in Sr[(HPAA)$(H_2O)_3$]·H_2O (left). Layered structure of Sr[(HPAA)$(H_2O)_3$]·H_2O along the *c*-axis, with lattice water molecules as exaggerated spheres separating the Sr-HPAA layers (right). Reproduced with permission from Ref. [70]. © American Chemical Society.

Å, whereas the closest Sr ··· Sr interlayer contact is 7.150 Å. Neighboring layers are connected through extensive hydrogen-bonding interactions that involve the water molecule of crystallization (Figure 9.19). Furthermore, the neighboring "double layers" interact additionally via hydrogen bonds that include two noncoordinating –(O)P–O–H moieties from two adjacent layers.

9.3.17.2 M(HPAA)$(H_2O)_2$ (M = Sr, Ba)

These materials are isostructural [70]. In their 3D structure, each HPAA links to five symmetry-equivalent, nine-coordinated M^{2+} centers through its phosphonate group (bridging three M^{2+} ions), the carboxylate group (bridging two M^{2+}), and the protonated hydroxyl moiety (coordinating terminally one M^{2+} center), Figure 9.20, for the Ba analog. There are no waters of crystallization, but only two Ba-coordinated waters (O(7) and O(8)) in the structure. The Ba–OH_2 bond distances are 2.782(4) and 2.817(5) Å. Ba–$O_{carboxylate}$ bond distances are 2.971(4) Å (monodentate), 2.858(4) Å (bridging), and 2.914(4) Å (bridging). The C–O bond lengths (C(1)–O(1) 1.273(4) Å and C(1)–O(2) 1.235(5) Å) suggest delocalization of the negative charge over both O carboxyl atoms. The monodeprotonated $-PO_3H^-$ group is coordinated to three Ba in a bridging μ_3-mode, with bond distances Ba–O(4) 2.706(4) Å, Ba–O_H(5) 2.855(4) Å, and Ba–O(6) 2.724(4) Å. O(5) bears the H atom. This is supported by the P–O bond lengths, P–O(4) 1.513(3) Å, P–O(5) 1.572(3) Å, and P–O(6) 1.484(3) Å. The near equivalency of the two P–O bond lengths (P–O(4) and P–O(6)) points to delocalization of the negative charge over the O(4)–P–O(6) moiety. The hydroxyl group remains protonated with its oxygen atom found at a 2.831(4) Å from Ba. The coordination environment of Ba could be best described as a *distorted bicapped trigonal antiprism*. The hydroxyl (O(3)) and one carboxyl oxygen (O(2)) create a five-member ring with Ba, whereas one phosphono-oxygen (O(6)) and the other carboxyl oxygen (O(1)) create a six-member ring with a neighboring

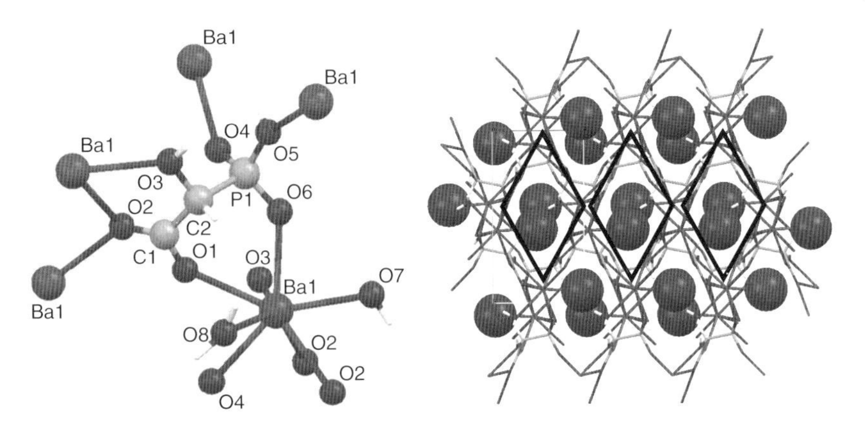

Figure 9.20 Coordination environment of Ba^{2+} centers in $Ba(HPAA)(H_2O)_2$ showing the multiple bridging of HPAA. 3D structure of $Ba(HPAA)(H_2O)_2$, showing packing along the *c*-axis and the Ba-coordinated waters (exaggerated O8) filling the "pores" (shown schematically as rhombs). Reproduced with permission from Ref. [70]. © American Chemical Society.

Ba. The multiple coordination ability of the $-PO_3H^-$ moiety in these structures can be contrasted to the monodentate coordination mode of the $-PO_3H^-$ group in $Sr[(HPAA)(H_2O)_3]{\cdot}H_2O$. In the latter, the absence of the bridging capability of the $-PO_3H^-$ group creates a layered structure, whereas in the former the bridging μ_3-mode of the $-PO_3H^-$ group is responsible for the generation of a 3D motif.

In all structures, both *R* and *S* HPAA isomers are incorporated in the layers. In $Sr[(HPAA)(H_2O)_3]{\cdot}H_2O$, each "double" layer is *R*–*S*–*R*–*S*–... In $Ba(HPAA)(H_2O)_2$, the *R* and *S* isomers are interwoven into a complicated 3D arrangement (Figure 9.20).

9.3.18 Calcium-Phosphonomethylene-Imino-Diacetate, Ca-PMIDA

This is a 2D layered material. The Ca^{2+} center is found in an octahedral environment shaped by five phosphonate oxygens and one Ca-bound water molecule (Figure 9.21a). The phosphonate and one carboxylate moieties act as bridging groups between Ca^{2+} ions. The second carboxylate group is monodentate. Ca–O bond distances fall in the range of 2.259–2.412 Å. The Ca–O(H_2O) is 2.314 Å.

The structure can be more precisely described as *pillared*, with the PMIDA dianions (organic layer) acting as pillars between the Ca/O layer (inorganic layer). This can be clearly seen in Figure 9.21b.

9.4 Metal–Phosphonate Anticorrosion Coatings

The literature on corrosion inhibition is vast. Inhibition approaches vary according to the peculiarities of the system, its metallurgy being one of the most significant.

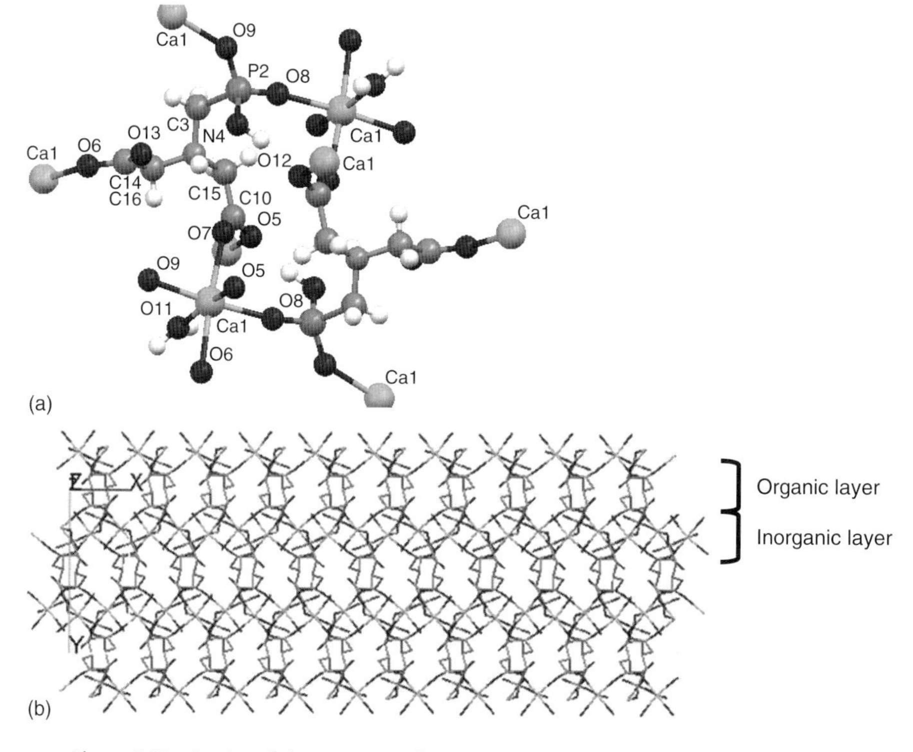

Figure 9.21 Portion of the structure of Ca-PMIDA, showing the numbering scheme (a). Packing diagram (along the *c*-axis) showing the 2D layered structure of the Ca-PMIDA material (b).

In this section, we briefly review the state of the art of the field of phosphonates as corrosion inhibitors, providing some characteristic examples. At the end of this section, we provide contributions put forth from our laboratory.

Mixtures of 2-aminothiophenol (ATP) and 1-hydroxyethylidene-*bis*(phosphonic acid) (HEDP) show a cooperative effect of inhibition of stainless steel (SS 41) [71]. The actual role of HEDP was explained in terms of its scale inhibition activity toward $FeCO_3$.

Kuznetsov *et al.* have studied the inhibitor efficiency of various phosphonate inhibitors of the aminomethylenephosphonate type (R-N-CH_2-PO_3H_2) [72]. The protective properties of aminophosphonic acids and their magnesium and calcium complexes were studied in soft water. 1,1-Hydroxycarboxypropane-3-amino-di(methylenephosphonic) and hexamethylenediamine-N,N-tetra(methylene-phosphonic) acids could suppress the corrosion of steel in water completely. Kouznetsov proposed that metal–phosphonate complexes are more effective than the corresponding acids; if the complexing agent is fixed, then the stability constants of the complexes become the major factor. For phosphonates of Mg^{2+} and Ca^{2+}, which are usually less stable than the corresponding iron complexes, the dependence of the protective concentration on stability constant passes through a maximum; the complexes of those acids whose own protective properties are weaker are more effective. Imino-*N*,*N*-diacetic-*N*-methylenephosphonic acid can serve as an example. By contrast, the relatively more stable complexes, for example, calcium nitrilo-tri(methylenephosphonate) (AMP), are much less effective than the acid itself.

Molybdate is a well-known corrosion inhibitor. Its combination with phosphonates enhances corrosion efficiency [73]. A level of 300 ppm MoO_4^{2-} had only 32% efficiency in inhibiting the corrosion of mild steel immersed in a neutral aqueous environment containing 60 ppm Cl^-, whereas a formulation of AMP (50 ppm) $-MoO_4^{2-}$ (300 ppm) – Zn^{2+} (50 ppm) exhibited 96% inhibition efficiency. The lower inhibition efficiency in the former case was due to the dissolution of the protective film formed on the metal surface, and getting precipitated in the bulk of the solution; this system controlled the anodic reaction only. The latter system controlled both the anodic and cathodic reactions; the dissolution of the protective film formed on the metal surface was reduced to a greater extent.

Protective layer formation of α,ω-diphosphono-alkane compounds on iron surface was studied [74]. Layer formation proved to be a spontaneous process on iron, and can be accomplished by simple immersion into an aqueous solution of phosphonate additives, resulting in a thin dense multimolecular adsorption layer with a high corrosion protection effect. It was concluded that the mechanism of inhibition was anodic type, hindering active iron dissolution because of the blocking of the metal surface. The protective layer formation completes after a few days of immersion, resulting in a thin but dense multimolecular adsorption layer. The "self-healing" effect of diphosphonates was demonstrated. On the basis of the high corrosion protection effect, surface modification by diphosphonates may be regarded as potential anticorrosive treatment.

Surface treatments have been carried out on carbon steel in solutions containing different phosphonates [75]. The compounds were dissolved in an ethanol/water mixture (80 : 20). Corrosion protection afforded by laurylphosphonic acid (LPA), ethyllaurylphosphonate (ELP; also called *lauryl phosphonic acid monoethylester*), and diethyllaurylphosphonate (DELP) was studied by steady-state current–voltage curves and electrochemical impedance measurements with a rotating disc electrode. Corrosion protection was only obtained for the ELP, which was able to form a relatively thick film on the carbon steel surface. Electron probe microanalysis corroborated that the film is thick and porous. Infrared spectroscopy indicated that the film was formed by reaction of the organic phosphonate with the steel surface to produce a metal salt.

The effect of phosphonates used in Russian heat-power engineering on the corrosion of carbon steel in deaerated delivery water at 90 °C was studied [76]. It was demonstrated that introduction of phosphonates reduced the susceptibility of steel to local corrosion. A Zn^{2+} complex of hydroxyethylidenediphosphonic acid (OEDP–zinc) was the most effective inhibitor of the anodic reaction.

For the protection of carbon steel from corrosion, AMP was more effective than HEDP, *N*,*N*-dimethylidenephosphonoglycine (DMPG), 1-ethylphosphonoethylidenediphosphonic acid (EEDP), and EDTMP [77]. A 20-min treatment in 1.0 M of AMP at pH 0.23 at 45 °C formed an anticorrosive complex film that was composed of 48.4% O, 28.6% P, 7.0% Fe, 4.3% N, and 11.7% C, based on X-ray photoelectron spectroscopy (XPS) and Auger electron spectroscopy. From differences in binding energies of Fe, N, and O, in the shift of C–N and P–O vibration, in the reflection FT-IR spectra, and in the change of P–OH and Fe–N vibration before and after film formation, it was deduced that N and phosphonate O in AMP were coordinated with Fe^{2+} in the film.

The corrosion inhibition of iron in 0.5 M sulfuric acid by *N,N*-dipropynoxy methyl amine trimethyl phosphonate was investigated by means of potentiodynamic polarization and electrochemical impedance spectroscopy (EIS) techniques [78]. *N,N*-dipropynoxy methyl amine trimethyl phosphonate was studied in concentrations from 40 to 320 ppm at a temperature of 298 K. The results revealed that the inhibition mechanism of *N,N*-dipropynoxy methyl amine trimethyl phosphonate is a combination of anodic and cathodic type. It was also found that this inhibitor obeys the Frumkin adsorption isotherm and the Flory–Huggins isotherm based on a substitutional adsorption process.

The role of pH and Ca^{2+} in the adsorption of an alkyl *N*-aminodimethyl phosphonate on mild steel (E24) surfaces was investigated by XPS [79]. Fe 2p3/2 and O 1s spectra showed that the oxide/hydroxide layer developed on the steel surface, immersed in the diphosphonate solution ($7 < pH < 13$, without Ca^{2+}) or in a filtered cement solution (pH 13, 15.38 M of Ca^{2+}), consists of Fe_2O_3, covered by a very thin layer of FeOOH (goethite). The total thickness of the oxide/hydroxide layer was ~3 nm and was independent of the pH and the presence/absence of Ca^{2+}. In the absence of Ca^{2+} ions, the N 1s and P 2p spectra revealed that the adsorption of the diphosphonate on the outer layer of FeOOH took place only for pH lower than the zero charge pH of goethite (7.55). At pH 7, the adsorbed diphosphonate layer was continuous and its equivalent thickness was ~24 Å (monolayer). In the presence of Ca^{2+} ions, the C 1s and Ca 2p signals indicated that Ca^{2+} was present on the steel surface as calcium phosphonate (and $Ca(OH)_2$, in very small amounts). The adsorption of the diphosphonate molecules on the steel surface was promoted in alkaline solution ($pH > 7.55$) by the divalent Ca^{2+} ions that bridged the O^{2-} of goethite and the $P{-}O^-$ groups of the diphosphonate molecules. The measured values for the Ca/P intensity ratio were in the range 0.75–1, which suggested that the diphosphonate molecules were adsorbed on steel forming a coordination polymer cross-linked by Ca^{2+} through their phosphonate groups. In the presence of Ca^{2+} ions in alkaline solution, the adsorbed diphosphonate layer was discontinuous and the surface coverage was found to be ~34%.

The effect of a new class of corrosion inhibitors, namely piperidin-1-yl-phosphonic acid (PPA) and (4-phosphono-piperazin-1-yl)phosphonic acid (PPPA) on the corrosion of iron in NaCl medium was investigated by electrochemical measurements [80]. Potentiodynamic polarization studies clearly revealed the type of the inhibitor. The addition of increasing concentrations of phosphonic acids caused a shift of the pitting potential (E_{pit}) in the positive direction, indicating the inhibitive effect of the added phosphonic acid on the pitting attack. The potential of corrosion was moved toward negative values and the corrosion current was reduced. The values of the current were lower in the presence of PPA and PPPA. This was explained by the fact that most of the surface of the electrode was covered by the molecules adsorbed. PPPA had a strongly inhibitive effect on chloride pitting corrosion. It was proposed that the addition of the $NCH_2PO_3H^-$ group (center adsorption) in the PPA para-position, giving PPPA, reinforced the active sites of this molecule and consequently increased its inhibition efficiency.

Phosphonate layer formation on a passive iron surface was investigated by electrochemical and atomic force microscopy techniques [81]. It was found that phosphonate groups bond more strongly to the oxide surface, while the metallic iron surface is disadvantageous for phosphonate layer formation in aqueous solutions. The rate of anodic dissolution was continually decreasing because of the time-dependent formation of a protective phosphonate layer. The kinetics of phosphonate layer formation on passive iron was determined by the potential applied for preceding passive film formation. The size and shape of iron oxide grains depended slightly on the potential of passivation. Changes in morphology because of the phosphonate layer formation were recorded by AFM.

The passivating ability of the zinc complex of the HEDP in a borate buffer solution was studied [82]. An *in situ* ellipsometric method was used to study the mechanism of formation of a protective film on iron in the presence of HEDP, Zn-HEDP, and $ZnSO_4$ in the course of the cathodic polarization of the electrode. The studies of Zn-HEDP adsorption on iron (at $E = -0.65$ V) in combination with XPS showed that on the metal surface a multilayer protective film formed consisting of an internal layer of $Zn(OH)_2$ and an outer layer consisting of HEDP complexes with Fe^{2+} and/or Zn^{2+}. It was found that the thickness of the passivating film does not exceed 60 Å, of which 7–10 Å correspond to the sparingly soluble $Zn(OH)_2$.

Electrochemical techniques were used for investigating the inhibitory activity on carbon steel of thiomorpholin-4-ylmethyl-phosphonic acid (TMPA) and morpholin-4-methyl-phosphonic acid (MPA) in natural seawater [83]. Measurement of the free corrosion potential indicated that the phosphonic acids tested inhibit the corrosion of carbon steel in seawater. Potentiodynamic polarization curves clearly proved that addition of these molecules is responsible for a decrease in the corrosion current density and a corresponding reduction of the corrosion rate (CR). The phosphonic acids tested as corrosion inhibitors of carbon steel in natural seawater are effective even at low concentrations. FT-IR spectroscopy was used to obtain information on the interaction between the metallic surface and the inhibitors. The morphology of the metal surface in the uninhibited and inhibited solution was examined using SEM/EDS.

New corrosion inhibitors, namely 3-vanilidene-amino-1,2,4-triazole phosphonate (VATP) and 3-anisalidene amino 1,2,4-triazole phosphonate (AATP) were synthesized and their action along with biocide on corrosion control of copper in neutral aqueous environment has been studied [84]. Potentiodynamic polarization measurements and EIS had been employed to analyze their inhibition behavior. VATP showed better protection over the other inhibitors used. The dissolution of copper in the presence of VATP and AATP with biocide mixture is negligible compared to blank. A combination of electrochemical methods and surface examination techniques is used to investigate the protective film and explain the mechanistic aspects of corrosion inhibition.

A thin-film coating consisting of Zr^{4+} and octadecyl phosphonate provided inhibition of O_2 reduction for Cu-rich aluminum alloys, such as AA2024-T3 [85]. The coating procedure produced films with thicknesses approaching that of a self-assembled monolayer. Using constant-potential experiments, the current density for the O_2 reduction reaction was shown to decrease by 2 orders of magnitude following five treatments. Auger electron spectroscopy of the coating showed that both Zr and P are present at the alloy surface following treatment. Characterization of the coating showed it to be hydrophobic, with the octadecyl chains being in a liquidlike environment.

The reaction of water-based solution of 1,5-diphosphono-pentane (DPP) and 1,7-diphosphonoheptane (DPH) with high-purity polycrystalline zinc surface was investigated at room temperature [86]. X-ray diffraction (XRD) and XPS studies confirmed the formation of a crystalline zinc–phosphonate film on the metal surface. The assembled layers gave hydrophilic properties to the surface. A conversion-type interaction of diphosphonic acid compounds with the oxidized zinc surface was unambiguously shown by XPS analysis. Conclusive results were obtained by synthesized zinc–diphosphonate model compounds modeling the ones deposited on the zinc surface, revealing a $\sim 1:1$ molar ratio of the phosphonate groups with zinc. Reactions with both diphosphonates resulted in significant protective effect of zinc against corrosion, although the structure and quality of the formed layers exhibited marked differences. The in-depth distribution of the composition and dissimilarity of the layer thickness were determined by glow discharge optical emission spectrometry (GD-OES). The corrosion inhibition was explained by the formation of insoluble zinc–phosphonate salt on the zinc surface, blocking the zinc dissolution process.

Efforts to understand the inhibition mechanism of 1,5-diphosphonic acid on mechanically polished zinc substrates were focused on the surface structure of layers formed in aqueous and ethanol solutions [87]. In spite of the differences noted in the protective layers formed in ethanol and in aqueous solution, their protection efficiency was identical after ~4 h of testing. On the basis of EIS, XPS, and GD-OES data, a simple oxide–hydroxide/diphosphonate model of the interface was proposed. Both XPS and GD-OES showed unambiguously that the surface film contained large amounts of Zn, comparable to the P content, especially when the layer deposition originated from an aqueous solution of 1,5-diphosphonic acid. On that basis, it was concluded that the surface film consisted of a zinc–phosphonate layer.

The cooperative effect of Ca^{2+} and tartarate ion on the corrosion inhibition of pure aluminum in an alkaline solution was investigated by hydrogen collection, electrochemical methods, and XPS [88]. The results of electrochemical experiments showed that the inhibition by Ca^{2+} was enhanced by the addition of tartarate ion, while the tartarate ion itself had only a little inhibition effect. Analysis by XPS showed that Ca^{2+} and tartarate ions did not contribute to the formation of the surface film on aluminum. This fact indicates that these two ions act as interface inhibitor. The cooperative effect of the two ions might be due to formation of a complex, which makes it easier for Ca^{2+} to adsorb on the aluminum surface.

The AMP layers were adsorbed on the surface of AA6061 aluminum alloy for improving the lacquer adhesion and corrosion inhibition as a substitute for chromate coatings [89]. The surface structure and features of the AMP layers on AA6061 aluminum alloy were investigated by means of XPS and ATR-FTIR analysis. The analyzed results showed that the AMP adsorption layers adsorbed on the surface of aluminum alloy via acid–base interaction in a bidentate conformation. After the AMP layers were coated with epoxy resin, the layers showed good adhesive strength and favorable corrosion resistance in contrast to chromate coatings.

A recent study investigated the surface modification of iron with self-assembled alkyl-phosphonic acid layers [90]. Passivated, native (covered with air-formed oxide), and metallic iron surfaces were used as substrates. The corrosion-protective effect of the phosphonate layer and the structure of the modified surfaces were investigated. The layer formation process was monitored directly in the phosphonate-containing solution. It was found that the phosphonate layer formation process depends on the conditions applied during iron passivation, and the oxide layer has an important role in the stability of the protective layer, while the bare metallic iron surface is disadvantageous for phosphonate bonding. Conversion electron Mossbauer spectroscopy (CEMS) and XPS proved the presence of the phosphonate monolayer on the passive iron surface.

Synergistic inhibition of corrosion of carbon steel in a low chloride environment using ascorbate as a synergist along with PBTC (see Figure 9.2) and Zn^{2+} was reported [91]. The synergistic effect of ascorbate has been established from the present studies. In the presence of ascorbate, lower concentrations of PBTC and Zn^{2+} are sufficient in order to obtain good inhibition, thus making this formulation more environmentally friendly. Potentiodynamic polarization studies inferred that this mixture functions as a mixed inhibitor, predominantly cathodic. Impedance studies revealed that an immersion period of 24 h is necessary for the formation of the protective film, with a very high charge transfer resistance. The surface analysis by XPS showed the presence of Fe, O, P, C, and Zn in the protective film. The XPS spectra indicated and FT-IR corroborated the presence of Fe^{3+}-oxides/hydroxides, $Zn(OH)_2$, and a proposed [Zn(II)–PBTC–ascorbate] complex on the film surface.

The system copper/benzotriazole or tolyltriazole in tap water was investigated as an example of interface inhibition. A quasi-polymeric film with a thickness of only a few molecular layers was formed from Cu(I) and inhibitor and is remarkably resistant even to air and solvents [92]. Phosphonic acids act as membrane inhibitors on iron in cooling water. Together with corrosion products and ions from the water, they form layers with a thickness of about 20–40 nm. As their growth then ceases, they cause no hindrance to heat transfer.

9.5
A Look at Corrosion Inhibition by Metal–Phosphonates at the Molecular Level

Phosphonate-based corrosion inhibitors are effective in decreasing metallic corrosion in nearly neutral conditions by forming weakly soluble compounds with the metal ion existing in the solution. These metal–phosphonate "complexes" precipitate on to the metallic surface to form a 3D protective layer. Such inhibitors (often called *interphase inhibitors*) for cooling water treatment technology in the past decades comprise different types of phosphonic acids [93]. In the previous section, it was shown that widely used phosphonic acids are HEDP, AMP, HPAA, and so on. Phosphonates are introduced into the system to be protected in the acid form or as alkali metal–soluble salts, but readily form more stable complexes with other metal cations found in the process stream (most commonly Ca, Mg, Sr, or Ba), depending on the particular application. It should be emphasized that research in this area has been stimulated by the need to develop inhibitor formulations that are free from chromates, nitrates, nitrites, inorganic phosphorus compounds, and so on. Phosphonates when blended with certain metal cations and polymers reduce the optimal inhibitor concentration needed for inhibition because of synergistic effects [94]. Synergism is one of the important effects in the inhibition process and serves as the basis for the development of all modern corrosion inhibitor formulations.

In spite of the significant body of literature, evidence regarding the molecular identity of the thin protective metal–phosphonate films lags behind. In this section, the corrosion inhibition performance of several metal–phosphonate materials is reported and a more systematic look at their corrosion inhibition mechanisms at the molecular level is presented. These exhibit dramatically different anticorrosion efficiencies, which are linked to their molecular structure. These metal–phosphonates are the following coordination polymers: Zn-AMP, $\{Zn(AMP)\cdot 3H_2O\}_n$; Zn-HDTMP, $\{Zn(HDTMP)\cdot H_2O\}_n$; Ca-PBTC, $\{Ca(PBTC)(H_2O)_2\cdot 2H_2O\}_n$; Ca-HPAA, $Ca_3(HPAA)_2(H_2O)_{14}$; M (Sr, Ba)-HPAA, $\{M(HPAA)(H_2O)_2\}_n$; M(Ca, Sr, Ba)-PMIDA, $\{Ca(PMIDA)\}_n$.

9.5.1
Anticorrosion Coatings by the Material $\{Zn(AMP)\cdot 3H_2O\}_n$

In this section, the CR is calculated from the equation:

$$CR = \frac{[534.57 \times (\text{mass loss})]}{[(\text{area})(\text{time})(\text{metal density})]}$$

Units: CR in millimeters per year, mass loss in milligrams, area in square centimeters, time in hours, metal density = 7.85 g/cm^3 (for carbon steel).

Synergistic combinations of 1 : 1 molar ratio Zn^{2+} and AMP are reported to exhibit superior inhibition performance than either Zn^{2+} or AMP alone [95]. However, no mention is made regarding the identity of the inhibitor species involved in corrosion inhibition. Therefore, a corrosion experiment was designed in order to verify the literature results and prove that the protective material acting as a corrosion barrier is an organic–inorganic hybrid composed of Zn and AMP [53]. A synergistic combination of Zn^{2+} and AMP in a 1 : 1 ratio (under identical conditions used to prepare crystalline Zn-AMP) offers excellent corrosion protection for carbon steel (see Figure 9.22). Although differentiation between the "control" and "Zn-AMP"-protected specimens is evident within the first hours, the corrosion experiment is left to proceed over a three-day period. On the basis of mass loss measurements, the CR for the "control" sample is 2.5 mm/year, whereas that for the Zn-AMP protected sample is 0.9 mm/year, which is a 270% reduction in CR. The filming material is collected and subjected to FT-IR, XRF, and EDS studies. These show that the inhibiting film is a material containing Zn (from added Zn^{2+}) and P (from added AMP) in an approximately 1 : 3 ratio, as expected. Fe was also present, apparently originating from the steel specimen. FT-IR showed multiple bands associated with the phosphonate groups that closely resemble those of an authentically prepared Zn-AMP material. For comparison, EDS and XRF spectra of a "protected" and an "unprotected" region show the presence of Zn and P in the former, but complete absence in the latter.

A characteristic example of a Zn-AMP film is shown in Figure 9.23 and is compared to a "bare" iron metal surface.

Figure 9.22 Corrosion inhibition by the material Zn-AMP at pH 3.0. The upper specimen is the "control" (no inhibitor present); the lower specimen is with Zn^{2+}/AMP combination present, both in 1 mM.

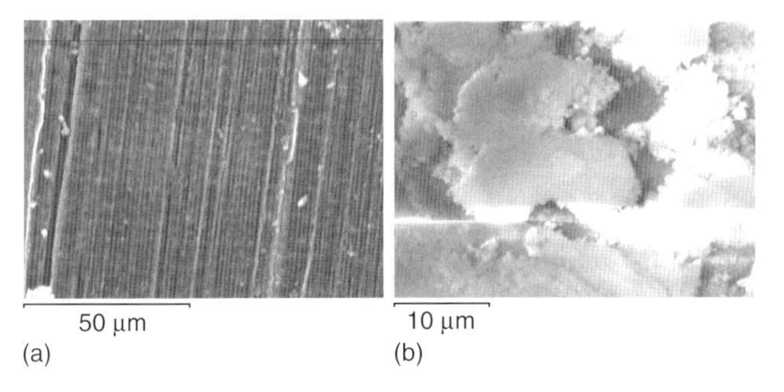

Figure 9.23 SEM images of a bare carbon steel surface (a, bar = 60 μm) and a Zn-AMP protected steel surface (b, bar = 10 μm). Deposition of an anticorrosive Zn-AMP material is obvious.

9.5.2 Anticorrosion Coatings by the Material $\{Zn(HDTMP)\cdot H_2O\}_n$

A combination of Zn^{2+} and HDTMP in a 1 : 1 ratio (under identical conditions used to prepare crystalline Zn-HDTMP) offers excellent corrosion protection for carbon steel (Figure 9.24) [58]. Although differentiation between the "control" and "Zn-HDTMP" protected specimens is profound within the first hours, the corrosion experiment is left to proceed over a three-day period. On the basis of mass loss measurements, the CR for the "control" sample is 7.28 mm/year, whereas that for the Zn-HDTMP protected sample is 2.11 mm/year, which is a ~170% reduction in CR. The filming material is collected and subjected to FT-IR, XRF, and EDS studies.

These show that the corrosion inhibiting film is a material containing Zn^{2+} (from externally added Zn^{2+}) and P (from added HDTMP) in an approximate 1 : 4 ratio. Fe was also present, apparently originating from the carbon steel specimen. FT-IR of the filming material showed multiple bands associated with the phosphonate groups in the 950–1200 cm^{-1} region that closely resemble those of the authentically prepared Zn-HDTMP material (Figure 9.25). For comparison, EDS and XRF spectra of a "protected" and an "unprotected" region show the presence of Zn and P in the former, but complete absence in the latter.

9.5.3 Anticorrosion Coatings by the Material $\{Ca(PBTC)(H_2O)_2\cdot 2H_2O\}_n$

A synergistic combination of Ca^{2+} and PBTC in a 1 : 1 molar ratio (under identical conditions used to prepare crystalline $Ca(PBTC)(H_2O)_2\cdot 2H_2O$ seems to offer excellent corrosion protection for carbon steel (Figure 9.26) based on visual observations. However, based on mass loss measurements, the CR for the "control" sample is 0.16 mm/year, whereas that for the Ca-PBTC protected sample is 1.17 mm/year, which is a ~10-fold increase in CR [68]. Therefore, PBTC essentially enhances the dissolution of bare metal, presumably forming soluble Fe–PBTC complexes. In contrast to aminomethylene-tris-phosphonate, AMP, PBTC does not

Figure 9.24 The anticorrosive effect of Zn-HDTMP films on carbon steel. The upper specimen is the "control," no inhibitor present. Corrosion inhibition in the lower specimen by a 1 mM Zn^{2+}/HDTMP synergistic combination is obvious.

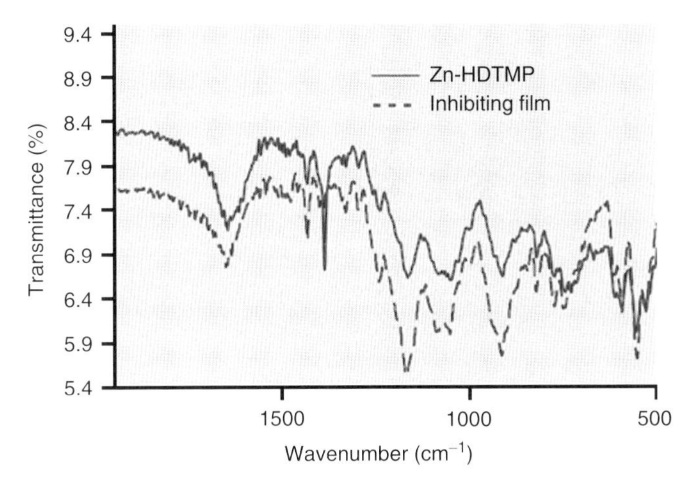

Figure 9.25 FT-IR spectra of "genuine" Zn-HDTMP and of the corrosion inhibiting film formed in situ from a 1 : 1 Zn^{2+}:HDTMP synergistic combination.

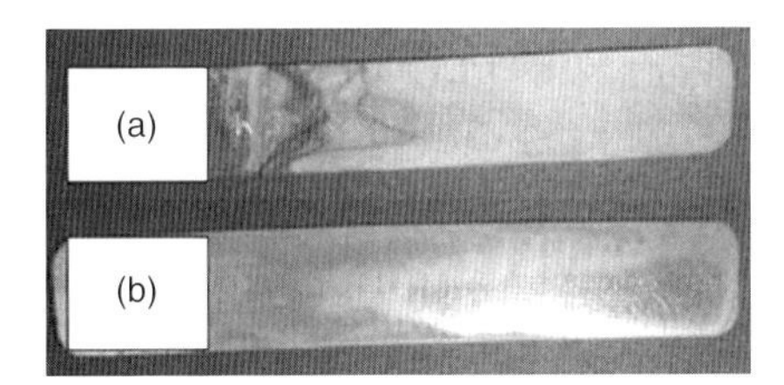

Figure 9.26 Phenomenology of the anticorrosive effect of Ca-PBTC films on carbon steel. The upper specimen is the "control" (a), no inhibitor present. Surface "cleanliness" in the lower specimen (b) by a 1 mM Ca^{2+}/PBTC synergistic combination is demonstrated, but metal loss is enhanced.

form stable metal–phosphonate protective films. This is consistent with the low complex formation constant for Ca-PBTC, 4.4 [36b].

9.5.4 Anticorrosion Coatings by the Material $Ca_3(HPAA)_2(H_2O)_{14}$

The assessment of the corrosion-inhibitory activity was based on mass loss measurements from carbon steel specimens, following a well-established NACE protocol. The effectiveness of corrosion protection by synergistic combinations of Ca^{2+} and HPAA in a 1 : 1 ratio is dramatically pH-dependent (Figure 9.27) [96].

At pH 2.0, mass loss from the steel specimens is profound resulting in high CRs (353×10^{-3} mm/year). However, specimen **2** (Figure 9.27) appears relatively clean from corrosion products, presumably because HPAA (either free or metal-bound) at the surface acts as a Fe-oxide-dissolving agent. At pH 7.3, CRs are appreciably suppressed (4×10^{-3} mm/year) in the presence of combinations of Ca^{2+} and HPAA (specimen **4**), reaching nearly quantitative inhibition.

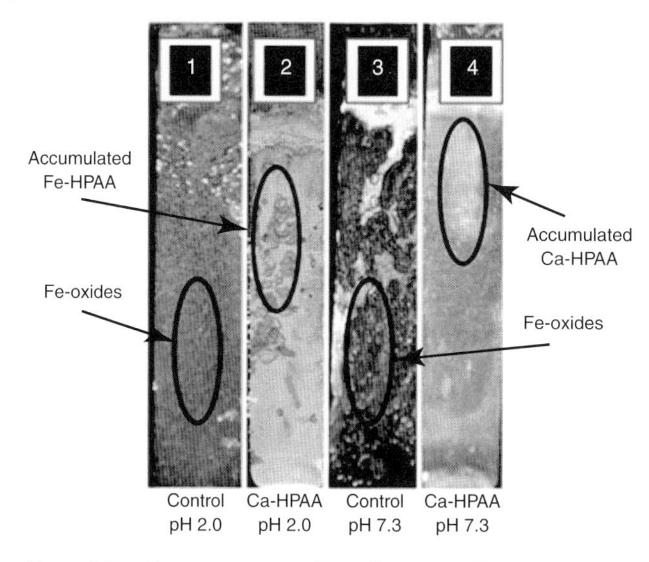

Figure 9.27 The anticorrosive effect of Ca-HPAA films on carbon steel. The effect of Ca-HPAA synergistic combinations is dramatically demonstrated in specimen 4.

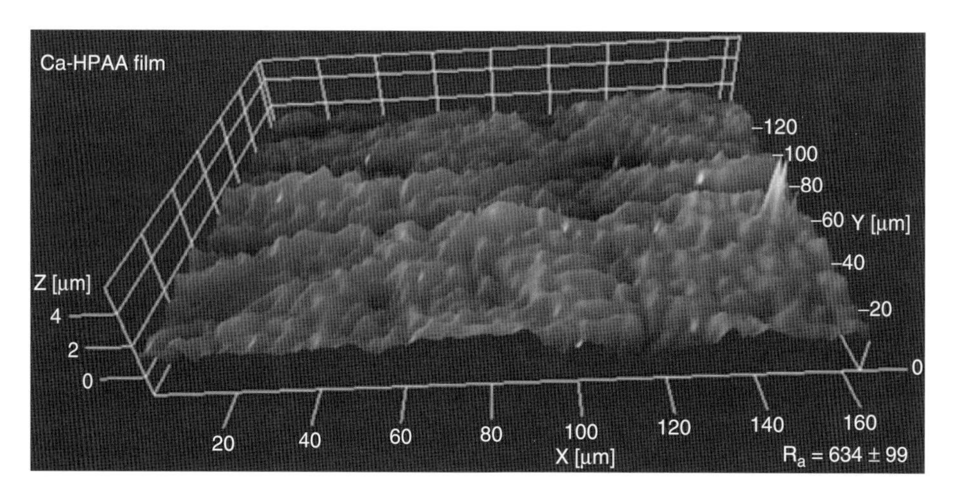

Figure 9.28 Anticorrosion film morphology viewed by VSI. Grid size is 9×10^4 μm^2. The height of z-axis is 4 μm. R_a is the average surface roughness, or average deviation, of all points from a plane fit to the test part of the surface.

The corrosion specimens (at pH 7.3) and film material were subjected to studies by vertical scanning interferometry (VSI, Figure 9.28), SEM, FT-IR, and EDS in order to fully characterize the protective coating. Data show that at pH 2.2 (ineffective protection) the only material identified on the steel specimens was iron oxide, while Ca^{2+} was completely absent or found in traces. Therefore, our interest focused on identifying the molecular identity of the inhibiting film at pH 7.3.

The inhibiting film (see Figure 9.28) is fairly uniform and contains Ca^{2+} and P (from externally added Ca^{2+} and HPAA), in an approximate 3 : 2 molar ratio (by EDS), suggesting a ratio of three Ca^{2+} and two $HPAA^{3-}$ ligands. FT-IR of the filming material showed multiple bands (in the 950–1200 cm^{-1} region) associated with the phosphonate groups and two bands at 1590 and 1650 cm^{-1} (assigned to the $\nu_{C=O}$ stretches), which closely match those of $Ca_3(HPAA)_2(H_2O)_{14}$. An XRD powder pattern of the Ca-HPAA anticorrosion film deposited on a carbon steel substrate is identical to that of a crystalline sample of $Ca_3(HPAA)_2(H_2O)_{14}$. This is unequivocal proof that the protective film on the steel surface is $Ca_3(HPAA)_2(H_2O)_{14}$.

9.5.5
Anticorrosion Coatings by the Materials $\{M(HPAA)(H_2O)_2\}_n$ (M = Sr, Ba)

Initial experiments were focused on exposure of carbon steel specimens to synergistic combinations of M^{2+} (Sr or Ba) and HPAA in oxygenated aqueous solutions, in a 1 : 1 ratio (under identical conditions used to prepare crystalline M-HPAA, at pH 2.0). Although the visual effect was at first encouraging (Figure 9.29, compare the "control" specimen **1** with specimens **2** and **3**), quantification of the CRs demonstrated that they were actually much higher than the "control" [70, 97].

Explanations for this lack of anticorrosion performance at low pH regions could be that the added HPAA first reacts preferentially with the Fe-oxide layer (formed almost instantaneously upon exposure of the carbon steel surface to oxygenated water) before it interacts with soluble Sr^{2+} or Ba^{2+}. Another possibility is that

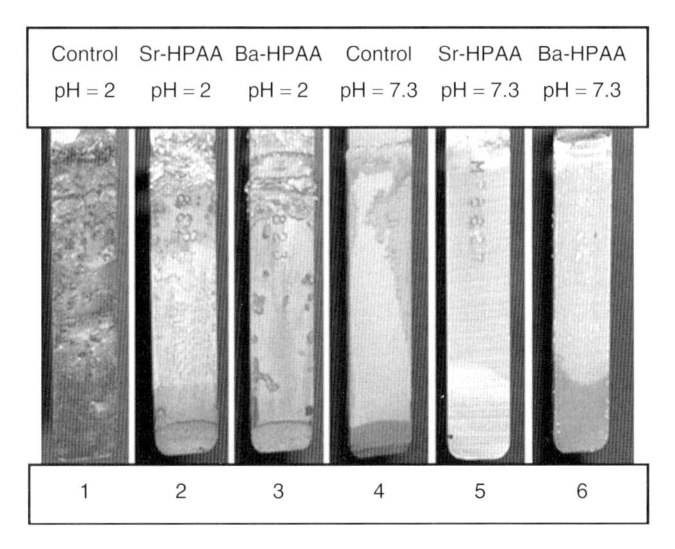

Figure 9.29 The anticorrosion effect of metal-HPAA films on carbon steel. Corrosion inhibition by metal-HPAA synergistic combinations is evident (specimens 2, 3, 5, and 6), compared to the "control" (specimens 1 and 4). Although specimens 2 and 3 (pH = 2.0) are free of iron oxides, corrosion rates are higher than the "control." Reproduced with permission from Ref. [70]. © American Chemical Society.

Sr-HPAA or Ba-HPAA compounds that may form in solution never reach the steel surface because they undergo bulk precipitation. We have discounted this scenario based on the following arguments. Indeed, we have observed white precipitates formed in the bulk in our corrosion experiments (pH 2.0) whose FT-IR, however, is distinctly different from those of authentically prepared Sr-HPAA or Ba-HPAA. These FT-IR spectra are the same as those of a Fe-HPAA material prepared at pH 2.0 using a Fe:HPAA ratio of 1 : 1, whose composition is consistent with the formula Fe(HPAA)·H_2O. A reasonable assumption is that HPAA at the surface acts as a Fe-oxide-dissolving agent. We have observed a similar behavior in similar experiments with M^{2+} (M = Ca, Zn) and 2-PBTC [68]. Owing to the ineffectiveness of the metal-HPAA materials to act as corrosion inhibitors at pH 2.0, no further experiments were pursued at that pH.

Hence, corrosion experiments were set up at higher pH 7.3. In general, CRs are lower as pH increases. This was confirmed in our "control" experiments (reduction of the CR by half, see specimens **1** and **4** in Figure 9.29). At higher pH (7.3) and in the presence of Sr^{2+} or Ba^{2+} and HPAA combinations, CRs are dramatically suppressed and corrosion inhibition reaches almost 100%. Although differentiation between the "control" and "metal-HPAA"-protected specimens is profound within the first hours, the corrosion experiments were left to proceed over a six-day period. Anticorrosion inhibitory activity was based on mass loss measurements. Anticorrosion performance was also demonstrated by visual inspection of specimens **5** and **6** that appear totally free of corrosion products. To further characterize the protective film, the corrosion specimens and film material were subjected to SEM, FT-IR, XRF, and EDS studies (Figure 9.30).

SEM images reveal a fairly uniform inhibiting film. This coating was found (by EDS, Figure 9.30, right) to contain M^{2+} (Sr or Ba from externally added salts) and P (from added HPAA), in an approximate 1 : 1 molar ratio. Fe was also present, apparently originating from the carbon steel specimen.

Furthermore, a complementary study of the inhibiting film was pursued by FT-IR spectroscopy. Figure 9.31 shows comparative FT-IR spectra of the filming material (from a corrosion experiment with Sr^{2+} and HPAA at pH 7.3) and a Sr-HPAA material that was synthesized at pH 7.3. A similar FT-IR spectrum was obtained for the Ba^{2+} + HPAA system. It is obvious that there is an excellent agreement between the two spectra.

The anticorrosion coatings composed of Sr-HPAA or Ba-HPAA function as corrosion inhibitors by reducing the cathodic current. This results in lower CRs. The films prevent oxygen diffusion toward the steel surface. This phenomenon is well known for phosphonate additives [98].

9.5.6 Anticorrosion Coatings by the Materials $\{M(PMIDA)\}_n$ (M = Ca, Sr, Ba)

PMIDA belongs to a family of phosphonates that contain both phosphonate and carboxylate moieties (see Figure 9.2). It can be considered as a close analog of AMP,

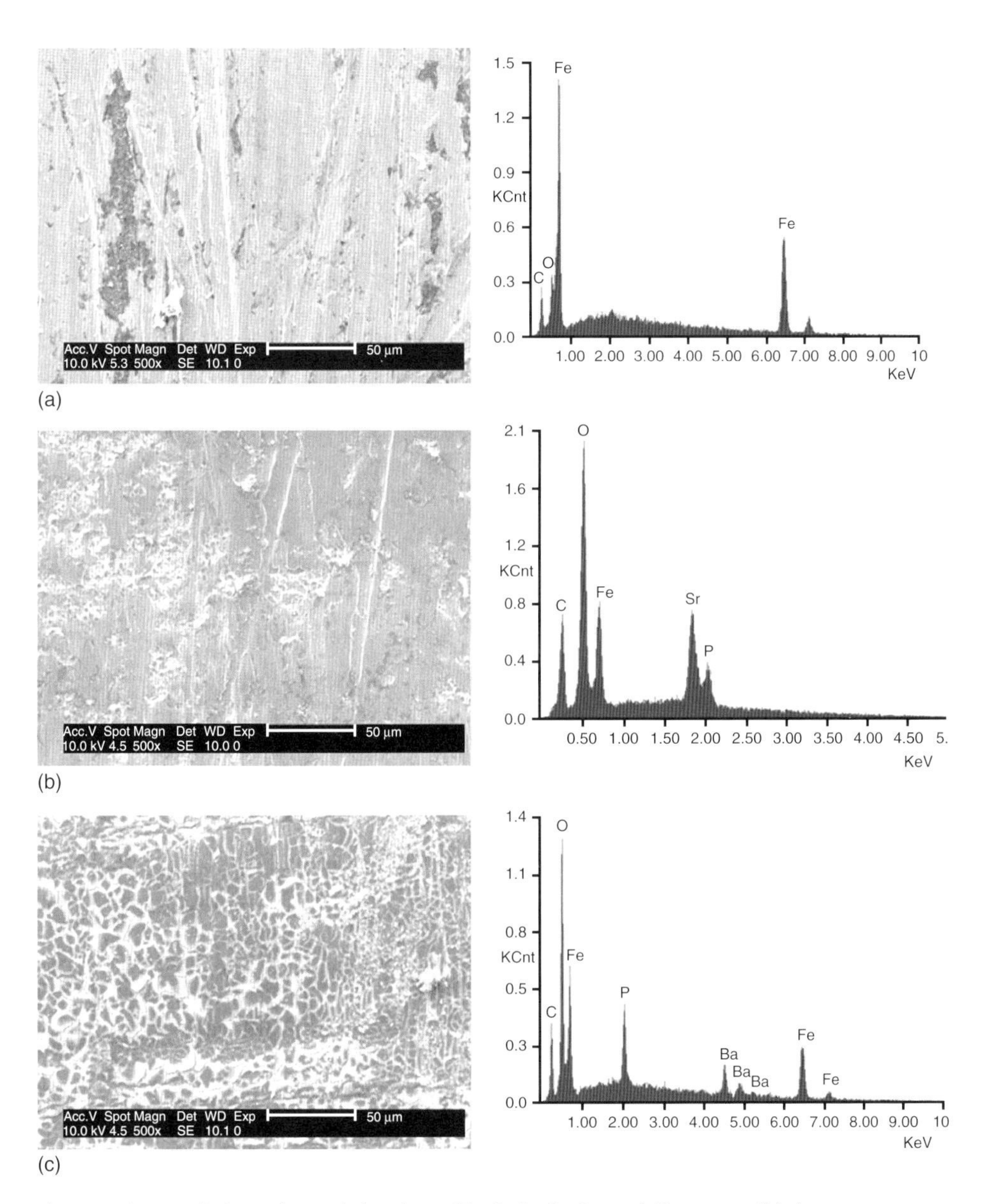

Figure 9.30 Morphology of corroded and metal-HPAA-protected steel surfaces by SEM: control (a), Sr-HPAA (b), and Ba-HPAA (c) at pH 7.3. Occasional film cracking is due to drying. Identification of film components (Fe, C, O, Sr, Ba, and P) was possible by EDS: control (a), Sr-HPAA (b), and Ba-HPAA (c). Reproduced with permission from Ref. [70]. © American Chemical Society.

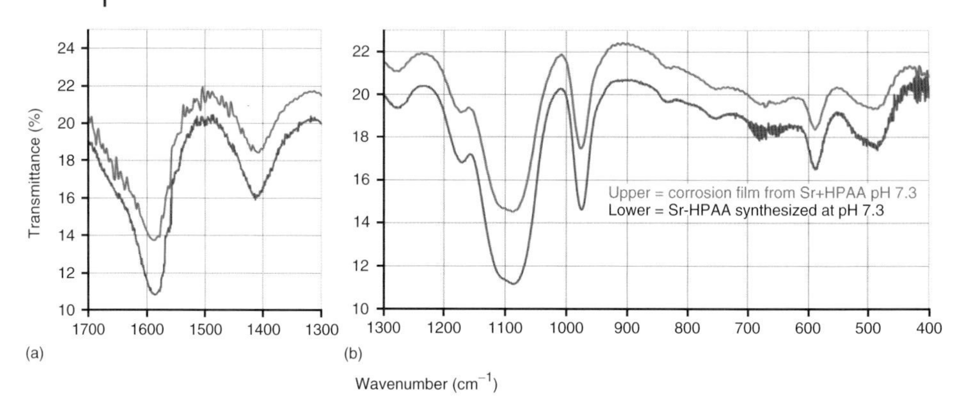

Figure 9.31 FT-IR of the anticorrosion protective film formed by combination of Sr^{2+} and HPAA at pH 7.3 and, for comparison, of Sr-HPAA synthesized at pH 7.3. (a) ν(C=O) asymmetric and symmetric stretching vibrations. (b) Vibrations associated with the phosphonate group. Reproduced with permission from Ref. [70]. © American Chemical Society.

but instead of possessing three phosphonate units, it has two carboxylate and one phosphonate groups.

A plethora of alkaline-earth metal PMIDA phosphonates (Mg, Ca, Sr, and Ba) were tested at two pH values (pH 3.4 and 6.0) and the measured CRs are shown in Table 9.1.

It is obvious that CRs are lower at the higher experimental pH of 6.0. However, upon examination of the CRs at pH 3.4, the inability of the inhibitors to reduce corrosion is evident. All CRs are higher than the control. Apparently, the inhibitor acts as a Fe-oxide remover in this case. Even at pH 6.0, only the inhibitors PMIDA (no metals), Ca-PMIDA, and Ba-PMIDA are somewhat effective. Continuation of this preliminary work will shed some light on the intricacies of this system.

Table 9.1 Corrosion rates measured with several metal-PMIDA inhibitors.

Inhibitor	**Corrosion rate (mm/yr)**	
	pH = 3.4	**pH = 6.0**
Control	0.228	0.124
PMIDA	0.437	0.089
Mg-PMIDA	0.448	0.158
Ca-PMIDA	0.387	0.087
Sr-PMIDA	0.550	0.118
Ba-PMIDA	0.414	0.084

9.5.7
A Comparative Look at the Inhibitory Performance by Metal–Phosphonate Protective Films

Metallic corrosion is a phenomenon affected by several factors. It is well established that pH is one of the major factors affecting CRs. The more acidic the fluid (water) in contact with the metal surface, the more aggressive is the corrosion. Therefore, pH plays a profound role in corrosion inhibition as well. In Table 9.2, the performance of several metal–phosphonate corrosion inhibitors has been systematically gathered, together with the pH of the inhibition experiments.

These data have been plotted in Figure 9.32. There are a number of important observations to be made. At higher pH regions (~7), most metal–phosphonate coatings perform well. At low pH regions (<3), corrosion inhibition is more challenging. This may have to do with incomplete formation of the protective coating. In addition, the high concentration of H^+ and the resulting high corrosion rates may be prevailing over the formation of the metal–phosphonate protective coating.

The results with Ca-PBTC and Zn-PBTC require further discussion. CRs in the presence of inhibitors are higher than those for the control (no inhibitor). This, at a first glance, is contrary to results obtained with several other inhibitors. This may be explained by several arguments. First, the metal–phosphonate film may not be robust, but porous in its microscopic nature. This, as mentioned before, would lead to localized attack and metal pitting. Such phenomena have not been observed upon examination of the metal specimens after the corrosion

Table 9.2 Comparative corrosion rates of carbon steel metal surfaces protected by metal–phosphonate corrosion inhibitors.

Corrosion inhibitor	pH	Corrosion rate (mm/yr)
Zn-AMP	3.0	0.900
Zn-PBTC	4.1	0.460
Ca-PBTC	4.1	1.170
Sr-HPAA	7.3	0.005
Ba-HPAA	7.3	0.005
Ca-HPAA	2.0	0.350
Ca-HPAA	7.3	0.004
Ca-EDTMP	7.0	0.028
Sr-EDTMP	7.0	0.031
Sr-HDTMP	2.2	2.200
Sr-HDTMP	7.0	0.230
Ba-HDTMP	2.2	3.100
Ba-HDTMP	7.0	0.060
Zn-HDTMP	2.2	2.110

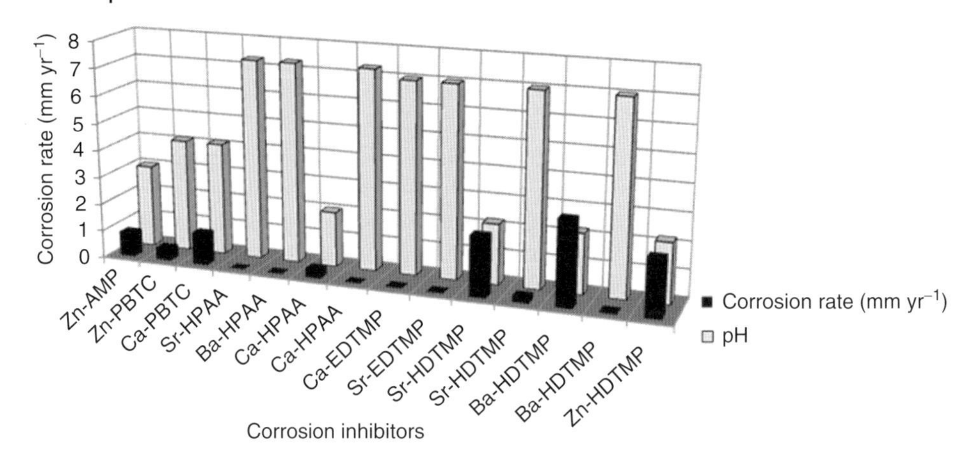

Figure 9.32 Corrosion rates of metal–phosphonate-protected surfaces as a function of pH.

experiments. Second, the metal–phosphonate (Ca or Zn-PBTC) is too soluble to deposit onto the metal surface, so it does not form a protective and anticorrosive thin film. This argument would be consistent with literature data on metal–PBTC complex formation constants (4.4 for Ca-PBTC and 8.3 for Zn-PBTC) that are considered to be very low [36b]. The difference in complex formation constants between Ca and Zn-PBTC would be consistent with the fact that Zn-PBTC is a more effective corrosion inhibitor than Ca-PBTC, as long as both inhibitors form films (albeit unstable) on the metal surface. If film formation does not take place, then CRs in the presence of Ca-PBTC or Zn-PBTC would be the same as the control, which is not the case.

Therefore, the results obtained with Ca-PBTC and Zn-PBTC indicate that these materials are soluble, and because of their acidic nature they act as metal dissolvers rather than corrosion inhibitors.

The two Zn-phosphonates have distinctly different crystal and molecular structures. The Zn-HDTMP material by virtue of its long chain linker between the two amino-bis(methylenephosphonate) moieties might be considered as a porous material. However, porosity measurements on this and the other phosphonates show absence of any porous structure. Therefore, differences in porosity cannot be invoked to explain the various anticorrosion properties of these metal–phosphonate materials.

Lastly, the ability of a metal–phosphonate corrosion inhibitor to adhere onto the metal surface plays a vital role in corrosion efficacy. Bulk precipitation of a metal–phosphonate complex will lead to loss of active inhibitors to precipitation, leading to insufficient levels for thin film formation. Surface adherence of the inhibitor films is a property that cannot be precisely predicted. However, it is a necessary condition for acceptable inhibition. In addition, the metal–phosphonate protective layer has to be robust and uniform.

9.6 Conclusions/Perspectives

Corrosion is a vast scientific field; however, it presents several facets that touch upon economics and people [99]. For example, the overall demand for corrosion inhibitors, after a rather constant rise of ~4.4%/year, reached US $1.6 billion in 2006 in the United States [3]. The petroleum-refining sector was expected to have a ~US $400 million share in this. Corrosion is an economical burden for several industry sectors. Research on the subject has been active for several decades. The solution to this complicated issue requires a multidisciplinary approach that unifies researchers from a diverse list of scientific and technological disciplines: chemistry, chemical engineering, electrochemistry, materials science and engineering, and many others.

In this chapter, our contribution to advancing solutions for corrosion issues relevant to industrial problems lies with the study of the corrosion event and its inhibition at the molecular level. In this context, we have shown that conveniently synthesized and structurally characterized organic–inorganic hybrid polymeric materials can act as protective corrosion inhibitors.

An ideal phosphonate corrosion inhibitor of the "complexing type" is required to possess the following features:

1) It must be capable of generating metal–phosphonate thin films on the surface to be protected.
2) It should not form very soluble metal complexes, because these will not eventually "deposit" onto the metal surface, but will remain soluble in the bulk.
3) It should not form sparingly soluble metal complexes because these may never reach the metal surface to achieve inhibition, but may generate undesirable deposits in the bulk or on other critical system surfaces.
4) Its metal complexes generated by controlled deposition on the metal surface must create dense thin films with robust structure. If the anticorrosion film is nonuniform or porous, then uneven oxygen permeation may create sites for localized attack, leading to pitting of the metal surface.

The popularization of Green Chemistry and Technology presents an opportunity to take corrosion-related research a step forward [100]. The quest for greener corrosion inhibitors that possess good performance characteristics, in addition to being nontoxic and environmentally friendly, is an ongoing endeavor that will certainly lead to great advances in the field [101].

References

1. Droffelaar, H. and Atkinson, J.T.N. (1995) *Corrosion and its Control*, NACE International, Houston.
2. (a)Sastri, V.S. (1998) *Corrosion Inhibitors: Principles and Applications*, John Wiley & Sons, Ltd, Chichester, p. 720; (b) Sastri, V.S. (2011) *Green corrosion inhibitors:*

Theory and practice, John Wiley & Sons, Ltd, Hoboken, NJ.
3. Editorial (2002) *Chem. Process.*, 11.B.
4. Healey, J.R. and Heath, B. (2006) BP spill highlights aging oil field's increasing problems; Environmentalists, activists call management into question. USA TODAY, August 11, B.1.
5. Javaherdashti, R. (2000) How corrosion affects industry and life. *Anti-Corros. Methods Mater.*, **47** (1), 30–34.
6. Kuznetsov, Yu.I. (1990) Role of complex-formation in corrosion inhibition. *Prot. Met.*, **26** (6), 736–744.
7. Popov, K., Rönkkömäki, H., and Lajunen, L.H.J. (2001) Critical evaluation of stability constants of phosphonic acids. *Pure Appl. Chem.*, **73** (10), 1641–1677.
8. Anastas, P.T. and Warner, J.C. (1998) *Green Chemistry: Theory and Practice*, Oxford University Press, New York.
9. Useful Information on Green Chemistry can be found in the U.S. Environmental Protection Agency, web site *http://www.epa.gov/greenchemistry* (accessed 30 July 2011).
10. Several Principles of Green Chemistry are Analyzed in the Canadian Green Chemistry Network, *http://www.greenchemistry.ca* (accessed 30 July 2011).
11. Einav, R., Harussi, K., and Perry, D. (2002) The footprint of the desalination processes on the environment. *Desalination*, **152** (1–3), 141–154.
12. Demadis, K.D., Mavredaki, E., Stathoulopoulou, A., Neofotistou, E., and Mantzaridis, C. (2007) Industrial water systems: problems, challenges and solutions for the process industries. *Desalination*, **213** (1–3), 38–46.
13. Abdel-Aal, N. and Sawada, K. (2003) Inhibition of adhesion and precipitation of $CaCO_3$ by aminopolyphosphonate. *J. Cryst. Growth*, **256** (1–2), 188–200.
14. Demadis, K.D., Neofotistou, E., Mavredaki, E., and Stathoulopoulou, A. (2005) Proceedings of the 10th European Symposium on Corrosion and Scale Inhibitors (10 SEIC), Annali dell Universita di Ferrara, N.S., Sezione V, Suppl. No. 12, August 29 –September 2, 2005, p. 451.
15. *http://www.ospar.org* (accessed 30 July 2011).
16. (a) Clearfield, A. (1998) Organically pillared micro- and mesoporous materials. *Chem. Mater.*, **10** (10), 2801–2810; (b) Maeda, K. (2004) Metal phosphonate open-framework materials. *Microporous Mesoporous Mater.*, **73** (1–2), 47–55; (c) Demadis, K.D. (2007) in *Solid State Chemistry Research Trends* (ed. R.W. Buckley), Nova Science Publishers, New York, pp. 109–172; (d) Sharma, C.V.K. and Clearfield, A. (2000) Three-dimensional hexagonal structures from a novel self-complementary molecular building block. *J. Am. Chem. Soc.*, **122** (18), 4394–4402; (e) Clearfield, A. (1998) Metal phosphonate chemistry. *Prog. Inorg. Chem.*, **47**, 371–510.
17. (a) Vioux, A., Le Bideau, L., Hubert Mutin, P., and Leclercq, D. (2004) Hybrid organic-inorganic materials based on organophosphorus derivatives. *Top. Curr. Chem.*, **232**, 145–174; (b) Mahmoudkhani, A.H. and Langer, V. (2002) Self-assemblies of extended hydrogen-bonded arrays using 1,4-butanebisphosphonic acid as a versatile building block. *Phosphorus, Sulfur Silicon*, **177** (12), 2941–2951; (c) Mahmoudkhani, A.H. and Langer, V. (2002) Phenylphosphonic acid as a building block for two-dimensional hydrogen-bonded supramolecular arrays. *J. Mol. Struct.*, **609** (1–3), 97–108; (d) Lazar, A.N., Navaza, A., and Coleman, A.W. (2004) Solid-state caging of 1,10-phenanthroline pi-pi stacked dimers by calix[4]arene dihydroxyphosphonic acid. *Chem. Commun.*, (9), 1052–1053; (e) Mahmoudkhani, A.H. and Langer, V. (2002) Supramolecular isomerism and isomorphism in the structures of 1,4-butanebisphosphonic acid and its organic ammonium salts. *Cryst. Growth Des.*, **2** (1), 21–25.
18. (a) Du, Z.-Y., Prosvirin, A.V., and Mao, J.-G. (2007) Novel Manganese(II) sulfonate-phosphonates with dinuclear, tetranuclear, and hexanuclear clusters. *Inorg. Chem.*, **46** (23), 9884–9894; (b) Du, Z.-Y., Xu, H.-B., and Mao, J.-G. (2006) Three novel zinc(II)

sulfonate-phosphonates with tetranuclear or hexanuclear cluster units. *Inorg. Chem.*, **45** (16), 6424–6430; (c) Yang, B.-P. and Mao, J.-G. (2005) New types of metal squarato-phosphonates: condensation of aminophosphonate with squaric acid under hydrothermal conditions. *Inorg. Chem.*, **44** (3), 566–571; (d) Lei, C., Mao, J.-G., Sun, Y.-Q., Zeng, H.-Y., and Clearfield, A. (2003) $\{Zn_6[MeN(CH_2CO_2)(CH_2PO_3)]_6(Zn)\}^{4-}$ anion: the first example of the oxo-bridged Zn-6 octahedron with a centered Zn(II) cation. *Inorg. Chem.*, **42** (20), 6157–6159.

19. (a) Cheetham, A.K., Ferey, G., and Loiseau, T. (1999) Open-framework inorganic materials. *Angew. Chem. Int. Ed.*, **38** (22), 3268–3292; (b) Forster, P.M. and Cheetham, A.K. (2003) Hybrid inorganic-organic solids: an emerging class of nanoporous catalysts. *Top. Catal.*, **24** (1–4), 79–86; (c) Merrill, C.A. and Cheetham, A.K. (2007) Inorganic-organic framework structures; M(II) ethylenediphosphonates (M = Co, Ni, Mn) and a Mn(II) ethylenediphosphonato-phenanthroline. *Inorg. Chem.*, **46** (1), 278–284; (d) Jhung, S.H., Yoon, J.W., Hwang, J.-S., Cheetham, A.K., and Chang, J.-S. (2005) Facile synthesis of nanoporous nickel phosphates without organic templates under microwave irradiation. *Chem. Mater.*, **17**, 4455–4460; (e) Merrill, C.A. and Cheetham, A.K. (2005) Pillared layered structures based upon M(III) ethylene diphosphonates: the synthesis and crystal structures of $M^{III}(H_2O)(HO_3P(CH_2)_2PO_3)$ (M = Fe, Al, Ga). *Inorg. Chem.*, **44** (15), 5273–5277.

20. (a) Gomez-Alcantara, M.M., Cabeza, A., Martinez-Lara, M., Aranda, M.A.G., Suau, R., Bhuvanesh, N., and Clearfield, A. (2004) Synthesis and characterization of a new bisphosphonic acid and several metal hybrids derivatives. *Inorg. Chem.*, **43** (17), 5283–5293; (b) Cabeza, A., Xiang, O.Y., Sharma, C.V.K., Aranda, M.A.G., Bruque, S., and Clearfield, A. (2002) Complexes formed between nitrilotris(methylenephosphonic acid) and M^{2+} transition metals: Isostructural organic-inorganic hybrids. *Inorg. Chem.*, **41** (9), 2325–2333; (c) Cabeza, A., Aranda, M.A.G., and Bruque, S. (1999) New lead triphosphonates: synthesis, properties and crystal structures. *J. Mater. Chem.*, **9** (2), 571–578.

21. (a) Turner, A., Jaffres, P.-A., MacLean, E.J., Villemin, D., McKee, V., and Hix, G.B. (2003) Hydrothermal synthesis and crystal structure of two Co phosphonates containing trifunctional phosphonate anions: $Co_3(O_3PCH_2NH_2CH_2PO_3)_2$ and $Co_3(O_3PCH_2NC_4H_7CO_2)_2{\cdot}5H_2O$. *J. Chem. Soc. Dalton Trans.*, (7), 1314–1319; (b) Hix, G.B., Wragg, D.S., Wright, P.A., and Morris, R.E. (1998) Synthesis and characterisation of $Al(O_3PCH_2CO_2){\cdot}3H_2O$, a layered aluminium carboxymethylphosphonate. *J. Chem. Soc. Dalton Trans.*, (20), 3359–3361; (c) Turner, A., Kariuki, B.M., Tremayne, M., and MacLean, E.J. (2002) Strategies for the synthesis of porous metal phosphonate materials. *J. Mater. Chem.*, **12** (11), 3220–3227; (d) Zakowsky, N., Hix, G.B., and Morris, R.E. (2000) Synthesis of a family of aluminium benzylphosphonates. *J. Mater. Chem.*, **10** (10), 2375–2380.

22. (a) Stock, N. and Bein, T. (2004) High-throughput synthesis of phosphonate based inorganic-organic hybrid compounds under hydrothermal conditions. *Angew. Chem. Int. Ed.*, **43** (6), 749–752; (b) Bauer, S. and Stock, N. (2007) Implementation of a temperature-gradient reactor system for high-throughput investigation of phosphonate-based inorganic-organic hybrid compounds. *Angew. Chem. Int. Ed.*, **46** (36), 6857–6860; (c) Forster, P.M., Stock, N., and Cheetham, A.K. (2005) A high-throughput investigation of the role of pH, temperature, concentration, and time on the synthesis of hybrid inorganic-organic materials. *Angew. Chem. Int. Ed.*, **44** (46), 7608–7611; (d) Bauer, S., Bein, T., and Stock, N. (2006) Inorganic-organic hybrid compounds: Synthesis and characterization of three new metal phosphonates with similar characteristic structural features. *J. Solid State Chem.*, **179** (1), 145–155.

23. (a) Breeze, B.A., Shanmugam, M., Tuna, F., and Winpenny, R.E.P. (2007) A series of nickel phosphonate-carboxylate cages. *Chem. Commun.*, (48), 5185–5187; (b) Langley, S., Helliwell, M., Raftery, J., Tolis, E.I., and Winpenny, R.E.P. (2004) Phosphonate ligands encourage a Platonic relationship between cobalt(II) and alkali metal ions. *Chem. Commun.*, (2), 142–143; (c) Baskar, V., Shanmugam, M., Sañudo, E.C., Shanmugam, M., Collison, D., McInnes, E.J.L., Wei, Q., and Winpenny, R.E.P. (2007) Metal cages using a bulky phosphonate as a ligand. *Chem. Commun.*, (1), 37–39; (d) Harrison, A., Henderson, D.K., Lovatt, P.A., Parkin, A., Tasker, P.A., and Winpenny, R.E.P. (2003) Synthesis, structure and magnetic properties of $[Cu_4(Hmbpp)_2(H_2NC(O)NH_2)_2(H_2O)_8]\cdot 4H_2O$. *Dalton Trans.*, (22), 4271–4274.
24. (a) Serre, C., Groves, J.A., Lightfoot, P., Slawin, A.M.Z., Wright, P.A., Stock, N., Bein, T., Haouas, M., Taulelle, F., and Ferey, G. (2006) Synthesis, structure and properties of related microporous N,N′-piperazine-*bis*methylene phosphonates of aluminum and titanium. *Chem. Mater.*, **18** (6), 1451–1457; (b) Ferey, G., Mellot-Draznieks, C., Serre, C., and Millange, F. (2005) Crystallized frameworks with giant pores: are there limits to the possible?. *Acc. Chem. Res.*, **38** (4), 217–225; (c) Serre, C., Lorentz, C., Taulelle, F., and Ferey, G. (2003) Hydrothermal synthesis of nanoporous metalofluorophosphates. 2. In situ and ex situ F-19 and P-31 NMR of nano- and mesostructured titanium phosphates crystallogenesis. *Chem. Mater.*, **15** (12), 2328–2337; (d) Barthelet, K., Nogues, M., Riou, D., and Ferey, G. (2002) Hydrothermal synthesis, structure determination, and magnetic properties of three new copper(II) methylenediphosphonates with hybrid frameworks (MIL-54, 55, 56), and of the Cu homologue of $Na_2Co(O_3PCH_2PO_3)\cdot(H_2O)$. *Chem. Mater.*, **14** (12), 4910–4918.
25. (a) Alberti, G., Casciola, M., Costantino, U., and Vivani, R. (1996) Layered and pillared metal(IV) phosphates and phosphonates. *Adv. Mater.*, **8** (4), 291–303; (b) Vivani, R., Alberti, G., Costantino, F., and Nocchetti, M. (2008) New advances in zirconium phosphate and phosphonate chemistry: structural archetypes. *Microporous Mesoporous Mater.*, **107** (1–2), 58–70; (c) Vivani, R., Costantino, F., Costantino, U., and Nocchetti, M. (2006) New architectures for zirconium polyphosphonates with a tailor-made open-framework structure. *Inorg. Chem.*, **45** (6), 2388–2399; (d) Costantino, U., Nocchetti, M., and Vivani, R. (2002) Preparation, characterization, and structure of zirconium fluoride alkylamino-N,N'-*bis*methylphosphonates: a new design for layered zirconium diphosphonates with a poorly hindered interlayer region. *J. Am. Chem. Soc.*, **124** (28), 8428–8434.
26. Demadis, K.D. and Katarachia, S.D. (2004) Metal-phosphonate chemistry: synthesis, crystal structure of calcium-amino-*tris*-(methylenephosphonate) and inhibition of $CaCO_3$ crystal growth. *Phosphorus, Sulfur Silicon*, **179** (3), 627–648.
27. Demadis, K.D. and Lykoudis, P. (2005) Chemistry of organophosphonate scale growth inhibitors: 3. Physicochemical aspects of 2-phosphonobutane-1,2,4-tricarboxylate (PBTC) and its effect on $CaCO_3$ crystal growth. *Bioinorg. Chem. Appl.*, **3** (3–4), 135–149.
28. Demadis, K.D. (2006) Chemistry of organophosphonate scale inhibitors, Part 4: stability of amino-*tris*(methylenephosphonate) towards degradation by oxidizing biocides. *Phosphorus, Sulfur Silicon*, **181** (1), 167–176.
29. Demadis, K.D. and Mavredaki, E. (2005) Green additives to enhance silica dissolution during water treatment. *Environ. Chem. Lett.*, **3** (3), 127–131.
30. (a) Dyer, S.J., Anderson, C.E., and Graham, G.M. (2004) Thermal stability of amine methyl phosphonate scale inhibitors. *J. Pet. Sci. Eng.*, **43** (3–4), 259–270; (b) Oddo, J.E. and Tomson, M.B. (1990) The solubility and stoichiometry of calcium-diethylenetriaminepenta

(methylenephosphonate) at 70 °C in brine solutions at 4.7 and 5.0 pH. *Appl. Geochem.*, **5** (4), 527–532; (c) Xiao, J.J., Kan, A.T., and Tomson, M.B. (2001) Prediction of $BaSO_4$ precipitation in the presence and absence of a polymeric inhibitor: phosphino-polycarboxylic acid. *Langmuir*, **17** (15), 4668–4673; (d) Friedfeld, S.J., He, S., and Tomson, M.B. (1998) The temperature and ionic strength dependence of the solubility product constant of ferrous phosphonate. *Langmuir*, **14**, 3698–3703.

31. (a) Tantayakom, V., Fogler, H.S., Charoensirithavorn, P., and Chavadej, S. (2005) Kinetic study of scale inhibitor precipitation in squeeze treatments. *Cryst. Growth Des.*, **5** (1), 329–335; (b) Browning, F.H. and Fogler, H.S. (1996) Fundamental study of the dissolution of calcium phosphonates from porous media. *AIChE J.*, **42** (10), 2883–2896; (c) Pairat, R., Sumeath, C., Browning, F.H., and Fogler, H.S. (1997) Precipitation and dissolution of calcium-ATMP precipitates for the inhibition of scale formation in porous media. *Langmuir*, **13** (6), 1791–1798; (d) Tantayakom, V., Fogler, H.S., de Moraes, F.F., Bualuang, M., Chavadej, S., and Malakul, P. (2004) Study of Ca-ATMP precipitation in the presence of magnesium ion. *Langmuir*, **20** (6), 2220–2226.

32. (a) Penard, A.-L., Rossignol, F., Nagaraja, H.S., Pagnoux, C., and Chartier, T. (2005) Dispersion of alpha-alumina ultrafine powders using 2-phosphonobutane-1,2,4-tricarboxylic acid for the implementation of a DCC process. *Eur. J. Ceram. Soc.*, **25** (7), 1109–1118; (b) Pearse, M.J. (2005) An overview of the use of chemical reagents in mineral processing. *Miner. Eng.*, **18** (2), 139–149.

33. (a) Sekine, I., Shimode, T., and Yuasa, M. (1992) Corrosion inhibition of structural-steels in CO_2 absorption process by organic inhibitor composed of 2-aminothiophenol, (1-hydroxyethylidene)bis(phosphonic acid), and diethanolamine. *Ind. Eng. Chem. Res.*, **31** (1), 434–439; (b) Mosayebi, B., Kazemeini, M., and Badakhshan, A. (2002) Effect of phosphonate based corrosion inhibitors in a cooling water system. *Br. Corros. J.*, **37** (3), 217–224.

34. (a) Kouznetsov, Yu.I. (2001) Role of the complexation concept in the present views on the initiation and inhibition of metal pitting. *Prot. Met.*, **37** (5), 434–439.

35. (a) Fang, J.L., Li, Y., Ye, X.R., Wang, Z.W., and Liu, Q. (1993) Passive films and corrosion protection due to phosphonic acid inhibitors. *Corrosion*, **49** (4), 266–271; (b) Paszternák, A., Stichleutner, S., Felhosi, I., Keresztes, Z., Nagy, F., Kuzmann, E., Vértes, A., Homonnay, Z., Peto, G., and Kálmán, E. (2007) Surface modification of passive iron by alkyl-phosphonic acid layers. *Electrochim. Acta*, **53** (2), 337–345.

36. (a) Nowack, B. and Van Briessen, J.M. (eds) (2003) *Biogeochemistry of Chelating Agents*, ACS Symposium Series, Vol. 910, ACS, Washington, DC ; (b) Knepper, T.P. (2003) Synthetic chelating agents and compounds exhibiting complexing properties in the aquatic environment. *Trends Anal. Chem.*, **22** (10), 708–724.

37. (a) Miyazaki, K., Horibe, T., Antonucci, J.M., Takagi, S., and Chow, L.C. (1993) Polymeric calcium-phosphate cements – setting reaction modifiers. *Dent. Mater.*, **9** (1), 46–50; (b) Atai, M., Nekoomanesh, M., Hashemi, S.A., and Amani, S. (2004) Physical and mechanical properties of an experimental dental composite based on a new monomer. *Dent. Mater.*, **20** (7), 663–668; (c) Nicholson, J.W. and Singh, G. (1996) The use of organic compounds of phosphorus in clinical dentistry. *Biomaterials*, **17** (21), 2023–2030; (d) Tschernitschek, H., Borchers, L., and Geurtsen, W. (2005) Nonalloyed titanium as a bioinert metal – A review. *Quintessence int.*, **36** (7–8), 523–530.

38. Cheng, F. and Oldfield, E. (2004) Inhibition of isoprene biosynthesis pathway enzymes by phosphonates, bisphosphonates, and diphosphates. *J. Med. Chem.*, **47** (21), 5149–5158.

39. Temperini, C., Innocenti, A., Guerri, A., Scozzafava, A., Rusconi, S., and

Supuran, C.T. (2007) Phosph(on)ate as a zinc-binding group in metalloenzyme inhibitors: X-ray crystal structure of the antiviral drug foscarnet complexed to human carbonic anhydrase I. *Bioorg. Med. Chem. Lett.*, **17** (8), 2210–2215.

40. Davini, E., Di Leo, C., Norelli, F., and Zappelli, P. (1993) Synthesis and applications of phosphonoacetic derivatives. *J. Biotechnol.*, **28** (2–3), 321–338.

41. (a) Bottrill, M., Kwok, L., and Long, N.J. (2006) Lanthanides in magnetic resonance imaging. *Chem. Soc. Rev.*, **35** (6), 557–571; (b) Finlay, I.G., Mason, M.D., and Shelley, M. (2005) Radioisotopes for the palliation of metastatic bone cancer: a systematic review. *Lancet Oncol.*, **6** (6), 392–400.

42. Kubicek, V., Rudovsky, J., Kotek, J., Hermann, P., Vander Elst, L., Muller, R.N., Kolar, Z.I., Wolterbeek, H.T., Peters, J.A., and Lukeš, I. (2005) A bisphosphonate monoamide analogue of DOTA: a potential agent for bone targeting. *J. Am. Chem. Soc.*, **127** (47), 16477–16485.

43. Kung, H., Ackerhalt, R., and Blau, M. (1978) Uptake of Tc-99m monophosphate complexes in bone and myocardial necrosis in animals. *J. Nucl. Med.*, **19** (9), 1027–1031.

44. (a) Padalecki, S.S. and Guise, T.A. (2001) The role of bisphosphonates in breast cancer: actions of bisphosphonates in animal models of breast cancer. *Breast Cancer Res.*, **4** (1), 35–41; (b) Stresing, V., Daubiné, F., Benzaid, I., Mönkkönen, H., and Clézardin, P. (2007) Bisphosphonates in cancer therapy. *Cancer Lett.*, **257** (1), 16–35; (c) Layman, R., Olson, K., and Van Poznak, C. (2007) Bisphosphonates for breast cancer: questions answered, questions remaining. *Hematol. Oncol. Clin. North Am.*, **21** (2), 341–367.

45. (a) Clearfield, A. (2002) Recent advances in metal phosphonate chemistry II. *Curr. Opin. Solid State Mater. Sci.*, **6** (6), 495–506; (b) Clearfield, A. (1996) Recent advances in metal phosphonate chemistry. *Curr. Opin. Solid State Mater. Sci.*, **1** (2), 268–278; (c) Barouda, E., Demadis, K.D., Freeman, S., Jones, F., and Ogden, M.I. (2007) Barium sulfate crystallization in the presence of variable chain length aminomethylenetetraphosphonates and cations (Na^+ or Zn^{2+}). *Cryst. Growth Des.*, **7** (2), 321–327.

46. See *http://www.dequest.com* (accessed 30 July 2011).

47. (a) Silvestre, J.-P., Dao, N.Q., and Salvini, P. (2002) Crystal structure of ethylenediammoniumhydroxyethylidenebisphosphonate. *Phosphorus, Sulfur Silicon*, **177** (4), 771–779; (b) Clearfield, A., Krishnamohan Sharma, C.V., and Zhang, B. (2001) Crystal engineered supramolecular metal phosphonates: crown ethers and iminodiacetates. *Chem. Mater.*, **13** (10), 3099–3112; (c) Dines, M.B., Cooksey, R.E., Griffith, P.C., and Lane, R.H. (1983) Mixed-component layered tetravalent metal phosphonates phosphates as precursors for microporous materials. *Inorg. Chem.*, **22** (6), 1003–1004; (d) Yang, H.C., Aoki, K., Hong, H.-G., Sackett, D.D., Arendt, M.F., Yau, S.-L., Bell, C.M., and Mallouk, T.E. (1993) Growth and characterization of metal(ii) alkanebisphosphonate multilayer thin-films on gold surfaces. *J. Am. Chem. Soc.*, **115** (25), 11855–11862; (e) Penicaud, V., Massiot, D., Gelbard, G., Odobel, F., and Bujoli, B. (1998) Preparation of structural analogues of divalent metal monophosphonates, using bis(phosphonic) acids: a new strategy to reduce overcrowding of organic groups in the interlayer space. *J. Mol. Struct.*, **470** (1–2), 31–38; (f) Serre, C. and Ferey, G. (1999) Hybrid open frameworks. 8. Hydrothermal synthesis, crystal structure, and thermal behavior of the first three-dimensional titanium(IV) diphosphonate with an open structure: $Ti_3O_2(H_2O)_2(O_3P(CH_2)PO_3)_2{\cdot}(H_2O)_2$, or MIL-22. *Inorg. Chem.*, **38** (23), 5370–5373; (g) Serpaggi, S. and Ferey, G. (1998) Hybrid open frameworks (MIL-n). Part 6 – Hydrothermal synthesis and X-ray powder ab initio structure determination of MIL-11, a series of lanthanide organodiphosphonates with three-dimensional networks, $Ln(III)H[O_3P(CH_2)_nPO_3](n = 1–3)$. *J. Mater. Chem.*, **8** (12), 2749–2755; (h) Distler, A., Lohse, D.L., and

Sevov, S.C. (1999) Chains, planes, and tunnels of metal diphosphonates: synthesis, structure, and characterization of $Na_3Co(O_3PCH_2PO_3)(OH)$, $Na_3Mg(O_3PCH_2PO_3)F{\cdot}H_2O$, $Na_2Co(O_3PCH_2PO_3){\cdot}H_2O$, $NaCo_2(O_3PCH_2CH_2CH_2PO_3)(OH)$, and $Co_2(O_3PCH_2PO_3)(H_2O)$. *J. Chem. Soc., Dalton Trans.*, (11), 1805–1812; (i) Poojary, D.M., Zhang, B., and Clearfield, A. (1997) Pillared layered metal phosphonates. Syntheses and X-ray powder structures of copper and zinc alkylenebis(phosphonates). *J. Am. Chem. Soc.*, **119** (51), 12550–12559; (j) Poojary, D.M., Zhang, B., Belling-Hausen, P., and Clearfield, A. (1996) Synthesis and x-ray powder structures of two lamellar copper arylenebis(phosphonates). *Inorg. Chem.*, **35** (17), 4942–4949; (k) Alberti, G., Vivani, R., and Murcia Mascaros, S. (1998) First structural determination of layered and pillared organic derivatives of γ-zirconium phosphate by X-ray powder diffraction data. *J. Mol. Struct.*, **470** (1–2), 81–92; (l) Alberti, G., Marcia-Mascaros, S., and Vivani, R. (1998) Pillared derivatives of γ-zirconium phosphate containing non-rigid alkyl chain pillars. *J. Am. Chem. Soc.*, **120** (36), 9291–9295.

48. (a) Klepetsanis, P.G. and Koutsoukos, P.G. (1998) Kinetics of calcium sulfate formation in aqueous media: effect of organophosphorus compounds. *J. Cryst. Growth*, **193** (1–2), 156–163; (b) Harmandas, N.G., Navarro Fernandez, E., and Koutsoukos, P.G. (1998) Crystal growth of pyrite in aqueous solutions. Inhibition by organophosphorus compounds. *Langmuir*, **14** (5), 1250–1255; (c) Zieba, A., Sethuraman, G., Perez, F., Nancollas, G.H., and Cameron, D. (1996) Influence of organic phosphonates on hydroxyapatite crystal growth kinetics. *Langmuir*, **12** (11), 2853–2858; (d) Reddy, M.M. and Nancollas, G.H. (1973) Calcite crystal growth inhibition by phosphonates. *Desalination*, **12** (1), 61–73; (e) Bochner, R.A., Abdul-Rahman, A., and Nancollas, G.H. (1984) Crystal growth of strontium fluoride from aqueous solution. *J. Chem. Soc. Faraday Trans. I*, **80**, 217–224; (f) Drela, I., Falewicz, P., and Kuczkowska, S. (1998) New rapid test for evaluation of scale inhibitors. *Water Res.*, **32** (10), 3188–3191.

49. Carter, R.P., Carroll, R.L., and Irani, R.R. (1967) Nitrilotri(methylenephosphonic acid), ethyliminodi-(methylenephosphonic acid), and diethylaminomethylphosphonic acid: acidity and calcium(II) and magnesium(II) complexing. *Inorg. Chem.*, **6** (5), 939–942.

50. Sergienko, V.S. (2001) Structural chemistry of 1-hydroxyethylidenediphosphonic acid complexes. *Russ. J. Coord. Chem.*, **27** (10), 681–710.

51. Bollinger, J.E. and Roundhill, D.M. (1993) Complexation of indium(iii), gallium(iii), iron(iii), gadolinium(iii), and neodymium(iii) ions with amino diphosphonic acids in aqueous-solution. *Inorg. Chem.*, **32** (13), 2821–2826.

52. Kortz, U. and Pope, M.T. (1995) Polyoxometalate-diphosphate complexes. 4. Structure of $Na_4[(O_3PCHN(CH_3)_2PO_3)W_2O_6]{\cdot}11H_2O$. *Inorg. Chem.*, **34** (14), 3848–3850.

53. Demadis, K.D., Katarachia, S.D., and Koutmos, M. (2005) Crystal growth and characterization of zinc–(amino–*tris*(methylenephosphonate)) organic–inorganic hybrid networks and their inhibiting effect on metallic corrosion. *Inorg. Chem. Commun.*, **8** (3), 254–258.

54. Demadis, K.D., Katarachia, S.D., Zhao, H., Raptis, R.G., and Baran, P. (2006) Alkaline earth metal organotriphosphonates: inorganic-organic polymeric hybrids from dication-dianion association. *Cryst. Growth Des.*, **6** (4), 836–838.

55. Sharma, C.V.K., Clearfield, A., Cabeza, A., Aranda, M.A.G., and Bruque, S. (2001) Deprotonation of phosphonic acids with M^{2+} cations for the design of neutral isostructural organic-inorganic hybrids. *J. Am. Chem. Soc.*, **123** (12), 2885–2886.

56. Daly, J.J. and Wheatley, P.J. (1967) The crystal and molecular structure of nitrilotrimethylene triphosphonic acid. *J. Chem. Soc. A*, 212–221.

57. Sagatys, D.A., Dahlgren, C., Smith, G., Bott, R.C., and Willis, A.C. (2000) Metal

complexes with N-(phosphonomethyl) glycine (glyphosate): the preparation and characterization of the group 2 metal complexes with glyphosate and the crystal structure of barium glyphosate dihydrate. *Aust. J. Chem.*, **53** (2), 77–81.

58. Demadis, K.D., Mantzaridis, C., Raptis, R.G., and Mezei, G. (2005) Metal–organotetraphosphonate inorganic–organic hybrids: crystal structure and anticorrosion effects of zinc–(hexamethylenediamine–*tetrakis* (methylene phosphonate)) on carbon steels. *Inorg. Chem.*, **44** (13), 4469–4471.

59. (a) Gomez-Alcantara, M., Cabeza, A., Martinez-Lara, M., Aranda, M.A.G., Suau, R., Bhuvanesh, N., and Clearfield, A. (2004) Synthesis and characterization of a new bisphosphonic acid and several metal hybrids derivatives. *Inorg. Chem.*, **43** (17), 5283–5293; (b) Song, H.-H., Zheng, L.-M., Wang, Z., Yan, C.-H., and Xin, X.-Q. (2001) Zinc diphosphonates templated by organic amines: syntheses and characterizations of $[NH_3(CH_2)_2NH_3]Zn(hedpH_2)_2{\cdot}2H_2O$ and $[NH_3(CH_2)_nNH_3]Zn_2(hedpH)_2{\cdot}2H_2O$ (n = 4, 5, 6) (hedp = 1-hydroxyethylidenediphosphonate). *Inorg. Chem.*, **40** (19), 5024–5029; (c) Drumel, S., Janvier, P., Deniaud, D., and Bujoli, B. (1995) Synthesis and crystal-structure of $Zn(O_3PC_2H_4NH_2)$, the first functionalized zeolite-like phosphonate. *J. Chem. Soc., Chem. Commun.*, (10), 1051–1052; (d) Hartman, S.J., Todorov, E., Cruz, C., and Sevov, S.C. (2005) Frameworks of amino acids: synthesis and characterization of two zinc phosphono-amino-carboxylates with extended structures. *Chem. Commun.*, (13), 1213–1214; (e) Mao, J.-G. and Clearfield, A. (2002) Metal carboxylate-phosphonate hybrid layered compounds: synthesis and single crystal structures of novel divalent metal complexes with N-(phosphonomethyl)iminodiacetic acid. *Inorg. Chem.*, **41** (9), 2319–2324; (f) Yang, B.-P., Mao, J.-G., Sun, Y.-Q., Zhao, H.-H., and Clearfield, A. (2003) Syntheses, characterizations, and crystal structures of three new metal phosphonocarboxylates with a layered and a microporous structure. *Eur. J. Inorg. Chem.*, (23), 4211–4217.

60. Bishop, M., Bott, S.G., and Barron, A.G. (2003) A new mechanism for cement hydration inhibition: solid-state chemistry of calcium nitrilotris(methylene)triphosphonate. *Chem. Mater.*, **15** (16), 3074–3088.

61. Stock, N., Stoll, A., and Bein, T. (2004) A new calcium tetraphosphonate containing small pores, $Ca[(HO_3PCH_2)_2N(H)CH_2C_6H_4CH_2N(H)(CH_2PO_3H)_2]{\cdot}2H_2O$. *Microporous Mesoporous Mater.*, **69** (1–2), 65–69.

62. Zheng, G.-L., Ma, J.-F., and Yang, J. (2004) Synthesis and crystal structure of a novel cobalt phosphonate containing the $[Co(H_2O)_6]^{2+}$ cation. *J. Chem. Res.*, (6), 387–388.

63. Demadis, K.D., Barouda, E., Zhao, H., and Raptis, R.G. (2009) Structural architectures of charge-assisted, hydrogen-bonded, 2D layered amine···tetraphosphonate and zinc···tetraphosphonate ionic materials. *Polyhedron*, **28** (15), 3361–3367.

64. Lee, B.H., Lynch, V.M., Cao, G., and Mallouk, T.E. (1988) Structure of $[Mg\{HO_3PCH(C_6H_5)_2\}_2]{\cdot}8H_2O$, a layered phosphonate salt. *Acta Cryst.*, **C44**, 365–367.

65. Demadis, K.D., Barouda, E., Stavgianoudaki, N., and Zhao, H. (2009) Inorganic-organic hybrid molecular ribbons based on chelating/bridging, "Pincer" tetraphosphonates, and alkaline-earth metals. *Cryst. Growth Des.*, **9** (3), 1250–1253.

66. Paz, F.A., Shi, F.-N., Klinowski, J., Rocha, J., and Trindade, T. (2004) Synthesis and characterisation of the first three-dimensional mixed-metal-center inorganic-organic hybrid framework with n-(phosphonomethyl) iminodiacetate. *Eur. J. Inorg. Chem.*, (13), 2759–2768.

67. Demadis, K.D. and Baran, P. (2004) Chemistry of organophosphonate scale growth inhibitors: two-dimensional, layered polymeric networks in the structure of tetrasodium 2-hydroxyethyl-amino-*bis*(methylenephosphonate). *J. Solid State Chem.*, **177** (12), 4768–4776.

68. Demadis, K.D., Lykoudis, P., Raptis, R.G., and Mezei, G. (2006) Phosphonopolycarboxylates as chemical additives for calcite scale dissolution and metallic corrosion inhibition based on a calcium-phosphonotricarboxylate organic-inorganic hybrid. *Cryst. Growth Des.*, **6** (5), 1064–1067.
69. Colodrero, R.M.P., Cabeza, A., Olivera-Pastor, P., Infantes-Molina, A., Barouda, E., Demadis, K.D., and Aranda, M.A.G. (2009) "Breathing" in adsorbate-responsive metal tetraphosphonate hybrid materials. *Chem. Eur. J.*, **15** (27), 6612–6618.
70. Demadis, K.D., Papadaki, M., Raptis, R.G., and Zhao, H. (2008) Corrugated, sheet-like architectures in layered alkaline-earth metal R,S-hydroxyphosphonoacetate frameworks: applications for anticorrosion protection of metal surfaces. *Chem. Mater.*, **20** (15), 4835–4846.
71. Sekine, I., Shimode, T., Yuasa, M., and Takaoka, K. (1992) Corrosion inhibition of structural-steels in co_2 absorption process by organic inhibitor composed of 2-aminothiophenol, (1-hydroxyethylidene)bis(phosphonic acid), and diethanolamine. *Ind. Eng. Chem. Res.*, **31** (1), 434–439.
72. Kuznetsov, Yu.I., Kazanskaya, G.Yu., and Tsirulnikova, N.V. (2003) Aminophosphonate corrosion inhibitors for steel. *Prot. Met.*, **39** (2), 120–123.
73. Rajendran, S., Apparao, B.V., Periasamy, V., Karthikeyan, G., and Palaniswamy, N. (1998) Comparison of the corrosion inhibition efficiencies of the ATMP-molybdate system and the ATMP-molybdate-Zn^{2+} system. *Anti-Corros. Methods Mater.*, **45** (2), 109–112.
74. Felhosi, I. and Kálmán, E. (2005) Corrosion protection of iron by alpha,omega-diphosphonic acid layers. *Corros. Sci.*, **47** (3), 695–708.
75. To, X.H., Pebere, N., Pelaprat, N., Boutevin, B., and Hervaud, Y. (1997) A corrosion-protective film formed on a carbon steel by an organic phosphonate. *Corros. Sci.*, **39** (10–11), 1925–1934.
76. Balaban-Irmenin, Yu.V., Rubashov, A.M., and Fokina, N.G. (2006) The effect of phosphonates on the corrosion of carbon steel in heat-supply water. *Prot. Met.*, **42** (2), 133–136.
77. Fang, J.L., Li, Y., Ye, X.R., Wang, Z.W., and Liu, Q. (1993) Passive films and corrosion protection due to phosphonic acid inhibitors. *Corrosion*, **49** (4), 266–271.
78. Du, T., Chen, J., and Cao, D. (2001) N,N-Dipropynoxy methyl amine trimethyl phosphonate as corrosion inhibitor for iron in sulfuric acid. *J. Mater. Sci*, **36** (16), 3903–3907.
79. Frateur, I., Carnot, A., Zanna, S., and Marcus, P. (2006) Role of pH and calcium ions in the adsorption of an alkyl N-aminodimethylphosphonate on steel: an XPS study. *Appl. Surf. Sci.*, **252** (8), 2757–2769.
80. Amar, H., Benzakour, J., Derja, A., Villemin, D., and Moreau, B. (2003) A corrosion inhibition study of iron by phosphonic acids in sodium chloride solution. *J. Electroanal. Chem.*, **558**, 131–139.
81. Paszternák, A., Felhosi, I., Keresztes, Zs., and Kálmán, E. (2007) Formation and structure of alkyl-phosphonic acid layers on passive iron. *Mater. Sci. Forum*, **537-538**, 239–538.
82. Kuznetsov, Yu.I., Zinchenko, G.V., Kazanskii, L.P., Andreeva, N.P., and Makarychev, Yu.B. (2007) On the passivation of iron in aqueous solutions of zinc 1-hydroxyethane-1,1-diphosphonate. *Prot. Met.*, **43** (7), 648–655.
83. Amar, H., Braisaz, T., Villemin, D., and Moreau, B. (2008) Thiomorpholin-4-ylmethyl-phosphonic acid and morpholin-4-methyl-phosphonic acid as corrosion inhibitors for carbon steel in natural seawater. *Mater. Chem. Phys.*, **110** (1), 1–6.
84. Ramesh, S., Rajeswari, S., and Maruthamuthu, S. (2004) Corrosion inhibition of copper by new triazole phosphonate derivatives. *Appl. Surf. Sci.*, **229** (1–4), 214–225.
85. Dufek, E.J. and Buttry, D.A. (2008) Inhibition of O_2 reduction on AA2024-T3 using a Zr(IV)-Octadecyl phosphonate

coating system. *Electrochem. Solid-State Lett.*, **11** (2), C9–C12.

86. Pilbath, A., Bertoti, I., Sajo, I., Nyikos, L., and Kálmán, E. (2008) Diphosphonate thin films on zinc: preparation, structure characterization and corrosion protection effects. *Appl. Surf. Sci.*, **255** (5), 1841–1849.
87. Pilbáth, A., Nyikos, L., Bertóti, I., and Kálmán, E. (2008) Zinc corrosion protection with 1,5-diphosphonopentane. *Corros. Sci.*, **50** (12), 3314–3321.
88. Shao, H.B., Wang, J.M., Zhang, Z., Zhang, J.Q., and Cao, C.N. (2002) The cooperative effect of calcium ions and tartrate ions on the corrosion inhibition of pure aluminum in an alkaline solution. *Mater. Chem. Phys.*, **77** (2), 305–309.
89. Wang, S.H., Liu, C.S., Shan, F.J., and Qi, G.C. (2008) Amino-tris-(methylenephosphonic acid) layers adsorption on AA6061 aluminum alloy. *Acta Metall. Sin. (Engl. Lett.)*, **21** (5), 355–361.
90. Paszternak, A., Felhosi, I., Paszti, Z., Kuzmann, E., Vertes, A., Kalman, E., and Nyikos, L. (2010) Surface analytical characterization of passive iron surface modified by alkyl-phosphonic acid layers. *Electrochim. Acta*, **55** (3), 804–812.
91. Appa Rao, B.V. and Srinivasa Rao, S. (2010) Electrochemical and surface analytical studies of synergistic effect of phosphonate, Zn^{2+} and ascorbate in corrosion control of carbon steel. *Mater. Corros.*, **61** (4), 285–301.
92. Holm, R., Holtkamp, D., Kleinstuck, R., Rother, H.-J., and Storp, S. (1989) Surface -analysis methods in the investigation of corrosion-inhibitor performance. *Fresenius Z. Anal. Chem.*, **333** (4–5), 546–554.
93. Gunasekaran, G., Natarajan, R., Muralidharan, V.S., Palaniswamy, N., and Appa Rao, B.V. (1997) Inhibition by phosphonic acids – an overview. *Anti-Corros. Methods Mater.*, **44** (4), 248–259.
94. Chougrani, K., Boutevin, B., David, G., Seabrook, S., and Loubat, C. (2008) Acrylate based anticorrosion films using novel bis-phosphonic methacrylates. *J. Polym. Sci. A: Polym. Chem.*, **46** (24), 7972–7984.
95. (a) Kalman, E., Lukovits, I., and Palinkas, G. (1995) A simple-model of synergism of corrosion-inhibitors. *ACH-Models Chem.*, **132** (4), 527–537; (b) Boffardi, B.P. (1993) in Corrosion and Electrochemistry. *Reviews on Corrosion Inhibitor Science and Technology* (eds A. Raman and P. Labine), NACE International, Houston, TX, pp. 6–16.
96. Demadis, K.D., Papadaki, M., and Cíšarová, I. (2010) Single-crystalline thin films by a rare molecular calcium carboxyphosphonate trimer offer prophylaxis from metallic corrosion. *ACS-Appl. Mater. Interfaces*, **2** (7), 1814–1816.
97. Demadis, K.D., Papadaki, M., Raptis, R.G., and Zhao, H. (2008) 2D and 3D alkaline earth metal carboxyphosphonate hybrids: anti-corrosion coatings for metal surfaces. *J. Solid State Chem.*, **181** (3), 679–683.
98. Benabdellah, M., Dafali, A., Hammouti, B., Aouniti, A., Rhomari, M., Raada, A., Sehaji, O., and Robin, J.J. (2007) The role of phosphonate derivatives on the corrosion inhibition of steel in HCl media. *Chem. Eng. Commun.*, **194** (10–12), 1328–1341.
99. Menke, J. (1999) Corrosion: a "people" problem. NACE Conference, Corrosion/99, Paper number 503.
100. Clark, J. and Macquarrie., D. (eds) (2002) *Handbook of Green Chemistry and Technology*, Blackwell Publishing, Oxford.
101. Hasson, D., Shemer, H., and Sher, A. (2011) State of the art of friendly "green" scale control inhibitors: A review article. *Ind. Eng. Chem. Res.*, **50** (12), 7601–7607.

10
Metal-Matrix Nanocomposite Coatings Produced by Electrodeposition

Caterina Zanella, Stefano Rossi, and Flavio Deflorian

10.1
Introduction

Various combinations of matrix materials (e.g., metals, alloys, and polymers) and dispersed phases (ceramics, metal particles, polymers, and encapsulated liquids) lead to a broad range of tailored properties such as improved wear resistance, hardness, or self-lubricating surfaces [1]. Recently, this idea has been developed to replace electrodeposited chromium coatings with nanostructured dispersion coatings [2–4].

The fact that electrochemical deposition can be used to synthesize composites has generated a great deal of interest in recent years. The advantages of this century-old process in comparison to other depositions techniques are as follows:

- low cost;
- free from porosity;
- high purity;
- industrial applicability;
- potential to overcome shape limitations or allows the production of free-standing parts with complex shapes;
- higher deposition rates;
- ability to produce coatings on widely differing substrates;
- ability to produce structural features with sizes ranging from nanometers to micrometers;
- easy to control alloy composition.

Electrodeposition or electroplating is the process by which an applied current or potential is used to deposit a film of metal or alloy by the reduction of metallic ions onto a conductive substrate. The electroplating process is a widely used industrial procedure and is applied in order to produce films that provide corrosion resistance and wear resistance, and could change thermal, magnetic, and optical characteristics of the surfaces. The interest in electrolytic and electroless composite coatings has increased substantially during the last decades as a result of the improved properties and the relatively low cost of these coatings [5]. The

Green Corrosion Chemistry and Engineering: Opportunities and Challenges, First Edition.
Edited by Sanjay K. Sharma.

electro-codeposition describes the embedding of a second phase, usually particles, in an electrodeposited metal matrix from a dispersion of those particles in the plating bath leading to the formation of a composite, and therefore to surface coatings with improved or sometimes completely new properties [6]. The final properties of the composite coatings depend on the combination of both particles and metal matrix, and ideally combine the best of the two worlds. Several metals have been studied for this application; however, nickel and copper are the most studied metal matrices [7] since they are the widest industrially applied coatings. Depending on the type of particles deposited, several properties such as abrasion, friction, and corrosion resistance can be achieved. The codeposition of ceramic particles such as Al_2O_3 [1, 8], SiC [9, 10], CeO_2 [11, 12], SiO_2 [13, 14], TiO_2 [15, 16], and WC [17, 18] leads to the hardening of the metal layer and to an improvement in the wear and abrasion resistance, while PTFE [19, 20] or MoS_2 [21] could be added in order to reduce the friction. On the other hand, the possibility of adding liquid-containing microcapsules enlarges further on the combination of composites: the addition of capsules containing lubricants or corrosion inhibitors that are released during abrasion or surface damage. These are just few example of the "smart coating" that could be produced by electrodeposition.

The unique functional properties of composite coatings are derived not only from the presence of the particles but also from the matrix microstructural changes induced by the interaction between the particles and the electrocrystallization process, [22] leading to a change in the matrix microstructure [23].

10.2
Electrodeposition of Composite Coatings–Theoretical Remarks

10.2.1
Suspension of Solid Particles in Electrolytes

Most materials exhibit a ζ-potential when immersed in water and its value is influenced by pH, chemical species in solution, and ionic strength of the medium [24]. The magnitude of the ζ-potential of the particles is a measure of the interaction of particles, and therefore can be used to predict the long-term stability of a suspension. If the suspended particles have a large negative or positive ζ-potential, then they will tend to repel each other and will not aggregate. On the contrary, if the particles have low ζ-potential values, that is, close to zero, then it is not possible to prevent the particles from approaching each other and aggregating. The limiting value between stable and unstable suspensions is generally 30 or -30 mV.

Many surfaces, such as an oxide dispersed in a diluted salt solution, will show a common pattern when the pH is varied: as the pH increases by adding alkali, the surface will become more negative, or at least less positive. Vice versa, when acid is added; ionization will cause the loss of hydroxyl ions that will make the surface more positive. In the ζ-potential versus pH plot, there may be a point where the curve passes through the value zero called the isoelectric point (IEP).

The charge can arise from ionization of surface groups, chemical bonding, or physical adsorption of ions from the liquid medium. Consequently, a charged particle suspended in an electrolyte solution tends to be surrounded by an ionic cloud.

10.2.2 Mechanism of Codeposition

Several theoretical models have been proposed to describe the codeposition phenomenon. However, the mechanism is not fully understood and the mathematical models to predict the amount of codeposited particles are verified only for a few systems and in controlled conditions; also, the equation cannot take into account all the variables of the process and therefore has been developed for particular experimental conditions. Moreover, these models have been developed from the investigation on the microsized particles and could not simply be applied to nanosized systems since in the nanorange all the forces act in a completely different way. Nevertheless, the physical description of the process can be considered to be valid for nanoparticles as well and could help in understanding the interaction between particles, electrolyte, electrodes, and process parameters such as pH, temperature, current conditions, and fluidodynamics.

A first model of the electrolyte codeposition was made by Guglielmi [25] and was based on a two-step, codeposition mechanism. Under the influence of an electric field, diffusion, and fluid flow, the particles migrate toward the cathode where Guglielmi assumed that the particles are first reversibly loosely adsorbed on the cathode surface and then embedded in the metal layer. The adsorption of metal ions on the surface of the particles forming an ionic cloud was subsequently considered [26] taking, therefore, into account the ζ-potential and electrophoretic forces acting on the solid–liquid interface. This model has been adopted for many years but the number of embedded particles was meant to be dependent only on the concentration of the particles and current density. To overcome the shortcoming in Guglielmi's model, Celis *et al.* [27] first and Fransaer *et al.* [28] later developed a model that also took into account the hydrodynamic effect and the characteristics of the particles. The basic assumption of this model is that all particles are surrounded by a cloud of adsorbed species, mainly metal ions, and can be incorporated into the metal matrix only if an efficient part of these chemical species is reduced at the same time with the metal ions on the cathode. Moreover, the model considers a current efficiency of 100% and spherical particles, and is based on the important influence of electrolyte stirring. The model consists of five consecutive steps as schematically described in Figure 10.1:

- formation of ionic clouds on the particles;
- convection toward the cathode surface;
- diffusion through a hydrodynamic boundary layer;
- diffusion through a concentration boundary layer;
- adsorption at the cathode where particles are entrapped within the metal deposit by the reduction of the ionic cloud.

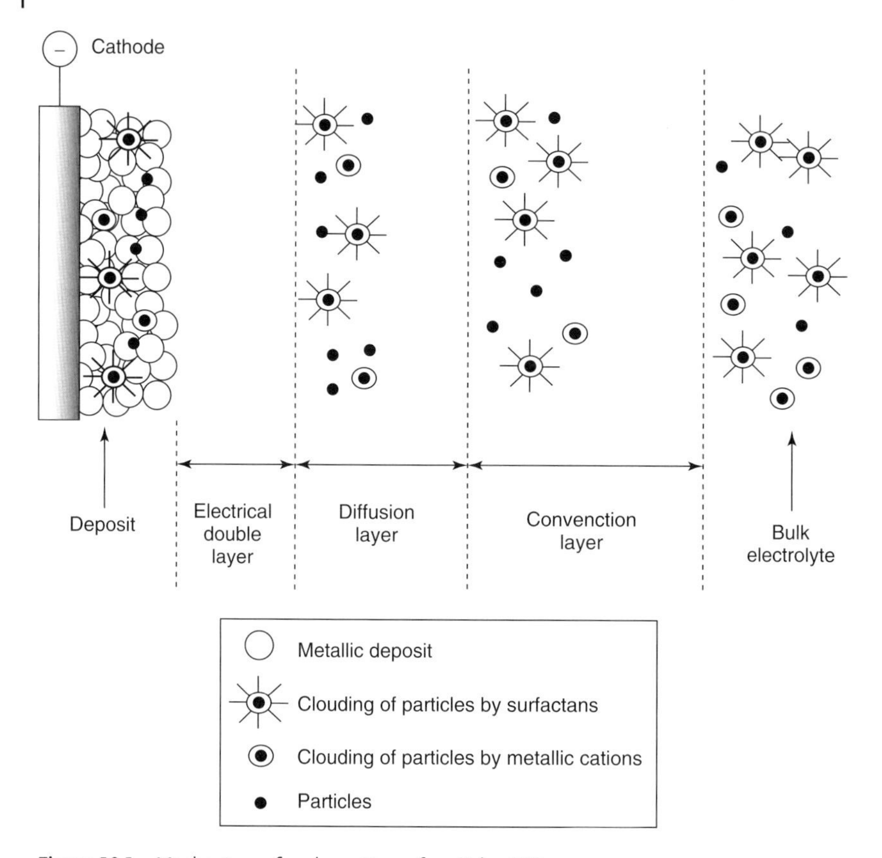

Figure 10.1 Mechanism of codeposition of particles [29].

The partial reduction of the ionic clouds on the particles was considered to be the dominating effect for the embedding of particles: in order to codeposit a particle, a specific quantity k of the total quantity K of the ions, which constitute the ionic adsorbed cloud of ions surrounding the particle, should be reduced on the cathode. This implies that not all the particles transported to the cathode are embedded into the metal deposit. The mass percentage of embedded particles is given by the expression:

$$a = \frac{W_\mathrm{p} \cdot N_\mathrm{p} \cdot P}{\frac{M \cdot i_\mathrm{p}}{n \cdot F} + W_\mathrm{p} \cdot N_\mathrm{p} \cdot P} \times 100 \tag{10.1}$$

where

W_p: the weight of one particle
N_p: the number of particles that diffuse through the diffusion layer per unit of time and surface area
P: the probability of one particle to be embedded into the metal matrix
n: the valence of the deposited metal

F: the Faraday constant
i_p: the current density
M: molecular weight.

The weight of the deposited metal is calculated using the Faraday law, while using the above-mentioned basic assumption, P depends on the probability $P(k/K,i)$ that at least k from the adsorbed ions K are reduced on the cathode, and so, if p_i is the probability that one ion is reduced for a current density i, then:

$$P_{(k/K,i)} = \sum_{z=k}^{K} C_z^K [1 - p_i]^{K-z} p_i^z \text{ and } P = H \cdot P_{(k/K,i)} \tag{10.2}$$

The probability p_i can be calculated from the Faraday's law data, while for its evaluation it is assumed that there is no differentiation between absorbed and free ions.

The factor H depends on the hydrodynamic conditions of the electrolytic bath, and from different experiments the following values have been given to it: $H = 1$ for regular flow, $0 < H < 1$ for a transient flow, and $H = 0$ for turbulent flow.

The number of particles N_p is correlated to both the number of ions N_m that pass through the diffusion layer per time and surface area units and to the kind of applied overpotential:

$$N_p = N_m \cdot \frac{C_p^*}{C_m^*} \cdot \left(\frac{i_{tr}}{i_p}\right)^{\alpha} \tag{10.3}$$

where

i_{tr}: the current density of the transition from the conditions of predominance of the charge-transfer overpotential to the conditions of predominance of the concentration overpotential.
C_p^*, C_m^*: the number of particles and the number of ions in the bulk electrolyte, respectively
α: parameter that expresses the interaction between the free and adsorbed ions. It has the value of zero when the charge-transfer overpotential is predominant and $\alpha \neq 0$ when the concentration overpotential is the predominant condition.

Also, this mathematical model, like the previous ones, did not succeed in describing all studied systems as it could not directly evaluate the percentage of codeposited particles. Moreover, the factors k, K, and α are recalculated at every single time using experimental data, while in the mathematical expression of the probability p_i, a time factor is missing (which should make this value dimensionless) and an evaluation error is therefore introduced.

All these models that take into account the electrostatic interaction between molecules adsorbed on the surface of the particles and the cathodic process assumed that only particles surrounded by positively charged ionic clouds can be codeposited since the cathodic surface is negative.

But it was empirically shown that negatively charged particles also could be codeposited into the metal deposit [29]. The electrostatic interaction between the surface of the particles and the cathodic surface is not very simple and definite as assumed in the mathematical models. The characteristic double layer of the particles is determined by the intrinsic surface charge, the adsorbed ions, and the ions compensating the charge in the double layer. Moreover, the cathodic surface is negatively charged, but could be covered by adsorbed positive ions and the double layer could compensate the charge. Moreover, in the presence of ionic surfactants, the ζ-potential of the metal layer must be considered among those of the particles. The electrostatic interaction between particles and electrodes is more complicated than assumed in the models and needs to be studied in more detail.

Other models that have been formulated afterward did not succeed in providing a general approach to the codeposition of inert particles into metallic coatings even if more factors were taken into consideration [30–33]. However, the flexibility and the reliability of each model to describe the behavior of a wide range of metallic coatings and particle types still require validation. Often, the mathematical relationship is strictly related to the experimental setup, and in all models available so far the effect of the codeposition of particles on the deposit electrocrystallization has not been considered. Moreover, when the particles have nanodimensions, all the forces act differently on the system and Brownian motion becomes more important. These models have therefore not been validated for the nanocomposites.

On the other hand, the properties of the metal-matrix composite coatings have been widely studied. Many publications are available regarding different types of microcomposite coatings using different kinds and sizes of particles as well as different kinds of metal or alloy matrices, while many studies on the nanocomposite coatings are limited to the codeposition optimization and understanding the mechanism; no systematic research has been carried out on the relationship between nanoparticle codeposition and both mechanical and protective protection.

10.2.3
Process Parameters Influencing the Incorporation

The process parameters influence the deposit characteristic and therefore need to be optimized for each electrodeposition process as a function of the final deposit's properties and application. In the electrodeposition of composite coating, the influence of process parameters is even more important since they affect the codeposition fraction. Since final properties of a composite coating depend mainly on the codeposited fraction of particles and their dispersion in the coating, the determination of the relationship between process parameters and fraction of embedded reinforcement leads to the possibility of tailoring the final coating and optimizing its properties for different applications.

The following parameters refer to the characteristics of particles: surface charge, type, shape, and size. Depending on the type and shape, the codeposited particles tend to be agglomerated or not, and the larger the particles, the higher is the codeposited fraction. For the change in the matrix microstructure and final properties, however, it is not the volume fraction but the number of efficient particles that is important. Therefore, if the particles are fine and well dispersed, a very low volume fraction is enough and even more effective than a big fraction of agglomerated or big particles.

Other parameters are determined by the choice of the galvanic bath: pH, temperature, additives, electrolyte composition, and concentration. It has been demonstrated that a low electrolyte concentration induces a higher particle codeposition rate since the ionic strength of the solution is reduced and therefore the suspension of particles is promoted.

Moreover, the choice of the geometry of the deposition cell is very important: both the fluidodynamics and the geometrical distribution of anodic and cathodic surfaces affect the deposition process. At the laboratory scale, the best condition in order to control the hydrodynamic is the rotating disk electrode, but considering the industrial application this configuration could not be reproduced.

While the mentioned parameters depend on the industrial plant and on the geometry of the device, the main factors that could be varied in order to tailor the deposition process are the current density and type, stirring, and concentration of particles.

- **Current conditions**: the current density is a very important parameter: the codeposition rate increases by increasing the current density up to a maximum value that depends on the particle-electrolyte system.
  Many different galvanostatic techniques can be used in order to optimize the deposit as, for example, direct current, pulse current, superimposed pulse current, and pulse reverse current. The differences in the pulse current conditions are more or less limited in the value of the current in the "off-time" that could be zero or anodic or even cathodic. (Figure 10.2)
  Pulse electrodeposition has been found to be effective in perturbing both the adsorption–desorption phenomena and electrocrystallization process, thus leading to an increase in the limit current density and to an opportunity to control the microstructure of electrodeposits [34, 35]. However, to achieve a significant grain refinement, the off-times should be longer than the on-times [36, 37]. Regarding the electro-codeposition, the use of pulse current has been efficient in increasing the codeposition rate, especially for nanosized reinforcement. The pulse reverse current condition induces a further increment in the concentration of particles and a more homogeneous dispersion of the particles in the metal matrix because of the partial dissolution of the metal deposit during the anodic period and the desorption of big agglomerates from the cathodic surface [38, 39]. The use of pulse current improves the number of the system variables. For example, pulse frequency and duty cycle (rate of off-time and on-time) can be varied to modify the system response, thus inducing a change in the adsorption and crystallization processes.

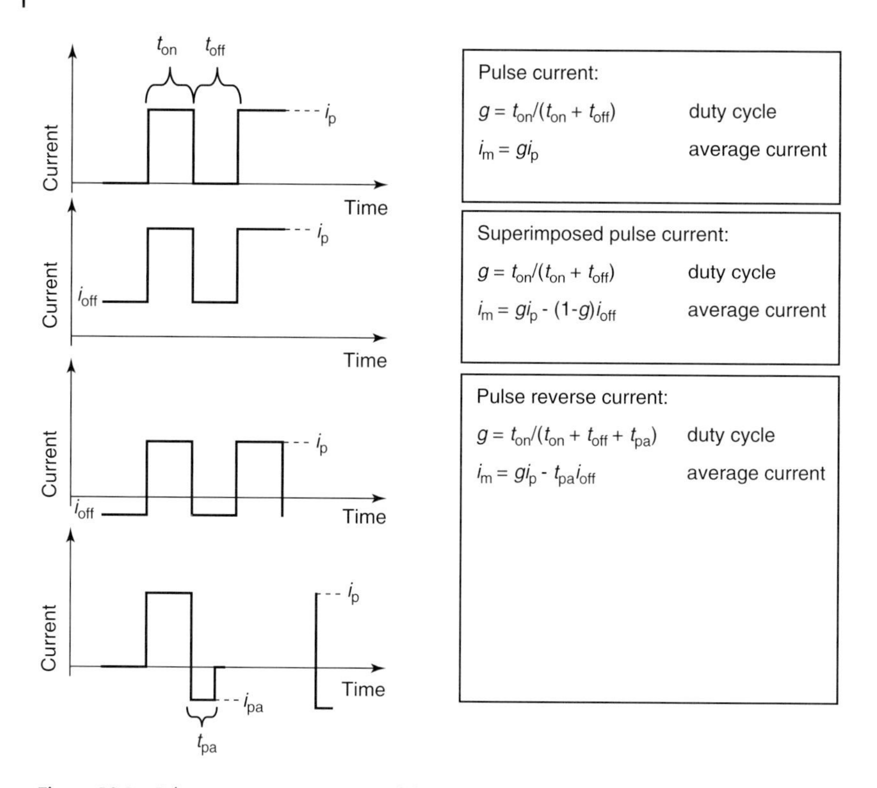

Figure 10.2 Schematic representation of the wave form and parameters of pulse current.

- **Stirring:** the transport of particles to the cathode is mainly determined by convection, and the hydrodynamic of the galvanic cell is very important for the codeposition process. If the electrolyte flow is too high near the cathodic surface, it is possible that it can remove the particles from the surface, thus leading to a decrease in the codeposition rate. On the contrary, a light stirring induces flocculation of the suspension and a smaller particle motion, thus producing a lower particle fraction and a higher heterogeneity in the codeposition of particles.
- **Concentration of particles:** the probability of codeposition and adsorption of particles is of course proportional to the concentration of particles in the galvanic bath. The experimental results show that the codeposited content of particles in the metal matrix increases by increasing the content of the particles in the galvanic bath with a decreasing rate and reaches a plateau that depends on the bath, temperature, and current condition. This behavior was confirmed by other studies for a wide range of codeposition systems [40–42]. There is an approximate saturation value of codeposited particles in the metal layer. This saturation value is different for each system of particle electrolytes and should be determined empirically.

10.3
Electrodeposition of Composite Coatings

The first paper on the electrodeposition of metal-matrix composite coatings studied the copper–graphite system for self-lubricating surfaces in car engines [43], but the development of hardened films started only in the 1960s [44].

The first electroplating baths containing particles appeared at the industrial scale for the production of Ni/SiC and Co/Cr_2O_3 coatings used for car engines, aircrafts construction, and printed circuits. Specifically, the Ni/SiC deposits have been used as internal coating of the aluminum cylinders of car engines [6]. The wear properties of Ni/SiC as well as Ni/Al_2O_3 composite coatings have been studied by many research groups as these coatings could be good candidates as substitutes for the hard chromium coatings.

Different particles sizes, current conditions, and different electroplating baths have been studied. The wear resistance of the coatings increases with the addition of micro- or submicro particles because of the high amount of incorporated particles and their high intrinsic hardness. On the other hand, the possible detachment of the partially embedded particles from the metal-matrix surface (Figure 10.3) during the abrasion leads to an improvement in the abrasion damage caused because they could act as a third body in the abrasion process.

Moreover, the presence of a high amount of relatively big particles could worsen the corrosion properties of the coating because of the presence of voids between the particles and the metal matrix, which are the preferred path of propagation for the corrosive solutions.

The researchers started to concentrate on the codeposition of smaller particles that could enhance the mechanical properties without penalizing the corrosion

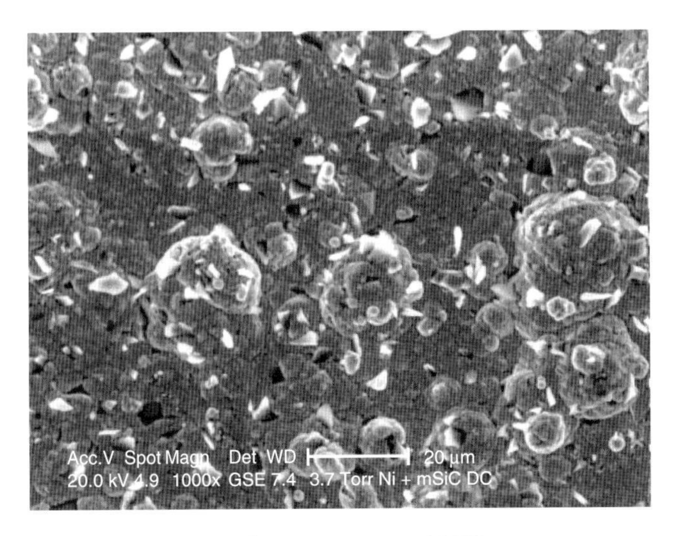

Figure 10.3 Top view of a microcomposite Ni/SiC.

resistance of the coating. Moreover, the presence of a smaller hard phase could avoid the formation of an abrasive third body during the wear process. Ding *et al.* [45] in 1998 used submicro particle sizes of α-alumina (0.11 and 0.4 μm) and showed that there is a linear increase in microhardness in both Ni and Cu with particle volume fraction in the deposit. The smaller particles produced a larger hardening effect, and the strengthening mechanism was explained by a combination of the Orowan-type strengthening and Hall–Petch effects.

In the last 10 years, the interest in the metal-matrix composite coatings has been revived as a result of the development of new nanotechnological methods for the production of nanoparticles of carbide or oxides.

Generally, the volume fraction of codeposited particles is limited for nanoparticles, and it is usually inversely proportional to their size [46]. For example, Shaou *et al.* [47] studied the rate of incorporation of two different sizes of Al_2O_3 nanoparticles (50 and 300 nm) into a nickel deposit. Using similar operating parameters (1000 rpm, 20 mA/cm), it was found that the percentage volume fraction of 300 nm Al_2O_3 in the nickel deposit was much higher compared to that in 50 nm Al_2O_3.

The presence of nanosized particles in a metal deposit may induce changes in the crystalline structure of the metallic coating. It has been shown that 20 nm SiC nanoparticles influence the competitive formation of nickel nuclei and crystal growth [48], and that the presence of nanoparticles will perturb the crystalline growth of a metal deposit, resulting in an increased number of defects in the crystal structure, facilitating a nanocrystalline structure [16, 49]. The hardening effect is therefore not only related to the presence of the second hard phase but also to the changes induced in the metal matrix.

The number of embedded nanoparticles depends on the operational parameters such as current density, pH, bath temperature, additives type, concentration, and stirring; but the relationship between operating parameters and codeposition amount is not clear and often the results are inconsistent.

Erler *et al.* demonstrated that the content of alumina nanoparticles in the deposit increases when the deposition current is decreased [50]. On the contrary, Bund *et al.* showed that the embedded fraction of nanoalumina is not affected by the current density in acidic nickel deposition bath, while it increases by increasing the current in the alkaline deposition bath [51].

The microstructure of the composite deposits depends on many factors but in general the addition of nanoparticles induces a refinement in the metal grain size. In Figures 10.4–10.6, an example of the microstructure of nickel coatings with different nanopowders is compared with the pure nickel ones [52].

The microstructure obtained under DC condition is very different in the three systems, Figures 10.4–10.6. The columnar grains of the pure coating are destroyed by the addition of the ceramic powders and SiC seems to be more effective in refining the microstructure than Al_2O_3, but Al_2O_3 leads to a more planar and flatter surface. However, both powders lead to the formation of a more compact layer, avoiding the formation of gaps between the grains.

Figure 10.4 Top view of Ni deposited under DC.

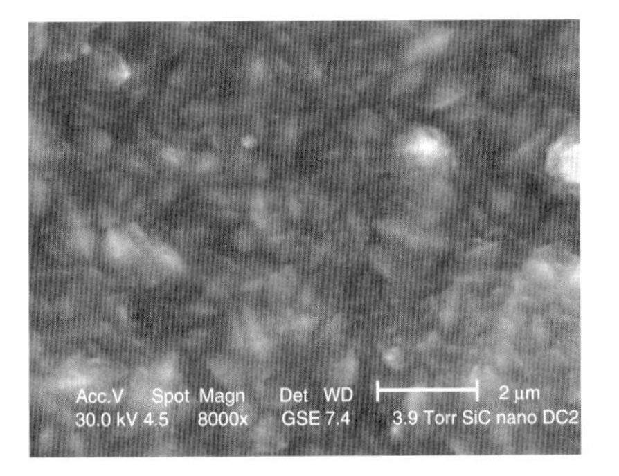

Figure 10.5 Top view of Ni/nanoSiC deposited under DC.

Figure 10.6 Top view of Ni/nanoAl_2O_3 deposited under DC.

In Figures 10.7–10.9, the microstructure analyses of the etched samples' cross section give interesting information about the coating growth and how it is affected by the codeposition or by the current condition [52]. Generally, nickel films present vertical columnar grains grown parallel to the applied electric field. The codeposition of the well-dispersed SiC nanopowder partially interrupts the growth

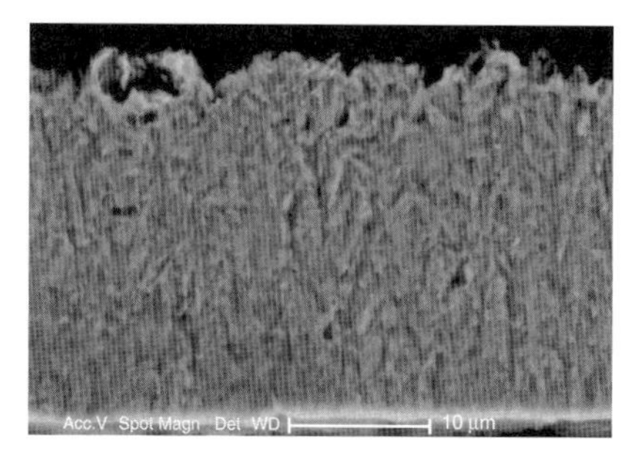

Figure 10.7 Cross section of Ni.

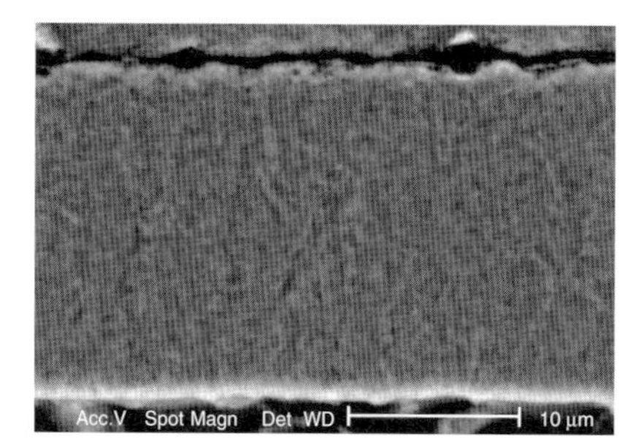

Figure 10.8 Cross section of Ni/nanoSiC.

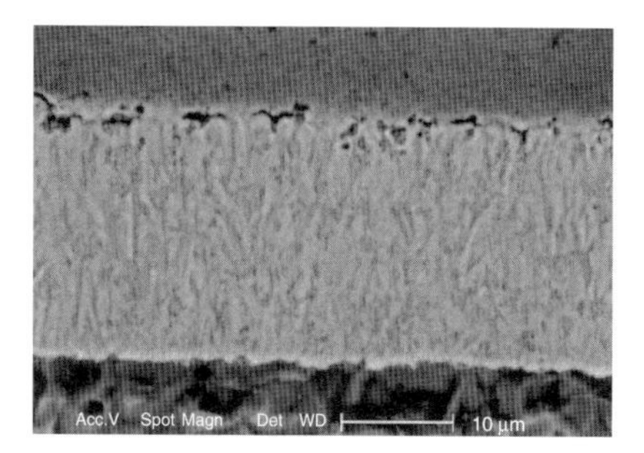

Figure 10.9 Cross section of Ni/nanoAl_2O_3.

of the columns, while the addition of nanoalumina does not induce the same refinement because the powder is strongly agglomerated as better highlighted in Figure 10.10.

The agglomeration degree of codeposited particles is an important property of the composite coating since smaller agglomerates are better distributed in the whole matrix compared to bigger ones and the reinforcement could be more homogeneous. Moreover, the smaller the particles, the higher is their influence on the electrocrystallization process and therefore on the final microstructure of the coating.

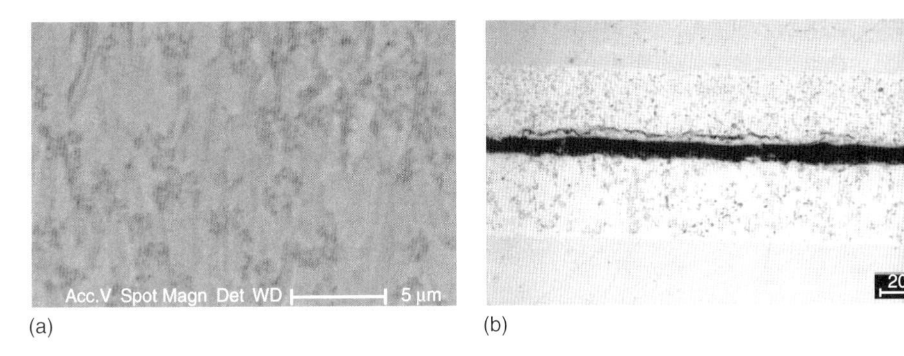

Figure 10.10 Etched cross section of Ni/nanoAl_2O_3 SEM image (a) and LOM image (b).

The results showing the effect of incorporation of particles on the final properties are often contradictory because they depend on many parameters and on both the embedded amount and the dispersion of the particles. The hardening effect caused by the codeposition of nanoparticles has been reported by many authors [53–55]: the increase in the hardness of the nickel nanocomposite film is associated with the presence of the intrinsic hardness of the codeposited ceramic phase, with a dispersion hardening, and with the refinement of the microstructure. The hardening efficiency is, however, strongly connected with the dispersion of reinforcement particles. As an example, the hardening effect of the composite systems is shown in Figures 10.8 and 10.9 and the relationship between microhardness and embedded fraction of particles is plotted in Figure 10.11 [52].

From this graph, it is clear that the codeposition of silicon carbide particles leads to harder coatings even if the content of embedded particles is comparable to the alumina powder. Both the graphs show that a linear proportionality exists between the final hardness and the content of ceramic particles. However, this relationship is different for the two systems. The microhardness is related not only to the hard phase content but also to its hardness, its dispersion degree, and to the microstructure and microstructure modifications that the particles induce in the metal matrix.

Fitting the points with a straight line, the Ni/SiC systems show a higher rate of hardening compared to Ni/Al_2O_3. The silicon carbide powder is not only harder than alumina, but the codeposition leads to a finer and more homogeneous dispersion of the particles into the metal layer. All these characteristics improve the effectiveness of the nanopowder in hardening, thus leading to a high rate of the hardness–particle content relationship. However, the alumina powder also shows a linear proportionality between the content and hardness of particles, but because of its agglomeration the room for improvement is smaller and the increase in the particle content leads to a smaller increase in the hardness.

As a consequence, the wear resistance is improved [9]. Garcia *et al.* [56] demonstrated that the wear resistance of Ni/SiC composites is higher for the SiC size of 300 nm than microparticles because larger particles could improve the wear load because of the abrasion performed by the pulled-out particles. Zimmerman *et al.*

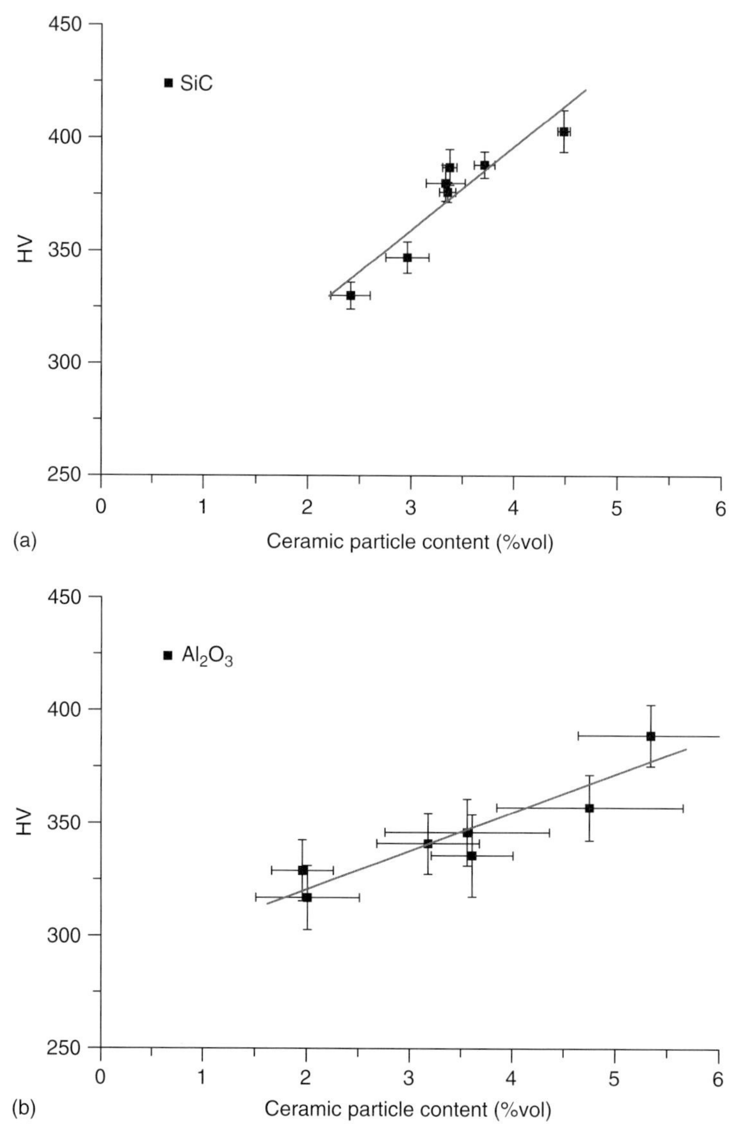

Figure 10.11 Relationship between hardness and content of particles for Ni/SiC (a) and Ni/Al_2O_3 (b).

[57] demonstrated that not only hardness but also the yield and tensile strengths are improved by codeposition of nanoparticles of SiC.

Regarding the protective properties of the nanocomposite coatings, there is no certain result shared by all researchers. Of course, the protection properties of the composite coatings strongly depend on the metal matrix. If the corrosion protection is a requirement, zinc or nickel are the most used coatings. The effect of the codeposition of nanoparticles on the electrochemical behavior of the metal matrix is quite interesting, but is still not defined in detail.

According to some groups, the incorporation of submicron SiC particles in the nickel matrix increases the corrosion resistance of the coatings, mostly due to microstructural modifications [58, 59]. In these cases, the codeposition of small particles leads to the formation of nanocrystalline, more compact, pore-free coatings, thus having higher corrosion resistance to both uniform corrosion and pitting corrosion.

According to other studies [60], the codeposition of SiC particles decreases the current density but only because of the fact that the coating surface is covered by inert ceramic particles and the free Ni surface in contact with the electrolyte is lower than in the case of pure nickel coatings. Others [61] demonstrated that nanocomposite coatings offer less protection, but this is mainly due to the presence of pores and defects caused by the agglomeration of the powder. Nevertheless, it is widely believed that the corrosion properties of composite coatings should increase if the grain size decreases and dispersion of particles is maximized. For this reason, many different attempts have been made by using different types of surfactants in order to avoid the agglomeration of the particles [62, 63], or by using pulse or triangular current in order to break the columnar structure of Ni deposits, to avoid pores and defects, and to create finer and more homogeneous microstructures [64, 65]. These attempts lead to finest deposits but the corrosion properties of these coatings have not been systematically studied.

In order to highlight the effect of powder dispersion on the protection properties, Figures 10.12 and 10.13 show the polarization curves of pure nickel and nickel-matrix nanocomposites. The Ni/nanoSiC layer is very well dispersed and homogeneous, while the Ni/nanoAl_2O_3 composite is quite agglomerated. These samples are produced under the effect of ultrasonic vibrations that lead to the production of pore-free coatings and permit the testing of electrochemical properties of the layers and are not influenced by the substrate.

The passive oxide layer grown on the Ni/SiC surface is more protective since the passive current density is almost 1 order of magnitude lower, while all the other parameters are comparable to the ones of pure nickel. On the other hand, the polarization curve of the Ni/Al_2O_3 layer is almost overlapped to the one of pure nickel and the only difference is the higher critical current density.

The codeposition of SiC nanopowder and the changes induced in the metal-matrix microstructure make the surface on Ni/SiC codeposits more resistant to uniform corrosion and lead to the formation of a more stable oxide.

From these curves, it can be deduced that, in environments containing chlorides, the presence of the nanoparticles neither affects the corrosion rate nor the passive current density, but the silicon carbide particles enhance the resistance to pitting corrosion. This is due not to the type of particles, but rather to the dispersion degree of the particles in the metal matrix.

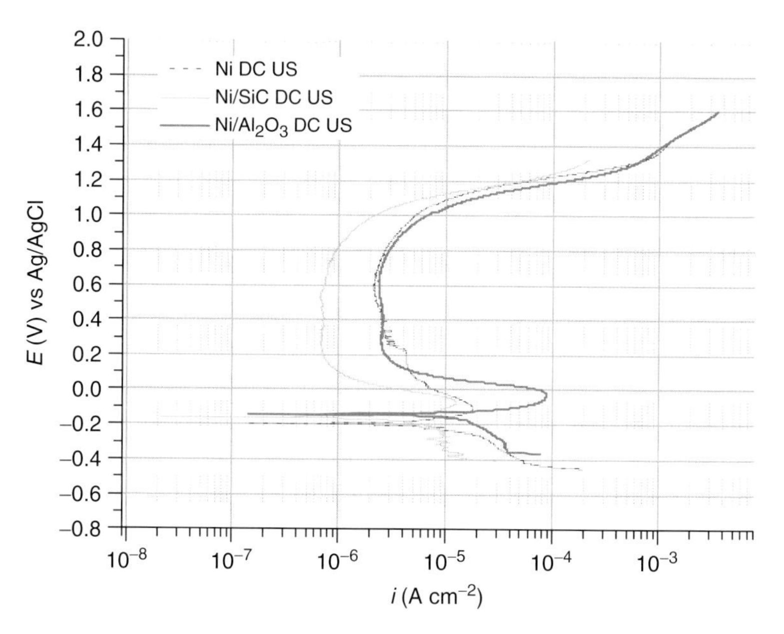

Figure 10.12 Potentiodynamic curves of pure nickel and nanocomposites in an acidic environment.

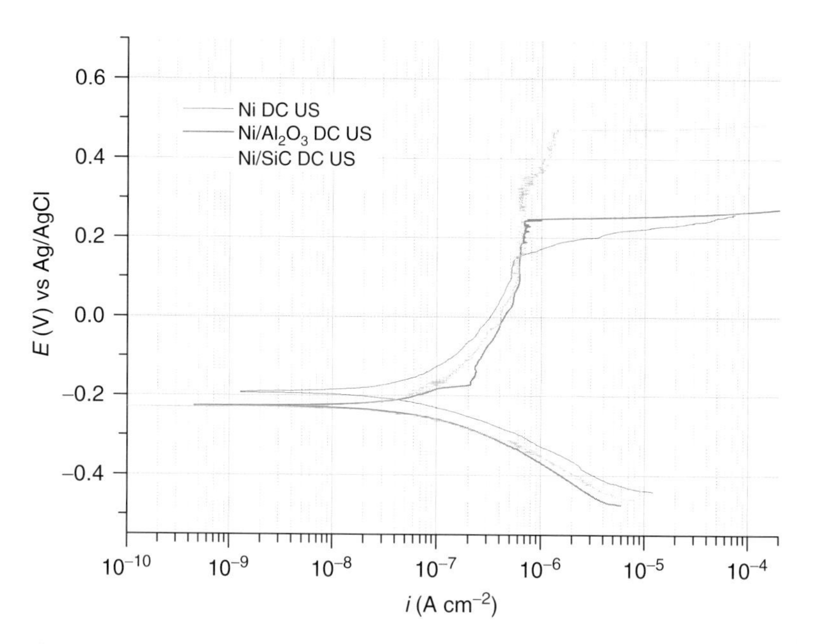

Figure 10.13 Potentiodynamic curves of pure nickel and nanocomposites in 3.5% NaCl electrolyte.

10.4
New Insight in the Electrodeposition of Composite Coatings

10.4.1
Ultrasonic Vibrations

Over the last few years, a large number of articles have been published that describe the variety of applications of power ultrasound in chemical processes [66]. These include synthesis, polymer chemistry, and some aspects of catalysis. Moreover, ultrasounds already have a wide diffusion and applications in the chemical engineering process, such as for dispersion of solids, crystallization, and degaussing processes.

Ultrasound was also shown to affect metal electrodeposition with benefit to the quality of the deposit, its adhesion and morphology, and also the diminution of brighteners and other additives needed in silent systems [67].

The effects of ultrasound in a liquid are to cause "acoustic streaming" and/or the formation of cavitation bubbles, depending upon the parameters of ultrasonic power, frequency, sonic source characteristics, and solution phenomena such as viscosity, volatility, and the presence of dissolved gases or other nucleation sites [68]. The power of ultrasound produces its effect via cavitation bubbles. These bubbles are generated during rarefaction cycle of the wave when the liquid structure is literally torn apart to form tiny voids, which collapse in the compression cycle [69].

Much work has been carried out from an industrial point of view empirically, in particular, on metals important in the electroplating technology such as zinc [70], iron [71], chromium [72, 73], copper [74], and nickel [75]. Hyde and Compton [76] in their studies on the influence of ultrasound on electrodeposition of metals conclude that, in electrodeposition under the influence of ultrasound, the critical effect is the increase in mass transport, which may be high enough to change a diffusion-controlled system into a charge-transfer-controlled system. Moreover, the charge-transfer reaction is promoted by the use of ultrasound because the pressure waves induce an activation of electrodes surfaces. The use of ultrasound during electroplating of pure metals may improve the final properties of the films, refining the microstructure and thus leading to a harder, more compact layer [77, 78].

Few studies have been performed on the use of the ultrasounds during the electro-codeposition of ceramic nanoparticles. Firstly, the ultrasonic vibrations were meant to be used only for the better dispersion of the particles in the galvanic bath to reduce the agglomeration and therefore to improve the suspension stability. Owing to the positive effects of ultrasounds on pure metal deposition, some researchers started to apply the ultrasonic vibration during the deposition process. Qu *et al.* [79] demonstrated that ultrasonic vibration promotes the uniform distribution of Al_2O_3 whiskers but decreases the ceramic content embedded in the nickel matrix. Rezrazi *et al.* [80] used ultrasounds to improve the codeposition rate of PTFE particles in gold matrix; Zheng and An [81] used the ultrasonic treatment to improve the content and uniformity of nano Al_2O_3 in NiZn alloy matrix. Lee *et al.* [82] found

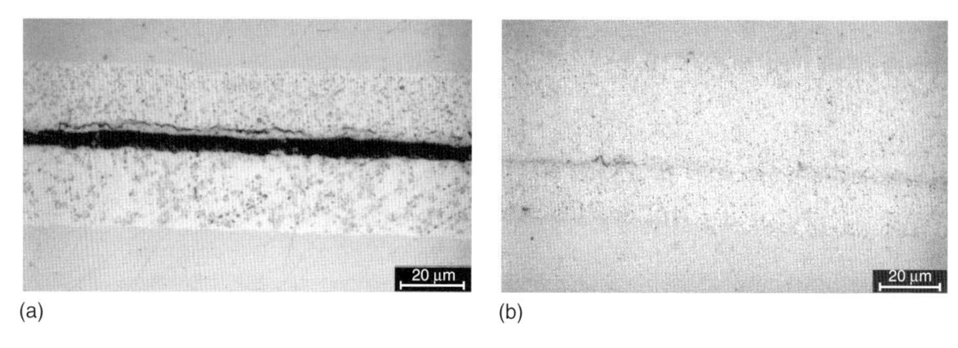

Figure 10.14 LOM cross sections of Ni/nanoAl_2O_3 deposited under silent (a) and sonicated (b) conditions.

out that ultrasonic vibration enhances the de-agglomeration of Al_2O_3 and CeO_2 in Cu matrix, especially in low concentrated baths.

Considering the very agglomerated composite system shown in Figure 10.10, the application of the ultrasounds during the deposition under the same conditions induces a sharp decrease in the agglomeration degree of the embedded ceramic powder as shown in Figure 10.14.

Besides this very important improvement, there is some more enhancement in the process efficiency. The application of the ultrasounds also has a beneficial influence on the codeposition rate, increasing the number of embedded particles in the metal layers. And, it has also been proved that the ultrasounds are a very successful alternative to surfactants in pitting control and allowing the production of high-quality and pore-free, and therefore, protective coatings.

10.4.2
Magnetic Field

The application of a magnetic field to a deposition cell can influence the kinetic of the ionic species inside the electrochemical cell. This effect is mainly caused by the Lorentz force and is called the magnetohydrodynamic (MHD) effect, which induces fluid flow in the cell and thus affects the transport rate of electroactive species toward the electrode.

The Lorentz f_L force acts on charges (ions) moving in a magnetic field:

$$\vec{f}_L = \vec{j} \times \vec{B} \tag{10.4}$$

where

J: the current density
B: the magnetic flux density.

This force becomes maximum when the magnetic field is aligned perpendicular to the current density and is zero when the magnetic field is parallel to the current density.

The magnetic field can increase the deposition rate, but decrease the current efficiency when the hydrogen evolution takes place since MHD not only enhances the mass transport process but also improves the discharge reaction [83]. Regarding the electrodeposition of metals and alloys, this enhanced mass transport can directly affect the electrocrystallization [84]. A superimposed magnetic field can change the crystallization behavior and induce a more uniform current distribution, and therefore a more uniform deposited layer. As a consequence, the hardness of the deposit can increase, the stress of the layers can change, and their corrosion properties can be improved [85].

Considering the deposition of composite coatings, few works studied the effect of the magnetic field on nonmagnetic particles. Dash *et al.* [86] and Vidrine and Podlaha [8] firstly applied a magnetic field into the preparation of $Cu-Al_2O_3$ composite coatings and found that the magnetic field was not only helpful to the suspension of particles in solution but also to improve the incorporated content of the particles in coatings. Peipmann *et al.* [87] demonstrated that the perpendicular magnetic field could not only improve the current efficiency but also the incorporation rate of the nanoscale alumina; the improvement might be due to the convection induced by MHD [88].

References

1. Bahrololoom, M.E. and Sani, R. (2005) *Surf. Coat. Technol.*, **192**, 154.
2. Aslanyan, I.R., Bonino, J.P., and Celis, J.P. (2006) *Surf. Coat. Technol.*, **200**, 2909.
3. Aal, A.A., Ibrahim, K.M., and Hamid, A.A. (2006) *Wear*, **260**, 1070.
4. Brooman, E.W. (2005) *Galvanotechnik*, **12**, 2843.
5. Di Bari, G. (2002) *Met. Finish.*, 35.
6. Roos, J.R. *et al.* (1990) *JOM*, 60–63.
7. Hovestad, A. and Jansenn, L.J.J. (1995) *J. Appl. Electrochem.*, **25**, 519.
8. Vidrine, B. and Podlaha, E.J. (2001) *J. Appl. Electrochem.*, **31**, 461.
9. Hou, K.H., Ger, M.D., Wang, L.M., and Ke, S.T. (2002) *Wear*, **253**, 994.
10. Orlovskaja, L., Periene, N., Krtinaitiene, M., and Surviliene, S. (1999) *Surf. Coat. Technol.*, **111**, 234.
11. Aruna, S.T., Bindu, C.N., Ezhil Selvi, V., Grips, V.K., and Rajam, K.S. (2006) *Surf. Coat. Technol.*, **200**, 6871.
12. Qu, N.S., Zhu, D., and Chan, K.C. (2006) *Scr. Mater.*, **54**, 1421.
13. Socha, R.P., Nowak, P., Laajalehto, K., and Vayrynen, J. (2004) *Colloids Surf. A*, **235**, 45.
14. Fransaer, J. and Celis, J.P. (2001) *Galavanotechnik*, **92**, 1544.
15. Li, J., Jiang, J., He, H., and Sun, Y. (2002) *J. Mat. Sci. Lett.*, **21**, 939.
16. Li, J., Sun, Y., Sun, X., and Qiao, J. (2005) *Surf. Coat. Technol.*, **192**, 331.
17. Stroumbouli, M., Gyftou, P., Plavlatou, E.A., and Spyrellis, N. (2005) *Surf. Coat. Technol.*, **195**, 325.
18. Jugovic, B., Stevanovic, J., and Maksimovic, M. (2004) *J. Appl. Electrochem.*, **34**, 175.
19. Ramesh, G.N.K. and Mohan, S. (1995) *Plat. Surf. Finish.*, **4**, 86.
20. Losiewiez, B., Stepien, A., Gierlotka, D., and Budniok, A. (1999) *Thin Solid Film*, **349**, 43.
21. Cang, Y.-C., Chang, Y.Y., and Lin, C.-I. (1998) *Electrochim. Acta*, **43**, 315.
22. Nowak, P., Socha, R.P., Aisheva, M. K., Fransaer, J., and Celis, J.-P. (2000) *J. Appl. Electrochem.*, **30**, 429.
23. Zimmerman, A.F., Clark, D.G., Aust, K.T., and Erb, U. (2002) *Mater. Lett.*, **52**, 85.

24. The Measurement of zeta potential using an autotitrator: Effect of pH, Application Note by Malvern Instruments, 2005.
25. Guglielmi, N. (1972) *J. Electrochem. Soc.*, **119**, 1009.
26. Foster, J. and Kariapper, A.M.J. (1973) *Trans. Inst. Met. Finish.*, **54**, 27.
27. Celis, J.P., Roos, J.R., and Buelens, C. (1987) *J. Electrochem. Soc.*, **13**, 1402.
28. Fransaer, J., Celis, J.P., and Roos, J.R. (1992) *J. Electrochem. Soc.*, **139**, 413.
29. Wunsche, F., Bund, A., and Plieth, W. (2004) *J. Solid State Electrochem.*, **8**, 209.
30. Valdes, J.L. (1987) *J. Electrochem.Soc.*, **134**, 223C.
31. Hwang, B.J. and Hwang, C.S. (1993) *J. Electrochem. Soc.*, **140**, 979.
32. Vereecken, P.M., Shao, I., and Searson, P.C. (2002) *J. Electrochem. Soc.*, **147** (7), 2572.
33. Bercot, P., Pena-Munoz, E., and Pagetti, J. (2002) *Surf. Coat. Technol.*, **157**, 282.
34. Ibl, N., Puipe, J.C., and Angerer, H. (1978) *Surf. Coat. Technol.*, **6**, 287.
35. Ibl, N. (1980) *Surf. Coat. Technol.*, **10**, 81.
36. Toth-Kadar, E., Bakonyi, I., and Pogany, L. (1996) *Surf. Coat. Technol.*, **88**, 57.
37. Qu, N.S., Zhu, D., Chan, K.C., and Lei, W.N. (2003) *Surf. Coat. Technol.*, **168**, 123–128.
38. Podlaha, E.J. (2001) *Nano Lett.*, **1** (8), 413.
39. Panda, A. and Podlaha, E.J. (2003) *Electrochem. Solid-State Lett.*, **6** (11), C149.
40. Chang, Y.S. and Lee, J.Y. (1998) *Mater. Chem. Phys.*, **20**, 309.
41. Graydon, J.W. and Kirk, D.W. (1990) *J. Electrochem. Soc.*, **137**, 2061.
42. Ramesh Bapu, G.N.K. and Yusuf, M.M. (1993) *Mat. Chem. Phys.*, **36**, 134.
43. Fink, C.G. and Pince, J.D. (1928) *Trans. Am. Electrochem. Soc.*, **54**, 315.
44. Williams, R.V. (1966) *Electroplat. Met. Finish.*, **19**, 92.
45. Ding, X.M., Merk, N., and Ilschner, B. (1998) *J. Mater. Sci.*, **33**, 803.
46. Maurin, G. and Lavanant, A. (1995) *J. Appl. Electrochem.*, **25**, 1113.
47. Shao, I., Vereecken, P.M., Cammarata, R.C., and Searson, P.C. (2002) *J. Electrochem. Soc.*, **149**, C610.
48. Benea, L., Bonora, P.L., Borello, A., and Martelli, S. (2002) *Wear*, **249**, 995.
49. Zhang, Y., Peng, X., and Wang, F. (2004) *Mater. Lett.*, **58**, 1134.
50. Erler, F., Jacobs, C., Romanus, H., Spiess, L., Wielage, B., Ampke, T. L., and Steinhauser, S. (2003) *Electrochim. Acta*, **48**, 3063.
51. Bund, A. and Thiemig, D. (2007) *Surf. Coat. Technol.*, **201**, 7092.
52. Zanella, C. (2010) Nanocomposite coatings produced by electrodeposition from additive-free bath: the potential of the ultrasonic vibrations. Doctorate thesis. University of Trento.
53. Pavlatou, E., Stroumbouli, M., Gyftou, P., and Spyrellis, N. (2006) *J. Appl. Electrochem.*, **36**, 385.
54. Ferkel, H., Muller, B., and Riehemann, W. (1997) *Mater. Sci. Eng. A*, **234**, 474.
55. Du, L., Xu, B., Dong, S., Yang, H., and Wu, Y. (2005) *Surf. Coat. Technol.*, **192**, 311.
56. Garcia, I., Fransear, J., and Celis, J.-P. (2001) *Surf. Coat. Technol.*, **148**, 171.
57. Zimmerman, A.F., Palumbo, G., Aust, K.T., and Erb, U. (2002) *Mater. Sci. Eng. A*, **328**, 137.
58. Garcia, I., Conde, A., Langelaan, G., Fransaer, J., and Celis, J.P. (2003) *Corros. Sci.*, **45**, 1173–1189.
59. Medeliené, V. (2002) *Surf. Coat. Technol.*, **154**, 104–111.
60. Malfatti, C.F., Ferreira, J.Z., Santos, C.B., Souza, B.V., Fallavena, E.P., Vaillant, S., and Bonino, J.P. (2005) *Corros. Sci.*, **47**, 567–580.
61. Lampke, T., Leopold, A., Dietrich, D., Alisch, G., and Wielage, B. (2006) *Surf. Coat. Technol.*, **201**, 3510.
62. Shrera, N., Masuko, M., and Saji, T. (2003) *Wear*, **254**, 555–564.
63. Ger, M.D. (2004) *Mater. Chem. Phys.*, **87**, 67–74.
64. Hu, F. and Chan, K.C. (2005) *Appl. Surf. Sci.*, **243**, 251–258.
65. Pavlatou, E.A., Stroumbouli, M., Gyftou, P., and Spyrellis, N. (2006) *J. Appl. Electrochem.*, **36**, 385–394.
66. Manson, T.J. (1986) *Ultrasonics*, **24**, 245.
67. Walker, R. (1993) in *Advances in Sonochemistry*, vol. 3 (ed. T.J. Manson), JAI Press.

68. Manson, T.J. and Lorimer, J.P. (1989) *Sonochemistry*, Ellis Horwood.
69. Margulis, M.A. (1976) *J. Phys. Chem.*, **50**, 1.
70. Prasad, P.B.S.N., Vasudevan, R., and Seshadri, S.K. (1993) *Trans. Indian Met.*, **46**, 247.
71. Walker, R. and Halagan, S.A. (1985) *Plat. Surf. Finish.*, **72**, 68.
72. Dereska, J., Jaeger, E., and Hovorka, F. (1957) *J. Acoust. Soc. Am.*, 69.
73. Namgoong, E. and Chun, J.S. (1984) *Thin Solid Films*, **120**, 153.
74. Walker, R.T. and Walker, C.T. (1975) *Ultrasonic*, **13**, 79.
75. Klima, J.K., Bernard, C., and Degrand, C. (1994) *J. Electroanal. Chem.*, **367**, 297.
76. Hyde, M.E. and Compton, R.G. (2002) *J. Electroanal. Chem.*, **531**, 19.
77. Ball, J.C. and Compton, R.G. (1999) *Electrochemistry*, **67**, 912.
78. Martins, L., Martins, J.I., Romeira, A.S., Costa, M.E., Costa, J., and Bazzaoui, M. (2004) *Mater. Sci. Forum*, **455–456**, 844.
79. Qu, N.S., Chan, K.C., and Zhu, D. (2004) *Scr. Mater.*, **50**, 1131.
80. Rezrazi, M., Doche, M.L., Bercot, P., and Hihn, J.Y. (2005) *Surf. Coat. Technol.*, **192**, 124.
81. Zheng, H.Y. and An, M.Z. (2008) *J. Alloys Compd.*, **459**, 548.
82. Lee, D., Gan, Y.X., Chen, X., and Kysar, W.J. (2007) *Mat. Sci. Eng. A*, **447**, 209.
83. Wang, C., Zhong, Y., Wang, J., Wang, Z., Ren, W.-L., Lei, Z., and Ren, Z. (2009) *J. Electroanal. Chem.*, **630** (1–2), 42–48.
84. Bund, A., Ispas, A., and Mutschke, G. (2008) *Sci. Technol. Adv. Mater.*, **9**, 024208.
85. Tacken, R. and Janssen, L. (1995) *J. Appl. Electrochem.*, **25**, 1.
86. Dash, J., Anderton, J., Litzenberger, B., and Trzynka, A. (1985) 168th Society Meeting, the Electrochemical Society, Las Vegas, p. 327.
87. Peipmann, R., Thomas, J., and Bund, A. (2007) *Electrochim. Acta*, **52**, 5808.
88. Hu, F., Chan, K., and Qu, N. (2007) *J. Solid State Electrochem.*, **11**, 267.

11
Adsorption Studies, Modeling, and Use of Green Inhibitors in Corrosion Inhibition: an Overview of Recent Research

Sanjay K. Sharma, Ackmez Mudhoo, and Essam Khamis

11.1
Introduction

Corrosion is the damage of material resulting from exposure and interaction with the environment. It is a major problem that must be confronted for safety, environment, and economic reasons [1]. Safety concerns are of utmost importance in every society and in all situations. Environmental concerns include pollution caused by corrosion, depletion of resources such as those needed for replacement of the corroded structures, and disposal of the corroded structures. To combine the technological progress with environmental safety is one of the key challenges of the millennium. Cleaner technology is a new dimension that is emerging rapidly at both the national and international level. Cleaner production has been identified as a key method for reconciling environment and economic development. The basic idea of cleaner production is to increase production efficiency while at the same time eliminate, or at least, minimize waste generation and emissions at their source rather than treat them at the end of the pipe only after they have been generated. The concept of cleaner production, pollution prevention, or waste reduction is still relatively new, although the concepts involved are much older. Both cleaner production and sustainability came into focus with the publication of "Our Common Future" [2]. This report provided a focused definition of the concept of sustainable development: "a process of change in which the exploitation of resources, the direction of investment, and the orientation of the technological development and instituted change are all in harmony and enhance both current and future potential to meet human need and aspiration." According to the World Commission on Environment and Development and the Brundtland Commission on Environment and Development, Brundtland Commission 1987, sustainable development is "development that meets the needs of the present without compromising the ability of future generations to meet their own needs." Sustainable development [3, 4] demands change, and requires doing more with lesser resource input and less waste generation. Instead of the end-of-pipe technology, it requires a pollution-prevention philosophy, which is "first and foremost, reduce waste at the origin through improved housekeeping and maintenance, and

Green Corrosion Chemistry and Engineering: Opportunities and Challenges, First Edition.
Edited by Sanjay K. Sharma.

modification in product design, processing, and raw material selection. Finally, if there is no prevention option possible, treat and safely dispose off the waste."

In the efforts to move toward "Sustainable Development," chemistry nowadays is at the forefront of the development of clean production processes and products. Chemistry is no doubt beginning to understand and protect our environment as the world's future is strongly dependent on the chemical processes adopted. Chemistry plays an integral part in our lives, and is all around us in the clothes we wear, the food we eat, the air we breathe, the buildings we use, and so on. Sustainability, eco-efficiency, and green chemistry are new principles that are guiding the development of next generation of products and processes [5]. Green Chemistry is considered to be an essential piece of a comprehensive program and also an alternating research methodology that is more eco- as well as economy friendly [6]. In its essence, Green Chemistry [7–11] is a science-based, and not a regulatory and economically driven, approach to achieving the goals of environmental protection and sustainable development.

In order to be eco friendly, or *green*, organic synthesis [12–15] must meet, if not all, at least some of the following requirements: avoid waste, be atom efficient, avoid use and production of toxic and dangerous chemicals, produce compounds that perform better or equal to the existing ones and are biodegradable, avoid auxiliary substances, reduce energy requirements, use renewable materials, and use catalysts rather than stoichiometric reagents. These requirements can be easily met by the concept of green chemistry. Safety and environmental concerns tend to be very difficult to translate and quantify in terms of monetary value. However, economic concerns tend to sensibly affect the cost estimates. According to a study completed in 1998 sponsored by the Federal Highway Administration (FHWA), corrosion of metals costs the United States in excess of US$276 billion per year. This loss to the economy is more than the entire gross national product of many countries around the world. This loss for 1975 was estimated at US$82 billion or 4.9% of the gross national product. However, corrosion not only affects the economy but it can also present a threat to life through the collapse of a structure or to the environment through the leakage of toxic chemicals.

Several efforts have been made for using corrosion-preventive practices, and the use of corrosion inhibitors is one of them. *Green Chemistry* provides many environment-friendly corrosion inhibitors, called "*green inhibitors*." The use of inhibitors for the control of corrosion of metals [16] and alloys, which are in contact with aggressive environment, is an accepted practice [17, 18]. Large numbers of organic compounds have been studied and are being studied to investigate their corrosion-inhibition potential. All these studies have revealed that organic compounds, especially those with N, S, and O, show significant inhibition efficiency. However, most of these compounds are not only expensive but also toxic, hence harmful to living beings [19]. It is needless to point out the importance of cheap, safe inhibitors of corrosion. Plant extracts and organic species have therefore become important as an environmentally acceptable, readily available, and renewable source for a wide range of inhibitors [20–29]. They are rich sources of ingredients that have very high inhibition efficiency [19, 30] and are hence termed "*green inhibitors*"

[24]. These green inhibitors are nonhazardous and eco friendly. This paper reviews and discusses the use of and research on corrosion inhibitors and green inhibitors recently reported in corrosion literature.

11.2
Adsorption Mechanisms in Corrosion Inhibition

Elayyachy *et al.* [31] have synthesized and tested two new telechelic compounds as inhibitors for the corrosion of steel in 1 M HCl solution. Weight loss measurements, potentiodynamic polarization, and electrochemical impedance spectroscopy (EIS) methods were used. Elayyachy *et al.* [31] found that the inhibiting action increased with the concentration of methyl 4-{2-[(2-hydroxyethyl)thio]ethyl}benzoate (T2) and 11-[(2-hydroxyethyl)thio]undecan-1-ol (T3) to attain 92% at 10^{-3} M and 90% at 10^{-4} M, respectively. The investigated adsorption of T3 on the steel surface was found to obey the Langmuir adsorption model. Elewady *et al.* [32] studied the role of some surfactants in the corrosion of Al in 1 M HCl using weight loss and galvanostatic polarization techniques. The results of Elewady *et al.* [32] showed that the inhibition occurred through adsorption of the inhibitor molecules on the metal surface. The inhibition efficiency was found to increase with increasing inhibitor concentration and decreased with increasing temperature, which is due to the fact that the rate of corrosion of Al is higher than the rate of adsorption. The latter observation was contrary to that observed by Elayyachy *et al.* [31] whereby the study of the effect of temperature between 35 and 80 °C showed that the inhibition efficiency had remained almost constant. Elewady *et al.* [32] also deduced, contrary to many other researchers, that the adsorption of these compounds on the metal surface was found to obey the Freundlich adsorption isotherm. Further research by Elewady *et al.* [32] showed that the inhibiting action of these compounds is considerably enhanced by the addition of KI, due to the increase in the surface coverage and therefore indicates the joint adsorption of these compounds and iodide ions. The thermodynamic parameters for adsorption and activation processes were determined, and the subsequent galvanostatic polarization data indicated that these compounds acted as mixed-type inhibitors.

On the other hand, Fuchs-Godec [33] came to a different conclusion in the study of mixed-type inhibitors. Fuchs-Godec [33] performed electrochemical measurements to investigate the effectiveness of cationic surfactants of the *N*-alkyl quaternary ammonium salt type, that is, myristyltrimethylammonium chloride (MTACl), cetyldimethylbenzylammonium chloride (CDBACl), and trioctylmethylammonium chloride (TOMACl), as corrosion inhibitors for type X4Cr13 ferritic stainless steel in a 2 M H_2SO_4 solution. The potentiodynamic polarization measurements made by Fuchs-Godec [33] showed that these surfactants hindered both anodic and cathodic processes, that is, they acted as mixed-type inhibitors. It was found that the adsorption of the *N*-alkyl ammonium ion in 2 M H_2SO_4 solution followed the Langmuir adsorption isotherm as observed by other researchers in their study [24, 34, 35]. The calculated values of the free energy of adsorption ΔG_{ads} by Fuchs-Godec

[33] were, in cases when the charge on the metal surface was negative with respect to the point of zero charge (PZC), relatively high, which is characteristic for the chemisorption. On the other hand, for positive metal surfaces, Fuchs-Godec [33] assumed that SO_4^{2-} anions were adsorbed first, so the cationic species would be limited by the surface concentration of anions. Accordingly, ΔG_{ads} values were lower in this case, and the adsorption was attributed to electrostatic attraction only, which was characteristic of physisorption.

In their study, Lebrini *et al.* [24] studied the corrosion inhibition of mild steel (MS) in perchloric acid by 3,5-bis(*n*-pyridyl)-4-amino-1,2,4-triazoles (*n*-PAT, $n = 2$, 3, and 4) at 30 °C using gravimetric and EIS techniques. Protection efficiencies of 95 and 92% were obtained with 12×10^{-4} Mof 3-PAT and 4-PAT, respectively; while 2-PAT reached only 65%. The inhibiting properties of *n*-PAT were found to depend on the concentration, and the order of increasing inhibition efficiency was correlated with the modification of the position of the nitrogen atom in the pyridinium substituent. It was shown by Lebrini *et al.* [24] that adsorption of 4-aminotriazole derivatives on the steel surface was consistent with the Langmuir adsorption isotherm, and the obtained standard free energy of adsorption values indicated that the corrosion inhibition of the MS in 1 M $HClO_4$ depends on both physi- and chemisorption, as opposed to the conclusions of sole physisorption from Fuchs-Godec [33]. Still, with respect to research on the adsorption mechanism of corrosion inhibition, Yurt *et al.* [35] investigated the effect of newly synthesized three Schiff bases – 2-[2-aza-2-(5-methyl(2-pyridly))vinyl]phenol, 2-[2-aza-2-(5-methyl(2-pyridly))vinyl]-4-bromophenol, and 2-[2-aza-2-(5-methyl(2-pyridly))vinyl]-4-chlorophenol – on the corrosion behavior of aluminum in 0.1 M HCl using potentiodynamic polarization, EIS, and linear polarization methods. Polarization curves obtained by Yurt *et al.* [35] indicated that all studied Schiff bases act as mixed-type inhibitors [33]. All measurements made by Yurt *et al.* [35] showed that inhibition efficiencies increased with an increase in inhibitor concentration [34, 36, 37]. This revealed that inhibitive actions of inhibitors were mainly due to adsorption on the aluminum surface. Adsorption of these inhibitors followed the Langmuir adsorption isotherms, and the thermodynamic parameters of adsorption (K_{ads}, ΔG_{ads}) of studied Schiff bases were calculated using the Langmuir adsorption isotherm. However, in the study of corrosion inhibition of copper in aerated, nonstirred 3% NaCl solutions in the temperature range 15–65 °C using sodium oleate as an anionic surfactant inhibitor, Amin [36] found that the sigmoidal shape of the adsorption isotherm confirmed the applicability of Frumkin's equation to describe the adsorption process rather than the much observed Langmuir isotherms.

The variation in inhibition efficiency values was found to depend on the type of functional groups substituted on benzene ring. It was also found by Yurt *et al.* [35] that the presence of bromine and chlorine atoms in the molecular structure of studied Schiff bases facilitates the adsorption of molecules on the aluminum surface, thus partially agreeing with the observations made by Elewady *et al.* [32] whereby the latter showed that the inhibiting action of telechelic compounds would be considerably enhanced by the addition of KI (iodide species). Yurt *et al.*

[35] also studied the correlation between the inhibition efficiencies of studied Schiff bases and their molecular structure using quantum chemical parameters obtained by modified neglect of differential overlap (MNDO) semiempirical self-consistent field – molecular orbital (SCF–MO) methods. Their results indicated that adsorption of studied Schiff bases depended on the charge density of the adsorption centers and their dipole moments. On the same line of discussion, El-Ashry *et al.* [38] also recently studied the correlation of the quantum chemical SCF calculations of some parameters of benzimidazoles with their inhibition efficiency in case of steel in an aqueous acidic medium. Geometric structures, total negative charge (TNC) on the molecule, highest occupied molecular orbital (E_{HOMO}), lowest unoccupied molecular orbital (E_{LUMO}), dipole moment (μ) and linear solvation energy terms, molecular volume (Vi), and dipolar polarization (π^*) were all found to be correlated to the corrosion-inhibition efficiency. The correlation between quantum parameters obtained calculation and experimental inhibition efficiency was validated by single-point calculations for the semiempirical AM1 structure using B3LYP/6–31G* as a higher level of theory and equations (with high R^2 values) were successfully proposed using linear regression analysis to calculate the corrosion-inhibition efficiency. It was ultimately established by El-Ashry *et al.* [38] that an increase in the orbital energies E_{HOMO} favored the inhibition efficiency toward steel corrosion.

11.3 Hybrid Coatings

Another field of extensive research on corrosion inhibition is the application of hybrid coatings preloaded with the inhibitors, especially sol–gel-derived organic–inorganic hybrid coatings [39]. In 2004, Pepe *et al.* [40] reported that chromates are among the most common substances used as corrosion inhibitors. However, these compounds are highly toxic, and an intense effort is being required (and is being undertaken) to replace them. According to Pepe *et al.* [40], cerium compounds seemed to fulfill the basic requirements for consideration as alternative corrosion inhibitors. The aim of the work of Pepe *et al.* [40] was therefore to study the effect of the incorporation of cerium ions in silica sol–gel coatings on aluminum alloys as potential replacement of chromate treatments. The main idea was to combine the "barrier" effect of silica coatings with the "corrosion inhibitor" effect of the cerium inside the coatings. Thin (below 1 μm for a single layer) and transparent cerium-doped silica sol–gel coatings were prepared by dipping 3005 aluminum alloys in sol–gel solutions. Ultraviolet–visible (UV–vis) spectra showed that cerium ions, Ce^{3+} and Ce^{4+}, were always present in the coatings, independently of the cerium salt or firing atmosphere used. Active protection with single- and two-layer coatings prepared with Ce(IV) salt seemed to improve corrosion protection of the coated aluminum, while coatings prepared with Ce(III) salt only entailed a protection when applied as a two-layer coating, possibly due to sealing of preexistent defects in the first layer. The improvement in active protection with immersion time would imply that corrosion is inhibited by cerium ions that migrate through the coating to the site of the attack (a defect in the coatings) and then react to passivate the site.

Khramov *et al.* [41] have developed sol–gel-derived organo-silicate hybrid coatings preloaded with organic corrosion inhibitors in order to provide active corrosion protection when the integrity of the coating is compromised. The incorporation of organic corrosion inhibitors into hybrid coatings has been achieved as a result of physical entrapment of the inhibitor within the coating material at the stage of film formation and cross-linking. Entrapped corrosion inhibitors become active in the corrosive electrolyte and can then slowly diffuse out of the host material. To ensure continuous delivery of the inhibitor to corrosion sites and long-term corrosion protection, a sustained release of the inhibitor is achieved by a reversible chemical equilibrium of either ion exchange of the inhibitor with the coating material or through cyclodextrin-assisted molecular encapsulation. Several organic compounds, such as mercaptobenzothiazole, mercaptobenzimidazole, mercaptobenzimidazolesulfonate, and thiosalicylic acid, have been selected and tested by Khramov *et al.* [41] to evaluate the effectiveness of these two approaches. Still, Lamaka *et al.* [42] studied the active corrosion protection of AA2024-T3 alloy provided by an environment-friendly, well-adhering pretreatment system consisting of an inhibitor-loaded titanium oxide porous layer and a sol–gel-based thin hybrid film. As a matter of fact, Lamaka *et al.* [42] proposed a novel approach aimed at developing a nanoporous reservoir for storing of corrosion inhibitors on the metal/coating interface. The nanostructured porous TiO_2 interlayer was prepared on the aluminum alloy surface by controllable hydrolysis of titanium alkoxide in the presence of a template agent. The morphology and the structure of the TiO_2 film were characterized with transmission electron microscopy (TEM), energy dispersive spectroscopy (EDS), scanning electron microscopy (SEM), and atomic force microscopy (AFM) techniques. Lamaka *et al.* [42] found that, in contrast to direct embedding of the inhibitors into the sol–gel matrix, the use of the porous reservoir eliminated the negative effect of the inhibitor on the stability of the hybrid sol–gel matrix. Hence, TiO_2/inhibitor/sol–gel systems showed enhanced corrosion protection and self-healing ability confirmed by EIS and scanning vibrating electrode technique (SVET) measurements.

Zheludkevich *et al.* [43] presented a new contribution to the development of a new protective system with self-healing ability composed of hybrid sol–gel films doped with nanocontainers that released entrapped corrosion inhibitors in response to pH changes caused by the corrosion process. A silica–zirconia-based hybrid film was used in this work as an anticorrosion coating deposited on a 2024 aluminum alloy. Silica nanoparticles covered layer by layer with polyelectrolyte layers and layers of inhibitor (benzotriazole) were randomly introduced into the hybrid films. The hybrid film with the nanocontainers revealed significantly enhanced long-term corrosion protection in comparison with the undoped hybrid film. The SVET also showed an effective self-healing ability of the defects. This effect was obtained due to regulated release of the corrosion inhibitor triggered and self-controlled by the pH feedback by the corrosion processes started in the cavities. This concept of feedback was further studied by Shchukin and Möhwald [44] wherein nanocontainers with the ability to release encapsulated active materials in a controlled way could be employed to develop a new family of self-repairing multifunctional coatings, which would not only possess passive functionality but also rapid feedback activity in response to changes in the local environment. Several approaches to fabricate self-repairing coatings on plastic and metal substrates were surveyed by Shchukin and Möhwald [44]. The release of the active materials was established to occur only when triggered by specific properties of the corrosion process, which prevented leakage of the active component out of the coating and increased coating durability.

Further research was conducted by Moutarlier *et al.* [45] to improve the self-repair properties of sol–gel films on the aluminum alloy. Moutarlier *et al.* [45] studied chromium(III), molybdate, permanganate, and cerium (III) by polarization resistance (Rp) in a chloride medium and compared it to the standard corrosion inhibitor, that is, hexavalent chromium [40]. The evolution of the composition of sol–gel coatings, during the corrosion test, was examined by glow discharge optical emission spectroscopy (GDOES). Moutarlier *et al.* [45] showed that the morphology of sol–gel and the solubility of the additive played a determinant role in the effectiveness of corrosion protection for a long term. Additives such as molybdate and permanganate ions decreased the sol–gel network stability and were too soluble (they were rapidly lost from the sol–gel films, in an aggressive medium), decreasing the power to prevent corrosion. Incorporation of Ce(III) was not efficient for a long time, as opposed to the findings of Pepe *et al.* [40] due to its high solubility.

Sol–gel films containing Cr(VI) and Cr(III) seemed to provide adequate corrosion protection, due to the sol–gel stability and their low solubility. Still maintaining that the chromium (III) and (IV) species were less suited for inhibition, Poznyak *et al.* [46] developed titania-containing, organic–inorganic hybrid sol–gel films as an alternative to chromate-based coatings for surface pretreatment of aluminum alloys. Stable hybrid sols were prepared by hydrolysis of 3-glycidoxypropyltrimethoxysilane and different titanium organic compounds in a 2-propanol solution in the presence of small amounts of acidified water. Different diketones were used as complexing agents in this synthesis for controllable hydrolysis of titanium organics. The properties of the obtained coatings were compared with those of zirconia-containing films. EIS measurements and standard salt spray tests have been performed by Poznyak *et al.* [46] to investigate the corrosion protection performance of the hybrid coatings. It was revealed that their protective properties depended significantly on the nature of metalorganic precursors and complexing agents used in the process of sol preparation. The best anticorrosive protection of AA2024 in chloride solutions was provided by the titania-containing sol–gel films prepared with titanium(IV) tetrapropoxide and acetylacetone as starting materials. In the case of zirconia-containing films, better protective properties were found when applying ethylacetoacetate as a complexing agent. Similarly, for AA2024, Yasakau *et al.* [47] have derived a certain number of corrosion inhibitors for AA2024 as additives to the hybrid sol–gel formulations in order to confer active corrosion protection without damaging the coating. 8-Hydroxyquinoline, benzotriazole, and cerium nitrate were added at different stages of the synthesis process to understand the role of the possible interaction of the inhibitor with components of the sol–gel system. SVET and EIS were employed as two main techniques to characterize the corrosion protection performance of the hybrid sol–gel films doped with inhibitors and to understand the mechanisms of corrosion protection. Yasakau *et al.* [47] have demonstrated that 8-hydroxyquinoline and cerium nitrate do not affect the stability of sol–gel films but would confer an additional active corrosion protection effect. In contrast to this, benzotriazole led to deterioration in the corrosion protection properties of hybrid sol–gel films.

Green algae were tested as natural additives for a paint formulation based on a vinyl chloride copolymer (VYHH) to evaluate its efficiency for protection of steel against corrosion in seawater using spectrophotometry, AC and DC electrochemical measurements, visual inspection, and surface analysis by Mansour *et al.* [27]. Both suspended and extracted forms of algae are utilized to achieve optimum performance of the algae-contained coatings. Poorest performance was obtained when algae were added in their suspended form, whereas the extracted form exhibited better performance based on impedance measurements. The data demonstrated that highest protection was obtained at an algae threshold concentration of about 1.4 wt%. The SEM and energy dispersive X-ray (EDX) analysis together with visual inspection of coated specimens exposed to the marine environment gave very good support to the electrochemical data [27].

11.4 Modeling Aspects

The value of mathematical and computer-aided models to organize data, to consider interactions in complex systems in a rational way, to correct the conventional wisdom, and to understand essential qualitative, and at times quantitative, features of biological, electrochemical, nuclear, physical, physicochemical, mathematical, and microscopic systems has been reasonably clearly documented in prior research for each of the latter fields of science and engineering.

The impact of modeling research specific to corrosion inhibition [48, 49] analysis discovery, however, has so far been moderate, but this will change in the future if we become adept at recognizing emerging opportunities and in integrating new concepts and tools into our research methodology. Mathematical structures and methods, allied with extraordinary contemporary computing power, are essential for the emerging field of metallurgical behavior prediction and analysis. In this quest, a hierarchy of powerful modeling, analysis, and computational tools, which can capture essential quantitative features of available experimental data and use these effectively for analysis and design of corrosion resistant structures, is therefore important. The few applications of computer modeling in corrosion inhibition discussed below attempt at providing an overview of the level of complexity involved in the corrosion process to a point to justifiably warrant the use of powerful modeling techniques.

Isgor and Razaqpur [50] developed a robust and comprehensive finite element model for predicting the rate of steel corrosion in concrete structures. The model of Isgor and Razaqpur [50] consisted of initiation and propagation stages, which were cast in the same time and space domains, that is, processes that commenced in the initiation stage, such as temperature, moisture, chloride ion, and oxygen transport within the concrete, maintained continuity into the propagation stage while active corrosion occurred contemporaneously. The model of Isgor and Razaqpur [50] was very innovative in the sense that it allowed the model to include the effects of changes in exposure conditions during the propagation stage on corrosion and the effects of the corrosion reactions on the properties of concrete. Moreover, as a novel approach to corrosion-inhibition analysis, Isgor and Razaqpur [50] calculated the corrosion rates on a steel surface by solving the intricate Laplace's equation for electrochemical potential with appropriate boundary conditions. These boundary conditions included the relationship between overpotential and current density for the anodic and cathodic regions. Owing to the nonlinear nature of these boundary conditions, a nonlinear solution algorithm was used. According to Isgor and Razaqpur [50], their model, being reasonably robust, can successfully enable designers to carry out comprehensive sensitivity analyses and to gauge the significance of variations in the values of certain parameters on the rate of corrosion in a concrete structure, so that ultimately better, corrosion-resistant concrete structures can be developed and cast for building and structural applications.

Kubo *et al.* [51] developed a mathematical model for the simulation of the changes in the pore solution phase chemistry of carbonated hardened cement paste when aqueous solutions of organic base corrosion inhibitors are applied to the surface of the material and constant current densities in the range of 1–5 A/m^2 are passed between anodes placed within the inhibitor solutions and steel mesh cathodes embedded within the paste. The model, derived from the Nernst–Planck equation, was used to predict the concentration profiles of electrochemically injected inhibitors and the major ionic species present within the pore electrolyte as the corrosion reactions proceeded. For their part of scientific contribution in the mathematical modeling of corrosion inhibition, Colorado-Garrido *et al.* [52] presented a predictive model for corrosion polarization curves using the artificial neural network (ANN). This proposed model obtained predictions of current in base of a corrosion inhibitor concentration and potential. The model of Colorado-Garrido *et al.* [52] was significantly different from other models developed to study corrosion inhibition in that the model took into account the variations of inhibitor concentration over steel by thermomechanical processing to decrease the corrosion rate. For the ANN, the Levenberg–Marquardt learning algorithm, the hyperbolic tangent sigmoid transfer function, and the linear transfer function were dexterously used by Colorado-Garrido *et al.* [52]. The best-fitting training data set was obtained with five neurons in the hidden layer, which made it possible to predict efficiency with accuracy, at least as good as that of the theoretical error, over the whole theoretical range. On the validation data set, simulations and theoretical data test were in very good positive agreement with an $R > 0.985$. Hence, the robustness of the model of Colorado-Garrido *et al.* [52] resided in its accurate predictive capacity of the current in short simulation times. With a different scenario of corrosion inhibition, Jingjun *et al.* [53] recently studied the flow-induced corrosion mechanisms for carbon steel in high-velocity flowing seawater and attempted to explain the corrosive phenomena. The design methodology of Jingjun *et al.* [53] comprised the derivation of an overall mathematical model for flow-induced corrosion of carbon steel in high-velocity flow seawater in a rotating disk apparatus using both extensive numerical computer simulation and test methods. By studying the impact of turbulent flow using the kinetic energy of a turbulent approach and the effects of the computational near-wall hydrodynamic parameters on corrosion rates, the corrosion behavior and mechanism were satisfactorily explored and explained by Jingjun *et al.* [53]. Jingjun *et al.* [53] held that their overall modeling and simulation approach could enable to understand the synergistic effect mechanism of flow-induced corrosion afresh and in depth. Jingjun *et al.* [53] also maintained that it was indeed scientific and reasonable to investigate carbon steel corrosion through correlation of the near-wall hydrodynamic parameters, which could accurately describe the influence of fluid flow on corrosion. The subsequent computational corrosion rates obtained by the model of Jingjun *et al.* [53] were in good agreement with the actual measured corrosion data. Jingjun *et al.* [53] also succeeded in showing that serious, flow-induced corrosion is caused by the synergistic effect between the corrosion electrochemical factor and the hydrodynamic factor, while the corrosion electrochemical factor plays a dominant role in flow-induced corrosion.

11.5 Green Inhibitors

All synthetic processes or engineering methods involve the use of different chemicals and materials that may cause corrosion to metals during the specific process. Some of them are very hazardous and environmentally problematic in nature. To control corrosion with conventional "toxic" corrosion inhibitors, they pose significant adverse effects on the environment and, at the end of the chain, to mankind. Such corrosion inhibitors are called "*gray inhibitors*" as has been discussed in the earlier sections. However, Green Chemistry provides many environment-friendly corrosion inhibitors, called "*green inhibitors,*" which are nonhazardous and eco friendly [19, 30, 54–59].

11.5.1 Natural Derivatives as Green Inhibitors

The safety and environmental issues of corrosion inhibitors that have arisen in industries have always been a global challenge. Chromates, for example, are used in the pretreatment of aluminum alloys [60]. Since chromates are suspected to be toxic as well as carcinogenic [61], many alternative corrosion inhibitors are being developed to reduce the jeopardizing effects on humans, animals, and the environment. *Green inhibitors* are environment-friendly corrosion inhibitors and range from rare earth elements [62–68] to organic compounds [69–77].

11.5.2
Research Orientations

The following digest of only some of the umpteen researches being carried out on the development and testing of green corrosion inhibitors exemplifies potential for the very wide array of green derivatives that can be synthesized and extracted for the said purpose. Many of the reported researches (over 3000) in literature may be broadly classified into the following categories of green derivatives for corrosion inhibitors: organic, amino acids, plant extracts, and rare earth elements based.

11.5.2.1 Organic-Based Green Inhibitors

Khaled [18] synthesized a new safe corrosion inhibitor namely *N*-(5,6-diphenyl-4, 5-dihydro-[1,2,4]triazin-3-yl)-guanidine (NTG), and its inhibitive performance toward the corrosion of MS in 1 M hydrochloric acid and 0.5 M sulfuric acid has been investigated. Corrosion inhibition was studied by weight loss, the Tafel extrapolation method and EIS. These studies of Khaled [18] have shown that NTG was a very good inhibitor in acid media and the inhibition efficiency could be reached as high as up to 99 and 96% in 1 M HCl and 0.5 M H_2SO_4 respectively. Khaled [18] intensified the research and made polarization measurements reveal that the investigated inhibitor is cathodic in 1 M HCl and mixed type in 0.5 M H_2SO_4. The activation energies of the corrosion process in the absence and presence of NTG were obtained by measuring the temperature dependence of the corrosion current density. Khaled [18] successfully demonstrated that the adsorption of the NTG inhibitor on the metal surface in the acid solution was found to obey the Langmuir's adsorption isotherm as has been observed with many of the conventional, but toxic, corrosion inhibitors [24, 31, 34, 35]. Gao and Liang [78] tested β-amino alcohols compounds in the series of 1,3-bis-dialkyl (C_nH_{2n+1}) aminopropan-2-ols as volatile corrosion inhibitors for brass in simulated atmospheric water using potentiodynamic, potentiostatic current transient, EIS, gravimetric, and volatile inhibition ability measurements. The evolution of the inhibition effect of the investigated compounds has been monitored by Gao and Liang [78] according to the length of alkyl chain. The results obtained by Gao and Liang [78] indicated that the inhibition efficiency increased with increasing the alkyl chain length and also inhibitor concentration. Polarization curves deduced by Gao and Liang [78] clearly supported that these compounds acted as good anodic inhibitors in water. Gao and Liang [78] discussed the inhibition mechanism in light of the chemical structure of undertaken inhibitors and came to the same, but widely validated, conclusion that the adsorption of the alcohol-based inhibitors on the brass surface followed the Langmuir adsorption isotherm model.

11.5.2.2 Amino Acids–Based Green Inhibitors

Barouni *et al.* [79] studied the inhibition effect of five amino acids on the corrosion of copper in molar nitric solution by using weight loss and electrochemical polarization measurements. Valine (Val) and glycine (Gly) were found to accelerate the corrosion process; but arginine (Arg), lysine (Lys), and cysteine (Cys) inhibited the corrosion phenomenon, with cysteine being the best among the three inhibitors. Its efficiency increased with concentration to attain 61% at 10^{-3} M. Ismail [80] further investigated the efficiency of cysteine as a nontoxic corrosion inhibitor for copper metal in 0.6 M NaCl and 1.0 M HCl by electrochemical studies. Potentiodynamic polarization measurements and EIS were equally employed by Ismail [80] to study the effect of cysteine on the corrosion inhibition of copper. Ismail [80] observed that a higher inhibition efficiency of about 84% could be achieved in the chloride solutions. The potentiodynamic polarization measurements made by Ismail [80] showed that the presence of cysteine in acidic and neutral chloride solutions affected mainly the cathodic process and decreased the corrosion current to a great extent and then shifted the corrosion potential toward more negative values. The experimental impedance data obtained by Ismail [80] were analyzed according to a proposed equivalent circuit model for the electrode/electrolyte interface. The results obtained from potentiodynamic polarization and impedance measurements by Ismail [80] were in good agreement, and the adsorption of cysteine on the surface of Cu, in neutral and acidic chloride solutions, followed the Langmuir adsorption isotherm very well with an adsorption free energy of cysteine on Cu of 25 kJ/mol revealing a strong physical adsorption of the organic-based inhibitor on the metal surface.

11.5.2.3 Plant Extracts–Based Green Inhibitors

Bendahou *et al.* [55] evaluated the effect of natural rosemary oil as a nontoxic inhibitor on the corrosion of steel in H_3PO_4 media at various temperatures. The oil was initially hydrodistilled and used as an inhibitor in various corrosion tests with gravimetric and electrochemical techniques being used to characterize the corrosion mechanisms. Chromatographic analysis by gas chromatography showed that the oil was rich in 1,8-cineole. Bendahou *et al.* [55] demonstrated good agreement between the various methods explored for corrosion-inhibition analysis. The polarization measurements showed that rosemary oil acted essentially as a cathodic inhibitor. The efficiency of the oil increased with the concentration (to attain 73% at 10 g/l) but decreased with the rise of temperature in the 25–75 °C range. According to novel results of Bendahou *et al.* [55], the natural oil could thus be used in chemical cleaning and pickling processes, thereby validating the originality of their work in finding of a safe and cheap inhibitor from natural plants. In Bothi Raja and Sethuraman [19], the corrosion-inhibitive effect of the extract of black pepper on MS in 1 M H_2SO_4 media was evaluated by conventional weight loss studies (33–50 °C), electrochemical studies namely Tafel polarization, AC impedance, and SEM. Results of weight loss studies reveal that black pepper extract acts as a good inhibitor even at high temperatures. The inhibition is through adsorption, which is found to follow the Temkin adsorption isotherm. Polarization curves revealed the mixed mode inhibition of black pepper extract. Analysis of impedance data has been made with equivalent circuit with constant phase angle element for calculation of a double-layer capacitance value. SEM studies provide the confirmatory evidence for the protection of MS by the green inhibitor.

El-Etre *et al.* [81] tested the aqueous extract of the leaves of henna (*Lawsonia*) as a corrosion inhibitor of steel, nickel, and zinc in acidic, neutral, and alkaline solutions, using the polarization technique. El-Etre *et al.* [81] found that the extract acted as a very good corrosion inhibitor for the three tested electrodes in all tested media. The inhibition efficiency increased as expected with the concentration of extract when increased. El-Etre *et al.* [81] postulated that the degree of inhibition depended on the nature of the metal and the type of the medium. For C-steel and nickel, the inhibition efficiency increased in the order: alkaline < neutral < acid, while in the case of zinc it increased in the order: acid < alkaline < neutral, thereby reconciling with the much observed concept of the Lawsonia extract being a mixed inhibitor as has been the case with umpteen gray inhibitors. The inhibitive action of the extract was discussed in view of adsorption of the complex Lawsonia molecules onto the metal surface. El-Etre *et al.* [81] found that this adsorption followed the Langmuir adsorption isotherm in all tested systems. El-Etre *et al.* [81] proposed, as a fresh explanation, that the formation of a complex between the metal cations and lawsone was an additional inhibition mechanism of steel and nickel corrosion. In their research with plant extracts for corrosion inhibition, Oguzie *et al.* [82] appraised the inhibiting effect of *Ocimum basilicum* extract on aluminum corrosion in 2 M HCl and 2 M KOH solutions, respectively, at 30 and 60 °C. The corrosion rates were determined using the gas-volumetric technique, and the values obtained in absence and presence of the extract was used in the calculation of the inhibition efficiency by Oguzie *et al.* [82]. Oguzie *et al.* [82] estimated the mechanism of inhibition from the trend of inhibition efficiency with temperature. According to Oguzie *et al.* [82], the *O. basilicum* extract was believed to inhibit aluminum corrosion in both the acidic and alkaline environments. Inhibition efficiency increased with extract concentration but decreased with rise in temperature, again suggesting physical adsorption of the organic matter on the metal surface. These results were corroborated by kinetic and activation parameters for corrosion and adsorption processes evaluated from the experimental data at the temperatures studied by Oguzie *et al.* [82]. On further testing, Oguzie *et al.* [82] found that halide additives synergistically improved the inhibition efficiency of the extract. Hence, the research of Oguzie *et al.* [82] provided new information on the possible application of the *O. basilicum* extract as an environment-friendly corrosion inhibitor. Oguzie *et al.* [82] maintained that the mixed extract–iodide formulation provided an effective means for retarding aluminum corrosion even in highly aggressive alkaline environments. Yet, a "sweeter" research was undertaken by Radojčić *et al.* [25] in their study of the influence of natural honey (chestnut and acacia) and natural honey with black radish juice on the corrosion of tin in aqueous and sodium chloride solutions using weight loss and polarization techniques. Radojčić *et al.* [25] observed that the inhibition efficiency of acacia honey was lower than that of chestnut honey, while the addition of black radish juice increased the inhibition efficiency of both honey varieties. The mechanism of corrosion inhibition was explained from a new perspective by Radojčić *et al.* [25] in that the process of inhibition was attributed to the formation of a multilayer adsorbed film on the tin surface, still following the established Langmuir adsorption isotherm.

Sethuraman and Bothi Raja [83] have evaluated the corrosion-inhibition potential of *Datura metel* in acid medium on MS with a view to develop new green corrosion inhibitors. The methodology of Sethuraman and Bothi Raja [83] consisted in studying an acid extract of the *D. metel* for its corrosion inhibitive effect by electrochemical and weight loss methods. Using weight loss measurement data, an attempt was made by Sethuraman and Bothi Raja [83] to probe the mechanism of inhibitive action by fitting the different established adsorption isotherms. Sethuraman and Bothi Raja [83] were very convincing in showing the significant corrosion inhibitive effect in an acid medium on MS. Sethuraman and Bothi Raja [83] explained on the basis of their findings that inhibition was through adsorption of the phytoconstituents on MS following both the Temkin and Langmuir adsorption isotherms. According to Sethuraman and Bothi Raja [83], the *Datura metel* plant was being investigated for the first time for its corrosion-inhibitive properties, and the green inhibitor developed by them could possibly find use in the inhibition of corrosion in industries where MS is used as a material of choice for the fabrication of machinery.

Acid cleaning of MS by dissolution in 1 M sulfuric acid/10% methanol (test solution) in the presence of Arghel herb extract as a green inhibitor was monitored by potentiodynamic and electrochemical impedance techniques in the temperature range 30–60 °C by Khamis *et al.* [28]. At all temperatures, the corrosion rate decreased with increasing inhibitor concentration. Potentiodynamic polarization measurements indicated that the inhibitor has a strong effect on the corrosion behavior of the steel and behaves as a mixed-type inhibitor [28]. Impedance results indicate that the charge transfer controls the dissolution mechanism of steel across the phase boundary in the absence and presence of the inhibitor. The inhibition efficiency obtained from the various methods employed was in good agreement. Thermodynamic and activation parameters obtained from this study indicated that the presence of the Arghel increases the activation energy, and the adsorption process of the inhibitor on the metal surface is spontaneous [28].

Inhibition of aluminum corrosion in 2 M sodium hydroxide solution in the presence and absence of 0.5 M NaCl using damsissa (*Ambrosia maritime*, L.) extract has been studied employing different chemical and electrochemical techniques by Abdel-Gaber *et al.* [29]. The chemical gasometry technique showed that addition of chloride ions or damsissa extract to sodium hydroxide solution decreases the volume of the hydrogen gas evolved. Potentiodynamic results manifested that chloride ion retards the anodic dissolution of aluminum below the pitting potential in sodium hydroxide solution. Damsissa extract, in the presence or absence of chloride ion, influenced both the anodic dissolution of aluminum and the generated hydrogen gas at the cathode, indicating that the extract behaved as a mixed-type inhibitor. The decrease in the observed limiting current with increasing damsissa extract concentration indicated that the anodic process is controlled by diffusion. Nyquist plots present two capacitive semicircles at higher and lower frequencies separated by an inductive loop at intermediate frequencies [29]. The inductive loops were clarified by the occurrence of adsorbed intermediates on the surface. A proposed equivalent circuit was used to analyze the impedance spectra for aluminum in alkaline solutions. The results showed that the damsissa extract could serve as an effective inhibitor for the corrosion of aluminum in alkaline solutions. The impedance measurements verified the remarkable stability of the extracts during storage up to 35 days. The Damsissa extract was found to be more effective in the presence of chloride ions than in its absence. Inhibition was found to increase with increasing concentration of the extract but decreased with increasing temperature. The associated activation parameters were determined and discussed [29].

The inhibition of corrosion of aluminum, using a package composed mainly of the extract of *Lupine* and *Damsissa* plants in addition to surfactant, has been studied using potentiodynamic and impedance techniques. The inhibition effect of surfactant in the inhibitor package is shown to increase at composition around its critical micelle concentration. The extract of *Arghel* introduced into concrete during its mixing reduces the attack of chloride ions to the steel reinforcement. Polarization measurements illustrated that *Arghel* behaved as an anodic-type inhibitor in simulated seawater solutions. EIS data revealed that *Arghel* has a pronounced effect on the diffusion process [84].

Khamis *et al.* [85] investigated the dual function of *olive* (*Olea europaea* L) extract as antiscalent for $CaCO_3$ deposit and as a corrosion inhibitor for steel in alkaline $CaCl_2$ brine solution. The antiscalent properties of this extract were studied using conductivity measurements, EIS, and chronoamperometry techniques in conjunction with microscopic examination. The inhibitive characteristics were investigated using EIS and potentiodynamic polarization curves measurements. Mineral scales were deposited from the brine solution by cathodic polarization of the steel surface at −0.9 V. The extract was found to impede $CaCO_3$ supersaturation and decrease the time of nucleation [85]. The scaling steps of nucleation, crystal growth, and total coverage of the surface regions were characterized by the chronoamperometry technique and different impedance spectra. The surface area occupied by the scale deposits decreased with increasing extract concentrations. Potentiodynamic polarization curves indicated that plant-leaf extract inhibits the corrosion of steel by controlling the cathodic oxygen reduction process. The data were compared with those obtained previously using *fig* leaves (*Ficus carica* L.) extract. The results showed that *olive* and *fig* leaf extract can decrease corrosion and scale buildup under the conditions tested [85].

11.5.2.4 **Rare Earth Elements–Based Green Inhibitors**

Considerable research efforts have been made to develop nontoxic and equally performing green corrosion inhibitors from rare earth elements. In principle, the rare earth element is selected from cerium, terbium, praseodymium, or a combination thereof, and at least one rare earth element should be in the tetravalent oxidation state. An inorganic or organic material is then used to stabilize the tetravalent rare earth ion to form a compound that is sparingly soluble in water. Specific stabilizers are chosen to control the release rate of tetravalent cerium, terbium, or praseodymium during exposure to water and to tailor the compatibility of the powder when used as a pigment in a chosen binder system. Stabilizers may also modify the processing and handling characteristics of the formed powders. Many rare earth–valence stabilizer combinations have been shown to equal the performance of conventional hexavalent chromium systems.

Blin *et al.* [86] investigated the corrosion-inhibition mechanisms of new cerium and lanthanum cinnamate–based compounds through the surface characterization of the steel exposed to NaCl solution of neutral pH. Blin *et al.* [86] used attenuated total reflectance–Fourier transform infrared (ATR–FTIR) spectroscopy to identify the nature of the deposits on the metal surface and confidently demonstrated that after accelerated tests the corrosion product commonly observed on steel (which is lepidocrocite, γ-FeOOH) is absent. The cinnamate species were clearly present on the steel surface upon exposure to NaCl solution for short periods and appeared to coordinate through the iron. Later on, the rare earth metal oxyhydroxide species were proposed by Blin *et al.* [86] to have formed as identified through the bands in the 1400–1500 cm^{-1} regions. According to Blin *et al.* [86], the protection mechanism appeared to involve the adsorption of the (rare earth metal) REM–cinnamate complex followed by the hydrolysis of the REM to form a barrier oxide on the steel surface, this being a novel mechanism involved and observed in corrosion inhibition. Blin *et al.* [87] further intensified their research on rare earth elements by demonstrating through a combination of linear polarization resistance (LPR) and cyclic potentiodynamic polarization (CPP) measurements that the lanthanum–4 hydroxy cinnamate compound could inhibit both the cathodic and anodic corrosion reactions on MS surfaces exposed to 0.01 M NaCl solutions. However, the dominating response was shown to vary with inhibitor concentration. At concentrations for which the highest level of protection was achieved, both REM and 4-hydroxy cinnamate (REM being lanthanum and mischmetal) displayed a strong anodic behavior for MS and their inhibition performance, including their resistance against localized attack, improved with time.

11.6 Conclusions

The diverse set of research summarized and discussed advocates that intense research efforts are being deployed to tackle the much problematic phenomenon of corrosion. Although it is being realized that the preceding discussions are modest, the essence remains that a wide variety of chemical, biological, physical, mechanical, electrical, metallurgical, nuclear, electrochemical, and computational techniques are being employed to make progress in the current research to shift from gray corrosion inhibitors to green corrosion inhibitors. Green corrosion inhibitors are increasingly being synthesized from organic, amino acids, and plant extracts and tested in their inhibition efficiencies, while modeling research is attempting to address and elucidate the mechanism of the chemical reactions involved in the corrosion process and in corrosion inhibition with the use of modeling techniques such as finite element modeling and ANN. Gray corrosion inhibitions unanimously adhere to the Langmuir adsorption model while several findings of research on equally performing and inhibitive green inhibitors have supported additional adsorption mechanisms captured rightly by Frumkin's equation and Freundlich and Temkin adsorption isotherms. Latest corrosion protection techniques involve the use of nanomaterials but the use of green inhibitors remains a much safer and environmentally secure way of protection against corrosion. It is not only environment friendly but also a cost-effective method with the added advantage of waste minimization. Therefore, it may be inferred that the applications of green inhibitors have a wide scope of research and development yet to be fully explored.

Acknowledgments

This chapter is a revised version of a review paper that has previously been published in the Journal of Corrosion Science and Engineering, an entirely electronic journal published by the Corrosion and Protection Centre, University of Manchester, in collaboration with the International Corrosion Council. The journal can be accessed at *http://www.jcse.org*. Readers are requested to make any comments on this paper to JCSE through their online comment mechanism. In this respect, the authors are also thankful to Prof. Bob Cottis, Editor of the Journal of Corrosion Science and Engineering for having been gracious enough to grant us the permission to reproduce this review in the current form.

References

1. Thompson, N.G., Yunovich, M., and Dunmire, D. (2007) *Corros. Rev.*, **25** (3/4), 247–262.
2. Brundtland, G.H. (1987) *Our Common Future*, Oxford University Press, New York.
3. Hutzinger, O. (1999) *Environ. Sci. Pollut. Res.*, **6**, 123.
4. Desimone, L. and Popoff, F. (2000) *Eco–Efficiency: The Business Link to Sustainable Development*, MIT Press, Cambridge, MA.
5. Sanghi, R. (2000) *Corros. Sci.*, **79**, 1662.
6. Sharma, S.K., Chaudhary, A., and Singh, R.V. (2008) *Rasayan J. Chem.*, **1** (1), 68–92.
7. Anastas, P.T. and Farris, C.A. (1994) *Benign by Design: Alternative Synthetic Design for Pollution Prevention*, ACS Symposium Series, Vol. N557, ACS, Washington DC.
8. Tundo, P. and Selva, M. (1996) in *Green Chemistry: Designing Chemistry for the Environment*, ACS Symposium Series, Vol. 626 (ed. T.C. Williamson), American Chemical Society, p. 81.
9. Collins, T.J. (1997) Green chemistry, in *Macmillan, Encyclopedia of Chemistry*, Vol. 2, Simon and Schuster Macmillan, New York, pp. 691–697.
10. Wilkinson, S.L. *Chem. Eng. News*, **75**, 35.
11. Anastas, P.T. (1999) *Crit. Rev. Anal. Chem.*, **29**, 167.
12. Grieco, P.A. (ed.) (1988) *Organic Synthesis in Water*, Blackie, London.
13. Faber, K. (1997) *Biotransformations in Organic Chemistry: Text Book*, 3rd completely rev. edn, Springer-Verlag, Berlin.
14. Jessop, P.G. and Leitner, W. (1999) *Chemical Synthesis in Supercritical Fluids*, Wiley-VCH Verlag GmbH, Weinheim.
15. Sanghi, R. (2000) *Resonance*, **5** (3), 77.
16. Valdez, B., Cheng, J., Flores, F., Schorr, M., and Veleva, L. (2003) *Corros. Rev.*, **21** (5–6), 445–458.

17. Taylor, S.R. and Chambers, B.D. (2007) *Corros. Rev.*, **25** (5/6), 571–590.
18. Khaled, K.F. (2008) *Int. J. Electrochem. Sci.*, **3**, 462–475.
19. Bothi Raja, P. and Sethuraman, M.G. (2008) *Mater. Lett.*, **62** (1), 113–116.
20. Rajendran, S., Amalraj, A.J., Joice, M.J., Anthony, N., Trivedi, D.C., and Sundaravadivelu, M. (2004) *Corros. Rev.*, **22** (3), 233–248.
21. Mathiyarasu, J., Pathak, S.S., and Yegnaraman, V. (2006) *Corros. Rev.*, **24** (5/6), 307–322.
22. Mesbah, A., Juers, C., Lacouture, F., Mathieu, S., Rocca, E., François, M., and Steinmetz, J. (2007) *Solid State Sci.*, **9** (3/4), 322–328.
23. Okafor, P.C., Osabor, V.I., and Ebenso, E.E. (2007) *Pigment Resin Technol.*, **36** (5), 299–305.
24. Lebrini, M., Traisnel, M., Lagrenée, M., Mernari, B., and Bentiss, F. (2008) *Corros. Sci.*, **50** (2), 473–479.
25. Radojčić, I., Berković, K., Kovač, S., and Vorkapić-Furač, J. (2008) *Corros. Sci.*, **50** (5), 1498–1504.
26. Refaey, S.A.M., Abd El Malak, A.M., Taha, F., and Abdel-Fatah, H.T.M. (2008) *Int. J. Electrochem. Sci.*, **3**, 167–176.
27. Mansour, E.M.E., Abdel-Gaber, A.M., Abd-El Nabey, B.A., Khalil, N., Khamis, E., Tadros, A., Aglan, H., and Ludwick, A. (2003) *Corrosion*, **59**, 242.
28. Khamis, E., Hefnawy, A., and El-Demerdash, A.M. (2007) *Materialwiss. Werkstofftech.*, **38**, 227.
29. Abdel-Gaber, A.M., Khamis, E., Abo-Eldahab, H., and Adeel, S. (2008) *Mater. Chem. Phys.*, **109**, 297.
30. Little, B.J., Lee, J.S., and Ray, R.I. (2007) *Biofouling*, **23** (2), 87–97.
31. Elayyachy, M., Hammouti, B., and El Idrissi, A. (2005) *Appl. Surf. Sci.*, **249** (1–4), 176–182.
32. Elewady, G.Y., El-Said, I.A., and Fouda, A.A. (2008) *Int. J. Electrochem. Sci.*, **3**, 177–190.
33. Fuchs–Godec, R. (2006) *Colloids Surf.*, **280** (1–3), 130–139.
34. Quraishi, M.A. and Khan, S. (2006) *J. Appl. Electrochem.*, **36** (5), 539–544.
35. Yurt, A., Ulutas, S., and Dal, H. (2006) *Appl. Surf. Sci.*, **253** (2), 919–925.
36. Amin, M.A. (2006) *J. Appl. Electrochem.*, **36** (2), 215–226.
37. Fouda, A.S., Mostafa, H.A., Ghazy, S.E., and El-Farah, S.A. (2007) *Int. J. Electrochem. Sci.*, **2**, 182–194.
38. El-Ashry, E.S.H., El Nemr, A., Essawy, S.A., and Ragab, S. (2008) *Prog. Org. Coat.*, **61** (1), 11–20.
39. Pathak, S.S., Khanna, A.S., and Sinha, T.J.M. (2006) *Corros. Rev.*, **24** (5–6), 281–306.
40. Pepe, A., Aparicio, M., Ceré, S., and Durán, A. (2004) *J. Non-Cryst. Solids*, **348**, 162–171.
41. Khramov, A.N., Voevodin, N.N., Balbyshev, V.N., and Donley, M.S. (2004) *Thin Solid Films*, **447–448**, 549–557.
42. Lamaka, S.V., Zheludkevich, M.L., Yasakau, Y.A., Serra, R., Poznyak, S.K., and Ferreira, M.G.S. (2007) *Prog. Org. Coat.*, **58** (2–3), 127–135.
43. Zheludkevich, M.L., Shchukin, D.G., Yasakau, K.A., Möhwald, H., and Ferreira, M.G.S. (2007) *Chem. Mater.*, **19** (3), 402–411.
44. Shchukin, D.G. and Möhwald, H. (2007) *Small*, **3** (6), 926–943.
45. Moutarlier, V., Neveu, B., and Gigandet, M.P. (2008) *Surf. Coat. Technol.*, **202** (10), 2052–2058.
46. Poznyak, S.K., Zheludkevich, M.L., Raps, D., Gammel, F., Yasakau, K.A., and Ferreira, M.G.S. (2008) *Prog. Org. Coat.*, **62** (2), 226–235.
47. Yasakau, K.A., Zheludkevich, M.L., Karavai, O.V., and Ferreira, M.G.S. (2008) *Prog. Org. Coat.*, **63** (3), 352–361. Article in press.
48. Papavinasam, S., Revie, R.W., Friesen, W.I., Doiron, A., and Panneerselvan, T. (2006) *Corros. Rev.*, **24** (3–4), 173–230.
49. Zieliński, A. and Sobieszczyk, S. (2008) *Corros. Rev.*, **26** (1), 1–22.
50. Isgor, O.B. and Razaqpur, A.G. (2006) *Mater. Struct.*, **39** (3), 291–302.
51. Kubo, J., Sawada, S., Page, C.L., and Page, M.M. (2007) *Corros. Sci.*, **49** (3), 1205–1227.
52. Colorado-Garrido, D., Ortega-Toledo, D.M., Hernandez, J.A., and Gonzalez-Rodriguez, J.G. (2007) Proceedings of the Electronics, Robotics

and Automotive Mechanics Conference – CERMA, pp. 213–218.
53. Jingjun, L., Yuzhen, L., and Xiaoyu, L. (2008) *Anti-Corros. Methods Mater.*, **55** (2), 66–72.
54. Davó, B., Conde, A., and de Damborenea, J.J. (2005) *Corros. Sci.*, **47** (5), 1227–1237.
55. Bendahou, M., Benabdellah, M., and Hammouti, B. (2006) *Pigment Resin Technol.*, **35** (2), 95–100.
56. Craddock, H.A., Caird, S., Wilkinson, H., and Guzmann, M. (2006) *J. Pet. Technol.*, **58** (12), 50–52.
57. Deacon, G.B., Forsyth, M., Junk, P.C., Leary, S.G., and Moxey, G.J. (2006) *Polyhedron*, **25** (2), 379–386.
58. Taha, A.A. (2006) The 10th Annual Green Chemistry and Engineering Conference.
59. Scendo, M. (2007) *Corros. Sci.*, **49** (7), 2985–3000.
60. Hinton, B.R.W., Arnott, D.R., and Ryan, N.E. (1986) *Met. Forum*, **9** (3), 162–173.
61. Costa, M. and Klein, C.B. (2006) *Crit. Rev. Toxicol.*, **36** (2), 155–163.
62. Kilbourn, B.T. (1985) *Ceram. Eng. Sci. Processes*, **6** (9–10), 1331.
63. Bethencourt, M., Botana, F.J., Cauqui, M., Marcos, M., and Rodriguez, M. (1997) *J. Alloys Compd.*, **250**, 455–460.
64. Virtanen, S., Ives, M.B., Sproule, G.I., Schmuki, P., and Graham, M.J. (1997) *Corros. Sci.*, **39** (10–11), 1897.
65. Bethencourt, M., Botana, F.J., Calvino, J.J., and Marcos, M. (1998) *Corros. Sci.*, **40** (11), 1803–1819.
66. Powell, S.M., Mcmurray, H.N., and Worsley, D.A. (1999) *Corrosion*, **55** (11), 1040.
67. Arenas, M.A., Conde, A., and de Damborenea, J. (2002) *Corros. Sci.*, **44**, 511–520.
68. Hughes, A.E., Ho, D., Forsyth, M., and Hinton, B.R.W. (2007) *Corros. Rev.*, **25** (5–6), 591–606.
69. Choi, D.J., Kim, Y.W., and Kim, J.G. (2001) *Mater. Corros.*, **52**, 697–704.
70. Davis, G.D., von Fraunhofer, A., Krebs Lorrie, A., and Dacres, C.M. (2001) Paper 01558, Presented at Conference CORROSION 2001, Government work published by NACE International.
71. Cano, E., Pinilla, P., Polo, J.L., and Bastidas, J.M. (2003) *Mater. Corros.*, **54**, 222–228.
72. Aballe, L., Bethencourt, M., Botana, F.J., and Marcos, M. (2001) *J. Alloys Compd.*, **323/324**, 855–858.
73. El-Maksoud, A.S.A. (2003) *Mater. Corros.*, **54**, 106–112.
74. Karda, G. and Solmaz, R. (2006) *Corros. Rev.*, **24** (3–4), 151–172.
75. Moretti, G., Guidi, F., and Grion, G. (2004) *Corros. Sci.*, **46** (2), 387–403.
76. El-Sawy, S.M., Abu-Ayana, Y.M., Mohdy, A., and Fikry, A. (2001) *Anti-Corros. Method Mater.*, **48** (4), 227–234.
77. Khamis, E. and Al-Andis, N. (2002) *Materialwiss. Werkstofftech.*, **33**, 550–554.
78. Gao, G. and Liang, C. (2007) *J. Electrochem. Soc.*, **154** (2), 144–151.
79. Barouni, K., Bazzi, L., Salghi, R., Mihit, M., Hammouti, B., Albourine, A., and El Issami, S. (2008) *Mater. Lett.*, **62** (19), 3325–3327.
80. Ismail, K.M. (2007) *Electrochim. Acta*, **52** (28), 7811–7819.
81. El-Etre, A.Y., Abdallah, M., and El-Tantawy, Z.E. (2005) *Corros. Sci.*, **47** (2), 385–395.
82. Oguzie, E.E., Onuchukwu, A.I., Okafor, P.C., and Ebenso, E.E. (2006) *Pigment Resin Technol.*, **35** (2), 63–70.
83. Sethuraman, M.G. and Bothi Raja, P. (2005) *Pigment Resin Technol.*, **34** (6), 327–331.
84. Khamis, E., Abdel-Gaber, A.M., Abd-El Nabey, B.A., Hefnawy, A., Aglan, H., and Ludwick, A. (2005) First World Congress on Corrosion in the Military: Cost Reduction Strategies, Grand Hotel Vesuvio –Sorrento, Italy.
85. Khamis, E., Abd-El-Nabey, B.A., Abdel-Gaber, A.M., and Abd-El-Khalek, D.E. (2008) 26th Water Treatment Technology Conference from 7–9 June 2008 in the Four Seasons Hotel at Alexandria San Stefano, Egypt.
86. Blin, F., Leary, S.G., Deacon, G.B., Junk, P.C., and Forsyth, M. (2006) *Corros. Sci.*, **48** (2), 404–419.
87. Blin, F., Koutsoukos, P., Klepetsianis, P., and Forsyth, M. (2007) *Electrochim. Acta*, **52** (21), 6212–6220.

12
Indian Initiatives for Corrosion Protection

Anand Sawroop Khanna

12.1
Introduction

Corrosion leads to the deterioration of materials, losing its esthetics, resulting in metal loss, and thereby decreasing the load-bearing capability of the supporting structure. Corrosion is a widespread process. It ranges from household materials to chemical, petrochemical plants, power plants, and offshore structures including ships. Corrosion cannot be avoided but can be minimized by a number of corrosion protection methods.

There is heated discussion in many national and international conferences/symposia and seminars in the country today to know the exact loss due to corrosion in India. Everyone has a different figure to offer. Why is this so? This is because there is hardly any study, carried out in India, to estimate how much our industry looses as a result of corrosion. The approximate values, ranging from Rs. 50 000 to 100 000 crore a year, are based on the independent interpretation from the 3% of GNP of any developed country – the figure given by American study in the year 2002.

The NACE India Section directly calculates the annual losses in India by just giving 3% of Indian GNP, which is ~Rs. 100 000 crore. It is on the very high side. However, if one carefully studies the growth of Indian GNP in the last decade, it cannot be mainly due to very strong industrial activity as is there in many other developed countries or in China. A big contribution to Indian GNP is by the IT industry, which contributes less toward corrosion losses. Hence, while calculating corrosion losses, it will be wrong to assume 3% of Indian GNP blindly, but a more conservative figure, such as half of this value, will be more reasonable. The Society for Surface Protective Coatings (SSPC) India proposes a value of Rs. 50 000 crore, which appears to be more reasonable than the 100 000 crore proposed by the NACE India Section. However, it is the need of the hour to verify these estimates by carrying out a systematic study on corrosion losses. In 2003, I submitted a proposal to the Council of Scientific and Industrial Research (CSIR), but they turned down the proposal without understanding its importance. It is high time that a national funding be made available to carry out a systematic study on corrosion losses.

Green Corrosion Chemistry and Engineering: Opportunities and Challenges, First Edition.
Edited by Sanjay K. Sharma.

What advantage will it give? Indian industries will know what mistakes they are making in not implementing the proper corrosion protection policies, and they will intend to use latest methods of corrosion protection, corrosion monitoring, and so on, which will result in minimizing corrosion losses and in turn accidents and plant shut downs. Figure 12.1 depicts an American study that shows the annual cost of corrosion for various industries.

Corrosion scenario in India is not very good. If we talk about education, there are only limited number of institutes in India, who have dedicated master's programs in Corrosion Science and Engineering. Although it has great demand, there are efforts to merge them or close them, which is unfortunate. In some universities and other IITs, corrosion is taught just as an elective course to fourth year metallurgical students only. Is it not surprising that we are lacking in propagating the knowledge of the subject because of which more than 50 000 crore is lost every year?

That is why knowledge of corrosion is mostly imparted by professional bodies or by experts who organize courses for industries and other users. NACE India Section and SSPC India are the main professional bodies who organize these courses for industry. There are, however, several institutes where R&D in corrosion is pursued. The most common are Atomic Energy, DRDO's, ISRO, and a few CSIR laboratories such as Karaikudi and NML Jamshedpur. Corrosion laboratories at Karaikudi are excellent; unfortunately, because of its remote location, many industries are unable to utilize it. Out of the various IITs, IIT Bombay has the best corrosion facilities on all aspects. And research on large areas, such as coatings, aqueous corrosion, high-temperature corrosion, and surface modification, is going on.

There are three main components of corrosion – control, monitor, and maintenance. Today, a better term is used called *corrosion management*, which basically involves three stages corrosion control by coatings, cathodic protection, or the use of inhibitors; corrosion monitoring by various online methods; and periodic maintenance. There is an additional component here, that is, failure analysis, and understanding its input is utilized to improve corrosion protection.

Many oil and gas industries are spending millions of rupees to control the corrosion at offshore, in refineries, and during the transportation of crude/gas to the respective sites. Millions of rupees are spent to protect underground pipelines from soil corrosion by monitoring the pipes with intelligent pigs from inside and cathodic potential from outside. Many of the coatings used for these pipelines are imported. In the same way, the monitoring pigs are rented by spending a huge cost. This situation is similar in many refineries, offshore structures, and petrochemical plants.

Hence, the need of the hour is to develop several institutions for corrosion education, supplemented with industrial courses and training programs by professional bodies, and for the development and manufacturing of highly corrosion-resistant materials such as special alloys, duplex stainless steels, Hastelloy, special plastics, fiber reinforced plastics (FRPs), new corrosion control coatings, and cathodic protection monitoring systems.

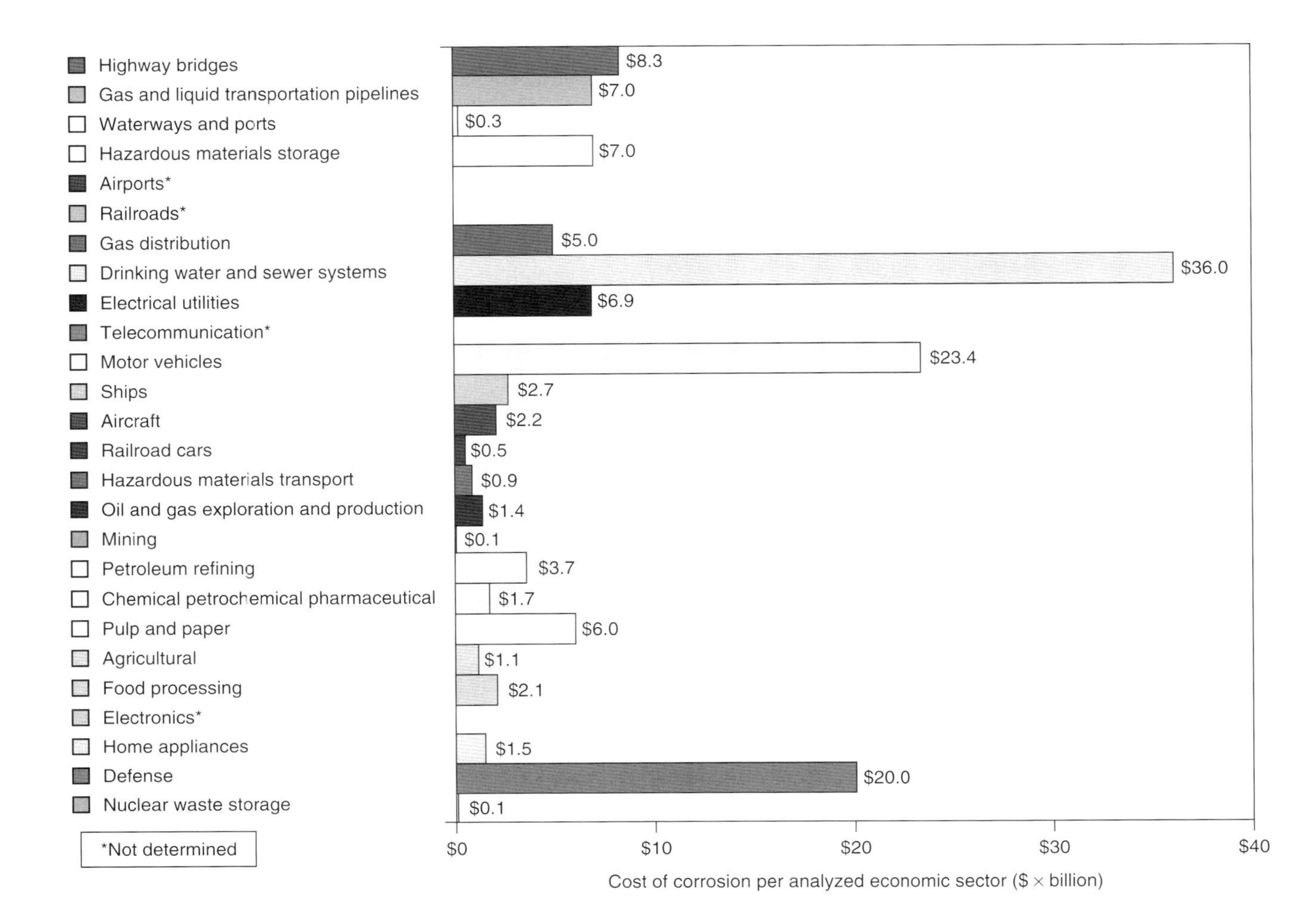

Figure 12.1 Cost of Corrosion in various sectors.

12.2
Scenario of the Indian Industry

Indian industry today is quite widespread and multifaceted. There is hardly any sector where it is not expanding. There are a number of refineries in various parts of the country spread from Kashmir to Kanyakumari and Assam to Mumbai, with the biggest refinery now in Jamnagar owned by Reliance Industries. These refineries convert the crude oil extracted from offshore into petrol, diesel, and many other important hydrocarbons and petroleum products. Petrochemical and chemical processing plants are the other category of industries working successfully in the country. Fertilizer plants and a large number of power plants are the heartbeat of the country, which provide the important ingredient (urea) to our farmers and electricity in the country. Another category of plants in India are the steel plants and copper, zinc, and aluminum plants. All these industries mentioned are corrosion prone. A rough statistics of some of these industries and the type of problems encountered are given in Table 12.1.

As per Chem-Pro-Tech India, the size of the Indian chemical industry is over Rs. 1750 billion, with a growth rate of 12.5% and is the 12th largest in world and 3rd largest in Asia. We are definitely not using the best corrosion protection technology. The concept of corrosion management in chemical industry is still a far dream. Hence, it is accepted that the annual losses can be over Rs. 50 billion (~3% of the total value).

Another most corrosion-prone area is the oil and gas exploration and distribution, shipping, ports and coast surveillance, and safety of ports. Today, India has more than 250 offshore platforms, 25 000 km of underground pipelines, perhaps more

Table 12.1 Statistics of various corrosion-prone industries and the type of corrosion observed [1–4].

Type of industry	Approximate number	Total capacity	Corrosion types observed
Refineries *www.petroleum.nic.in/refi.htm* (till 2005)	18	127.37 MMTPA	U, L, S, M
Petrochemical plants *business.mapsofindia.com/india-petroleum-industry/indian-petrochemical-plants.html*	56	5 MTPA	U, L, S, C, HTC
Power plants (fossil fuel based) *http://www.cpcb.nic.in/divisionsofheadoffice/pci2/ThermalpowerPlants.pdf*	83	147 000 MW	H, HTC, U
Fertilizer plants *http://www.energymanagertraining.com*	115	430 08 500 MTPA	U, L, SCC, HTC
Steel plants *http://www.energymanagertraining.com*	11	241 97 000 MTPA	U, L

Abbreviations: U, uniform; L, localized; C, carburization; S, sulfidation; H, hot corrosion; HTC, high-temperature corrosion; M, metal dusting; S, stress corrosion cracking.

than a dozen of ports, and hundreds of ships including luxury cruise ships, warships, barges, coast guard, and cargo ships. In order to meet the demands of these industries, we have various sophisticated industries such as L&T, BHEL, who are masters in machine design and execution. Knowledge of corrosion-resistant materials, their evaluation, and performance optimization are of great concern. Many industries such as Wellspun, Jindal, PSL, and Punj Llyods are manufacturing pipes, applying coatings, and laying down large lengths of pipelines.

Infrastructure is another area that is growing at a very strong pace. Constant developments of roads, flyovers, bridges, dams, airports, multistoried buildings, malls, multiplexes, housing complexes, and railway stations have resulted in better corrosion-resistant materials. Use of best steels and stainless steels in building construction, coatings of steel-reinforced bars, and concrete structures is a specialized area.

Transportation is perhaps another area where India is growing very fast. Metros are coming in all major cities, which require sophisticated vibration-free design with the best concrete technology. Better materials for rails and racks is one of the requirements. Superfast trains and multipurpose railway stations with marketing facilities are being planned. Increased passenger requirement has necessitated more airplanes and bigger airports with better design and that are long lasting and safe, requiring little maintenance. Steel structures, well protected by a long-lasting coating, and long-life claddings such as polyvinylidene fluoride (PVDF)-coated aluminum composites are being used. Airplanes with very high speeds and long traveling time required a host of better and corrosion-resistant materials.

The metal and alloy manufacturing industry is also quite big and is also corrosion prone. In India, at present, we have several steel plants, under both government and private sectors. SAIL and Vizag are perhaps the important steel plants in the government sector, and TATA Steel, Jindal Steel, Essar, and Ispat are well-known names in the private sector. Hindustan Zinc, Hindalco, and Hindustan Copper are among the important plants that manufacture nonferrous metals such as Zn, copper, and aluminum. All these industries are corrosion prone. We get several of their corrosion problems to solve.

Thus, today, India is an industrial state in the true sense, and most of the industries mentioned are corrosion prone. Hence, a concept of corrosion management is very much required to minimize corrosion losses; to avoid accidents, plant shut down, and death of workers; and finally to sustain growth. Where do we stand today?

12.3 Corrosion Protection Scenario in India

The four important methods of corrosion protection are

- using better materials;
- protection by coatings;

- cathodic protection;
- protection by inhibitor chemicals.

Except for the first method, the development of new materials for corrosion protection, there is a significant presence of the other three. Protection by coatings is perhaps well represented in the country. According to Frost and Sullivon, the total coating market in India is about Rs. 13 000 crore. Out of which, 50% is decorative, and the balance is industrial. The industrial coating is further subdivided into auto, protective, and marine. The share in auto is again about Rs. 2500 crore, and the balance is in protective and marine. The sector that is still lacking in India is the high performance coating market, especially for functional applications such as splash zone, underwater paints, insulation paints, paints for high-temperature applications, fluoride-based paint systems. Powder coatings for underground cross-country pipelines are fully imported. Many adhesives, used in pipe coatings, are imported. There are more than 500 paint manufacturing companies in India, out of which Asian, Berger, Shalimar, and Nerolac are the top manufacturers. Another group of companies are several multinational companies such as International Paints, PPG, Dupon't, Chugoku, and KEC International that transfer a lot of good protective coating systems in the Indian market.

Cathodic protection is another area that is owned by a group of private sector companies. There is expertise in developing basic equipment such as rectifiers and anodes and in installing them properly at site. The prominent companies in cathodic protection system are Universal Corrosion Prevention INDIA, De Nora India Limited, Electro Protection Services India Private Limited, Adhunik Power Systems Pvt. Ltd., Alpha Solar Instrumentation Company, Anjani Solar Power Company, Caltech Engineering Services, Corrosys India Pvt Ltd., Electro Corr Damp, Flascan CORCON Systems, Golconda Corrosion Control (P) Ltd., Kothari Refractory & Acid Proof Tiles Co., Rajasthan Electronics & Instruments Limited, Sai Titanium Products Private Limited, Sargam Metals Pvt. Ltd., Ti Anode Fabricators Pvt Ltd, TIFAB ENGINEERING, Titanium Tantalum Components Industries, Titanium Tantalum Product, and Titanium Tantalum Products Ltd. More details can be seen at the site in [5].

Protection by inhibitors is a quite discrete business. One set of business deals only with selling the inhibitor chemicals to the user industries, and the other set takes turnkey order of supplying and maintenance of the inhibitor systems for the plants. There are hundreds of companies in India that manufacture various inhibitor chemicals and take turnkey projects to control corrosion in the plant.

Many corrosion-resistant materials, especially special metals and alloys, are generally imported. Even a large amount of stainless steel used in India is imported. Very limited manufacture of stainless steels is made in India by Jindal Group. Many high-performance alloys, such as duplex stainless steels, super ferritics, and super austenitic, are fully imported. Special corrosion-resistant alloys such as Incoloy 625, Hastelloy, and C-276 are fully imported. MIDHANI is the only company which makes special superalloys and titanium alloys.

12.4 Corrosion Education

In a survey of many Indian engineering and science colleges, it was found that corrosion engineering is not a prescribed course in the curriculum of most bachelor's degree programs. It is available only in those institutions where metallurgy/materials science is available as the core discipline and is not available in other institutes, in spite of the fact that all branches of engineering are affected by corrosion degradation. Even in materials and metallurgy departments, corrosion is not a core subject but an elective and is mostly taught in the last semester. Thus, most undergraduates have an inadequate background in corrosion engineering principles and practices. Many industries, especially chemical, petrochemical, and oil and gas related, need employees with competence in corrosion engineering, but they are not finding it in today's graduates. Their principal concern is that those making design decisions do not know much about corrosion. This lack of knowledge and awareness ultimately jeopardizes the health, wealth, and security of a plant and, in one sense, the security of our country. The other matter of great concern is the availability of experts to teach the subject. This in turn depends on the quantum of the corrosion research community. If corrosion engineering education is to flourish, the number of materials science engineering faculty specializing in corrosion will need to increase. Industry, therefore, will need to support university-based corrosion specialists, who will become teachers. For a country like India – which is going in a big way in oil and gas exploration, distribution, and refining; chemical and petrochemical industry; power plants; and shipping – a dedicated effort is needed by both industry and academics to work together to create better awareness of basic corrosion and to promote R&D in latest corrosion protection technology and corrosion modeling and management.

12.4.1 Why Corrosion Education Is the Need of the Hour?

All materials degrade with time in their environments. A basic understanding of this process is crucial to the education of the nation's scientists and engineers. Any industry requires corrosion engineers right from the design stage and for the selection of materials, the corrosion allowance and corrosion protection technique to be applied, the proper strategy of corrosion monitoring, and periodic maintenance. Along with this, a planned strategy of R&D and failure analysis is required. Envisaging the continued need of Corrosion Engineers, University of Akron, Ohio, has started a bachelor's course on Corrosion science & engineering. I think India must follow this approach and start such kind of specialized programs in corrosion science and engineering and/or surface engineering, which impart a basic knowledge of engineering with focus on material design, protection and failure, corrosion management, the methodology for oil drilling, and making corrosion-free component plants or structures. Along with postgraduate programs on corrosion modeling, corrosion-related problems in oil drilling, undersea crude transportation, and development of newer materials for highly aggressive environments are also included.

In conclusion, it can be stated that corrosion education is perhaps the only way to create a better awareness of corrosion protection, which in turn can help minimize the losses due to corrosion. It will also help in overcoming a very common myth that nothing can be done about corrosion and in the following recommendations.

- Preventive strategies in nontechnical areas:
  - increase awareness of the significant corrosion costs and the potential savings;
  - change policies, regulations, standards, and management practices to increase cost savings through sound corrosion management;
  - improve education and training of staff in the recognition and control of corrosion.
- Preventive strategies in technical areas:
  - advance design practices for better corrosion management;
  - advance life prediction and performance assessment methods;
  - advance corrosion technology through research, development, and implementation.

12.4.2 Pioneers of Corrosion Education in India

Although corrosion research is being done in India from the past four to five decades, the Central Electrochemical Research Institute (CECRI) is perhaps the pioneer in corrosion research, a systematic study of corrosion, leading to a degree in corrosion science started in 1982 at IIT, Bombay. In 1994, systematic activities on corrosion started with the formation of NACE India Section, though, there were other professional bodies such as the National Council of Corrosion and the Society for Advancement in Electrochemical Science and Technology (SAEST). Also, there are several societies related to corrosion protection by paints, such as the Indian Paint Association (IPA), the Colour Society, the Indian Small Scale Paint Association (ISSPA), and the most latest SSPC, India.

12.4.3 Corrosion Science and Engineering, IIT Bombay

The corrosion science and engineering program at IIT Bombay (1982) is the unique academic program in India, specializing in corrosion and its control. It is a two year program giving a degree in Master of Technology (M. Tech.) in corrosion science and engineering and also a Ph. D. program. The course consists of about 54 credits of courses and a year long project on some basic or applied corrosion problem. In addition to teaching, there is a strong interaction with industry in the form of sponsored projects, consultancy projects, and training courses for industries.

12.4.4
The Central Electrochemical Research Institute (CECRI)

It is a CSIR laboratory working only on corrosion-related problems, covering all facets of electrochemical science and technology: corrosion science and engineering, electrochemical materials science, functional materials and nanoscale electrochemistry, electrochemical power sources, electrochemical pollution control, electrochemicals, electrodics and electrocatalysis, electrometallurgy, industrial metal finishing, and computer networking and instrumentation. The institute provides a single and unique canopy under which all aspects of electrochemistry and related areas are researched in their dimensions. CECRI's activities are directed toward development of new and improved products and processes as well as novel innovations in electrochemical science and technology. CECRI runs several projects in collaboration with laboratories from within and outside India. CECRI assists the Indian industry by conducting surveys and undertaking consultancy projects.

12.4.5
NACE India Section

NACE India Section was established in 1992 to promote corrosion awareness. Since its inception, India Section has made significant contributions through international and national conferences, conventions, technical programs, training courses, workshops, and so on. The activities of India Section provide a unique platform for engineers and professionals to deliberate on corrosion and benefit by sharing experiences and knowledge. Over the years, it has organized more than a dozen international conferences, several seminars/workshops, and more than 50 corrosion awareness programs on cathodic protection, paint and coating inspection, pipeline corrosion, fundamentals of corrosion, industrial cooling water treatment management, corrosion control in refining process, and so on. These programs are tailored to meet the need of the professionals in Indian industries and are regularly updated to include new inputs of technologies and their applications with a mission to minimize corrosion losses. It also organizes certification courses such as the Coating Inspector Program of NACE International.

12.4.6
The Society for Surface Protective Coatings India

The society was established in 2003 with an aim to create a common platform for paint manufacturers, suppliers, contractors, applicators, R&D personnel's from industry, and educational institutions to discuss the latest in paint technology, new formulations, and new applications as well as to discuss how to use international standards for various industrial paint applications. So far, SSPC India has carried out seven international symposia on surface protective coatings with good participation from paint industry, user industry, educational institutes, and contractor lobby. It has also organized several technical courses on paint coating inspection and quality control, which is one of the most effective coating course that is cost effective and affordable by the Indian public and much better than the NACE International CIP courses that are charged exorbitantly and provide a very standard mundane training program.

12.4.7 The National Corrosion Council of India (NCCI)

The National Corrosion Council of India (NCCI) started functioning from 1990. It is located in the premises of the CECRI, Karaikudi, Tamil Nadu. However, it only nominated institutional members from various major industries/institutions and public sector undertakings. The NCCI organizes the National Congress on Corrosion Control every year, covering all aspects of corrosion science and engineering at important venues for enlightening and increasing awareness about the corrosion of metals and the appropriate remedial measures to control it. The NCCI provides a forum for the presentation of industrial problems and ideas derived from current research and various topics on corrosion. It provides an excellent opportunity to share knowledge, solve industrial problems, and promote budding entrepreneurs.

12.5 An Overview of Highly Corrosion-Prone Industries in India

12.5.1 Oil and Gas Industry

The oil and gas industry is perhaps one of the biggest and is spread from the length to breadth of the country, with more than 300 onland/offshore drilling rigs, wells, and process platforms, in six to seven basins ranging from Mumbai offshore to Rajasthan basin in desert, producing 1 million barrels of oil equivalent per day, thus taking care of 80% of domestic requirements. With more than 20 000 km of cross-country pipelines and more than 4500 km of subsea pipelines, India has the biggest pipeline network. In addition, it has a large network of refineries spread all over India. The important oil and gas industries are ONGC, Cairn Energy, Reliance Industries, Indian Oil Company Ltd. (IOCL), Bharat Petroleum Company Ltd. (BPCL), Hindustan Petroleum Company Ltd. (HPCL), Kochi Refineries, Chennai Petroleum Corporation, Oil India, Gas Authority of India Ltd. (GAIL), and Gujarat Gas.

The oil and gas industry utilizes one of the major shares of various corrosion protection technologies, whether it is paint coatings, inhibitors, or cathodic protection. Many of the offshore structural components and refineries use the most advanced alloys such as duplex stainless steels, superalloys, and FRPs. The most latest paint technologies include the use of glassflake coatings for splash zone, underwater paint systems, fire-resistant coatings, and fusion bond epoxy coatings for underground pipelines. Transportation pipelines for crude use tons of inhibitors to control the corrosion of internal pipeline. The outside pipeline needs a combination of highly sophisticated coatings and cathodic protection. Also, the need for corrosion monitoring pipelines by coupons, LPR, and ER probes is well known. Intelligent pigging and other NDT methods are used to monitor the health of the internal pipelines, which transport oil and gas.

Thus, all control and monitoring methods are being used sparingly at some places. Different kinds of coatings for repair and rehabilitation are being used. Failures are seen quite often. The reason for this is the lack of qualified corrosion professionals, lack of knowledge about the latest control and monitoring methods, ignorance of analyzing failure case histories, and, in general, lack of corrosion management approach. What is corrosion management approach? As per *Successful Health and Safety Management*, manual, HS(G) 65, HSE Books 1991, ISBN 0-11-882055-9, the framework of a successful corrosion management approach is

- the overall policies adopted by an organization;
- *the role and responsibilities of managers and staff within the organization*, including the development and maintenance of appropriate strategies;
- *the development of plans and procedures, as well as the means of* implementation of various corrosion control measures;
- *the methods adopted for performance measurement of the system against* predetermined criteria;
- *the use of systematic and regular reviews of system performance;*
- *the use of periodic audits of the management and monitoring systems.*

It simply means the use of corrosion knowledge right from design stage and for the selection of materials, the use of proper corrosion control strategy, and periodic maintenance and monitoring. Monitoring, databank of all corrosion failures, R&D and its implication on improving design. However, as per Figure 12.2, an organization's willingness to adopt this approach is one of the biggest hurdle. Unless qualified personnel with knowledge of corrosion are employed and enough finance is allocated for monitoring, maintenance, and R&D, it is difficult to stop frequent failure and hence minimize corrosion losses.

12.5.2 Process Chemical and Petrochemical Industry

The second most corrosion-prone industry is the chemical and petrochemical industry. Indian chemical industries are one of the fastest growing industries and contribute 13% to the national Gross Domestic Product (NGDP). Petrochemical industry with basic products such as ethylene, propylene, benzene, and xylene; intermediates such as MEG, PAN, and LAB; synthetic fibers such as nylon, PSF, and PFY; and polymers such as LDPE/HDPE, polyvinyl chloride (PVC), polyester, and PET is a fastest sector with growth rate at 13%, and major petrochemical

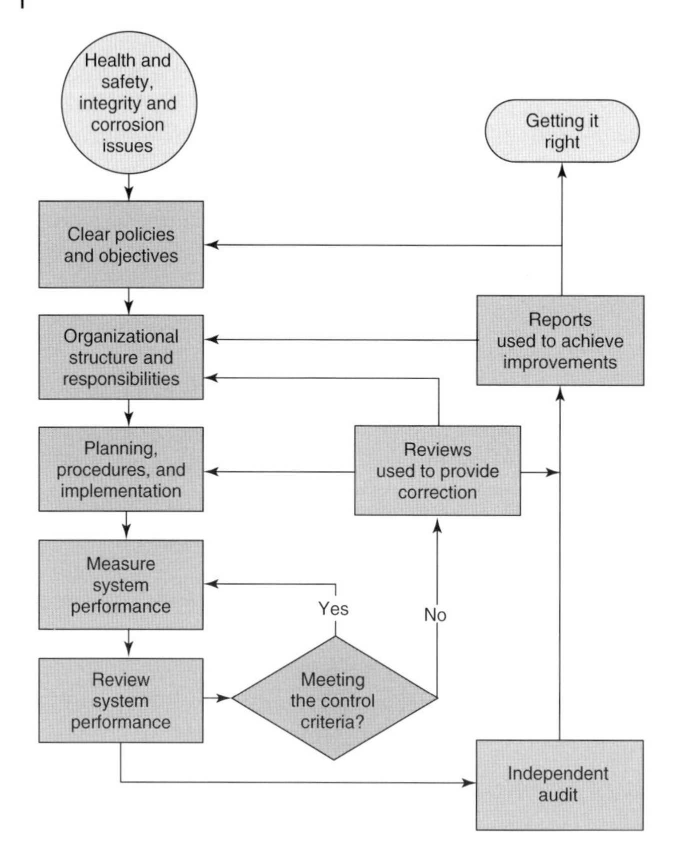

Figure 12.2 Schematic of Corrosion Management.

industries are IPCL, NOCIL, REL, and GAIL. Other chemicals such as inorganic chemicals and organic chemicals have $2.5 and $1 billion industry in India. Several failures due to corrosion are seen regularly, and many severe accidents have taken place in the past. In these industries, not only ambient temperature corrosion but also high-temperature corrosion is a problem. Thus, selection of high-temperature materials and high-temperature coatings is needed.

12.5.3 Pulp and Paper Industry

Paper production consists of a series of processes and can be roughly divided according to the five major manufacturing steps: pulp production, pulp processing and chemical recovery, pulp bleaching, stock preparation, and paper manufacturing. Each manufacturing step has its own corrosion problems related to the size and quality of the wood fibers, the amount and temperature of the process water, the concentration of the treatment chemicals, and the materials used for machinery construction. Examples of corrosion affecting production are

- corrosion products polluting the paper;
- corrosion of rolls scarring the sheets of paper.

Corrosion of components may also result in fractures or leaks in the machines, causing production loss and safety hazards.

12.5.4 Power Plants

Power plants are the lifeline of any country. Power is the basic need of life today. However, power can be obtained from various types of power plants, fossil fuels, nuclear, hydro, windmills, and solar. The power obtained from fossil fuel power plants has enormous corrosion problems. A schematic of a coal-based power plant is shown in Figure 12.3.

The main function of a power plant is to generate a stream of clean steam to run the turbine and generate power. Exhaust steam enters the condenser, where cooling water converts it into condensate. The condensate collects in the hotwell and is pumped through a series of feedwater heaters back into the boiler. To replace the small amount of water constantly lost from the system, a demineralizer provides high purity makeup water to the hotwell.

Sound water treatment and chemical control minimize the effects of corrosion. Thus, continuous monitoring of key parameters such as pH, conductivity, oxygen,

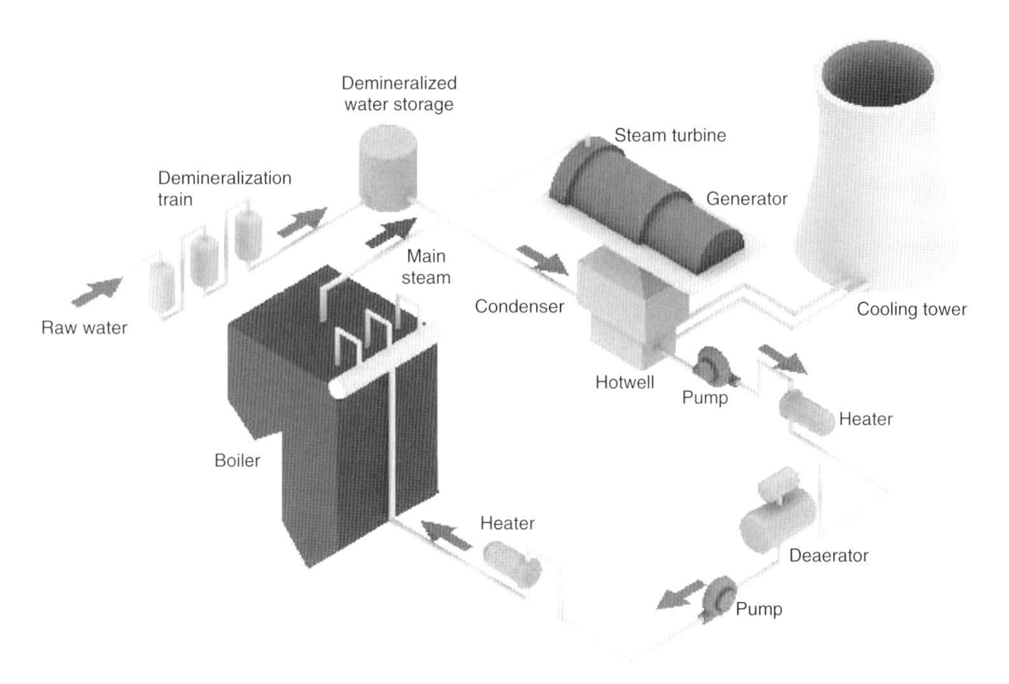

Figure 12.3 Schematic of a Power Plant.

sodium, phosphate, and silica indicates the health of the power plant. Since water and steam are in constant contact with metal surfaces, possibility of corrosion is always there; water purification and chemical treatment greatly reduce and control corrosion in the plant.

Corrosion and scaling are further minimized by the addition of chemical scale and corrosion inhibitors. On the fireside, corrosion problems are even more severe and they depend on the quality of sulfur and salt impurities in the coal. In case the coal has lots of sulfur, there is no other way than to protect the superheater tubes with a chromium-rich high-temperature coating.

Thus, in power plants, corrosion protection by coatings and by the addition of inhibitor chemicals is the most important. Thus, a knowledge of which inhibitors and at what optimized concentration they should be used and the knowledge of special coatings for various plant components, ranging from cooling towers to exhaust chimneys to heat exchanger tank and so on, are needed.

12.6 Conclusions

The conclusion that one can arrive from the above discussion is that it is high time we should stop neglecting the importance of corrosion control. Corrosion is a widespread menace, covering almost every industry, household items, and infrastructure and is indirectly responsible for polluting the environment. Alarming losses due to corrosion need to be curbed by creating general awareness among masses, modifying academic curricula, and introducing strict guidelines with penalty clauses for not properly controlling corrosion in industry. Predicting corrosion by various modeling approach must be practiced, and industries must be encouraged to follow corrosion management approach.

12.7 Recommendations

12.7.1 For Government or Relevant Ministry Responsible for Controlling Industrial Discipline

- Formation of strict guidelines for various industries to control corrosion.
- More governmental support for the cause of corrosion education and research and development.
- Directions to identify vulnerable components and structural parts, which can result in failure of the plant because of corrosion.
- Enforce regular refresher courses for imparting basic corrosion principles and update corrosion technology.
- Enforce regular interaction with corrosion agencies, professional bodies, and academic institutions for timely guidance.

- Emphasize corrosion management approach.
- Emphasize an effort to minimize corrosion losses by providing incentives such as awards and recognition.

12.7.2
For Academic Institutes and Research Organizations

- Creating a four year bachelor's course in corrosion engineering or surface engineering, covering the basics of corrosion, its control methodology, monitoring and prediction, and various techniques to modify and characterize is needed.
- An institute elective course in corrosion science and engineering should be made mandatory for all branches of engineering.
- A strong research and development effort is required not only from academic institutions but also from various concerned industries such as atomic energy, aerospace, oil and gas, chemical process industries, power plants, and infrastructure to study and implement the latest control and monitoring methods.

12.7.3
Roles and Responsibilities of the Industry

The industry should

- strengthen the provision of corrosion engineering education;
- develop corpus funds for promoting and supporting research and development in the field of corrosion science and engineering;
- provide incentives to the universities, such as endowed chairs in corrosion control, to promote the hiring of corrosion experts at the universities;
- fund the development of educational modules for corrosion courses;
- support faculty development by offering corrosion-related internships and sabbatical opportunities and supporting cooperative programs between universities and government laboratories to facilitate the graduate student research experience;
- increase support for the participation of their engineers in short courses when specific skill shortages are identified and are required to be filled in a short term.

References

1. Petrochemical Plants Details in India, *http://business.mapsofindia.com/national-fertilizers/*.
2. Indian Refinery Information, *http://petroleum.nic.in/refi.html*.
3. Indian Fertilizers Plants Information, *http://www.energymanagertraining.com/Plot/Fertilizer.html*.
4. Steel Plants in India, *http://www.energymanagertraining.com/Plot/Iron_Steel.html*.
5. *http://www.energy.sourceguides.com/businesses/byGeo/byC/India/byP/cathodic/cathodic.shtml*.
6. (2002) Corrosion Cost and Preventive Strategies in United States,

FHWA-RD-01-156, March 2002, US Department of Transportation.

Further Reading

(2008) *Assessment of Corrosion Education, Committee on Assessing Corrosion Education,* National Research Council, 180 p., 7 x 10, ISBN: 0-309-11975-8.

(2008) *Guidance for Corrosion Management in Oil and Gas Production and Processing,* Energy Institute, London, May 2008, p. 2.

Moloney, M.H. (2007) Proceedings of the Materials Forum 2007, Corrosion Education for the 21st Century, Editor ISBN: 0-309-10894-2.

13
Protective Coatings: Novel Nanohybrid Coatings for Corrosion and Fouling Prevention

S. Anandakumar and R. Savitha

13.1
Introduction

In recent days, there is a growing need to protect the environment against degradation for sustainable development and in order to meet the needs and aspirations of the present and future generations. Economic competitiveness and environmental concerns have driven the coating technologists to explore newer chemistry and approaches to improve the efficiency of organic coatings at minimum volatile organic components. Organic coatings on a substrate give an esthetic appearance as well as protection from the destructive phenomenon known as *corrosion*. Coatings can provide materials with the desired esthetic properties such as color and gloss and also protect them against environmental influences, including moisture, radiation, biological deterioration, and damage from chemical and mechanical origins. The effectiveness of protection of a substrate against natural deterioration depends on factors such as quality of the coatings, substrate characteristics, the properties of the coating–substrate interface, and the corrosiveness of the environment.

The tendency toward corrosion and fouling of most strategic metals such as steel, aluminum, and galvanized steel can be controlled in an effective manner by coatings. Surface coatings often consist of both inorganic particles such as TiO_2, $CaCO_3$, and clay and organic polymers such as epoxies, vinyl acetates, acrylics, and styrene butadiene. However, organic coatings, in general, have the advantageous properties that water can only slightly penetrate through the hydrophobic organic film. The adequate performance of organic coatings depends on two main factors. The coating must constitute a barrier to water, ions, and oxygen permeation. However, when water infiltrates through the coating, the inhibitor present in the coating must passivate the substrate. The disadvantage of organic coatings, however, is their high volatile content representing an environmental hazard. The volatile organic solvent is injurious to human health and also contributes to global warming and depletion of the ozone layer leading to the collapse of the Antarctic ice shelf and rise in sea level and flooding. Consequently, research is focussed towards developing novel coatings with tailored nanostructures. The

Green Corrosion Chemistry and Engineering: Opportunities and Challenges, First Edition.
Edited by Sanjay K. Sharma.

published scientific reports over the past decade have led to the knowledge-based concept of tuning coating properties according to the end-use application. With the expanding knowledge on the effect of structures by means of nanoscale modification, it is certainly possible to create new types of coatings that can take advantage of the size-tunable properties of nanosized components. The developing ability to synthesize uniform nanosized components and to include them in macromolecular structures will allow coating technologists to have a new level of control over the physicochemical properties of the components that makeup the macroscopic materials. Nanomaterials promote synergisms in structural integrity, functionality, versatility, and cost-effective fabrication as well.

13.2 Background

Steel is one of the major materials used for the construction of industrial structures, storage tanks, vehicles, pipelines, ships, and so on. However, research has shown that commercial coatings used for the protection of steel and marine structures can be rendered useless within one month of continued use [1]. This can result in direct and indirect losses to a country, which could be as high as 5% of Gross National Product [1]. However, coating appears to be one of the most convenient methods for preventing the corrosion of steel surfaces, storage tanks, and surfaces of pipelines, ships' hulls, and their superstructures, although there are other much more costly alternatives available [1, 2]. The phenomenon of biofouling poses another serious economic problem to the pursuance of maritime activities [1–3]. The literature contains descriptions of a variety of methods, including immobilization of marine bacteria in hydrogels [4, 5], piezoelectricity [6, 7], and electrolytic technologies [8], for the control of biofouling. However, none of these methods is able to totally prevent fouling, and the use of antifouling paints has historically been the method of choice in many instances [3]. But the common tributyltin (TBT) oxide and copper compounds used for the past 40 years were toxic and consequently their use in coating formulations has been banned in many countries [9–15]. For this reason, nontoxic alternatives are needed, and hence, an attempt has been made to formulate nanohybrid coatings that significantly improved the corrosion resistance of steel substrates, besides offering fouling prevention and decontamination of the surfaces of materials.

13.3 Fouling

Fouling refers to the accumulation of unwanted materials on solid surfaces, most often in an aquatic environment [1]. The fouling materials are of two types:

1) living organisms (biofouling) (e.g., marine fouling);
2) nonliving substances (inorganic or organic) (e.g., petroleum refineries).

13.4
Marine Fouling

Marine fouling is a perennial problem for ships, boats, ports, and anything kept in the sea for a specific period of time. The sea is teeming with the tiny larvae of marine organisms, which swim around until they find somewhere to settle and grow. Smooth surfaces are particularly attractive to many of these creatures and are quickly encrusted.

13.4.1
Stages of Marine Fouling

Fouling of immersed surface takes place in four stages:

1) Organic materials such as proteins, polysaccharides, and glycoprotein get physically adhered to the substrates.
2) Within 24 h, bacteria and unicellular algae are adsorbed; subsequently, the existence of the microbial film that has formed provides sufficient food to allow the fixing.
3) Spores of microalgae will constitute biofilm, which in turn will allow the increased capture of more particles and organisms.
4) After two or three weeks of immersion, the conditions are then set for the fixing and growth of either macroalgae or marine invertebrates.

13.4.2
Consequences of Marine Fouling

The deleterious effects of marine fouling severely affect oceangoing vessels to a greater extent. Fouling organisms such as barnacles, tube worms, and algae that accumulate on any submerged surfaces greatly increase drag and reduce speed and fuel economy of commercial and military vessels, which have a severe impact on the overall cost of the travel.

13.4.3
Methods Used for Fouling Prevention

The need to protect ship hulls from marine fouling is as old as man's use of ships as a means of locomotion, which summarizes the main antifouling products used before the mid-nineteenth century. Since ancient times, the use of natural products such as wax, tar, and asphalt has been practiced. According to Almeida *et al.* [16] the Phoenicians and Carthaginians seem to have been the first to use copper for this purpose. This technique was similarly adopted by the Greeks and Romans, who also investigated the use of lead sheathing. In the eighteenth century, it was common to use wooden sheathing covered with a mixture of tar, fat, and pitch and studded with numerous metal nails, whose heads, closely in contact with each other, seem to have formed a sort of second metallic sheath. Even in the nineteenth century, several countries returned to the use of copper sheathing, with copper and zinc nails, and experimented with sheathings of zinc, lead, nickel, galvanized steel, and other materials, as well as copper-coated wood sheathing. Nonmetallic sheathings were also suggested, namely, those made of rubber, vulcanite, cork, and others, which were eventually abandoned because of their high cost and/or difficulty in application. The first antifouling paints appeared in the mid-nineteenth century, containing copper, arsenic, or mercury oxide as toxicants dispersed in linseed oil, shellac, or rosin [15].

13.4.3.1 **Antifouling Paints**

The first antifouling paints emerged in the mid-nineteenth century were based on the idea of dispersing a powerful toxicant in a polymeric binder. These were followed by other paints with binders based on different bituminous products and natural resins, whose dilution was achieved with turpentine spirit, benzene, or naphtha. However, since the pigments used in these paints, which were applied in direct contact with the ship hull, caused corrosion on the first steel hulls, the application of a primer capable of protecting the hull was quick to appear. Meanwhile, new products were emerging, including *hot plastic paints* with natural binders and copper or other toxicants, *rust preventive compounds* that were shellac-based products containing toxicants, and, with the development of polymer chemistry, *cold plastic paints* that used different synthetic resins or natural products alone or in combination. The natural products, which were easier to apply by means of *airless* spraying, were also developed around that time, allowed dry docking intervals of up to 18 months, of antifouling paints used on steel hulls before 1960 [3].

The first organometallic paints (with tin, arsenic, mercury, and others) appeared around 1950 and gave rise, after numerous and successive developments, to TBT-based antifouling paints, which became famous because of their great efficiency and versatility. TBT has been referred to as perhaps the most effective antifouling biocide ever developed, but the International Maritime Organization (IMO) banned TBT after 2003 because of its environmental impact on nontarget organisms (e.g., it affects the shell growth of oysters and also leads to the development of male characterization in the female genitalia of dog whelks). Thus, the paint industry has been urged to develop TBT-free products, which can cause less harmful effects on the environment. Fouling may be controlled by foul-release coatings that have low surface energies and do not contain harmful biocides. Attached organisms are removed from vessels as a result of frictional forces caused by their motion through the water [16, 18]. While effective on fast-moving vessels, these coatings are not suitable for use on the majority of vessels. Electrochemical technologies have also been examined by Nakasono *et al.* in 1993 but were expensive to apply to large structures [19]. Natural products from plant extracts, which have been proven for their antimicrobial activity, are used as a biocide in antifouling paint formulations and studied for their antifouling and corrosion rate as well [19, 20].

13.5
Corrosion

Corrosion, a worldwide problem, is the disintegration of a material into its constituent atoms because of the chemical reactions with the surroundings. *Corrosion* may also be defined as the loss of useful properties of a material as a result of chemical or electrochemical reaction with its environment [1].

13.5.1
Consequences of Corrosion

Corrosion is a serious problem in all fields of application of metals. The loss of metal resources whose abundance is limited poses a danger to conservation and causes serious economic problems (e.g., in UK, 1 ton of steel is converted into rust for every 90 s).

1) Plant shutdown because of failure (e.g., nuclear reactor during decontamination process).
2) Replacement of corroded equipment, resulting in heavy expenditure.
3) High cost of preventive maintenance, such as pickling.
4) Loss of efficiency.
5) Loss of product from corroded container.
6) Safety requirement measures for fire hazards or explosion.
7) Health problems due to chemicals escaping from corroded equipment.
8) Necessity of overdesign to prevent for corrosion.

13.5.2
Methods Used for the Prevention of Corrosion

Various methods available for corrosion prevention are

- design improvement;
- change of metal;
- change of environment;
- change of metal electrode potential;
- use of protective coatings.

Among these methods, coating appears to be the most convenient method for the prevention of corrosion of various substrates.

13.5.3
Characteristics of a Good Coating

1) It should easily cover the base surface and possess high covering power.
2) It should produce a tough, adherent, impervious, glossy, and stable film.

3) It should be easily coated with a brush or spraying device to produce a smooth and uniform surface.
4) It should not crack, peel, or chalk out after its application and should function as a strong protective barrier against a corrosive atmosphere.
5) It must have good resistance to acid, alkali, and salts because these conditions exist in corrosive media.
6) It must have good moisture and UV resistance because high density and sunlight are very detrimental to most protective coatings.
7) It should have a fast drying property.

13.5.4 Evaluation of Corrosion Resistance of Coatings

Corrosion resistance of a material is defined as the ability to withstand corrosive attack by any medium in a given set of conditions. The materials used can generally be metals or alloys. From the corrosion measurement, inference could be made about the suitability of the material for a particular environment, and further, the remaining life of the material can also be assessed. Corrosion being an electrochemical process requires cathode and anode in electrical contact through an electrolyte. The process occurs through the flow of electrons between the anode and cathode. This electron flow rate corresponds to the reaction (oxidation and reduction) that occurs on the surface. Thus, the capability of assessing the kinetics of the corrosion process is monitored by the electron flow. Several methods have been used to evaluate the corrosion process. Electrochemical impedance spectroscopy (EIS) study is one among the important methods.

13.5.5 Electrochemical Impedance Studies (EISs)

A wealth of kinetic and mechanistic information about the study of electrochemical systems can be determined by AC impedance measurements [21]. EIS is an alternating current technique used for studying the spontaneous passivity of metals in an electrolyte. It also tells that the value of the coating resistance, which can be affected by the electrolyte and also by the resistance produced by the metal and ions in the solution. This resistance can be calculated by the values corresponding to the movement of metal ions from the metal into the solution using Bode and Nyquist plots [1, 21, 22].

13.6 Epoxy Resin Coatings

Epoxy resins are a broad class of versatile reactive compounds. The need for a high degree of control over both the network properties and resin processability make epoxy chemistry appealing to more applications. Epoxy resins having good mechanical and chemical properties can be processed relatively simply and safely to the level of the cured epoxy resin molded materials. Epoxy resins consist of a linear chain molecule with a reactive epoxy group at each end of the chain. They are cured using amines, which link the resin with their chains together, resulting in a chemical bond formation leading to a three-dimensional network. Wegmann [23] suggested that epoxy resins play a dominant role in protective coatings mostly because of their outstanding performance in terms of corrosion protection and chemical resistance. The actual properties of the final coatings are very much depending on the type of epoxy and hardener used. Epoxy resins have been known for some of their unique combination of properties including outstanding adhesion to various surfaces, lightweight, high strength, extreme durability, stability under UV exposure, and chemical resistance. They are used in aerospace, automobiles, land and marine transportation, chemical process industries, and electrical industries because of their mechanical properties and excellent processability [4]. This versatility in formulation made epoxy resins widely applicable in surface coatings, adhesives, laminates, composites, potting, painting materials, encapsulates for semiconductor, and insulating material for electrical devices [23]. However, in the case of civil and marine applications, epoxy resins require some modifications and predominantly need curing agents with amines [1].

13.6.1
Advantages of Epoxy Resin

Epoxy resin is chosen for this study owing to its exceptional combination of properties such as

- high safety;
- excellent solvent resistance;
- chemical resistance;
- toughness;
- low shrinkage;
- good electrical, mechanical, and corrosion resistance;
- excellent adhesion to many substrates.

13.6.2
Disadvantages of Epoxy Resin

They are

- poor impact strength;
- high rigidity;
- moisture absorbing nature;
- high brittleness;
- inadequate fouling and corrosion prevention.

These disadvantages restrict their use as high-performance fouling and corrosion-resistant coatings.

13.6.3
Justification

Epoxy resins with a unique combination of properties have occupied a dominant place in the development of high-performance coatings. At the same time, the more demanding requirements of the end users for greater resistance to corrosion and fouling of these materials push the existing technology and knowledge to their limits. Such challenging requirements have attracted the researchers into the relatively new field of organic–inorganic nanohybrid polymers. The amalgamation of organic–inorganic entity into a single polymer offers a unique combination of properties of both constituents such as improved toughness, flexibility, corrosion, fouling, and moisture resistance with good processability and thermo-oxidative stability [24]. Although numerous paint/coating systems based on epoxy resin are available for corrosion and fouling prevention, they are not completely satisfactory in the field, where high corrosion and fouling resistance are required. The demand for epoxy resin as corrosion-resistant/fouling-resistant coating is restricted mainly because of its inferior characteristics mentioned earlier. Hence, the development of modified epoxy resin with a better set of coating properties ideally suitable for corrosion and fouling prevention is inevitable. With all these ideas in mind, an attempt has been made in the present investigation to formulate tris(*p*-isocyanatophenyl)-thiophosphate-modified nanohybrid epoxy coatings containing natural biocides and a chemical biocide encapsulated in a nanozeolite container, with an unique self-polishing composition ideally suitable for marine substrates. These nanocoatings are capable of preventing algae, bacteria, and protozoa, besides offering excellent corrosion resistance to the steel surface through the presence of phosphorus and sulfur group.

13.6.4
Need for Nanotechnology

Nanotechnology has created a key revolution in the twenty-first century, exploiting the new properties, phenomena, and functionalities exhibited by matters when dealt at the level of few nanometers. At this level, the physical, chemical, and biological properties of materials differ in fundamental and valuable ways from properties of individual atoms and molecules or bulk matter. Nanoscience and technology is a young field that encompasses nearly every discipline of science and engineering. Nanophase and nanostructured materials constitute a new branch of material research and attract a great deal of attention because of their potential applications in areas such as coatings, electronics, optics, catalysis, ceramics, magnetic data storage, and polymer nanocomposites (PNCs) [25]. The unique properties and improved performance of nanomaterials are determined by their sizes, surface structures, and interparticle interactions. The role played by particle size is comparable to the role of the particle's chemical composition, adding another parameter for designing and controlling particle behavior. Research and development in nanotechnology is directed toward understanding and creating new coating materials, devices, and systems that exploit these new properties [25]. A significant effect in nanotechnology is the surface free-energy effect. The high surface area/volume ratio of nanoscale materials results in an increased interfacial area between the macromolecular binder and the nanoparticle. The interfacial region imparts new structural arrangements to the molecular scale, providing new properties intermediate to those of the organic and inorganic components, and is responsible for efficient stress transfer across the composite components. Thus, the surface or interface energy effects of nanomaterials result in higher elastic modulus and mechanical strength than conventional composite materials [26, 27].

13.6.5 Polymer Nanomaterials

Nanomaterials can be classified into nanostructured materials and nanophase/ nanoparticle materials [28]. The former usually refers to condensed bulk materials that are made of grains (agglomerates), with grain sizes in the nanometer size range, whereas the latter are usually the dispersive nanoparticles. To distinguish nanomaterials from the bulk, it is crucial to demonstrate the unique properties of nanomaterials and their prospective impacts on science and technology. The reinforcement of polymers using fillers, whether inorganic or organic, is common in modern plastics to result in PNCs exhibiting multifunctional, high-performance polymer characteristics beyond what traditional filled polymeric materials possess. PNCs or the more inclusive term, polymer nanomaterials (PNMs), represent a radical alternative to the traditional filled polymers or polymer compositions. Many examples of PNMs can be found in the literature demonstrating substantial improvements in physical and mechanical properties of coatings. The value of PNM technology is not based solely on mechanical enhancements of the neat resin. Rather, its value comes from providing value-added properties not present in the neat resin, without sacrificing the inherent processability and mechanical properties of the neat resin. Two main PNM processing methodologies have been developed: *in situ* routes and exfoliation. Currently, exfoliation of layered silicates, carbon nanofiber/carbon nanotubes (CNF/CNTs), and polyhedral oligomeric silsesquioxanes (POSSs) in commodity and high-performance resins is the most investigated PNM processing methodology by the government, academic institutions, and industries all over the world.

13.6.6 Different Types of Nanoparticles

The fabrication of nanoparticles is significant from a technological point of view. Nanoparticles exhibit size-dependent trends in structure and reactivity. In general, various geometries and sizes of nanoparticles can be produced during the manufacturing process, and the size and geometry affect the properties of the nanocomposites derived from them. The benefit of using nanosized particles in coating formulations includes obtaining improved [25] chemical, physical, and color performance [26]; thermal stability and flame retardancy; and [27, 28] weathering and corrosion stability with a low level of nanoparticle loading. Therefore, the incorporation of nanosized particles can reduce the pigment usage in the coating formulation. Many paint and coatings manufacturers are currently both using and evaluating the potential of nanomaterials in their formulations. There are different types of commercially available nanoparticles that can be incorporated into the polymer matrices to form PNCs. Depending on the application; the researcher must determine the type of nanoparticle needed to provide the desired effect.

- Montmorillonite (MMT) organoclays;
- CNFs;
- POSS;
- CNTs;
- nanosilica (N-silica);
- nanoaluminum oxide ($A1_2O_3$);
- nanotitanium oxide (TiC);
- others.

POSS and nanosilica (nanocontainer) alone are discussed here, as they are relevant to the research work carried out for this investigation.

13.6.7
Polyhedral Oligomeric Silsesquioxanes (POSSs)

POSS possesses two unique structural features: (i) the chemical composition is a hybrid, intermediate ($RSiO_{1.5}$) between that of silica (SiO_2) and siloxanes (R_2SiO) and (ii) POSS molecules are nanoscopic in size, ranging from approximately 1 to 3 nm (Figure 13.1 and 13.2). These materials are thermally and chemically more robust than siloxane, and their nanostructured shape and size provide unique properties by controlling polymer chain motion at the molecular level. The key purpose of POSS technology is to create hybrid coating materials that are rigid and tough and easy to process with the characteristics of high corrosion resistance.

13.6.8
Need for Nanocontainer (Nanozeolite)

As far as the antifouling coatings are concerned, they have to provide release of the biocides rapidly after changes in coating integrity. The active antifouling biocides can be introduced in different components of the coatings. The agents are effective only if their solubility in the defective environment is in the right range. Very low solubility leads to lack of active agent at the substrate interface and consequently to a weak feedback activity. If the solubility is too high, the substrate will, however, be protected for only a relatively short period since the

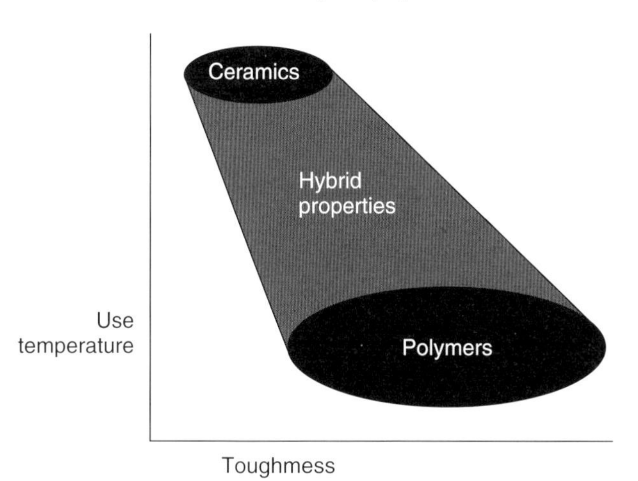

Figure 13.1 Property of POSS molecule.

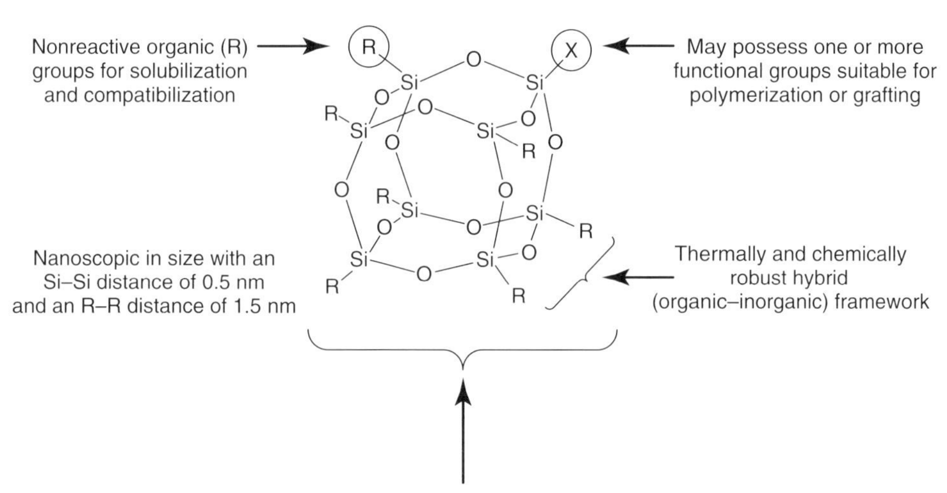

Figure 13.2 Anatomy of a POSS molecule.

active agent will be rapidly leached out from the coatings. Most developments in biocide technology aim to address environmental issues, as biocides are essentially toxic and the environmental impact of their release requires careful monitoring and control. Significant advances could be made if

- the distribution of the biocide could be concentrated at or near the exposed surface;
- the minimum inhibitory concentration of biocide could be sustained at an optimal level for an extended period;
- the biocide could be immobilized and released only on demand;
- the risks associated with safe handling (skin sensitization) could be minimized since many film biocides carry the skin sensitization label;
- the biocide could be shielded from any ill effect of environmental exposure during the time it remains in the coatings.

Nanocontainer could provide the means to satisfy the above criteria, whereby an active biocide is retained within its protective framework until some trigger mechanism stimulates the release of biocides. Such technology has already made an impact on other fields, for example, drug release and fragrance delivery [29].

13.6.9 Self-Repairing Multifunctional Coatings

Nanocontainers with the ability to release encapsulated active biocides in a controlled way can be employed to develop a new family of self-repairing multifunctional coatings [29], which possess not only functionality but also rapid feedback activity in response to change in local environment. The release of the active biocides occurs only when triggered, which prevents leakage of the active component out of the coating and enhances coating durability.

13.6.9.1 Fabrication of Nanocontainers

Silica is considered to be a polymer of silicic acid containing interlinked SiO_4 tetrahedron. This construction terminates by the formation of siloxane bridges (Si–O–Si) or a silanol group (SiOH). The surface chemistry of silica is dominated by the nature, distribution, and accessibility of these structures. By definition, silanol groups are weakly acidic, but variation in Si–O bond angles leads to a spectrum of activities. The siloxane bridges are nonpolar, and so calcinations increase the hydrophobicity of the silica surface. Hence, controlled calcinations provide a means to control the nature and availability of absorption sites [29, 30].

13.6.9.2 Biocide Release Mechanism from Nanocontainer

Usually, the biocidal coatings kill microbes and fungi by slowly and steadily releasing biocide from the dried film. Unfortunately, this mechanism is also responsible for the ultimate deactivation of the biocidal activity; once the biocide has leached out or washed out of the coating, all protection against microbes is lost. Typically, the biocidal activity lasts only around 18 months, and even less in hot, humid environments, and a new coating must be reapplied to regain antifungal protection. Hence, there is a need for the development of a new biocidal coating system with a prolonged biocidal activity by incorporating biocide additives on a nanoparticulate medium using advanced nanoparticle technologies. It is essential that the embedded biocides are released into the environment only when needed.

In this context, the proposed development of nontoxic, nanohybrid coatings with natural antifungal and biocidal agents incorporated into zeolite will be very valuable.

13.6.9.3 Selection of Natural Products as Biocides

The selection of natural products as biocides is very important because they possess certain essential qualities such as biodegradability, less toxicity, easy availability, and cost-effectiveness. Naturally available plant extracts have antimicrobial activity, which do not cause unexpected side effects on living organisms and environment. The selected vegetations for this work are

1) *Michelia champaca leaf;*
2) neem oil;
3) nontoxic chemical biocide.

M. champaca

Plant profile

Botanical name	:	*Michelia champaca*
Family	:	Magnoliaceae
Habit	:	Tree
Tamil name	:	Sembagam
Sanskrit	:	Swarnachampaka

A tall handsome, evergreen tree with straight trunk and found in South Asian region; wild in the Himalayan tract, Assam, Burma, Western Ghats, South India, and much cultivated in various parts of India.

Neem Oil

Plant profile

Botanical name	:	*Azadirachta indica*
Family	:	Meliaceae
Habit	:	Tree
Tamil name	:	Vembu
Sanskrit name	:	Nimba

Neem is a fast-growing tree that can reach a height of 15–20 m (about 50–65 ft), rarely 35–40 m (115–131 ft). It is evergreen, but in severe drought, it may shed most or nearly all of its leaves. The branches are widespread. The fairly dense crown is roundish or oval and may reach a diameter of 15–20 m in old, freestanding specimens.

Nontoxic Chemical Biocide – Bronopol (Commercially Known as Pandol) Bronopol is a highly active antimicrobial chemical compound whose chemical formula is 2-bromo-2-nitropropane-1,3-diol.

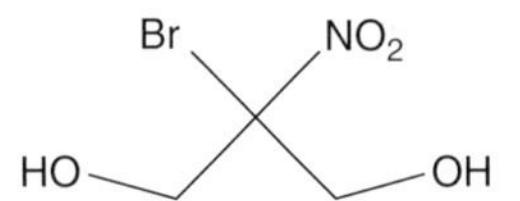

Bronopol was invented by The Boots Company PLC, Nottingham, England, in the early 1960s, and its first applications were as a preservative for pharmaceuticals. Bronopol's low mammalian toxicity (at in-use levels) and exceptional activity against bacteria (especially the troublesome gram-negative species) ensured that it became popular as a preservative in many consumer products such as shampoos and cosmetics. Bronopol was subsequently taken up as an effective antimicrobial in many industrial environments such as paper mills, oil exploration, and production facilities, as well as cooling water disinfection plants.

13.7
Scope and Objectives

- To develop tris(*p*-isocyanatophenyl) thiophosphate (Desmodur)-modified epoxy coating system.
- To synthesize the functionally terminated POSS nanoreinforcement.
- To synthesize the mesoporous nanocontainer.
- To encapsulate the natural products and Pandol into nanocontainer to be used as self-repairing coatings.
- To characterize the above coatings using IR and NMR analyses.
- To evaluate their corrosion resistance using advanced EIS studies and salt spray test.
- To understand the antifouling behavior of such coatings by seawater immersion.
- To predict the surface morphology using SEM surface analytical technique.

13.8
Experimental: Synthesis and Structural Characterization of the Nanohybrid Coatings

13.8.1
Materials

Epoxy resin GY250 representing diglycidyl ethers of bisphenol A (DGEBA) with an epoxy equivalent weight (EEW) of 180–190 and a viscosity of about 10 000 mPa was used for our study. Aradur 140 (polyamidoimidazoline) curing agent was obtained from Huntsman (India). The organophosphorus compound tris(*p*-isocyanatophenyl) thiophosphate (Desmodur) was supplied by Anabond India Pvt. Limited (Chennai, India). Incompletely condensed POSS and aminopropyltriethoxysilane were purchased from Alfa Aesar (Germany). Dimethyl formamide

(DMF), tetrahydrofuran (THF), acetone, and toluene purchased from Sisco Research Laboratories, India, were used after proper drying. DMF was dried using calcium hydride (1 g per 100 ml) as the drying agent followed by filtering off the hydride and distillation. Toluene and THF were dried with sodium and distilled before use. Acetone was dried using $KMnO_4$ followed by distillation. Nanocontainer MCM-41 was synthesized as per the reported procedure and was then calcinated at 550 °C for 6 h and used. *M. champaca* plant leaves were collected from Coimbatore and processed. Neem oil was purchased from Maruthi Pharmacy (Chennai, India), and Pandol was supplied by Gayathri Chemicals and Agencies (Chennai, India).

13.8.2 Surface Preparation of the Mild Steel Specimens

Mild steel specimens (of composition C, 0.04%; Si, 0.01%; Mn, 0.17%; P, 0.002%; S, 0.005%; Cr, 0.04%; Mo, 0.03%; Ni, 1.31%; and Fe, balance) were used for our study. The specimens were degreased with acetone to remove impurities from the substrate, after which the specimens were subjected to sandblasting at a pressure of 100 psi through the nozzle to get the appropriate crevices. The particle size of the sand was 80 meshes. The distance between the substrate and the blaster was maintained at approximately 2 ft. The specimens were then kept in the desiccators for conditioning.

13.8.3 Synthesis of Phosphorus-Containing Polyurethane Epoxy Resin

The tris(*p*-isocyanatophenyl) thiophosphate and DGEBA-type epoxy resin were flushed with nitrogen for 12 h at 80 °C. The synthesis is shown in Scheme 13.1. Tris (*p*-isocyanatophenyl) thiophosphate (Desmodur) and DGEBA (1 : 3 ratio) were together mixed with THF at 60 °C in a 50 ml round-bottomed flask. The contents of the flask were refluxed with a condenser under nitrogen atmosphere, which involved stirring for 2–3 h. The EEW of the resulting phosphorus-containing epoxy resin was determined to be 348.

13.8.4 Synthesis of Amine-Functionalized POSS (POSS-NH_2)

The synthesis of POSS-NH_2 was carried out as per the reported procedure. Stoichiometric amount of POSS-triol dissolved in dry toluene was mixed with aminopropyltriethoxysilane in a 50 ml round-bottomed flask and refluxed for 8 h at 90 °C. The filtrate was then subjected to solvent evaporation to obtain the desired product, which was confirmed by IR and NMR spectra. The synthesis is also shown in Scheme 13.1.

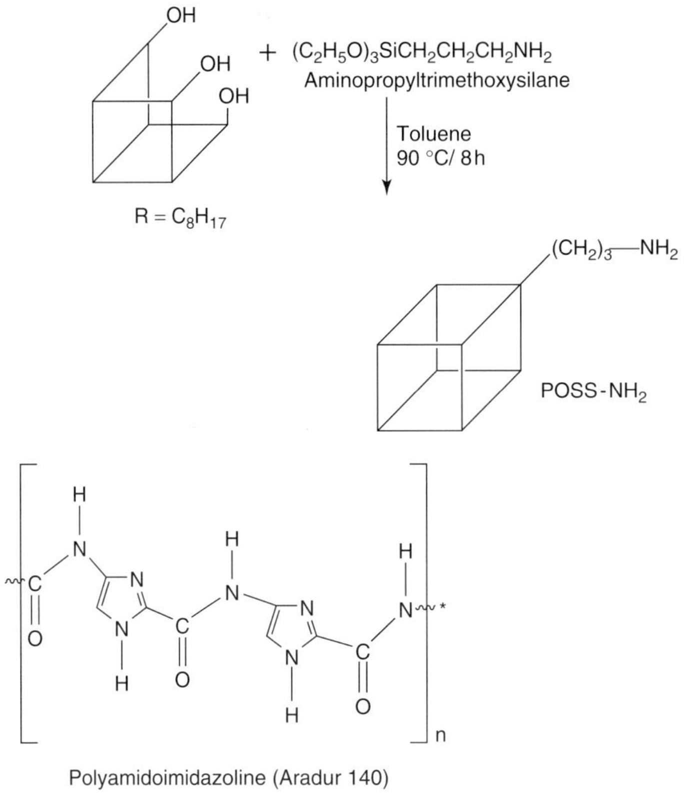

$$+ (C_2H_5O)_3SiCH_2CH_2CH_2NH_2$$

Aminopropyltrimethoxysilane

Toluene
90 °C/ 8 h

$R = C_8H_{17}$

$(CH_2)_3$—NH_2

POSS-NH_2

Polyamidoimidazoline (Aradur 140)

Scheme 13.1

13.8.5 Preparation of Biocides

The *M. champaca* plant leaves were collected and dried under controlled conditions (shadow drying) to avoid too many chemical changes. Five percent of extract was prepared by taking 5 g of powdered dry leaves in 100 ml water and refluxed for 3 h and then kept overnight and filtered next day and subjected to characterization. The neem oil was purchased and used as it is and subjected to characterization. Along with these natural biocides, a commercially available chemical biocide Pandol is also taken for the study. This was also chosen for our study to compare its antifouling and anticorrosive extent with the other two natural biocides.

13.8.6 Loading of Biocides

Mesoporous materials are of great interest to many researchers since they possess highly ordered periodic arrays of uniformly sized channels and large surface area

(1000 m^2 g^{-1}). MCM-41 is a well-known first-synthesized mesoporous molecular sieve. These voids are usually occupied by charge-neutralizing cations and water molecules. The cations are mobile and can usually exchange to varying degrees. The solvents used for loading the biocide into the MCM-41 porous material included deionized water, ethanol, methanol, pyridine, and acetone. The purpose of the organic solvent was to increase the miscibility of the biocide solution with zeolite since zeolite is immiscible with water, and then the organic solvent was added to obtain a final solution containing 10 wt% biocide. The biocide solution and the base zeolite were first thoroughly mixed to form a homogeneous dispersion and then heated at 150 °C for 4 h to totally remove the solvent. The gel-like material can be obtained after evaporating the solvent (organic solvent and water). The loading of biocides into zeolite followed by its surface functionalization and reaction with monomer is depicted in Scheme 13.2.

13.8.7
Preparation of Tris(*p*-Isocyanatophenyl)-Thiophosphate-Modified Epoxy Nanocoatings

Phosphorus-containing polyurethane epoxy resin was mixed with 5% of the flow control agent (urea formaldehyde) at 60 °C for 15 min with constant stirring. To this, a calculated amount of 5 wt% of POSS-NH_2 was dissolved at 60 °C under vigorous stirring. After complete dissolution, stoichiometric amount of Aradur 140 curing agent was added, and the agitation was continued at 60 °C until a homogeneous mixture was obtained. These phosphate-modified nanocoatings cured by Aradur 140 with and without biocides (encapsulated in nanozeolite) were coated on mild steel using a brush with a thickness of 200 ± 10 μm. The samples were kept for seven days to allow complete curing. Finally, the specimens were tested for their corrosion and fouling resistance performance by subjecting them to NaCl immersion, salt spray testing, and aggressive fouling environments. The nomenclature of coating systems studied for the investigation is given in Table 13.1.

13.8.8
Test Methods

FT-IR spectra of the monomers were recorded by Perkin–Elmer 781 infrared spectrometer using NaCl to confirm their structure. ^1H, ^{13}C, ^{31}P NMR spectra were also recorded on a Bruker 500 MHz NMR spectrometer with DMSO and $CDCl_3$ as solvents. Using Association of computing machinery (ACM) impedance testing equipment (Gill AC Serial No. 900 Sequencer Version 4), potentiodynamic polarization and impedance measurements were made for the virgin epoxy and tris(*p*-isocyanatophenyl)-thiophosphate-modified epoxy nanocoated specimens. Salt spray test for the coated specimens was carried out as per ASTMB-117 standard, and

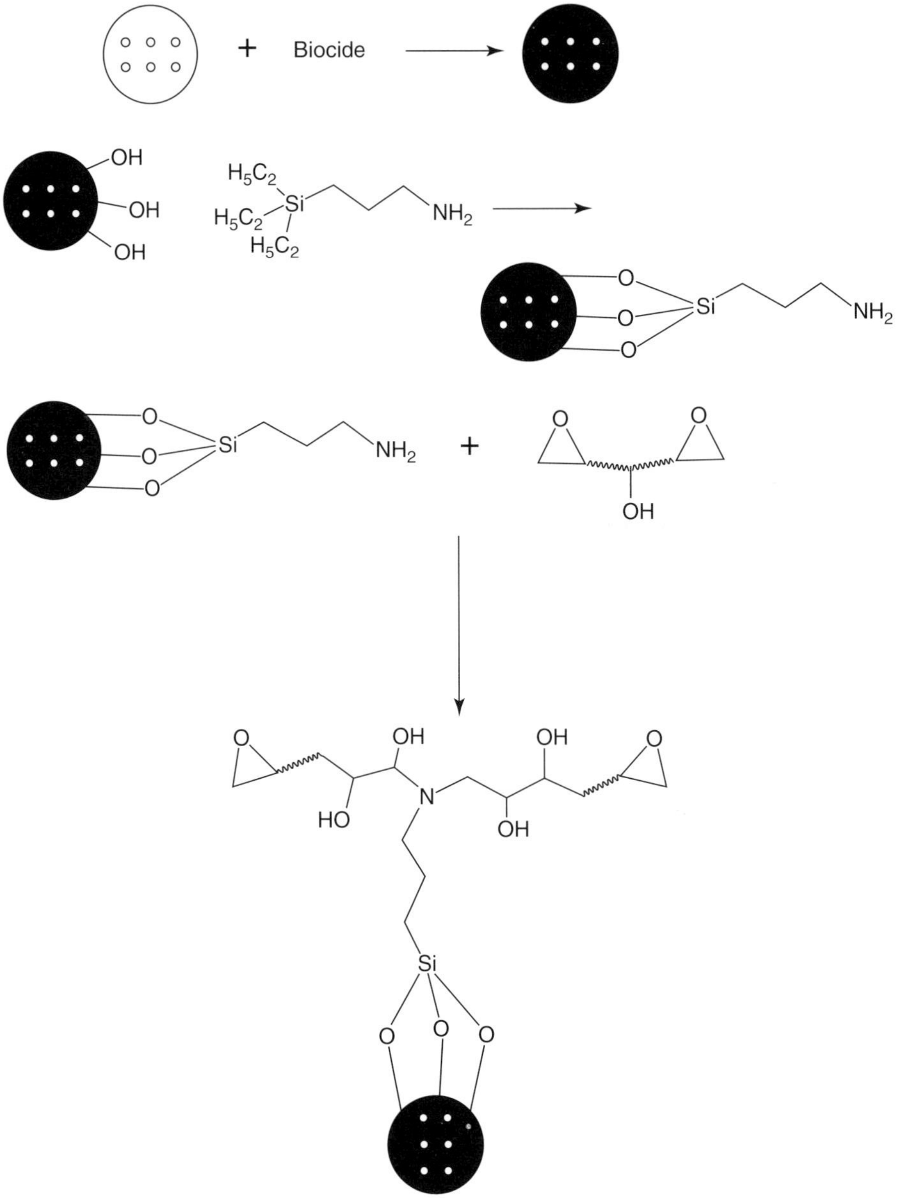

Scheme 13.2

the surface morphology was examined by means of SEM to evaluate their fouling resistance. X-ray diffraction (XRD) and Differential scanning calorimetry-thermal gravimetric analyzer (DSC-TGA) analyses were made for the nanozeolite and its incorporation with natural biocides and synthetic biocide.

Table 13.1 Coating systems under investigation.

Systems	Reactants	Curing agent	Nanocontainer	Additives
A	DGEBA + Desmodur	Aradur 140	MCM-41	*Michelia champaca*
B	DGEBA + Desmodur	Aradur 140	MCM-41	Neem
C	DGEBA + Desmodur	Aradur 140	MCM-41	Bronopol
D	DGEBA + Desmodur	Aradur 140	MCM-41	–

13.9 Results and Discussion

13.9.1 Structural Characterization of Tris(*p*-Isocyanatophenyl)-Thiophosphate-Modified Epoxy Resin

Figure 13.3 depicts the FT-IR spectrum of raw Desmodur with characteristic peaks as follows: IR (NaCl)–1186 and 930 cm^{-1} (P–O–Ph), 1210 cm^{-1} (–PS), and 2270–2273 cm^{-1} confirmed the presence of isocyanate (–NCO). Figure 13.4 illustrates the IR spectrum for DGEBA, which confirms the presence of secondary –OH of epoxy at 3300–3400 cm^{-1}. The completion of the reaction between –NCO of Desmodur and secondary –OH of DGEBA was ascertained by the disappearance of –NCO peak and appearance of –NH and –CN linkage peaks observed at 3300–3500 and 1400 cm^{-1}, respectively, as shown in Figure 13.5.

Figure 13.6 illustrates the NMR spectrum of virgin DGEBA. The oxirane protons appear at δ 2.5–3.5, methylene protons at δ 4.0, secondary –OH of DGEBA at δ 5.5, and aromatic protons at δ 6.8–7.2 confirmed the structure of DGEBA.

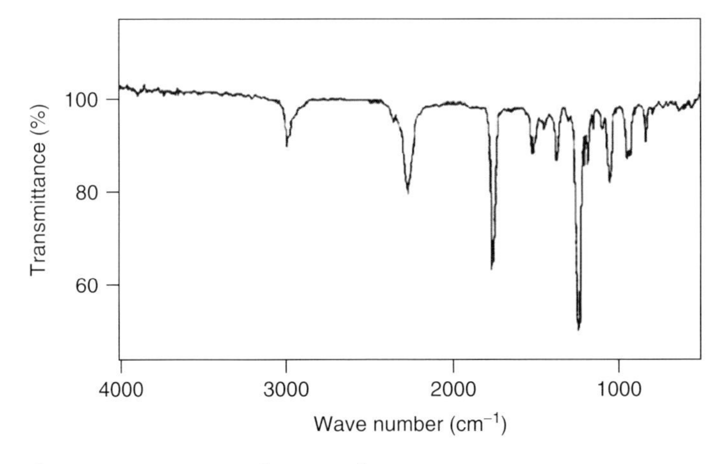

Figure 13.3 IR spectrum for Desmodur.

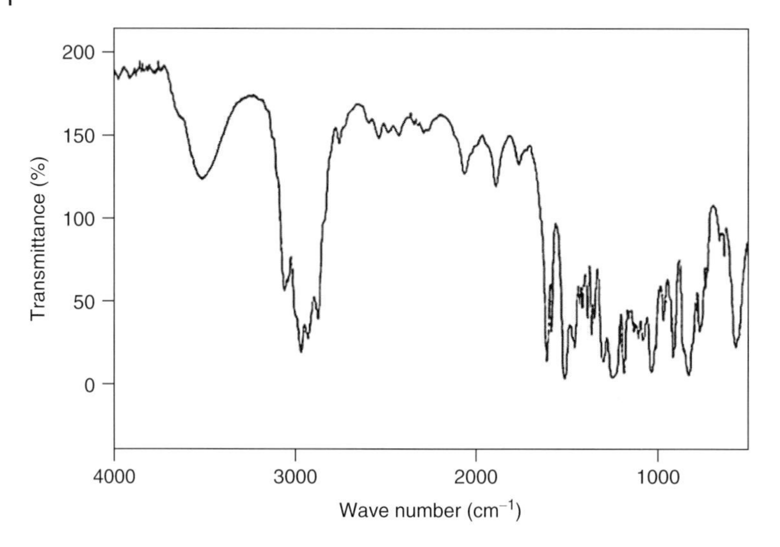

Figure 13.4 IR spectrum for DGEBA.

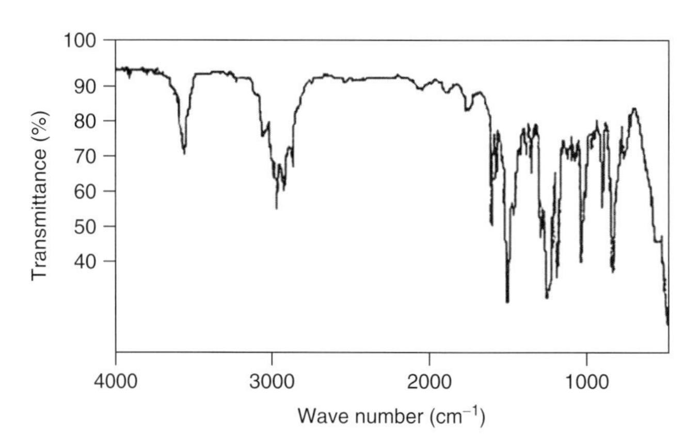

Figure 13.5 IR spectrum of Desmodur reaction with epoxy resin.

The reaction between the –NCO group in Desmodur and secondary –OH of DGEBA was ascertained by ^{1}H NMR spectrum as shown in Figure 13.7. The –NH proton appears at δ 8.2 and –CH–NH– protons appear at δ 3.35, confirming the completion of the reaction. Furthermore, the chemical shifts at δ 2.5–3.5, δ 4.0, and δ 6.8–7.5 confirmed the presence of the oxirane ring, methylene group, and aromatic ring of phosphate-modified epoxy resin, respectively.

The ^{13}C NMR spectrum of phosphate-modified resin is depicted in Figure 13.8. The signals at δ 174 and 150 correspond to –CO group and –C–N. The oxirane carbon at δ 100–160, $-CH_2$ carbon at δ 60, and CH_3-C-CH_3 appearing at δ 30 confirmed the structure of phosphate-modified epoxy resin. Figure 13.9 illustrates the ^{31}P NMR spectrum of phosphate-modified epoxy resin. A single signal at

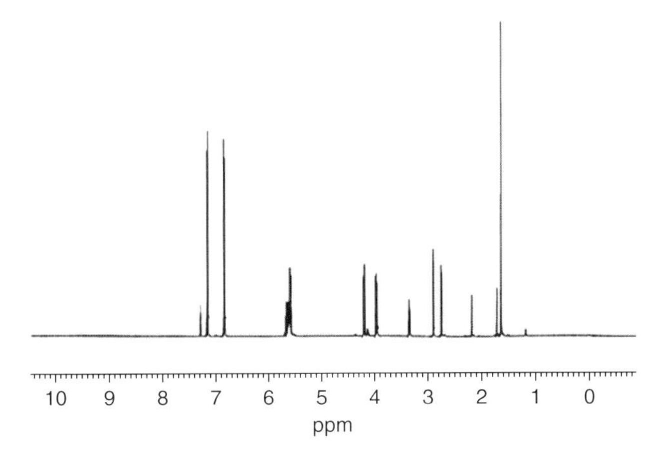

Figure 13.6 ^1H NMR spectrum for DGEBA.

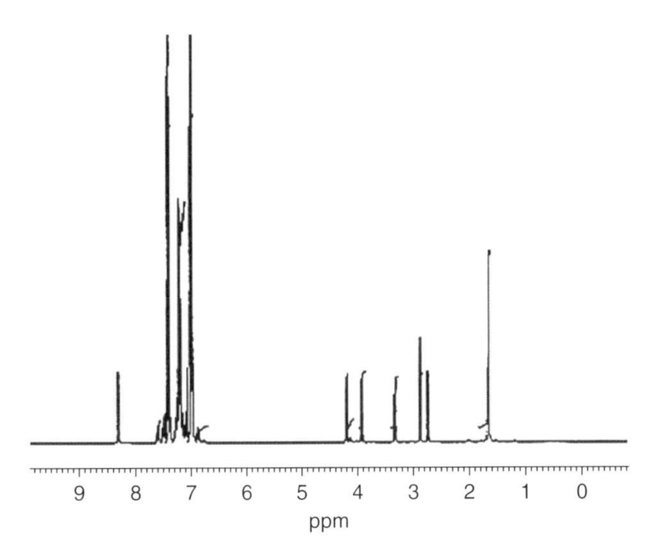

Figure 13.7 ^1H NMR spectrum for isocyanate-reacted epoxy resin.

δ 14.75 indicated the presence of phosphorus. The ^1H NMR spectrum signals (Figure 13.10) corresponding to POSS-NH_2 are assigned as follows: δ 1.25 for aliphatic amine and δ 0.52 for Si–CH_2 protons.

13.9.2 Structural Characterization of MCM-41

IR spectroscopy and XRD have been used extensively to characterize mesoporous molecular sieves. The IR spectrum of the calcined MCM-41 is depicted

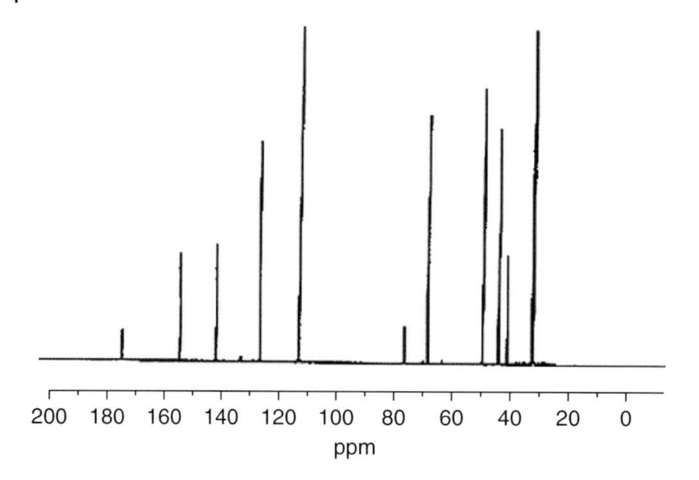

Figure 13.8 ^{13}C NMR spectrum of phosphate-modified resin.

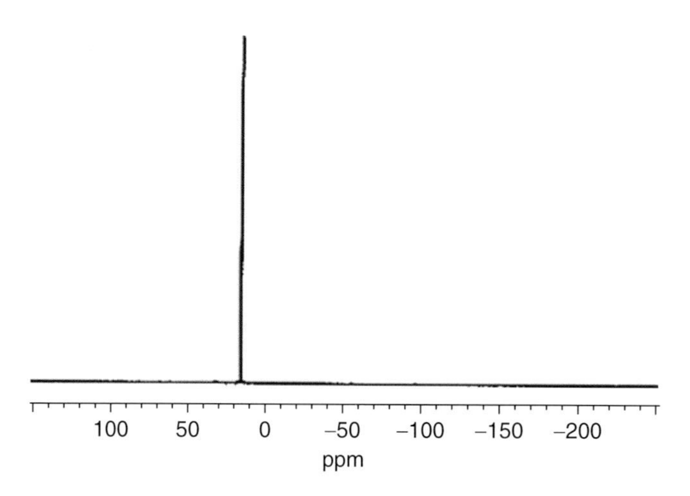

Figure 13.9 ^{31}P NMR spectrum of phosphate-modified resin.

in Figure 13.11. The broad envelop in 3448 cm^{-1} is due to –OH stretching of surface hydroxyl and bridged hydroxyl groups. There were less intense peaks just below 2400 cm^{-1} in the spectrum, which were assigned to symmetric and asymmetric stretching modes of the $-CH_2$ group of the template. Their corresponding bending modes were observed at 1641 cm^{-1}. The peaks between 500 and 1200 cm^{-1} were assigned to framework vibration. XRD provides direct information of pore architecture and phase purity of the MCM-41 material. The XRD pattern of calcined MCM-41 is shown in Figure 13.12. The diffraction patterns have reflection peaks only in the low-angle range because of its mesoporous nature.

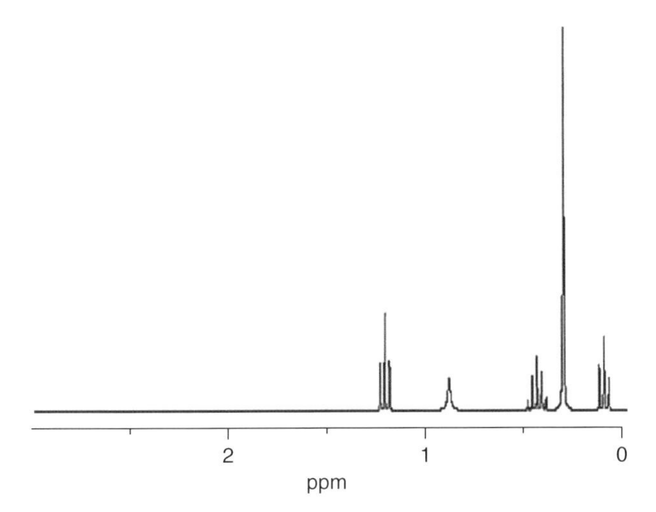

Figure 13.10 ^{1}H NMR spectrum of amine-functionalized POSS.

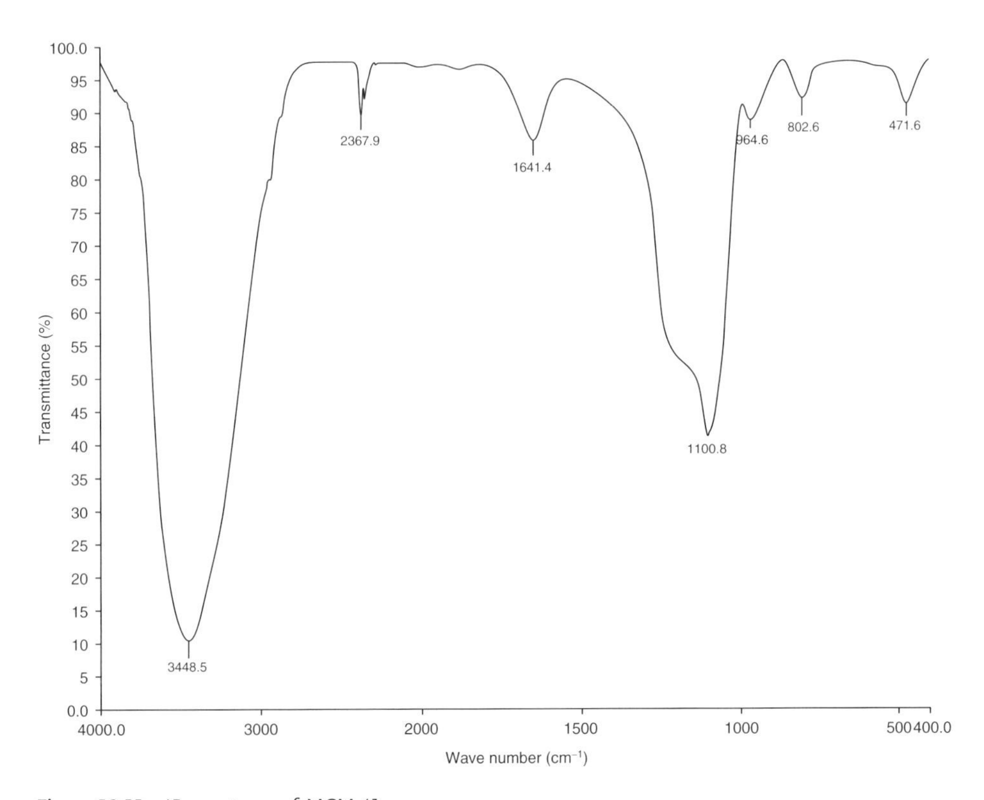

Figure 13.11 IR spectrum of MCM-41.

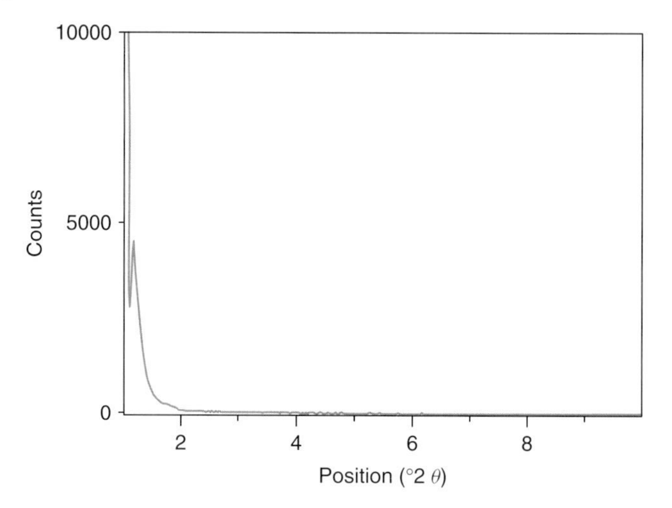

Figure 13.12 XRD of MCM-41.

13.9.3
Composition of Natural Products

13.9.3.1 M. champaca

The extract of *M. champaca* leaf was subjected to IR and ^{1}H NMR (Figures 13.13 and 13.14), and with the help of previous literatures, its composition is given in Table 13.2.

The ^{1}H NMR spectra of natural products, such as *M. champaca* leaf, are complicated, and their composition with respect to time keeps changing, which was observed by Logo *et al.* [31]. The ^{1}H NMR spectrum obtained by us for the *M. champaca* leaf (Figure 13.13) was in good agreement with his reported work.

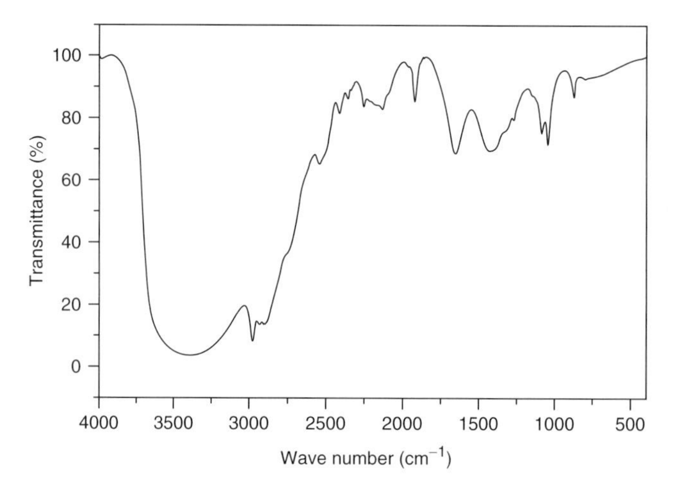

Figure 13.13 IR spectrum of *Michelia champaca*.

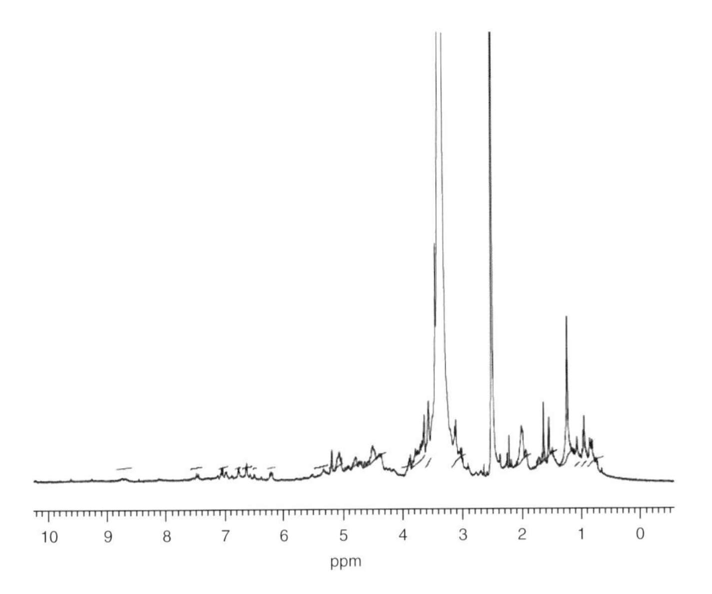

Figure 13.14 ^{1}H NMR of *Michelia champaca*.

Table 13.2 Composition of *Michelia champaca* leaf.

S. no.	Composition	Percentage (%)
1	Monoterpenes	10–14
2	Sesquiterpene hydrocarbon	47–69
3	Oxygenated sesquiterpene	15–20

13.9.3.2 Neem Oil

The IR and ^{1}H NMR of commercially available neem is depicted in Figures 13.15 and 13.16, respectively, and the composition of neem reported in previous studies by Logo *et al.* [32] is given in Table 13.3.

13.9.4 Confirmation of Loading of Biocide

The biocides before and after loading in MCM-41 can be confirmed by DSC-TGA studies. Figure 13.17 illustrates the DSC-TGA thermogram of MCM-41 loaded with and without *M. champaca* leaf extract. The percentage of residue before loading was found to be 40%, while after loading it was enhanced to 84–86% clearly indicating that *M. champaca* leaf extract was successfully loaded in the pores of MCM-41.

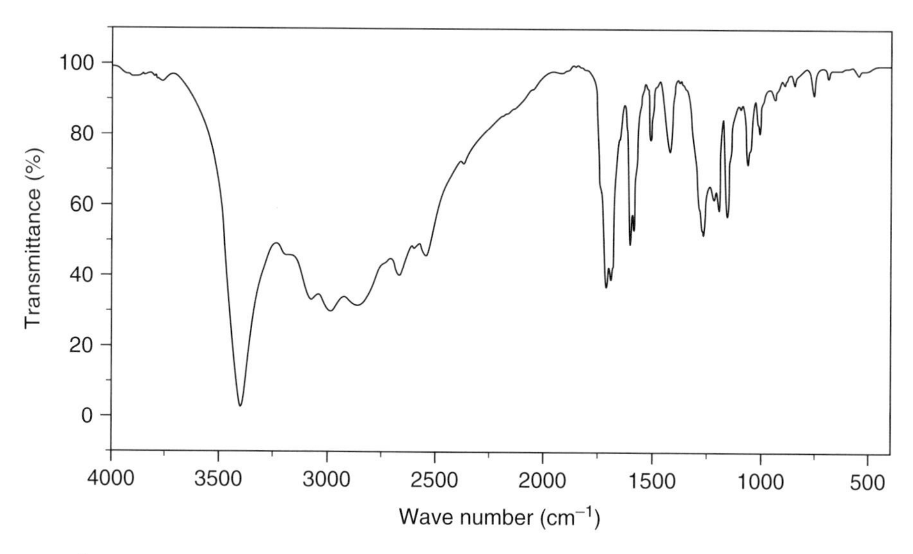

Figure 13.15 IR spectrum of neem oil.

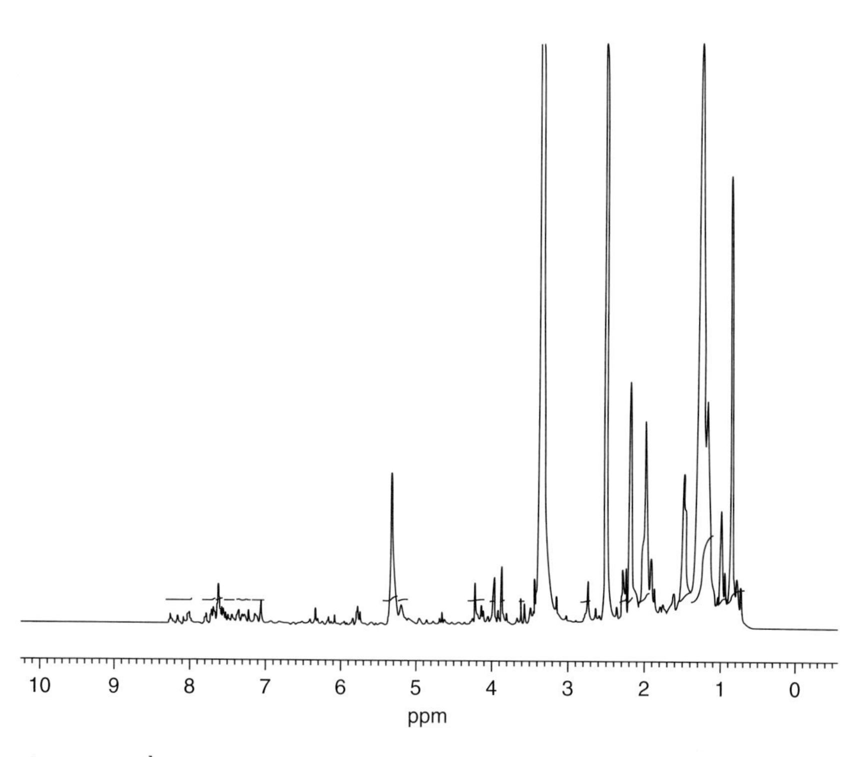

Figure 13.16 ^{1}H NMR spectrum of neem oil.

Table 13.3 Composition of neem oil.

S. no.	Composition	Percentage (%)
1	Oleic acid	25–54
2	Linoleic acid	6–16
3	Hexadecanoic acid	16–33
4	Octadecanoic acid	9–24

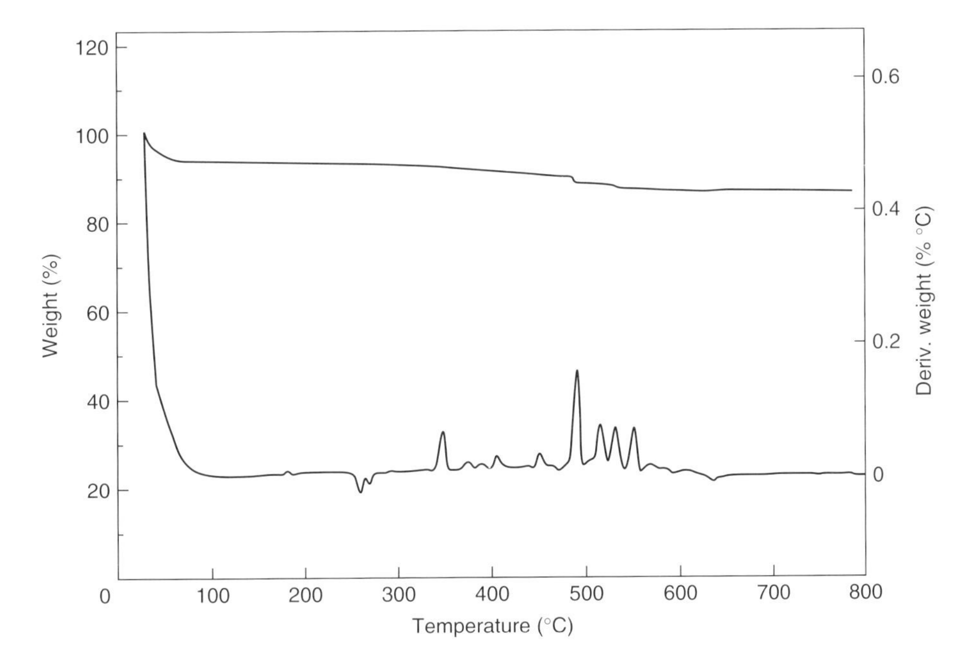

Figure 13.17 DSC-TGA of loaded MCM-41.

13.9.5 Evaluation of Corrosion Resistance by EIS

EIS studies were carried out on mild steel panels coated with phosphate-modified epoxy resin containing two natural biocides (*M. champaca* leaf extract and neem oil), containing one chemical biocide (Pandol), and without biocide in seawater. The results thus obtained from EIS were illustrated in Figure 13.18. It was interesting to note that except the neem-oil-loaded coating, all coated samples with and without biocides showed very good corrosion resistance up to $10^9\ \Omega\ cm^2$, indicating no corrosion taking place initially. In contrast to this observation, a different impedance pattern was obtained for the same coated samples after 15, 30, 45, and 60 days

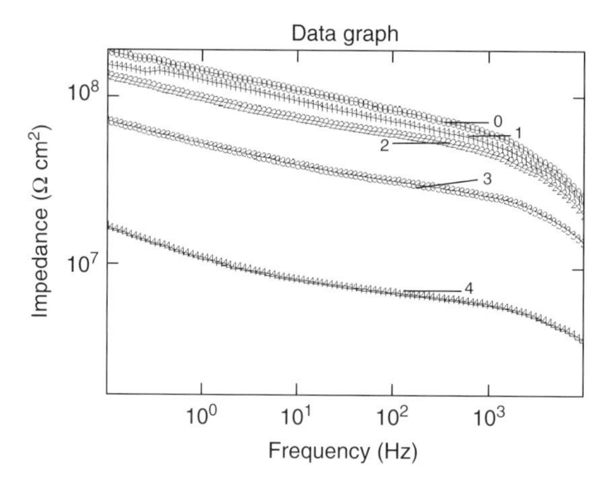

Figure 13.18 Bode plot of panel A coated with *Michelia champaca* for (0) initial, (1) 15 days, (2) 30 days, (3) 45 days, and (4) 60 days of immersion in seawater.

of immersion in seawater, and the results are depicted in Figures 13.18–13.21, respectively. From the figures, it was clearly evident that the coating without biocide (D) and an excellent corrosion resistance is superior to other coatings that contain biocides (A, B, and C), and the order of corrosion resistance is found to be as follows:

$$D > C > A > B$$

This superior corrosion resistance offered by the coating system D may be due to the molecular structure of Aradur 140 (polyamidoimidazoline), which was able to adsorb on the metal surface through the $\prod$ electrons of the aromatic ring and lone pair of electrons of nitrogen and sulfur atoms. Similar observation was made by He *et al.* [33] who explained the relationship between molecular structures and revealed that the corrosion-resistant properties improve with increasing number of nitrogen atoms, thereby increasing the coordination capacity of the heterocyclic molecules that influence the stability and protective properties of the cured film.

The reason for the decreased corrosion resistance observed in the case of other coatings, namely, C, B, and A, is not clear and seem to be complicated. The reason may, however, be due to the presence of porous structure of such coatings paving an avenue for the entry of corrosive species to the substrate when they are subjected to the accelerated corrosion testing by means of EIS. It was also interesting to note that the corrosion resistance offered by the neem-oil-loaded coating B seems to be peculiar, that is, its corrosion resistance was very low at the beginning ($10^5\ \Omega\ cm^2$) and found to be maximum as the days of immersion reached 30 ($10^9\ \Omega\ cm^2$). However, the value of corrosion resistance decreased to $10^7\ \Omega\ cm^2$ at the end of 60 days. The maximum corrosion resistance observed in the case of coating B on the thirtieth day of immersion may be due to the barrier coating, which might have resulted from the reaction of leached neem biocide with electrolyte.

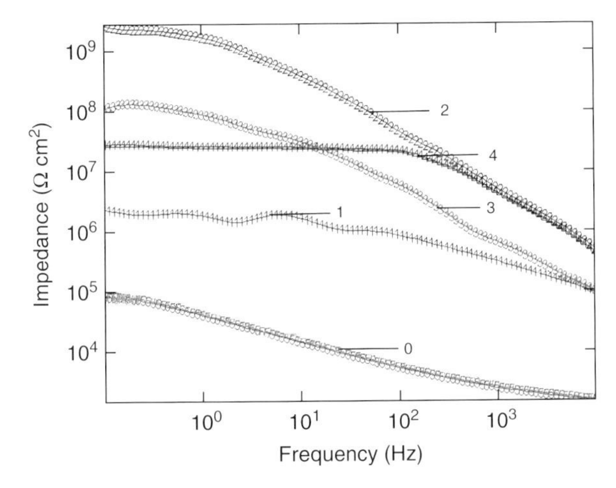

Figure 13.19 Bode plot of panel B with neem oil biocide for (1) initial, (2) 15 days, (3) 30 days, (4) 45 days, and (5) 60 days of immersion in seawater.

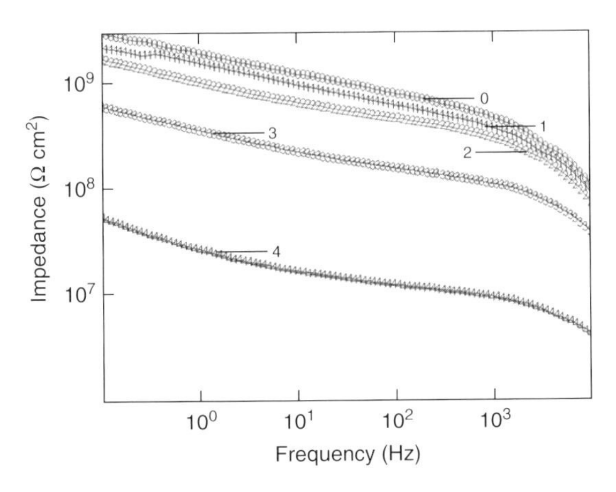

Figure 13.20 Bode plot of panel C with bronopol as biocide for (0) initial, (1) 15 days, (2) 30 days, (3) 45 days, and (4) 60 days of immersion in seawater.

Its decreased corrosion resistance after 30 days of immersion may be due to the decomposition of barrier coating. Similar observation was made by Ananda Kumar *et al.* in the year 2001 for zinc-containing epoxy and siliconized epoxy coatings.

13.9.6 Colorimetric and Gravimetric Analyses

The data obtained from EIS were correlated with the data resulted from colorimetric and gravimetric analyses. The iron content of the corrosive media in which A-, B-,

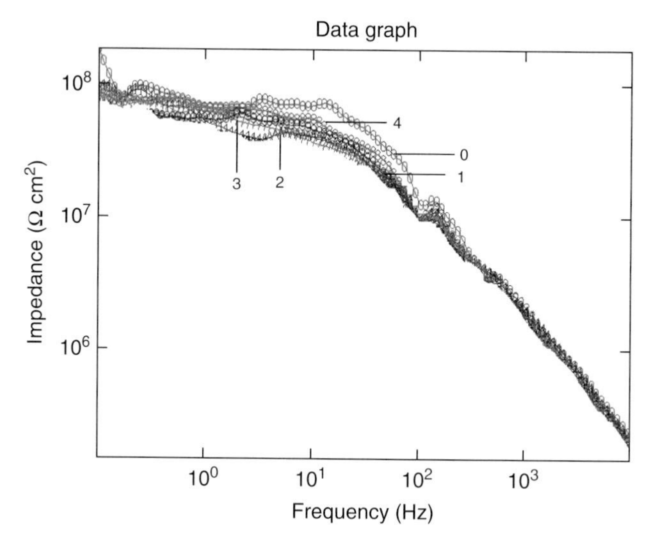

Figure 13.21 Bode plot of the panel D without biocide for (0) initial (1) 15 days (2) 30 days (3) 45 days, and (4) 60 days of immersion in sea water.

C-, and D-coated samples were immersed for corrosion study was measured by means of colorimetric and gravimetric analyses. It was interesting to observe that the data resulted from colorimetric and gravimetric analyses were found to be in very good agreement with each other and also with the data obtained from EIS. The iron and iron oxide content determined by colorimetric and gravimetric analyses are given in Tables 13.4 and 13.5 and also shown as graph in Figures 13.22a,b.

Table 13.4 Data resulted from colorimetric analysis.

Systems	Fe values (ppm)
A	23.4
B	33.4
C	22.3
D	3.4

Table 13.5 Data resulted from gravimetric analysis.

Systems	Fe_2O_3 values (ppm)
A	46.2
B	67.5
C	40
D	5.5

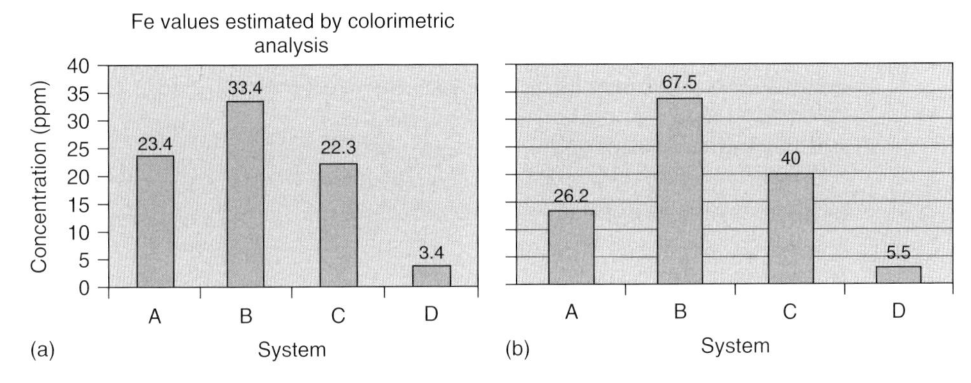

Figure 13.22 (a) Bar diagram of Fe values of corrosive medium. (b) Bar diagram of Fe_2O_3 of corrosive medium.

From the tables, it was clearly evident that the values of iron and iron oxide content exhibited by sample D were found to be minimum and those of sample B were found to be maximum. The values of other two samples fall in between these two extremes. The minimum value obtained for sample D indicates its superior corrosion resistance with less metal dissolution in corrosion media. The order of corrosion resistance D > C > A > B observed from this study exactly matches with the order of corrosion resistance seen in the case of EIS studies.

13.9.6.1 pH Analysis

The pH values of the corrosive medium (in which A-, B-, C-, and D-coated samples were immersed for corrosion test) were noted periodically, and the data obtained is given in Table 13.6. The difference in pH value of the coating system D during the initial and final stages was found to be minimum with respect to the number of days of immersion. For example, its pH measured during the first day of immersion was found to be 6.47 and after 30 days 6.21, whereas in the case of other samples (A, B, and C), the variation between the pH values of first-day immersion and after 30 days was more, which can be seen in Table 13.6. This clearly indicates the superior corrosion resistance of sample D in comparison with the pH values

Table 13.6 pH values of corrosive medium.

Coating systems	First-day immersion	After 10 d of immersion	After 20 d of immersion	After 30 d of immersion
A	7.42	6.41	6.36	6.23
B	6.98	6.11	6.31	6.38
C	6.92	6.66	6.49	6.42
D	6.47	6.39	6.24	6.21

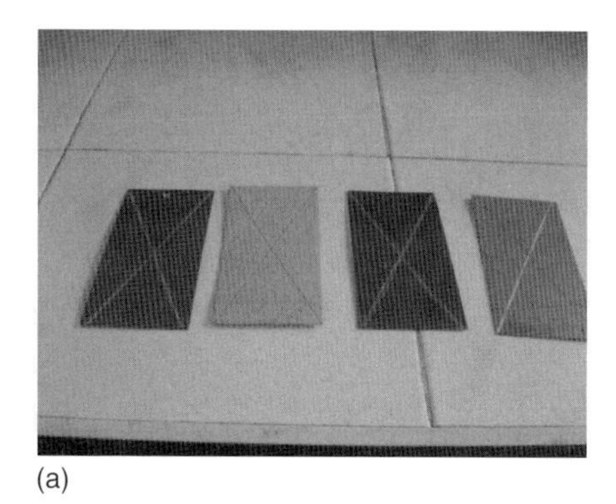
(a)

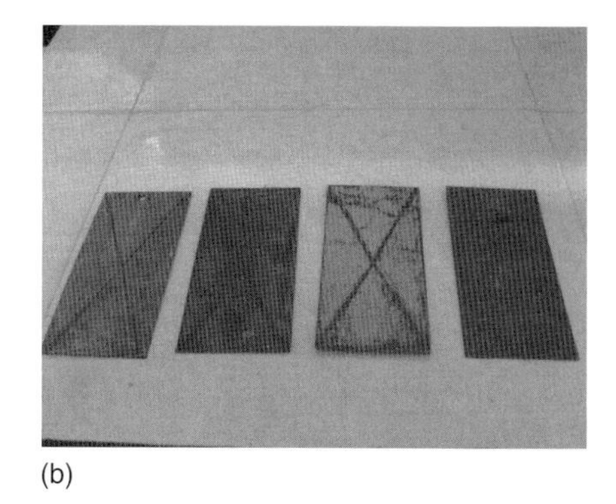
(b)

Figure 13.23 (a) Scratched samples before exposure to salt spray. (b) Scratched samples after 1000 h of exposure to salt spray.

of other samples, namely, A, B, and C. This data supports the data obtained from EIS, salt spray test, and colorimetric and gravimetric analyses.

13.9.6.2 Salt Spray Test Results

The coated specimens were subjected to a diagonal scratch with the help of a sharp knife in order to expose the base metal to the continuous salt fog chamber containing 3.5% NaCl solution. The 1000 h salt spray test supports the results obtained from EIS studies. The salt spray results of coated panels before and after exposure to salt fog atmosphere are schematically given in Figure 13.23a & b. At the end of the salt spray test, no visible corrosion products were seen on the surface of the unscratched area of panels coated with nanohybrid epoxy resin with biocides A, B, and C and without biocide D. As observed in the case of EIS and colorimetric and gravimetric analyses, sample D containing no biocide offered superior corrosion resistance than the other three samples (A, B, and C) chosen for the study. This behavior may be again due to the molecular structure of Aradur 140 and nano-cross-linking effect of POSS offering excellent corrosion resistance to demonstrate their synergistic effect.

13.9.6.3 Seawater Immersion Test

Fouling resistance of these coatings was determined by antifouling studies by subjecting the coated samples in sea for a period of 180 days at the east coast of India (Chennai, Tamil Nadu). SEM photographs of the A-, B-, C-, and D-coated specimens having attached fouling organisms were taken and compared with the similar work reported by Soren Kill *et al.* [34]. The surface morphology of A-, B-, C-, and D-coated samples after immersion in sea for the evaluation of fouling was examined by SEM, and the results are illustrated in Figures 13.24–13.27. From the fouling study, we observe that the attachment of fouling organisms was more on panel D coated with phosphate-modified epoxy without biocide and the intensity of fouling organisms attached to A-, B-, and C-coated panels was minimum.

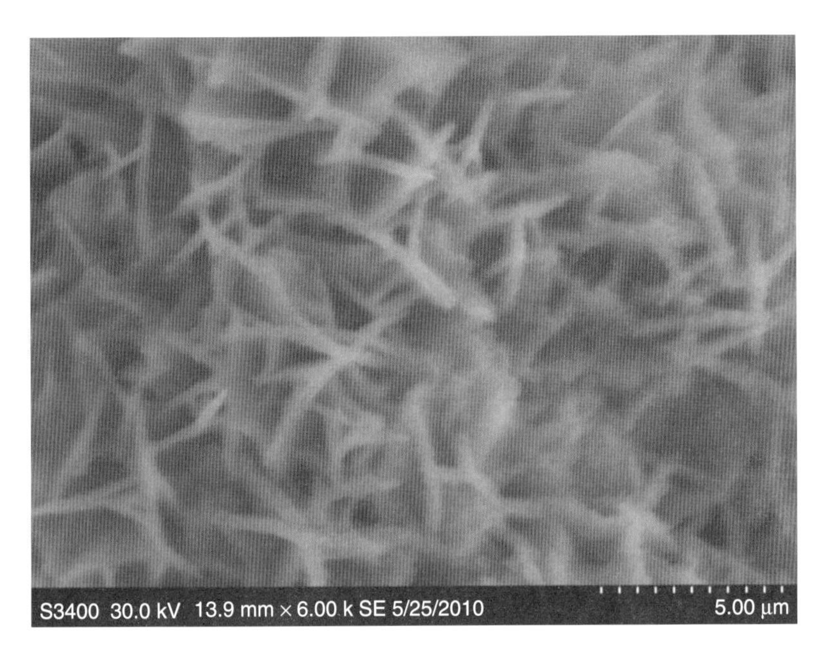

Figure 13.24 SEM picture of sample A.

Figure 13.25 SEM picture of sample B.

Figure 13.26 SEM picture of sample C.

Figure 13.27 SEM picture of sample D.

Among the biocide-loaded coated panels, panel coated with C (Pandol as biocide) showed least fouling indicating its excellent antifouling ability (Figure 13.26) than panels coated with A (Figure 13.24) and B (Figure 13.25). This may be due to the superior antimicrobial activity offered by Pandol than the other two nontoxic natural biocides. Although Pandol appears to be toxic, it has been found to have good activity in selectively low concentrations of 0.01–0.2% as an antimicrobial preservative and antiseptic in tropical pharmaceutical formulations, cosmetics, and toiletries and is safe in concentrations that we have used in this study. The order of fouling resistance of all the coatings is

$$C > B > A > D$$

13.9.6.4 Antifouling Studies by Scanning Electron Microscopy (SEM)

Photomicrograph of the coating system D exhibits heavy bacterial colonization and biofilms attached (Figure 13.27) to the surface of a mild steel coupon, indicating its inferior antifouling activity. In contrast, there was a marked inhibition of bacterial adhesion on the mild steel surface coated with coating systems C, B, and A. SEM Figures 13.24–13.26 show the surface of well-protected mild steel coupons with few bacterial colonies and a very thin biofilm formation, indicating their superior fouling resistance. The fouling attack appeared to proceed at a lower rate on the surfaces coated with coating systems C, B, and A. Furthermore, coating C reveals its poor adhesive strength towards adsorbed seawater components (including proteins released by barnacles, fungi, and alga) on it, affirming its controlled release and self-polishing nature [34]. The superior fouling resistance offered by the coating system C may be due to the low surface energy of POSS and combined effect of Pandol as an effective biocide against fouling. The mechanism of foul release from POSS-NH_2 surfaces is still not completely understood [35]. However, it is generally believed that the antifouling properties are because of their low surface energies. A similar observation was made by Chen *et al.* [36] for the low surface energy nontoxic organosilicon nano-SiO_2 antifouling coatings, which showed less adhesion of the biofouling organisms on the coating films.

13.10 Summary and Conclusion

The main goal of this work is to develop nontoxic nanohybrid coatings based on nanohybrid epoxy resin, which contain antifungal agents and nontoxic/chemical biocides incorporated in novel nanozeolite molecular sieves as nanocontainer. The embedded biocides were designed to be released into the environment only as needed, thus extending the lifetime of the biocidal activity. These imbedded biocides, therefore, would eliminate the undesired contamination of the surrounding area and remain within a coating.

In the proposed work, an epoxy resin is reinforced with POSS nano-cross-linker, using Desmodur having phosphorus and sulfur moieties as modifier by means of an inter-cross-linking polymer network (ICN) to formulate nanohybrid coating having improved properties ideally suitable for high-performance applications. Use of modifiers improved the environmental and corrosion-resistant properties compared to common commercially available resins that are currently used in the field where high corrosion and fouling resistance are essential. Furthermore, this invention also involved the synthesis of a novel nanozeolite, which served as a nanocontainer medium for incorporating nontoxic antifouling agent and naturally available biocides. Nontoxic biocide/natural antifouling agents chosen from naturally available neem oil, *M. champaca* leaf, and a chemical biocide Pandol were encapsulated in mesoporous silica followed by activation of its surface by 3-aminopropyltriethoxysilane to achieve a coating with controlled release of biocidal/antifungal agents for the prevention of fouling. The purpose of zeolite is to promote the longevity of biocides and antifouling agents by shielding them from direct contact with enzymes and microorganisms, with a controlled dispersion of biocide from the polymer to the fouling environment. The anticorrosive and antifouling efficiencies of the developed nanohybrid coatings were evaluated by standard test methods to determine their suitability to and efficacy in a range of environments.

The panels were evaluated for their corrosion resistance by means of EIS studies, salt spray test, and immersion tests (in 3.5% NaCl). The fouling resistance of these coatings was determined by antifouling studies by subjecting the coated specimens in seawater for a period of 180 days at the east coast of India (Chennai, Tamil Nadu).

The data resulted from corrosion and fouling studies clearly indicate that the chemical biocide Pandol and natural products used as biocide showed better fouling rate than the coating without biocide. It can be concluded that the natural products and Pandol may be used as an effective antifouling biocide in nanohybrid coatings than the conventional epoxy coatings currently used, for better performance and longevity. As far as corrosion resistance is concerned, the nanohybrid coating without biocide showed the maximum corrosion resistance. The versatility of the nanohybrid coatings thus achieved will then be exploited for their commercialization.

Acknowledgment

We gratefully acknowledge the financial assistance provided by DRDO (Defence Research Development Organization, Government of India) (Grant No. **ERIP/ER/0503520/M/01** dated 20/07/06) for this work, and the authors thank the Department of Chemistry, Anna University, Chennai, and the National Metallurgical Laboratory, Chennai, for their help in carrying out this work. Furthermore, the authors sincerely thank Mrs. A. Jaya, Former Principal DKM College (W), Vellore, TN, India and Mr. D. Duraibabu, Doctoral candidate of Chemistry department, Anna University, Chennai for their timely help in proof correction.

References

1. Anandakumar, S., Balakrishnan, T., Alagar, M., and Denchev, Z. (2006) Development and characterization of silicone/phosphorus modified epoxy materials and their application as anticorrosion and antifouling coatings. *Prog. Org. Coat.*, **55** (2), 160–167.
2. Ananda Kumar, S., Alagar, M., and Mohan, V. (2002) Studies on corrosion-resistant behaviour of siliconized epoxy interpenetrating coatings over mild steel surface by electrochemical methods. *J. Mater. Eng. Perform.*, **11** (2), 123–129.
3. Yebra, D.M., Kiil, S., and Dam-Johansen, K. (2004) Antifouling technology: past, present and future steps towards efficient and environmentally friendly anti-fouling coatings. *Prog. Org. Coat.*, **50** (2), 75–104.
4. Rascio, V.J.D., Giudice, C.A., and Delamo, H. (1988) Biocidal performance of acrylated glyphosphate in a model photopolymerisable coating formulation. *Corros. Rev.*, **8** (1–2), 87–153.
5. Clarke, A.S. (1995) Natural ways of banish barnacles. *New Sci.*, **18**, 38–41.
6. Nair, K.V.K. (1999) Marine biofouling and its control with particular reference to condenser cooling circuits of power plants. *J. Indian Inst. Sci.*, **17**, 497–511.
7. Gatenholm, P., Kellberg, S., and Marita, J.S. (1992) *Proceedings of the ACS Division of Polymeric Materials Science and Engineering*, ACS, Washington, DC, p. 490.
8. Usani, M., Tomoshige, K., and Marita, H. (1994) Proceedings of the International Offshore and Polar Engineering Conference, ISOPE, Golden, CO, p. 644.
9. Rahmoune, M., and Latour, M. (1996) Application of Mechanical Waves Induced by Piezofilms to Marine Fouling Prevention. *J. Intelligent Mater. Sys. Struct.,*, **7** (1), 33–43.
10. Rahmoune, M. and Latour, M. (1995) *Smart Mater. Struct.*, **4** (3), 195–204.
11. Dalley, R. (1989) *Biofouling*, **1**, 363–366.
12. Maguire, R.J. (1992) *Water Sci. Technol.*, **25** (11), 125–132.
13. Van Slooten, K.B. and Tarradellas, J. (1994) *Environ. Toxicol. Chem.*, **13** (5), 755–762.
14. Shugui, D., Guolan, H., and Yong, C. (1995) *Water Pollut. Res. J. Can.*, **30** (1), 33.
15. Tas, J.W., Keizer, A., and Opperhuizen, A. (1996) Bioaccumulation and lethal body burden of four triorganotin compounds. *Bull. Environ. Contam. Toxicol.*, **57** (1), 146–154.
16. Almeida, E., Teresa, C., and Orlando de Sousa, D. (2007) Marine paints particular case of antifouling paints. *Prog. Org. Coat.*, **59**, 2–20.pp.
17. Bultman, J.D., Griffith, J.R., Thompson, M.F., Nagabhushanam, R., Sarojini, R., Fingerman, M., and Balkema, A.R. (eds) (1994) Fluoropolymer and silicone fouling-release coatings international, in *Recent Developments in Biofouling Control*, pp. 383–389.
18. Kavanagh, C.J., Schultz, M.P., Swain, G.W., Stein, J., Truby, K., and Darkangelo Wood, C. (2001) Variation in adhesion strength of Balan.us eburneus, Crassosterea virginica and Hydroides dianthus to fouling release coatings. *Biofouling*, **17**, 155–167.
19. Nakasono, S., Burgess, J.G., Takahashi, K., Murayama, C., Nakamura, S., and Matsunaga, T. (1993) Electrochemical prevention of biofouling with a carbon-chloroprene sheet. *Appl. Environ. Microbiol.*, **59**, 2757–2762.
20. Bakus, G.J., Schulte, B., Jhu, S., Wright, M., Green, G., and Gomez, P. (1991) Antibiosis and antifouling in marine sponges laboratory versus field study, in *New Perspective in Sponge Biology*, Smithsonian Institution Press, Washington, DC, pp. 102–108.
21. Rittschof, D., McClintock, J.B., and Baker, B.J. (eds) (2001) *Natural Product Antifoulants and Coatings Development in Marine Chemical Ecology*, Chapter 17, CRC Press, Boca Raton, FL.
22. Murray, J.N. (1997) Electrochemical test methods for evaluating organic coatings on metals, introduction and generalities

regarding electrochemical testing of organic coatings. *Prog. Org. Coat.*, **30**, 225.
23. Wegmann, A. (1997) Freeze-thaw stability of epoxy resin emulsions. *Pigment Resin Technol.*, **26** (3), 153–160.
24. Bayramolu, E.E. (2007) Unique biocide for the leather industry, essential oil of *oregano. J. Am. Leather Chemists Assoc.*, **102**, 347–351.
25. Anandakumar, S., Denchev, Z., and Alagar, M. (2008) Development and characterization of phosphorus containing epoxy resin coatings. Proceedings of Coatings Science International Conference (COSi 2008), p. 93.
26. Chattopadhyay, D.K. and Raju, K.V.S.N. (2007) Structural engineering of polyurethane coatings for high performance applications. *Prog. Polym. Sci.*, **32**, 352–418.
27. Tian, L. and Rajapakse, R.K.N.D. (2007) Finite element modeling of nanoscale inhomogeneities in an elastic matrix. *Comput. Mater. Sci.*, **41**, 44–53.
28. Crosby, A.J. and Lee, J.Y.J. (2007) Polymer Nanocomposites: The "Nano" Effect on Mechanical Properties. *Polym. Rev.*, **47** (2), 217–219.
29. Koo, J.H. (2006) *Polymer Nanocomposites: Processing, Characterization and Application*, Nanoscience and Technology Series, 1st edn, McGraw-Hill, 26–28.
30. Edge, M., Allen, N.S., Turner, D., Robinson, J., and Seal, K. (2001) The enhanced performance of biocidal additives in paints and coatings. *Prog. Org. Coat.*, **43** (1–3), 10–17.
31. Logo, J.H.G., Avila, P. Jr., Moreno, P.R.H., Limberger, R.P., Apcl, M.A., and Henriques, A.T. (2003) Analysis comparison and variation in the chemical composition from the leaf volatile oil of Xylopia Aromatica. *Biochem. Syst. Ecol.*, **31**, 669–672.
32. Logo, J.H.G., Favero, O.A., and Romoff, P. (2006) Microclimatic factors and phenology influences on chemical composition of essential oils from Pettosporum Undulatum Vent leaves. *J. Brazil Chem. Soc.*, **17**, 1334–1338.
33. He, Z., Rao, W., Ren, T., Liu, W., and Xue, Q. (2002) The tribochemical study of some nitrogen-containing heterocyclic compounds as lubricating oil additives. *Tribol. Lett.*, **13**, 87–93.
34. Svendsen, J.R., Kontogeorgis, G.M., Kill, S., Weinell, C.E., and Grønlund, M. (2007) Adhesion between coating layers based on epoxy and silicone. *J. Colloid Interface Sci.*, **316**, 678–686.
35. Meyer, A., Baier, R., Wood, C.D. *et al.* (2006) Contact angle anomalies indicate that surface-active elutes from silicone coatings inhibit the adhesive mechanisms of fouling organisms. *Biofouling*, **22**, 411–423.
36. Chen, M., Qu, Y., Yang, L., and Gao, H. (2008) Structures and antifouling properties of low surface energy non-toxic antifouling coatings modified by nano-SiO_2 powder. *Sci. China, Ser. B-Chem.*, **51**, 848–852.

Further Reading

Kim, J., Nyren-Erickson, E. *et al.* (2008) Release characteristics of reattached barnacles to non-toxic silicone coatings. *Biofouling*, **24**, 313–319.

Index

a
acid catalysts *186, 188*
AC impedance 66–67
acousting streaming 313
acrylic urethanes 107
adsorption
– mechanism, in corrosion inhibition 321–323
– theory 127–129
airless spraying 358
aliphatic isocyanates 106
alkyd resins 99–100, *100*
ambiodic and mixed inhibitors 135
amino acids-based green inhibitors 330
aminoazophenylene (AAP) 136
anesthetics and narcotics 141
3-anisalidene amino 1,2,4-triazole phosphonate (AATP) 273
anodic (passivating and film-forming) inhibitors 126–127
– adsorption theory 127–129
– generalized film theory 127
anodic metal dissolution 4–6
anodic oxide 8
anodic Tafel constant 59
anticorrosion rust 28
antifouling paints 358
antifouling studies, by scanning electron microscopy 389
artificial neural network (ANN) 327
asphyxiants 141
atmospheric corrosion
– anticorrosion rust 28
– chemistry 24–26
– weathering steel corrosion 26–27

b
barium-(amino-tris-(methylenephosphonate)) (Ba-AMP) 255, *256*
barium-phosphonomethylene-imino-diacetate (Ba-PMIDA) 260, *262*
barrier coatings 108
base catalysts *187, 190*
benzotriazole 133
bicapped octahedron 267
biofouling 356
bronopol 367–368
Butler–Volmer equation 58

c
calcium-(amino-tris-(methylenephosphonate)) (Ca-AMP) 252–253
calcium-hexamethylene-diamine-*tetrakis* (methylenephosphonate) (Ca-HDTMP) 264, *265*
calcium-hydroxyphosphonoacetate (Ca-HPAA) 264, *266*, 267
$Ca_3(HPAA)_2(H_2O)_{14}$ 279–281
calcium-phosphonobutane-1,2,4-tricarboxylate (Ca-PBTC) 263, *264*
$\{Ca(PBTC)(H_2O)_2{\cdot}2H_2O\}_n$ 278–279
calcium-phosphonomethylene-imino-diacetate (Ca-PMIDA) 269, *270*
carcinogens 142
cathodic (adsorption-type) inhibitors 129
– mechanism 129–130
cathodic oxidant reduction 6
cathodic protection 340, 344, 347, 348, 349
cathodic Tafel constant 59
Central Electrochemical Research Institute (CECRI) 346, 347
chemical gasometry technique 332
chemisorbed films 128
chemisorption 134, 163
chloride-breakdown of passive films 12–13
coincidence site lattice boundaries (CSLBs) 227

Green Corrosion Chemistry and Engineering: Opportunities and Challenges, First Edition.
Edited by Sanjay K. Sharma.

cold plastic paints 358
colorimetric and gravimetric analyses 383–389
concentration polarization 61–68
condensation reaction
– with acid catalysts *186*
– with base catalysts *187*
– metal surface *197*
contact adsorption 49, *50* 163
conversion electron Mossbauer spectroscopy (CEMS) 275
corrosion cost, in various sectors *341*
corrosion education, India 345
– Central Electrochemical Research Institute (CECRI) 347
– corrosion science and engineering in IIT (Bombay) 346
– NACE India Section 347
– National Corrosion Council of India (NCCI) 348
– pioneers in India 346
– reasons 345–346
– Society for Surface Protective Coatings India 347–348
corrosion inhibitors 125, 143–147, 244. *See also individual entries*
– classification of
– – anodic (passivating and film-forming) inhibitors 126–129
– – cathodic (adsorption-type) inhibitors 129–130
– – mixed inhibitors 130–131
– – precipitating inhibitors 131
– – vapor phase inhibitors 131–139
– phosphonate-based *246*
corrosion loss 339–340, 343, 347, 349, 353
corrosion management 340, *350*
corrosion potential 3–4, 55
corrosion-prone industries, in India
– oil and gas industry 348–349
– power plants 351–352
– process chemical and petrochemical industry 349–350
– pulp and paper industry 350–351
corrosion rust 19–20
– electron-selective rust 22–24
– ion-selective rust 20–22
– redox rust 24
corrosion science and engineering, in IIT (Bombay) 346
coupling agent 182, 191, 203–204
crevice corrosion 16–18
crevice protection potential 16
critical crevice corrosion temperature 18
critical pitting temperature 16

d

dangerous inhibitors. *See* anodic (passivating and film-forming) inhibitors
diethyllaurylphosphonate (DELP) 271
dimethylaminomethylene-bis(phosphonic acid) (DMABP) 249, *250*
dimethylbenzyl bromide 137
dip coating 192–194
1,5-diphosphono-pentane (DPP) 274
1,7-diphosphonoheptane (DPH) 274
distorted bicapped trigonal antiprism 268
dynamic electrochemical processes 49–61

e

eco-friendly corrosion inhibitors 125–126. *See also* green corrosion inhibitors
– anodic (passivating and film-forming) inhibitors 126–127
– – adsorption theory 127–129
– – generalized film theory 127
– cathodic (adsorption-type) inhibitors 129
– – mechanism 129–130
– mixed inhibitors 130–131
– precipitating inhibitors 131
– toxicity of inhibitors 139–140
– – anesthetics and narcotics 141
– – asphyxiants 141
– – carcinogens 142
– – irritants 140–141
– – mutagens 142
– – sensitizers 142
– – systemic poisons 141–142
– – teratogens 142–147
– vapor phase inhibitors 131–139
electrical double layer 49
electrochemical cell *67*, 68
electrochemical impedance spectroscopy (EIS) 272
electrochemistry 2, 3, 4, 25, 29, 33–39
– concentration polarization 61–68
– dynamic electrochemical processes 49–61
– electrode potential measurements 44–45
– equilibrium electrode potentials 45–48
– free energy and electrode potential 41–44
– Pourbaix diagrams, use of 49
– thermodynamics and stability of metals 40–41
electrodeposition 297–298
– of composite coatings 305–312
– – codeposition mechanism 299–302
– – magnetic field 314–315

– – process parameters influencing incorporation 302–304
– – solid particle suspension in electrolytes 298–299
– – ultrasonic vibrations 313–314
electrode potential 2
electron acceptors 134
electron backscatter diffraction (EBSD) 229
electron configuration theory. *See* adsorption theory
electron-selective rust 22–24
electron-sink area 37
electron transfer processes 2
energy dispersive X-ray spectroscopy (EDS) 226
energy profile
– for anode at equilibrium and for anodic activation polarization *58*
– for copper in equilibrium with divalent ion solution *57*
– of copper oxidation *56*
epoxy resin coatings 101–105, *102*, *103*, *104*, 360–361
– advantages 361
– biocide release mechanism from nanocontainer 366–367
– disadvantages 361–362
– justification 362
– nanocontainer fabrication 366
– nanocontainer need 364–366
– nanoparticle types 363–364
– natural product selection as biocides 367–368
– need for nanotechnology 362–363
– polyhedral oligomeric silsesquioxanes (POSSs) 364, *365*
– polymer nanomaterials 363
equivalent circuit 66
ethylenediamine-*tetrakis*(methylene-phosphonic acid) (EDTMP) 249, *250*
ethyllaurylphosphonate (ELP) 271
Evans model 26
exchange current 57
exfoliation 216, 223, 225, 226, 227, 230, 232, 235, 238, 239

f
Faraday's Law 55, 56
Fermi level 2
film breakdown potential 12
fire-resistant coatings 119
Frumkin isotherm 158
fusion bonded epoxy (FBE) 91

g
generalized film theory 127
glassflake epoxies 111
glow discharge optical emission spectrometry (GD-OES) 274
grain boundary engineering (GBE) 227, 229–230, 233–234, 236, 237, 239
gray inhibitors 328
green chemistry
– and corrosion control 245
– and sustainable development 320
green corrosion inhibitors 157–160, 320–321. *See also* eco-friendly corrosion inhibitors
– amino acids-based 330
– in developing countries
– – metal usage and present corrosion management 173–174
– – researchers' work to develop green inhibition science 174–176
– natural derivatives as 328–329
– natural products as 166–169
– organic-based 329
– plant extracts–based 330–333
– protection against corrosion 160–161
– rare earth elements-based 333–334
– research and progress 169–170
– – proposed mechanism of extract inhibitory behavior 170–173
Gum Arabic (GA) 168
Guoy–Chapman analysis 52

h
hard acid 5
hard base 5, 6
heat-affected zone (HAZ) 73, 74, 77, 78
Helmholtz double-layer model 49–52, 62
hematological system 140
herbal extracts, as corrosion inhibitors 143–146, 164–169, 170
high-performance coatings 109
– cent percent solventless epoxies 110–111
– fire-resistant coatings 119
– organic–inorganic hybrid waterborne coatings 119–121, *120*
– polysiloxane coatings 114, 117–118
– polyvinylidenedifluride coatings 113–114
– underwater coatings 111, 113
hot plastic paints 358
hybrid coatings, in corrosion inhibition 323–326
hydrolysis reaction
– in acidic conditions *184*
– in alkaline conditions *184*

hydrolysis reaction (*contd.*)
– and condensation with the pH of the solution *187*
– metal surface *197*
– for organically modified alkoxide *192*
– for silicon tetralkoxide *185*

i

Indian initiatives, for corrosion protection 339–340, 343–344
– corrosion education 345
– – Central Electrochemical Research Institute (CECRI) 347
– – corrosion science and engineering in IIT (Bombay) 346
– – NACE India Section 347
– – National Corrosion Council of India (NCCI) 348
– – pioneers in India 346
– – reasons 345–346
– – Society for Surface Protective Coatings India 347–348
– highly corrosion-prone industries in India
– – oil and gas industry 348–349
– – power plants 351–352
– – process chemical and petrochemical industry 349–350
– – pulp and paper industry 350–351
– industry scenario 342–343
– recommendations 352–353
inhibition efficiency (IE) 129
inhibitor. *See also individual inhibitors*
– choice of 163–166
– definition of 161
– inhibition mechanism 162–163
inorganic zinc-rich coatings 108
interfacial potential 2
interpenetrating polymer network (IPN) 199, *201*
interphase inhibitors 276
intumescent coatings 119
ion-selective rust 20–22
ion transfer process 2
irritants 140–141
isocyanates 106
isoelectric point (IEP) 20, 28, 298

k

Kelvin potential 91, *92*, 93–94

l

laurylphosphonic acid (LPA) 271
Levich–Landau–Derjaguin equation 193
Lewis acid–base 5, 19, 29
limiting current density 62
linear polarization technique 59
local cell theory of corrosion 38
localized corrosion
– crevice corrosion 16–18
– pitting corrosion 13–16
– potential–dimension diagram 18–19, *18*
localized electrochemical impedance spectroscopy (LEIS) 204
– local electrochemical activity of precracked steel specimen, characterization of 88–89, *90*
– measurement *81*, *85*, *89*
– microscopic metallurgical electrochemistry of pipelines steels 86–88
– steel corrosion at coating defect base 82–86
– technique and principle 81

m

magnetohydrodynamic (MHD) effect 314, 315
malignant cells 142
marine fouling 357
– consequences 357
– fouling prevention methods 357–358
– stages 357
MCM-41 structural characterization 375–378
metal–AMP (amino-tris-(methylenephosphonate)) organic–inorganic hybrids 251–252
metallic corrosion 132
– anodic metal dissolution 4–6
– basic processes 1–2
– cathodic oxidant reduction 6
– corrosion potential 3–4
– potential-pH diagram 2–3, *3*
metallic passivity
– chloride-breakdown of passive films 12–13
– passivation of metals 9–10
– passive films 11
– passivity of metals 7–9
metal-matrix coatings 298, 302, 303, 304, 305–306, 309, 310–311
metal–phosphonate anticorrosion coatings 243, 269–275
– comparative look at inhibitory performance by protective films 285–286
– green chemistry and corrosion control 245
– metal–phosphonate materials 247
– – barium-(amino-tris-(methylenephosphonate)) (Ba-AMP) 255, *256*

– – barium-phosphonomethylene-imino-diacetate (Ba-PMIDA) 260, *262*
– – calcium-(amino-tris-(methylene-phosphonate)) (Ca-AMP) 252–253
– – calcium-hexamethylene-diamine-*tetrakis*(methylenephosphonate) (Ca-HDTMP) 264, *265*
– – calcium-hydroxyphosphonoacetate (Ca-HPAA) 264, *266*, 267
– – calcium-phosphonobutane-1,2,4-tricarboxylate (Ca-PBTC) 263, *264*
– – calcium-phosphonomethylene-imino-diacetate (Ca-PMIDA) 269, *270*
– – dimethylaminomethylene-bis(phosphonic acid) (DMABP) 249, *250*
– – ethylenediamine-*tetrakis*(methylene-phosphonic acid) (EDTMP) 249, *250*
– – metal–AMP (amino-tris-(methylene-phosphonate)) organic–inorganic hybrids 251–252
– – $M(HPAA)(H_2O)_2$ (M = Sr, Ba) 268–269
– – phosphonobutane-1,2,4-tricarboxylic acid (PBTC) 247–249, *248*
– – strontium-(amino-tris-(methylene-phosphonate)) (Sr-AMP) 253–255, *254*
– – strontium and calcium-ethylene-diamine-*tetrakis*-(methylene phosphonate) (Sr-EDTMP and Ca-EDTMP) 259–260, *261*
– – strontium/barium-hexamethylene-diamine-*tetrakis*(methylenephosphonate) (Sr/Ba-HDTMP) 257, *259*
– – $Sr[(HPAA)(H_2O)_3]\cdot H_2O$ 267–268
– – tetrasodium-hydroxyethyl-amino-bis(methylenephosphonate) (Na_4-HEABMP) 260, 262–263
– – zinc-hexamethylene-diamine-*tetrakis* (methylenephosphonate) (Zn-HDTMP) 255, 257, *258*
– – zinc-tetramethylene-diamine-*tetrakis* (methylenephosphonate) (Zn-TDTMP) 259, *260*
– at molecular level 276
– – $Ca_3(HPAA)_2(H_2O)_{14}$ 279–281
– – $\{Ca(PBTC)(H_2O)_2\cdot 2H_2O\}_n$ 278–279
– – $\{M(HPAA)(H_2O)_2\}_n$ (M = Sr, Ba) 281–282, *283*, *284*
– – $\{M(PMIDA)\}_n$ (M = Ca, Sr, Ba) 282, 284
– – by $\{Zn(AMP)\cdot 3H_2O\}_n$ 276–277
– – by $\{Zn(HDTMP)\cdot H_2O\}_n$ 278, 279
– perspectives 287
metal protection, employing green inhibitors *175*
metastasis 142
methyl resins 118
$\{M(HPAA)(H_2O)_2\}_n$ (M = Sr, Ba) 268–269, 281–282, *283*, *284*
Michelia champaca 378, *379*
microelectrochemical technique application, in corrosion research 71
– localized electrochemical impedance spectroscopy
– – local electrochemical activity of precracked steel specimen, characterization of 88–89, *90*
– – microscopic metallurgical electrochemistry of pipelines steels 86–88
– – steel corrosion at coating defect base 82–86
– – technique and principle 81
– scanning Kelvin probe
– – coating disbondment monitoring 91–94
– – technique and principle 89, 91
– scanning vibrating electrode technique
– – local dissolution of welding zone of pipeline steel 73–79
– – mill scale and corrosion product deposit effects on steel corrosion 79–81
– – and principle 72–73
mixed inhibitors 130–131
modeling aspects, for corrosion inhibition 326–328
modified alkyds 101
moisture-cured polyurethanes 107–108
molybdate 271
morpholin-4-methyl-phosphonic acid (MPA) 273
$\{M(PMIDA)\}_n$ (M = Ca, Sr, Ba) 282, 284
mutagens 142

n

NACE India Section 347
nanohybrid coatings synthesis and structural characterization
– amine-functionalized POSS (POSS-NH_2) 369, *370*
– biocide loading 370–371
– biocide preparation 370
– materials 368–369
– mild steel specimen surface preparation 369
– phosphorus-containing polyurethane epoxy resin synthesis 369
– test methods 371–372

nanohybrid coatings synthesis and structural characterization (*contd.*)
– tris(*p*-isocyanatophenyl)-thiophosphate-modified epoxy nanocoatings preparation 371
National Corrosion Council of India (NCCI) 348
National Council of Corrosion 346
natural products, as corrosion inhibitors. *See* herbal extracts, as corrosion inhibitors
neem oil 379, *380*, *381*
Nernst equation 61
nickel coating microstructure *305*, 306, *307*, *308*, *309*, 311
Nyquist diagrams 83, *84*, *85*, *88*
Nyquist plot 66

o
open-circuit potentials 54
organic-based green inhibitors 329
organic zinc-rich coatings 108
organofunctional alkoxysilanes 189–192, *191*, *192*
organofunctional sol–gel coating corrosion properties 202–203
OSPAR Commission (Oslo and Paris Commission) 245
overpotential 57
overvoltage. *See* overpotential
oxide film theory 127
oxygen overvoltage 63

p
pandol. *See* bronopol
passivation 273, 275
passivation–depassivation pH 17
passivation–depassivation potential 15
passivation of metals 9–10
passivation potential 7
passivators. *See* anodic anodic (passivating and film-forming) inhibitors
passive films 8, 11
passivity of metals 7–9, 48
pH analysis 385–386
phenyl resins 118
phosphonic acids 244
phosphonobutane-1,2,4-tricarboxylic acid (PBTC) 247–249, *248*
(4-phosphono-piperazin-1-yl)phosphonic acid (PPPA) 272, 273
photo-excitation 23
physically absorbed films 128
physisorption 134, 163
piperidin-1-yl-phosphonic acid (PPA) 272, 273
pit-repassivation potential 14, 15, 19
pitting corrosion 13–16
pitting potential 12, 13, 16
plant extracts–based green inhibitors 330–333
polarization 57
– concentration 61–68
– curves 3
– of electrodes 54
– experimental resistance plot *60*
– types 56
polarization diagram 54, *55*, *65*
– cathodic and anodic branches of 55
polyamines 104
polyhedral oligomeric silsesquioxanes (POSSs) 364, *365*
– amine-functionalized (POSS-NH_2) 369, 370
polymer nanocomposites (PNCs) 363, 364, 364
polymethylimines 136
polyols 107
polysiloxane coatings 114, 117–118
polyurea coatings 122
polyvinylidenedifluride coatings 113–114
potential–dimension diagram 18–19, *18*
potential-pH diagram 2–3, *3*, *41*
potentiodynamic polarization studies 64, 272, 275, 321, 332
Pourbaix diagrams 45–48, *46*, *48*
– use of 49
precipitating inhibitors 131
primary inhibition 137
protective coatings 97, 355–356
– background 356
– classification 98–99
– corrosion
– – consequences 359
– – electrochemical impedance studies (EIS) 360
– – good coating characteristics 359–360
– – prevention methods 359
– – resistance of coatings evaluation 360
– epoxy resin coatings 360–361
– – advantages 361
– – biocide release mechanism from nanocontainer 366–367
– – disadvantages 361–362
– – justification 362
– – nanocontainer fabrication 366
– – nanocontainer need 364–366
– – nanoparticle types 363–364

– – natural product selection as biocides 367–368
– – need for nanotechnology 362–363
– – polyhedral oligomeric silsesquioxanes (POSSs) 364, *365*
– – polymer nanomaterials 363
– fouling 356
– high-performance coatings 109
– – cent percent solventless epoxies 110–111
– – fire-resistant coatings 119
– – organic–inorganic hybrid waterborne coatings 119–121, *120*
– – polysiloxane coatings 114, 117–118
– – polyvinylidenedifluride coatings 113–114
– – underwater coatings 111, 113
– marine fouling 357
– – consequences 357
– – fouling prevention methods 357–358
– – stages 357
– MCM-41 structural characterization 375–378
– – biocide loading confirmation 379, *381*
– – colorimetric and gravimetric analyses 383–389
– – corrosion resistance evaluation 381–383
– – natural product composition 378–379
– nanohybrid coatings synthesis and structural characterization
– – amine-functionalized POSS (POSS-NH_2) 369, 370
– – biocide loading 370–371
– – biocide preparation 370
– – materials 368–369
– – mild steel specimen surface preparation 369
– – phosphorus-containing polyurethane epoxy resin synthesis 369
– – test methods 371–372
– – tris(*p*-isocyanatophenyl)-thiophosphate-modified epoxy nanocoatings preparation 371
– paint application 121–122
– paint coatings selection 97–98
– resin chemistry
– – acrylic urethanes 107
– – aliphatic isocyanates 106
– – alkyd resins 99–100, *100*
– – epoxy resin 101–105, *102*, *103*, *104*
– – isocyanates 106
– – modified alkyds 101
– – moisture-cured polyurethanes 107–108
– – polyols 107
– – urethanes 105–106
– – zinc-based coatings 108–109
– scope and objectives 368
– supervision, inspection, and quality control 122–123
– surface preparation 121
– training and certification courses 123
– tris(*p*-isocyanatophenyl)-thiophosphate-modified epoxy resin structural characterization 373–375
proton acceptors 134
1,4-bis[2-pyridyl]-5H-pyridazino[4,5-b] indole (PPI) 139

q
quinolines 137

r
rare earth elements-based green inhibitors 333–334
redox rust 24
reference electrode (RE) 44, *45*
relative humidity (RH) 98
resin chemistry
– acrylic urethanes 107
– aliphatic isocyanates 106
– alkyd resins 99–100, *100*
– epoxy resin 101–105, *102*, *103*, *104*
– isocyanates 106
– modified alkyds 101
– moisture-cured polyurethanes 107–108
– polyols 107
– urethanes 105–106
– zinc-based coatings 108–109
rust preventive compounds 358

s
salt spray test results 386
scanning Kelvin probe (SKP) 204
– coating disbondment monitoring 91–94
– measurement 91, *93*, 94
– technique and principle 89, 91
scanning vibrating electrode technique (SVET) 204, 324, 326
– local dissolution of welding zone of pipeline steel 73–79
– measurement *74*, *75*, *76*, *77*, 80
– mill scale and corrosion product deposit effects on steel corrosion 79–81
– and principle 72–73
seawater immersion test 386, 387, *388*, 389
secondary inhibition 137
sensitizers 142
shot peening 231–232, 239

silanes-based pretreatment for organic coatings adhesion 182, *200*
– corrosion protection by sol–gel coatings 202
– – investigation methods of silicon alkoxide sol–gel coating properties 203–204
– – organofunctional sol–gel coating corrosion properties 202–203
– – practical examples 204–207
– hybrid silane sol–gel coatings 182
– – dip coating 192–194
– – interaction between silicon alkoxides and metallic substrates 194–199
– – silicon alkoxides and organofunctional silicon alkoxides 183–192
silicon oils 118
Society for Advancement in Electrochemical Science and Technology (SAEST) 346
Society for Surface Protective Coatings India 347–348
soft acid 5
soft base 5, 6
sol–gel coatings 182, 202
– dip coating 192–194
– interaction between silicon alkoxides and metallic substrates 194–199
– investigation methods of silicon alkoxide sol–gel coating properties 203–204
– organofunctional sol–gel coating corrosion properties 202–203
– practical examples 204–207
– silicon alkoxides and organofunctional silicon alkoxides 183–192
specific adsorption 163
spray and walk coatings 122
standard hydrogen electrode (SHE) 44
steel corrosion, weathering 26–27
Stern model, of double layer 52–53, *53*
strontium-(amino-tris-(methylene-phosphonate)) (Sr-AMP) 253–255, *254*
strontium/barium-hexamethylene-diamine-*tetrakis*(methylenephosphonate) (Sr/Ba-HDTMP) 257, *259*
strontium and calcium-ethylene-diamine-*tetrakis* (methylene phosphonate) (Sr-EDTMP and Ca-EDTMP) 259–260, *261*
$Sr[(HPAA)(H_2O)_3]\cdot H_2O$ 267–268
supercritical water (SCW)
– alloy oxidation thermodynamics 216–220
– and applications 211–214
– austenitic stainless steels and Ni-base alloys 214–215
– – general corrosion behavior 215–216
– – oxide layer structure 225–227
– – surface morphology 223–225, *224*
– – weight change 221–223
– cooled reactor (SCWR) 213, 216, 217
– factors influencing corrosion
– – grain size effect 237–239
– – microstructure effect 236–237
– – test conditions 234–235
– – thermodynamics and kinetics 235–236
– gasification (SCWG) 212
– novel corrosion control methods
– – grain size refinement 231–233
– – microstructural optimization 227–231
– – performance comparison 233–234
– oxidation (SCWO) 212, 213
sustainable development 319–320
synergism 131
synergistic effect 131
systemic poisons 141–142

t

Tafel equation 59
teratogens 142–147
tetrasodium-hydroxyethyl-amino-bis (methylenephosphonate) (Na_4-HEABMP) 260, 262–263
thin film 244, 274, 286, 287
thiomorpholin-4-ylmethyl-phosphonic acid (TMPA) 273
thiourea (TU) 138
toxicity of inhibitors 139–140
– anesthetics and narcotics 141
– asphyxiants 141
– carcinogens 142
– irritants 140–141
– mutagens 142
– sensitizers 142
– systemic poisons 141–142
– teratogens 142–147
transpassive state 8, 9
tris(*p*-isocyanatophenyl)-thiophosphate-modified epoxy
– nanocoatings preparation 371
– resin structural characterization 373–375

u

underwater coatings 111, 113
urethanes 105–106
3-vanilidene-amino-1,2,4-triazole phosphonate (VATP) 273

v

vapor phase inhibitors 131–139
vinyl chloride copolymer (VYHH) 326

volatile corrosion inhibitors (VCIs). *See* vapor phase inhibitors

x

X-ray diffraction (XRD) 226, 274
X-ray photoelectron spectroscopy (XPS) 227, 274, 275
$\{Zn(AMP)\cdot 3H_2O\}_n$ 276–277

z

zinc-based coatings 108–109
$\{Zn(HDTMP)\cdot H_2O\}_n$ 278, 279
zinc-hexamethylene-diamine-*tetrakis* (methylenephosphonate) (Zn-HDTMP) 255, 257, *258*
zinc-tetramethylene-diamine-*tetrakis* (methylenephosphonate) (Zn-TDTMP) 259, *260*